PHANEROZOIC STROMATOLITES II

PHANEROZOIC STROMATOLITES II

Edited by

Janine Bertrand-Sarfati
University of Montpellier, C.N.R.S., France

and

Claude Monty
University of Nantes, France

SPRINGER-SCIENCE+BUSINESS MEDIA, B.V.

Library of Congress Cataloging-in-Publication Data

Phanerozoic stromatolites II / edited by Janine Bertrand-Sarfati and
  Claude Monty.
      p.   cm.
  Includes index.
  ISBN 978-94-010-4491-2   ISBN 978-94-011-1124-9 (eBook)
  DOI 10.1007/978-94-011-1124-9

    1. Stromatolites.   I. Bertrand-Sarfati, Janine.   II. Monty,
  Claude, 1937-   . III. Title: Phanerozoic stromatolites 2.
  QE955.P47   1994
  552'.58--dc20   94-4221

ISBN 978-94-010-4491-2

Printed on acid-free paper

TABLE OF CONTENTS

Preface   vii

Introduction   ix

PART I
MODERN STROMATOLITES FROM FRESHWATER AND MARINE
ENVIRONMENTS

The Modern Thrombolites of Lake Clifton, Western Australia
    L. S. Moore and R. V. Burne   3

Bacterially Controlled Calcification of Freshwater *Schizothrix*-Stromatolites: An Example from the Pieniny MTS, Southern Poland
    J. Szulc and B. Smyk   31

Stromatolitic Mats in Antarctic Lakes
    R. A. Wharton, Jr.   53

Recent Fresh-Water Lacustrine Stromatolites, Stromatolitic Mats and Oncoids from Northeastern Mexico
    B. M. Winsborough, J-S. Seeler, S. Golubic, R. L. Folk and B. Maguire Jr.   71

Peritidal Potential Stromatolites – A Synopsis
    G. Gerdes and W. E. Krumbein   101

Stromatolite and Serpulid Bioherms in a Holocene Restricted Lagoon (Sabkha El Melah, Southeastern Tunisia)
    E. Davaud, A. Strasser and Y. Jedoui   131

PART II
CENOZOIC STROMATOLITES IN LAKES AND MARINE STROMATOLITIC
PHOSPHORITES

Microstructures in Tertiary Nonmarine Stromatolites (France). Comparison with Proterozoic
    J. Bertrand-Sarfati, P. Freytet and J. C. Plaziat   155

Stromatolites From the East African Rift: A Synopsis
J. Casanova   193

Lacustrine Stromatolites and Oncoids: Manuherikia Group (Miocene), New Zealand
J. K. Lindqvist   227

Stromatolitic Phosphorites in the Eocene of the Negev (Southern Israel)
D. Soudry and G. Panczer   255

PART III
MESOZOIC DEEP MARINE STROMATOLITES AND BACTERIAL MARINE PHOSPHORITES

Amino Acids in the Pelagic Stromatolites of the Rosso Ammonitico Veronese Formation (Middle-Upper Jurassic, Southern Alps, Italy)
L. Ballarini, F. Massari, S. Nardi and L. Scudeler Baccelle   279

Deep-Marine Microbial Structures in the Upper Jurassic of Western Tethys
G. Dromart, C. Gaillard and L. F. Jansa   295

Mesozoic Stratigraphic Breaks and Pelagic Stromatolites in the Betic Cordillera, Southern Spain
J. A. Vera and A. Martin-Algarra   319

Mesozoic Pelagic Phosphate Stromatolites From the Penibetic (Betic Cordillera, Southern Spain)
A. Martin-Algarra and J. A. Vera   345

PART IV
PALEOZOIC STROMATOLITES AND THROMBOLITES

Siliciclastic-Carbonate Stromatolite Domes, in the Early Carboniferous of the Ajjers Basin (Eastern Sahara, Algeria)
J. Bertrand-Sarfati   395

Thrombolitic-Stromatolitic Cycles of the Cambro-Ordovician Boundary Sequence, Precordillera Oriental Basin, Western Argentina
C. Armella   421

Thrombolites and Stromatolites Within Shale-Carbonate Cycles, Middle-Late Cambrian Shannon Formation, Amadeus Basin, Central Australia
J. M. Kennard   443

PREFACE

Precambrian stromatolites have received in depth, consideration from geologists and paleontologists; they were indeed searching for biosedimentary structures that were sufficiently characteristic and widely distributed to be considered as useful tools for stratigraphic correlation. Silicified stromatolites are also of interest as they contain preserved traces of ancient life.

Calcareous Phanerozoic stromatolites have not received very much attention from geologists. Logan's too schematic morphological classification of 1964, was not so helpful to the knowledge of Phanerozoic stromatolites because neither their morphology nor their microstructure were studied in the same detail in which Proterozoic stromatolites have now been described. We therefore know little about the Phanerozoic stromatolites which, do, however, show an interesting range of diversification. A major questions still remaining to be answered include the history of stromatolite development and wether their morphology has "evolved" in addition to detailed information concerning Cenozoic nonmarine stromatolites which precipitate carbonate and the Recent giant stromatolites which trap particles.

For these reasons Claude Monty, in 1981, launched the first volume of what was going to be a series on "Phanerozoic stromatolites" in order to describe their morphology, microstructure and paleoecology and to present them in their stratigraphic context.

Thanks to Kluwer, the second volume of the series is now ready in spite of problems which have delayed its completion. The book contains 17 contributions spanning the entire Phanerozoic up to the Recent, comprising four parts in which stromatolites are dealt with in relation to their environment. Emphasis is placed on fresh-water and deep marine settings.

Janine Bertrand-Sarfati, Montpellier
Claude Monty, Nantes

INTRODUCTION

PART I: MODERN STROMATOLITES FROM FRESHWATER AND MARINE ENVIRONMENTS

Modern and quaternary stromatolites (laminated) and thrombolites (non-laminated), all related to microbial activity, occur in diverse fresh water environments (four papers) while only stromatolites occur in marine settings (two papers). New data emphasize the presence of stromatolites in subtidal environments.

L. S. MOORE and R. V. BURNE The occurrence of thrombolite and stromatolite, in southern Australian modern lakes, is discussed. Their study indicates that modern unlaminated thrombolite fabric occurs in communities of filamentous cyanobacteria and are not due to grazing and burrowing by metazoans as proposed for explaining ancient thrombolites. The clotted fabric is definitely a growth fabric.

J. SZULC and B. SMYCK Fresh water stromatolites, described in Polish streams, are built by filamentous cyanobacteria. The processes of calcification are controlled by bacteria and the polyhedral-like habit of the calcified colonies results from competition between biological versus physicochemical mechanisms.

R. A. WHARTON Under the ice cover of Antarctic lakes, complex communities of microorganisms are building stromatolites. As in other environments, the morphology of each particular mat results from a combination of biological, geochemical and sedimentological processes.

B. M. WINSBOROUGH, J-S. SEELER, S. GOLUBIC, R. L. FOLK and B. MAGUIRE Stromatolites and other microbial accretions are reported from spring-fed lakes in northern Mexico. Growth is influenced by light, depth and current. In the light zone a laminated fabric is induced by the growth of two alternating filamentous cyanobacteria while in the shade (older mat stages), numerous varieties of cyanobacteria and endolith strongly modify the calcifying stromatolite.

G. GERDES and W. KRUMBEIN In peritidal settings, emphasis is placed on tidal range and climate as controls on potential stromatolite growth, together with mechanisms producing biolamination and especially the amount of sediment. Two topics valuable for interpreting pre-Cambrian record are discussed : a) thick biogenic varvites are capable to form by self-burial via mat-by-mat overgrowth, in ponds with almost zero energy and in the absence of sediment input and b) bioherm growth is related to deeper waters.

E. DAVAUD, A. STRASSER and Y. JEDOUI Sea level high stand, in a southeastern Tunisian Holocene lagoon, allowed serpulid and then stromatolite bioherm to grow, related to increasing salinity of the environment. Aragonite precipitated within the primary porosity may be under microbial control. Lower Carboniferous and Triassic serpulid-stromatolite bioherms are discussed.

PART II: CENOZOIC STROMATOLITES IN LAKES AND MARINE STROMATOLITIC PHOSPHORITES

Cenozoic stromatolites are widespread in nonmarine environments where they have built extensive beds. Emphasis is placed on the microstructure description (3 papers) as the biological framework is often striking. Phosphatic stromatolites and nodules (1 paper) are described from marine settings.

J. BERTRAND-SARFATI, P. FREYTET and J.C. PLAZIAT Nonmarine stromatolites from France are dominated by precipitation processes. In an attempt to clarify descriptions of the stromatolite microstructure a nongenetic classification is proposed according to the size of filaments, cocoids or other microorganisms and the laminae fabric. Analogs are proposed both with modern stromatolite building cyanobacteria and algae and with Proterozoic stromatolite microfabrics.

J. CASANOVA Morphology and microstructure of stromatolites in East African lakes are essentially under environmental control : hydroclimatic variations, nature of the substrate, bathymetry and hydrodynamism. In these rift lakes, they provide a good isotopic record of the paleohydrology, the bathymetry and the climatic seasonality.

J. LINDQVIST Stromatolites, oncolites and ooids are the only carbonates found in a New Zealand lacustrine siliciclastic sequence. The microstructures are well preserved and some spheroidal cells, usually attributed to insect eggs or cocoid algae, are recognized, for the first time, as fungal spores.

D. SOUDRY and G. PANCZER Phosphatic stromatolites, well known in pre-Cambrian and Cambrian settings, are reported from Eocene marine environments in southern Israel. Ministromatolites and microstromatolites accreted on the sheltered margins of a topographic high, while nodules are reported from nearby depressions. Diverse capsule-like or globose cells are interpreted as microbial primary producers and bacteria-decomposers responsible for the laminated stromatolite. Only part of the early phosphatisation may be due to microbial activity (see also § III).

PART III: MESOZOIC DEEP MARINE STROMATOLITES AND BACTERIAL MARINE PHOSPHORITES

Mesozoic stromatolites reported grew in marine environments. Carbonated, in very deep water, or phosphatic and related to condensed facies, they are generally very small and attributed to bacteria rather than to cyanobacteria.

L. BALLARINI, F. MASSARI, S. NARDI and L. SCUDELER-BACELLE
Stromatolites and oncolites are found in carbonate pelagic environments of the Italian
Ammonitico-Rosso. The amino-acids of these laminated fabrics are consistent with the
composition of the cell wall of modern and fossil cyanobacteria.

G. DROMART, C. GAILLARD and L.F. JANSA Very deep marine stromatolitic
carbonates are reported from France and Canada. Relatively small, diverse columnar and
planar stromatolites encrust hard substrates in environments of low sedimentation rates.
Their growth in basins of considerable depth (up to 1000m) indicates
non-photoautotrophic microorganisms.

J.A. VERA and A. MARTIN-ALGARRA Pelagic (i.e. open marine versus neritic)
stromatolites have been largely ignored. In the Spanish Betic Cordillera, numerous
discontinuity surfaces are covered by phosphatic stromatolites growing at the beginning
of sea level rises.

A. MARTIN-ALGARRA and J.A. VERA These phosphatic stromatolites and oncolites
whose morphologies are described in an unusual way, are attributed to bacterial growth.
The very thin phosphate laminae (about 2-4 μm thick) are evenly repeated and contain
some filamentous remnants. Phosphate-rich precursor colloidal substances may floculate
around bacterial threads and phosphate precipitation will then occur during very early
diagenesis.

PART IV: PALEOZOIC STROMATOLITES AND THROMBOLITES

In the Paleozoic record, stromatolites are widespread in cratonic basins from two
especially stable periods : Cambro-Ordovician and Carboniferous. In Cambro-Ordovician
they built reefs and bioherms in association with thrombolites, unknown or very rare in
more ancient and more recent strata. Stromatolites are abundant in the Middle
Carboniferous of Northeastern Africa and also in association with serpulids, in Europe
(see § I-6)

J.B. BERTRAND-SARFATI Middle Carboniferous fluvio-marine deposits cover a large
flat area in the eastern Sahara. Stromatolites survive there in brackish protected waters,
coping with a continuous supply of clay and silt sediment. Absence of filamentous
structures and thick laminae suggests a high content of coccoid cyanobacteria and
bacteria.

C. ARMELLA Thrombolite-stromatolite cycles from the Cambro-Ordovician in
Argentina are described in relation to high-frequence sea level rises. Thrombolites are
clearly controlled by microbial activity and their occurrence with stromatolites in an
environment deprived of fauna contradicts their interpretation as stromatolites disturbed

by grazing and burrowing matazoans.

J.M. KENNARD Thrombolite-stromatolite cycles from the Australian middle-upper Cambrian are treated in relation to the local environment. Microbial activity completely controlled the micro-structure, whereas the meso-structure and mega-structure are partially environmental. Cyclic arrangement is very much the same as that described in Argentina. Thrombolite clotted structure is attributed to activity of coccoid communities; however we saw in §1 that modern thrombolites are built by filamentous cyanobacteria.

PART I

Modern Stromatolites
from Freshwater
and Marine Environments

THE MODERN THROMBOLITES OF LAKE CLIFTON, WESTERN AUSTRALIA

L.S. MOORE[1] and R.V. BURNE[2]
1 Department of Microbiology, University of Western Australia ; WA 6009 Australia
2 Australian Geological Survey Organisation, PO Box 378, Canberra ;
ACT 2601 Australia

ABSTRACT

Thrombolites and stromatolites are microbialites with contrasting internal structures. The decline of stromatolites at the end of the Proterozoic and the rise of thrombolites during the Cambrian have been related to the evolution of burrowing and grazing metazoans, and it has been suggested that the thrombolites were the result of metazoan activity disrupting the original stromatolitic laminae. Some maintain that the thrombolitic structure is intrinsic, and due to penecontemporaneous mineralisation associated with coccoid-dominated BMCs. However, it is recognised that thrombolitic fabrics are complex, show a great variation, and may have originated in several ways. The interpretation of the genesis of thrombolites has been limited by the absence of well-documented modern examples. Thrombolitic and stromatolitic microbialites are presently forming in Lake Clifton, a marine-derived coastal lake in southwestern Australia. The thrombolites are by far the most predominant form of microbialite in Lake Clifton, with small stromatolites restricted to certain upper shore areas. The thrombolites exhibit a wide range of external morphologies including conical, domical, discoidal and tabular formations which vary considerably in size, as well as more irregular and columnar structures up to 1.3 m high. Many of the tabular and domical forms have coalesced to form an extensive reef-like formation over 6 km long. As documented by other studies of modern microbialites, external morphology appears to be primarily the product of the environmental setting. In Lake Clifton, seasonal fluctuations in water depth, regional variations in sedimentation rates and the effects of prevailing winds and currents are major controlling factors. By contrast, the internal morphology of the various thrombolites is remarkably similar, composed of a framework of aragonitic mesoclots. The aragonite is precipitated as a consequence of microenvironmental chemical changes induced by the metabolic activity of principally filamentous cyanobacteria. The Lake Clifton thrombolites represent an important modern analogue for understanding fossil thrombolites. The mesoclots appear to be a growth fabric and are not the result of disruption of pre-existing stromatolitic laminae. Moreover, Lake Clifton provides an important example of the coexistence between both

3

J. Bertrand-Sarfati and C. Monty (eds.), Phanerozoic Stromatolites II, 3–29.
© 1994 *Crown Copyright.*

stromatolitic and thrombolitic microbialites and an abundant and diverse metazoan fauna as well as demonstrating an apparent inability of this fauna to markedly influence microbialite development. Thus apart from changing chemical conditions, competition for space (and the resultant lack of suitable habitats) was probably a major factor restricting thrombolite distribution throughout the Phanerozoic following the middle Ordovician.

INTRODUCTION

Thrombolites (Aitken, 1967) and stromatolites (Kalkowsky, 1908) are two of several forms of organosedimentary structures considered to have accreted as a result of the activity of benthic microbial communities (BMCs). Collectively, these structures have been successively defined as *cryptalgal sedimentary rocks* or *rock structures* by Aitken (1967), *stromatolites* by Awramik and Margulis (1974), *microbial structures* by Kennard and James (1986), and finally as *microbialites* by Burne and Moore (1987).

Thrombolites and stromatolites may have similar external forms; it is their internal structure that sets them apart. Thrombolites are characterised by an internally clotted texture and lack fine laminations, whereas stromatolites are characterised by an internal structure of discrete laminations (see e.g. Aitken, 1967; Monty, 1976; Kennard and James, 1986; Burne and Moore, 1987).

The decline of stromatolites at the end of the Proterozoic and the rise of thrombolites during the Cambrian have been related to the evolution of burrowing and grazing metazoans (Garrett, 1970; Awramik, 1971, 1981, 1984, 1991; Walter and Heys, 1985). Walter and Heys (1985) suggested that fossil thrombolites, including those described by Aitken (1967), were the result of burrowing animals disrupting the formation of stromatolitic laminae, and that the early Palaeozoic thrombolites provided the stratal evidence for the disruptive influence of metazoan life on stromatolites. Monty (1973), however, pointed to a lack of stratal and evolutionary evidence for the supposed disruption to the stromatolitic laminae. Kennard and James (1986) also challenged the view of Walter and Heys (1985). They considered the internal structure of Cambrian thrombolites to be intrinsic, and caused by calcification associated with coccoid-dominated BMCs. Moreover, Monty (1976) described three modes of modern thrombolite formation, none of which involved the disruption of an initially laminated structure by burrowing or grazing animals.

As Monty (1976) and Burne and Moore (1987) concluded, it is clear that thrombolitic fabrics are complex, show great variation and might have originated in several ways. In this context, studies on certain modern ecosystems such as Lake Clifton can elucidate the mode of thrombolite formation and the role of a co-existing metazoan fauna.

Thrombolitic microbialites are presently forming in Lake Clifton, a saline coastal lake in southwestern Australia. Lake water salinity ranges from 14 g L^{-1} in winter to 35 g L^{-1} in summer. The lake has evolved from a marine embayment to a saline lake as a result of transgressive barrier building and relative sea level fall over the past 4 000 years. The thrombolites are therefore growing in an environment intermediate between an open-marine environment in which their early Palaeozoic counterparts flourished, and a continental setting from which most modern microbialites have been recorded (Monty, 1973; Golubic, 1991).

The marine-derived waters of this former coastal lagoon support an abundant and diverse fauna comprising marine, estuarine and lacustrine elements, a situation permitting studies on microbialite formation in the presence of a wide variety of benthic metazoans. Such studies have a bearing on those interpretations which suggest that the origin of thrombolites, as well as the decline in the abundance and diversity of marine stromatolites, was mainly the result of burrowing and grazing animals. Lake Clifton, therefore, provides an important modern analogue for improving our understanding of the development of Phanerozoic microbialites and, more importantly, the origin of thrombolites.

REGIONAL SETTING

Lake Clifton, which lies between 1.5 and 4 km from the Indian Ocean, is one of eleven interdunal lakes comprising the Clifton-Preston lakeland system on the south-west coast of Australia (32° 47'S, 115° 38'E; Fig. 1). This system lies on the western edge of the 25-30 km wide Swan Coastal Plain, an accumulation of Late Tertiary and Quaternary limestones, sands and clays. This coastal plain abuts the Precambrian Yilgarn Block to the east and rests on Cretaceous sandstones, siltstones and claystones some 20-25 m below sea level (Playford, Cockbain and Lowe, 1976; Commander, 1988).

All the Clifton-Preston lakes are less than 4 km from the Indian Ocean and act as groundwater sinks (Fig. 1). They occupy interdunal depressions between a series of linear coastal barriers that formed as a result of repeated sea level changes during the Quaternary (Searle and Semeniuk, 1985). Lake Clifton lies between two Pleistocene coastal barriers; to the east is the Clifton/Harvey barrier, a prominent ridge which attains elevations of up to 70 m, while to the west a low, narrow ridge separates the Lake from a narrow and discontinuous chain of nine lakes running from Swan Pond in the north to Lake Newnham in the south (Fig. 1). This barrier is 0.7 km wide at the south but has become so narrow at the north that it is breached at times of high lake levels, allowing exchange between Lake Clifton and Swan Pond. A third Pleistocene barrier separates the narrow chain of lakes from Lake Preston, which lies further to the southwest (Fig. 1).

The barrier dune systems are of composite beach and aeolian origin. Each marks the high point of a Quaternary marine transgression and derives predominantly from the shoreward transport of quartz and skeletal carbonate sands from the continental shelf by prevailing westerly and southwesterly swells (Playford et al., 1976; Semeniuk and Johnson, 1985). Pleistocene limestone, formed by the lithification of the carbonate sediments of the barrier systems, is common in the region with numerous outcrops occurring along the lake margins (Playford et al., 1976; Semeniuk and Johnson, 1985).

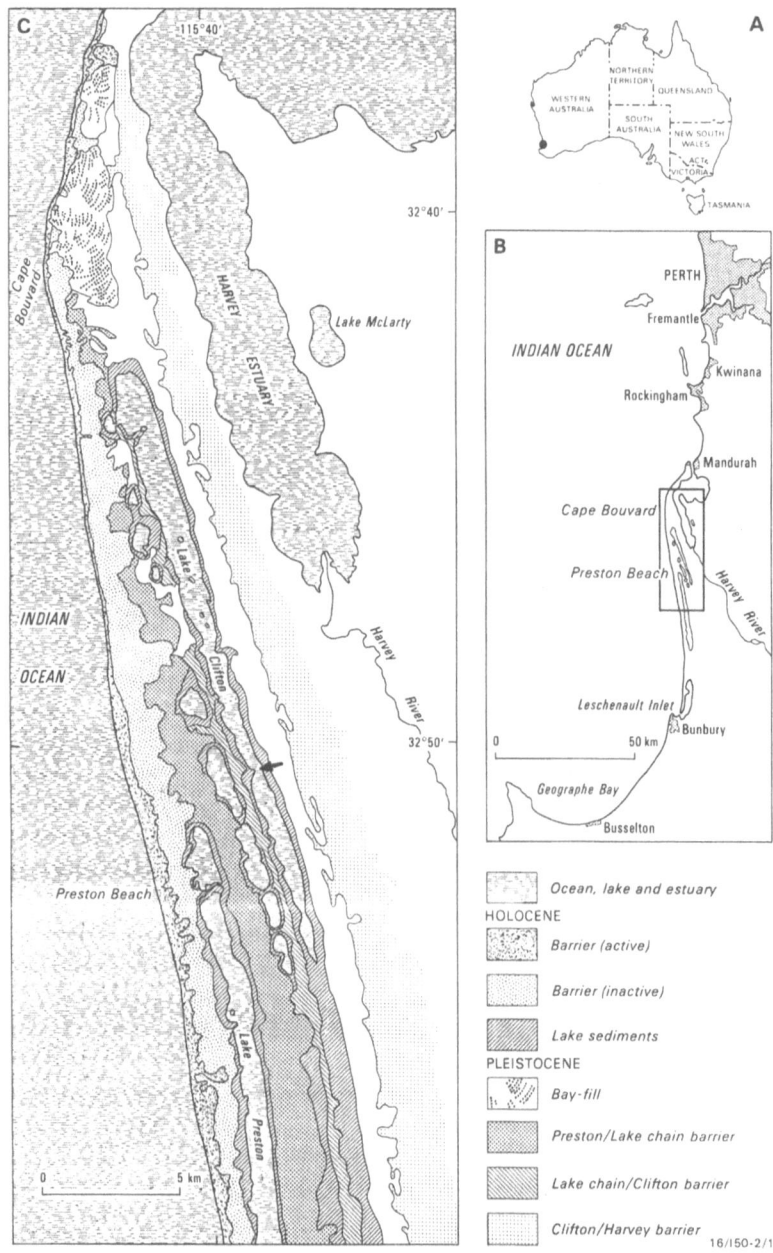

Figure 1: The location and morphostratigraphy of Quaternary coastal barriers and interdunal lakes forming the Clifton-Preston lakeland system between the Indian Ocean and Harvey Estuary, southwest Australia. Arrow indicates position of large sand spit and deep-water channel.

The present coastal barrier is 20-40 m high and 1-3 km wide, and was initiated in the Holocene about 7 800 years BP, some time after the start of the Flandrian Transgression (Semeniuk and Meagher, 1981a,b). At the height of this transgression (4 770-3 610 years BP), sea level rose to 3-4 m above its present position before subsequently falling to present-day levels by at least 2 800 years BP. The Holocene Barrier runs obliquely to the Pleistocene Barriers, stepping over them and their interdunal depressions until it abuts the western margin of Swan Pond (Fig. 1). Further north it forms the western boundary of the Clifton interdunal depression, which in this region is filled by a series of low arcuate ridges of uncertain age (Biggs, 1977). Thus although Lake Clifton lies between Pleistocene ridges, it was probably connected with the open ocean at its northern end until the Holocene Barrier reached its present position at, or shortly after, the height of the Flandrian transgression. This view is consistent with the results of a recent radiocarbon study on the adjacent Peel-Harvey system (Semeniuk and Semeniuk, 1991).

LAKE SEDIMENTS

Lake Clifton is surrounded by a low platform of Holocene sediments that relate to higher water levels possibly associated with the ~4 000 year BP maximum sea-level stand. The sediments of the Lake consist of carbonate mud, skeletal remains of ostracods and gastropods, calcified stems and oogonia of charophytes, peloids and small, irregular carbonate concretions. Debris from the microbialites consists of sand to gravel-sized chips that are most common in the vicinity of the more exposed structures. Short cores demonstrate that sediments similar to those of the present-day range in depth from 0.25 to over 1 m. At the base of these sediments, a deeper water facies is occasionally indicated. This facies is underlain by estuarine and marine sequences richer in quartz and containing bivalves such as *Katelysia, Mytilus* and *Brachidontes*. The richest marine sequence is found in the northern end of the Lake where it includes echinoid fragments, serpulids and foraminifera. Radiocarbon dates of the top of the estuarine sequence range between 4 670 and 3 890 years BP, supporting the view that Lake Clifton did not become isolated from the sea until at least the high Holocene sea-level stand between 4 770 and 3 610 years BP. Following its isolation, the Lake was somewhat higher than today, a feature consistent with a subsequent fall in sea level to its present position between 3 610 and 2 800 years BP. Thrombolites became established in the Lake Clifton during or after this time.

Prevailing southwesterly winds have led to the western shore of the Lake forming a protected environment characterised by fine grained sediments, whereas the eastern shore is more exposed and its sediments are coarser. Sediments in the 3 m deep basin at the north end of the Lake consist of a purple microbial ooze, while those in a 2-3m deep channel in the south-central portion of the Lake comprise a lag of the marine bivalve layer exposed by the erosion of more recent facies (Moore, 1993).

Differences in the character of the sediments in the lakes of the Clifton-Preston system reflect past variations to the lakewater chemistry of each basin. These are thought to be the result of temporal changes to the relative contribution of sea water influx, hypersaline and/or fresh water lenses, and regional groundwater discharge from unconfined aquifers lying to the east of the lakeland system (Moore, 1987). Thus the surface sediments (top 2 m) of the hypersaline Lake Preston contain a marine molluscan fauna consistent with salinities equal to or greater than that of sea water. In contrast to Lake Preston, the marine-like facies in the Clifton basin, which is also dominated by bivalves such as *Katelysia*, ceased to be deposited almost 4 000 years ago.

HYDROLOGY

Lake Clifton, the northernmost member of the interdunal lake system, is 21.5 km long and has a maximum width of 1.5 km. A wide, partially vegetated sand spit constricts the width of the lake to less than 200 m at a point approximately 7 km from its southern end (Fig. 1). While much of the lake is less than 1.5 m deep, certain areas, such as the deep basin in the north and the channel passing around the spit, are up to 3 m in depth. Both the basin and mean annual water level of Lake Clifton are lower than mean sea level. Like the other lakes of the system, Clifton is underlain by a body of hypersaline groundwater. A large unconfined aquifer of fresh water (1-2 g L^{-1} TDS) along the eastern side of the lake has a saturated thickness of 20 m and extends for 3 km to the east of the foreshore (Commander, 1988). This groundwater forms a lens above the hypersaline body, and seepage occurs predominantly along the eastern shore of Lake Clifton throughout the year (Moore, 1987). Although there are no streams feeding the lakeland system and there is little surface runoff, lake water levels fluctuate seasonally by up to 1 m; rising in winter with direct rainfall and increased groundwater input, and falling over summer by evaporation. Annual fluctuations vary considerably between years; for example during the winters of 1979, 1983 and 1984, water levels in Clifton rose by 0.61, 0.92 and 0.86 m respectively (Moore, Knott and Stanley, 1983; Moore, 1987).

While most lakes of the system are seasonally or permanently hypersaline, Lake Clifton remains at or below seawater salinity throughout much of the year and is the only lake which continues to support living microbialites. Salinity varies from 15 g L^{-1} in winter to 35 g L^{-1} in summer in the main part of the Lake, though salinities in the very shallow southern end of the Lake approach 40 g L^{-1} in summer (C. Burke, University of WA, pers.comm.). All the lakes are marine-derived, the waters dominated by sodium and chloride ions, with the general pattern of ionic dominance being $Na^+ > Mg^{2+} > Ca^{2+} > K^+$ and $Cl^- > SO_4^{2-} > HCO_3^-$ (Moore, 1987; Moore and Turner, 1988). The relative proportions of major ions are similar to those of standard sea water, and differ markedly

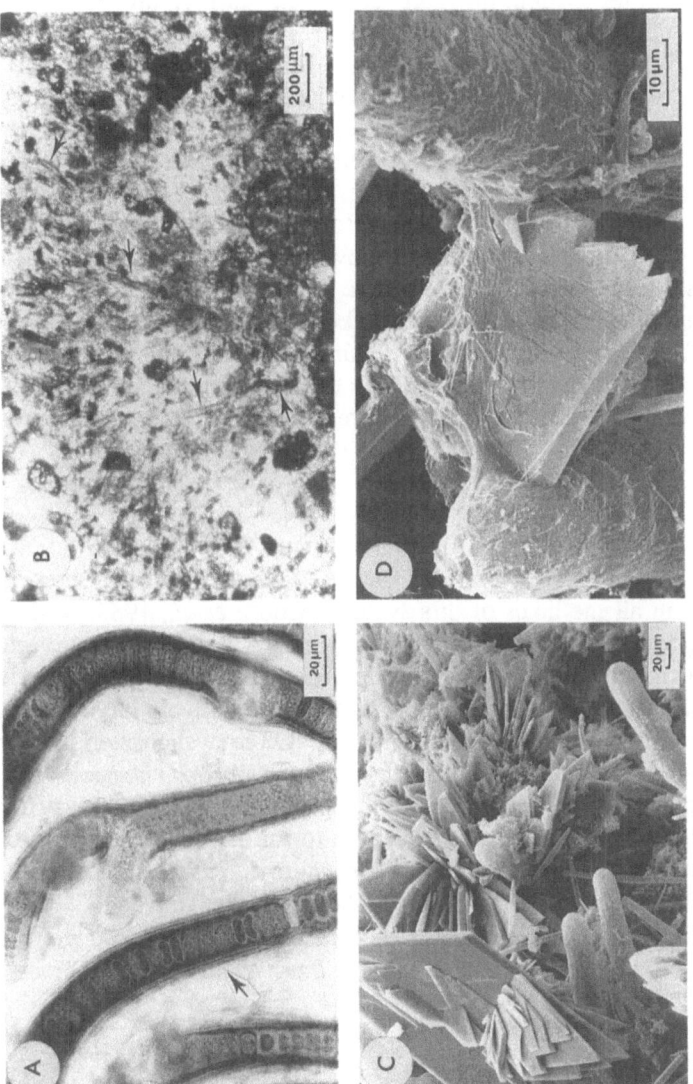

Figure 2: a) Live filaments of *Scytonema siculum* showing false branching, laminated sheath (arrowed) and heterocysts; b)Thin section (30 µm) showing remnants of *Scytonema* filaments (arrows) within aragonitic mesoclot; c) Scanning electron micrograph of the surface of a thrombolite, showing arrays of orthorhombic crystals of aragonite surrounding protruding filaments of *Scytonema*.; d) Scanning electron micrograph showing an aragonite crystal between two filaments of *Scytonema*.

from those of the groundwater seeping into Lake Clifton which is enriched in Ca^{2+} and HCO_3- (Moore, 1987).

Seepage rates up to 1800 mL m^{-2} h^{-1} have been recorded in the vicinity of the microbialites. There is a close association between the distribution of these living microbialites and areas of groundwater discharge. Intrusion of low salinity groundwater maintains reduced salinities within the lake sediments and at the sediment-water interface, reduces the effects of desiccation on the BMC during periods of exposure, and provides a primary source of Ca^{2+} and HCO_3^- which, interacting with the saline lakewaters, facilitates the production of calcified microbialites by the BMC (Moore, 1987).

Although groundwater discharge along the foreshores of lakes close to Lake Clifton has yet to be investigated in detail, the presence of wide, indurated carbonate platforms along the margins of many of these lakes points to the precipitation of $CaCO_3$ from groundwater seepage. While these platforms now preclude groundwater discharge along the immediate foreshore region, inactive "relict" microbialites (microbial framestones *sensu* Burne and Moore, 1987) in Lakes Pollard, Newnham, Preston and Martin's Tank (Fig. 1) indicate zones of former groundwater discharge and possibly less saline conditions in the past. Since radiocarbon dating suggests these framestones are only 2 540 to 1 230 years old, it is unlikely they relate to a period of higher sea level than present.

MICROBIAL AND ALGAL COMMUNITIES

Benthic microbial communities of varied species composition and preservation potential are common in all the lakes of this system (Moore *et al.*, 1983; Neil, 1984). The BMC associated with the Lake Clifton microbialites is composed of a variety of cyanobacteria and eukaryotic algae. By far the most abundant cyanobacterium is· *Scytonema* (*S. siculum*, Moore and Couté, unpublished), a relatively large filamentous organism with a sheath diameter of 20-40 μm (Fig. 2a). Other cyanobacteria within this BMC include *Oscillatoria, Dichothrix, Chroococcus, Gleocapsa, Johannesbaptista, Gomphosphaeria* and *Spirulina*. Periphytic and epiphytic diatoms are particularly numerous throughout the lake and are not exclusive to the BMC of the microbialites. The most common genera are *Amphora, Brachysira, Cymbella, Entomoneis, Mastogloia, Navicula, Nitschia*, and *Synedra* (J. John, Curtin University, WA, pers. comm.). Macroalgae also occur in the lake and include *Ruppia megacarpa* and recently *Cladophora vagabunda*, as well as the charophytes *Lamprothamnium papulosum* and *Nitella* sp.

METAZOAN COMMUNITIES

An abundant and diverse community of metazoans is associated with the Lake Clifton thrombolites (Table 1). The fenestrae within the structures provide an important habitat for isopods, amphipods, coleopteran and trichopteran larvae, shrimps and juvenile

CNIDARIA
    Anthozoa   Diadumenidae   *Haliplanella luciae*
ASCHELMINTHES
    Nematoda   Unident. species
    Rotifera   Unident. species
ANNELIDA
    Polychaeta   Neriidae   *Ceratonereis aequisetis*
                                          Unident. species
    Hirudinea   Glossiphonidae   ? *Placobdella* sp.
ARTHROPODA
    Crustacea
        Anostraca   1 unident. species
        Ostrocoda   Cyprididae   1 unident. species
        Amphipoda   Melitidae   *Melita zeylancia*
                          Talitridae   *Talorchestia* sp.
                          Corophiidae   *Paracorophium excavatum*
        Isopoda   Sphaeromatidae   *Sphaeroma* sp.
        Decapoda   Palaemonidae   *Palaemonetes australis*
                          Parastacidae   *Cherax plebejus*
    Insecta
        Coleoptera   Hydrophilidae   1 unident. species
                          Tabanidae   larvae of unident. species
        Trichoptera   Leptoceridae   ? *Symphitoneuria wheeleri*
MOLLUSCA
    Gastropoda   Hydrobiidae   *Coxiella striatula*
                                           Potamopyrgus sp.
    Bivalvia   Erycinidae   *Arthritica semen*
BRYOZOA
    Gymnolaemata   Anascae   **Conopeum aciculata*
CHORDATA
    Osteichthyes   Gobiidae   *Pseudogobius olorum*
                                         Favonigobius suppositus
                        Atherinidae   *Atherinsoma (?wallacei)*
    Reptilia   Cheloniidae   *Chelodina oblonga*

* formerly identified as *Membranipora* sp. New identification by Sprigg and Bone (1993).

Table 1 : Metazoan fauna recorded in and around the Lake Clifton thrombolites between 1979 and 1988 (after Moore et al. , 1983, with additions).

gobiid fishes *(Pseudogobius olorum).* Other fauna frequently encountered within, or close to, the thrombolites include nematodes, polychaetes, ostracods, copepods and two other species of teleost fish (Table 1). Abundance can be appreciated by the fact that several thousand animals have been extracted from single thrombolites less than 30 cm in diameter (B. Knott, University of WA, pers.comm.). In view of the grazing habits of much of this fauna, it is reasonable to assume that the thrombolites provide both a source of food as well as refuge from predation (Moore *et al.,* 1983).
Two species of gastropod are generally found grazing on the sediment in the shallow regions of the foreshore rather than on and within the microbialites. Sampling with 65

mm diameter cores to a depth of 30 mm showed that *Coxiella striatula* commonly occurs with maximum densities exceeding 100 animals per 100 cm^3 of sediment, while the maximum density recorded for *Potamopyrgus* sp. was 154 individuals (Moore *et al.*, 1983). In addition, colonial bryozoans have been found in a number of areas where they not only colonise the thrombolites but are also incorporated into the framework of some of these structures. The sea anenome, *Haliplanella luciae*, a species widespread in Australian estuaries, is common in the northern portion of Lake Clifton, where it colonises both the thrombolites and areas of coarse sediments.

THROMBOLITES AND STROMATOLITES

The Clifton thrombolites occur principally along the eastern margins of the lake with some minor developments along the north-western shore. They lie on or are partially buried in the sediments of the lake basin. Radiocarbon dating of submerged thrombolites show ages ranging from 1 950 years BP to modern. Serially-dated examples imply minumum net growth rates of the calcified structure of 10 cm per 100 years.

External Morphology

The thrombolites of Lake Clifton exhibit a wide range of external morphologies including conical, domical, discoidal and tabular formations which vary considerably in size, as well as more irregular and columnar structures up to 1.3 m high. As documented by other studies of modern microbialites such as those in Shark Bay (e.g. Logan, Hoffman and Gebelein, 1974; Playford, 1980), external morphology appears to be primarily the product of the environmental setting. In Lake Clifton, seasonal fluctuations in water depth, regional variations in sedimentation rates and the effects of prevailing winds and currents are major controlling factors.

Large tabular, discoidal and domical thrombolitic microbialites, which range in diameter from 20 to 150 cm, are found at the north end of the lake. Along the north-eastern shoreline they have coalesced to form an extensive wave-resistant structure over 6 km long and in several places up to 120 m wide (Fig. 3a). This can appropriately be termed a reef after both Heckel (1974) and Fagerstrom (1987). The tabular thrombolites comprise a platform area comparable with the 'reef flat' of coral reefs. Thus the 'rear margin' of the 'reef flat' is the highest part of the formation, most regularly emergent and least exposed to wave activity, whilst the 'reef platform' and 'reef front' (although often emergent) are regularly exposed to wave activity.

The tabular thrombolites typically consist of a series of concentric rings 3-18 cm wide surrounding a central core (Fig. 3b). On the platform the rings often abut or overlap those of adjacent thrombolites, with outer rings encompassing two or even three

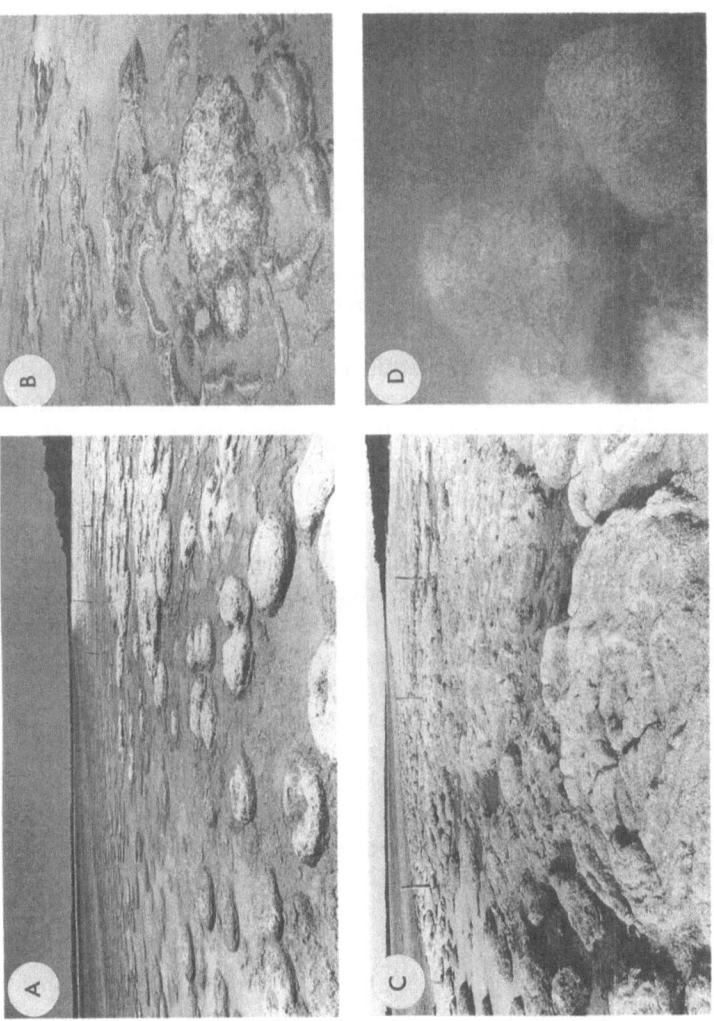

Figure 3: a) Part of the reef-like formation along north-eastern shore of Lake Clifton at the end of summer, showing the region where the tabular thrombolites forming the emergent platform give way to isolated domical structures; b) Tabular thrombolites with concentric rings. Some of the outer rings partially encircle two or more adjacent thrombolites ; c) Part of the platform area where the tabular thrombolites have coalesced to form this structure; d) Two permanently submerged, conical thrombolites (85 and 50 cm in height) lying in front of the platform where annual water depths typically range from 1 to 2 m.

formerly discrete structures. This growth pattern has clearly facilitated the development of a continuous and indurated platform (Fig. 3c). Processes thought to promote these Clifton ring formations are related to the lateral accretion of the structures where vertical growth is limited by shallow water. A similar process is involved in the formation of microatolls in modern corals (Stoddart and Scoffin, 1979; Scoffin and Stoddart, 1978). The marked delineations between the rings in the Lake Clifton structures are probably related to prolonged interruption and eventual resumption in BMC activity in a manner similar to the episodic growth of coral microatolls described by Woodroffe and McLean (1990) (see also Internal Morphology).

The tabular structures comprising the 'rear margin' are more widely spaced, lower in height, and have fragmented and incomplete rings. These features give this area a disorderly and rubbly appearance reminiscent of the rear margin of coral atoll reef flats. That the tabular microbialites might represent the eroded remains of once domical or conical structures which had formed when lake levels were much higher is difficult to accept for a variety of reasons. For example, their wide diameter would mean that 'giant' domical or conical thrombolites formed within a relatively short period of time. Moreover, there is no sign of relict or 'stranded' microbialites above the highest water mark of the present-day shoreline.

In front of the emergent platform, mainly domical thrombolites pass offshore into a deeper area (1-3 m), where the structures are conical, more isolated and interspersed with unconsolidated sediments (Fig. 3a). These permanently submerged conical structures are up to 1 m high (Fig. 3d). Gross 'external laminae' of these conical thrombolites are often present (Fig. 3d). They are semi-detached from the core of the structure, thereby producing a distinct outer shell or 'cloak' under which lakewater can readily circulate (Fig. 4c). Since the connections between the cloak and core are tenuous, the cloak is often fragile and easily dislodged (Burne and Moore, 1987). This is not the case, however, for the coalescent rings of the tabular thrombolites forming the platform (Fig. 3b,c). The fragility of the cloaks may in part be related to the steep sides and offshore location of the conical thrombolites, features which discourage sediment accumulation, infilling and the subsequent stabilisation of these layers.

To the south of the reef formation, there is a gradual transition to smaller and generally more widely separated thrombolites on a broad, gently-sloping shore of fine sands and silts. Discrete and mainly discoidal thrombolites (Fig. 4a) which range from 2 to 50 cm in diameter, are partially buried in unconsolidated sediment near and beyond the mean

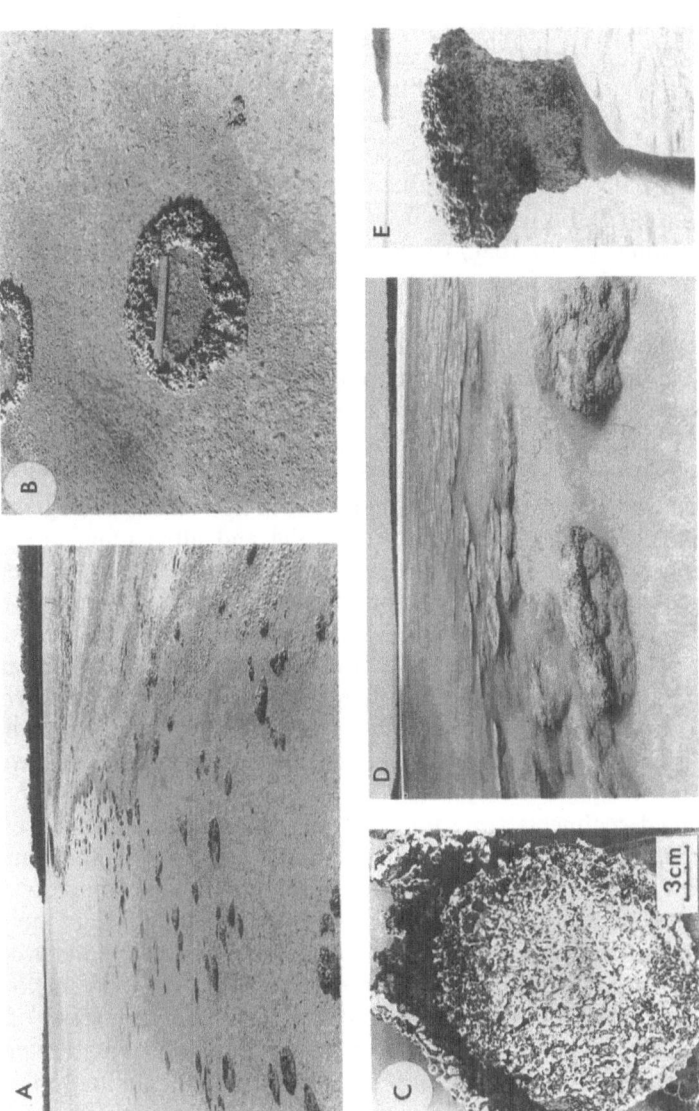

Figure 4: a) A stretch of the lower eastern shore of Lake Clifton south of the reef-like area, where discoidal thrombolites are distributed in a broad zone at and beyond the mean annual low water mark; b) Ring-like thrombolites at the same location as Fig. 4a, but lying just above the mean annual low water mark (ruler is 30 cm long); c) Cross-section of a small, permanently-submerged conical thrombolite with a single outer cloak enveloping the central core. Cavities and tubes (ie. the open fenestrae) occur predominantly near the surface of the core and cloak; d) View of the narrow deep-water channel showing some of the irregular tabular, sub-spheroidal and columnar thrombolites associated with this region of the lake. All thrombolites are submerged and refraction has exaggerated their tabular nature; e) Club-shaped thrombolite from the scalloped western edge of the narrow, deep-water channel. The stalk and lower half of the 'head' were buried in the unconsolidated fine sediments.

annual low water mark. However, many thrombolites, particularly those which lie at or just above the mean low water mark, are ring or partial-ring structures surrounding a necrotic centre of low relief. These have been described as "pustular doughnuts" (Fig. 4b; also Moore *et al.*, 1983). Grey and Thorne (1985) suggest that the formation of the necrotic centre may be due to emergence of the top of these thrombolites during periods of low water levels, a process similar to the genesis of microatolls. Subaerial exposure of the upper surface restricts growth to the margins of the structure producing a rim. The rim then serves to trap sediments which further inhibit growth and help produce the necrotic centre, thereby leading to a ring-like structure. It is likely, however, that a combination of processes may be involved, including wave action *per se*, as well as those processes operating during occasional periods of complete exposure. These include differential wicking (i.e. capillary draw) of groundwater and wetting by spray, both of which would tend to favour BMC growth on the perimeter of these thrombolites.

Towards the southern end of the lake and lying across the narrow, deep-water channel which passes around the sand spit, are discoidal, domical, sub-spheroidal and columnar thrombolites (Fig. 4d). These microbialites exhibit a wide range of sizes, with the largest (up to 1.3 m in height) being found in the deeper parts of the channel. Those lying along the scalloped edges of the sand spit are partially or almost completely buried by fine sediments, and many prove to be club-shaped upon removal (Fig. 4e).

The thrombolites in the channel are subjected to strong and alternating north/south currents over summer and autumn. These wind-driven currents result from afternoon south-westerly sea breezes giving way to nocturnal easterly winds. Compared to the structures on the north-eastern shore, the channel thrombolites are far more irregular in shape, have much flatter tops, and often have one or more holes up to 15 cm in diameter facing the direction of the wind-driven currents. It is not yet clear whether the irregular shapes of the channel thrombolites have arisen as a consequence of continual exposure to such currents, or whether they are the result of cyclical periods of burial, re-exposure and erosion.

The currents have eroded sediments from the base and sides of the channel, cutting down to the horizon of the former marine phase, exposing the band of shells dominated by *Katelysia*. Many of these shells form a substrate for thrombolite formation, with small structures 1-5 cm wide and 2-4 cm high attached to the upper surface of the shells (Fig. 5a). Since the surface sediments within the channel are continually disturbed by the currents, the tops of *Katelysia* shells exposed above the sediment provide the only stable substrate at the sediment-water interface that allows the formation of new thrombolites. Despite their location in such a dynamic environment, both the external and internal morphology of these small thrombolites are remarkably similar to those of their counterparts in much calmer areas of the lake (*cf.* Fig. 5a, 6d).

Apart from the small thrombolites forming on the *Katelysia* shells, the Clifton microbialites are not attached to hard substrates, but lie on or are partially buried in the unconsolidated sediments of the lake basin. Moreover, despite the number of artificial and natural hard substrates present in the lake, such as wooden fence posts, limestone

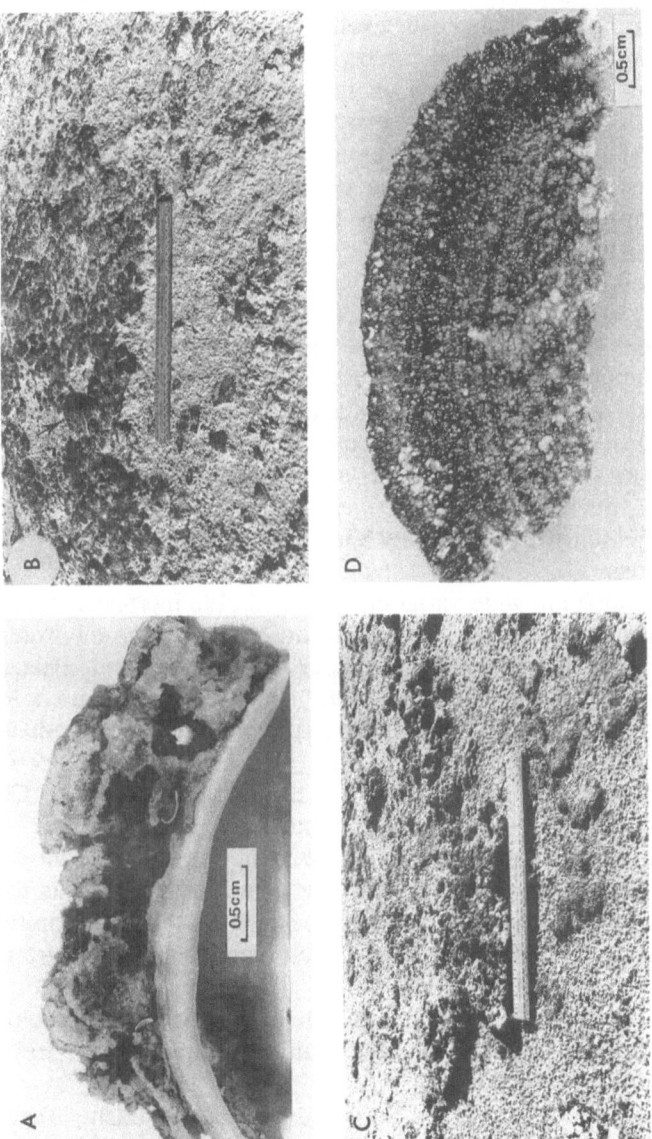

Figure 5: a) Cross-section of a small thrombolite formed on the upper surface of a *Katelysia* shell (taken from the centre of the deep-water channel); b) Aggregates of unlithified, pustular stromatolites (8-12 mm diameter) and a larger, domical stromatolitic microbialite (5 cm diameter; arrowed) on the upper eastern shore in an area of continuous groundwater resurgence. Ruler is 30 cm long in this and Figure 5c; c) Aggregates of partially lithified, pustular stromatolites in the mid-shore region of the same site as Figure 5b; d) Cross-section through one of the domical stromatolites (see Fig. 5b) showing alternating light and dark laminations reminiscent of 'pincushions' from Andros Island.

and clay bricks, PVC tubing, driftwood and outcropping limestone, the BMCs colonising these surfaces do not contain *Scytonema* and do not lithify. It is not yet clear why *Scytonema* is absent.

The three major types of thrombolite (ie. tabular, domical and conical/columnar) often comprise distinct, parallel bands along much of the shore. Since the height of the thrombolites is limited by the maximum lake level, it is not surprising that a recent study of external morphological variation along several transects perpendicular to the shoreline shows significant correlations between mean water depth and thrombolite height and shape (Moore, 1993). There is also a significant positive correlation between thrombolite height and energy of the environment, with taller thrombolites occurring in higher energy locations where fine sediments do not accumulate, and shorter, oblate or club-shaped structures occurring in lower energy areas where they protrude only slightly above the surrounding fine sediment.

In addition to the thrombolites, aggregates of small (less than 5 cm diameter) stromatolitic structures of low relief occur in Lake Clifton. These are confined to those regions of the upper eastern shoreline which, even when exposed by low lake levels during summer, remain saturated by low salinity groundwater seepage for most of this period (Fig. 5b). Although morphologically distinct from the thrombolites and often unlithified, these stratiform and domical microbialites are also dominated by *Scytonema*. In the mid-shore region, they are partially lithified (Fig. 5c). The unlithified stromatolitic structures (Fig. 5d) closely resemble the laminated "pincushions" described by Monty (1967), Monty and Hardie (1976) and Hardie (1977) from Andros Island, Bahamas.

Knowledge of the way in which the wide variation in thrombolite morphology in Lake Clifton reflects environmental factors may assist in understanding the environments which influenced the external morphology of both Proterozoic and Palaeozoic microbialites. For example, Grey and Thorne (1985) have drawn attention to the similarity between some fabrics of the Lake Clifton microbialites and stromatolites occurring in "upward-shallowing sequences" of the early Proterozoic Duck Creek Dolomite. They commented on a similarity between the fabric of the Lake Clifton "pincushions" and that of the microdigitate stromatolite *Asperia ashburtonia* Grey 1985. Likewise they suggested that the niche-like structures in the columnar form *Pilbaria perplexa* Walter, may have been formed by a mechanism similar to that controlling the formation of the necrotic centre of the 'pustular doughnut' thrombolites in Lake Clifton. From analogy with the Lake Clifton microbialites they suggested that a shallow-water, seasonally influenced environment dominated many of the upward-shallowing cycles observed in the Duck Creek Dolomite and in other geological sequences containing associations of niched and digitate stromatolites (Grey and Thorne, 1985).

The Cambrian thrombolites from the Shannon Formation (Amadeus Basin, Australia) and Petit Jardin Formation in Newfoundland represent even closer fossil analogues of the Lake Clifton thrombolites. For example, there are strong similarities between the large thrombolites from the deep-water channel in Lake Clifton and the thrombolite pillars of the Shannon Formation (Kennard and James, 1986, p. 493). According to Kennard and James (1986), these pillars also grew on a loose substrate and had high synoptic relief before becoming enveloped by wave-rippled sediments indicative of a relatively high-energy environment.

Internal Morphology

Although gross morphology varies between and within localities, the internal framework and dominant microbial components of the thrombolites are remarkably consistent throughout the lake. The internal structure consists of a calcified framework composed of mesoscopic clots (mesoclots) of microcrystalline aragonite, and an interframework (*sensu* Kennard and James, 1986) of fenestrae and detrital sediment (Fig. 6a,b).

Examination of thin sections of thrombolites from various parts of the Lake demonstrates that remnants of *Scytonema* filaments occur within the aragonitic mesoclots (Fig. 2b), indicating that this organism has been present since the genesis of the thrombolites. The filaments appear as radial arrays of golden-brown threads. This colour is most likely due to scytonemine, the persistent extracellular sheath pigment characteristic of *Scytonema*. The close association between living *Scytonema* filaments and the developing carbonate structure has been highlighted through the use of scanning electron microscopy (SEM) which shows the arrangement of dense clusters of aragonite crystals around individual filaments (Fig. 2c), as well as aragonitic crystals between filaments (Fig. 2d). However, the presence of *Scytonema* on and within the thrombolites does not necessarily imply any causal relationship with the genesis of the structure, and the above features alone do not provide sufficient evidence for microbially-influenced deposition of carbonate. The $\delta^{13}C$ values for the carbonate forming the mesoscopic framework of Lake Clifton thrombolites are markedly elevated ($+7.1$ $^{o}/_{oo}$) compared to those of the lake water (-7.0 $^{o}/_{oo}$) and groundwater (-11.7 $^{o}/_{oo}$). The fractionation evident in the Lake Clifton thrombolites is most probably the result of CO_2 uptake by the photosynthesising BMC, and indicates that the precipitation of the carbonate constituting the mesoclots is biologically-influenced (Moore, 1988; Moore and Turner, 1988). These data support the view that the major process involved in the formation of the Lake Clifton thrombolites is the precipitation of aragonite in a microenvironment determined by *Scytonema* and other members of the BMC, i.e. a process of biologically-influenced, non-skeletal mineralisation.

The clots are non-laminated and are the essential frame-building component of all thrombolites (Kennard and James, 1986; Burne and Moore, 1987). X-ray diffraction analyses and SEM with EDAX demonstrate that the mesoclots, which vary in width from 3-12 mm and exhibit a variety of geometrical shapes, arise from the accretion of arrays of 10-200 μm aragonitic crystal laths. These crystals are initially precipitated at the surface of the thrombolite, immediately around the mucopolysaccharide sheaths of *Scytonema* (Fig. 2c, 6c). The unusual form of the crystals closely resembles that of the

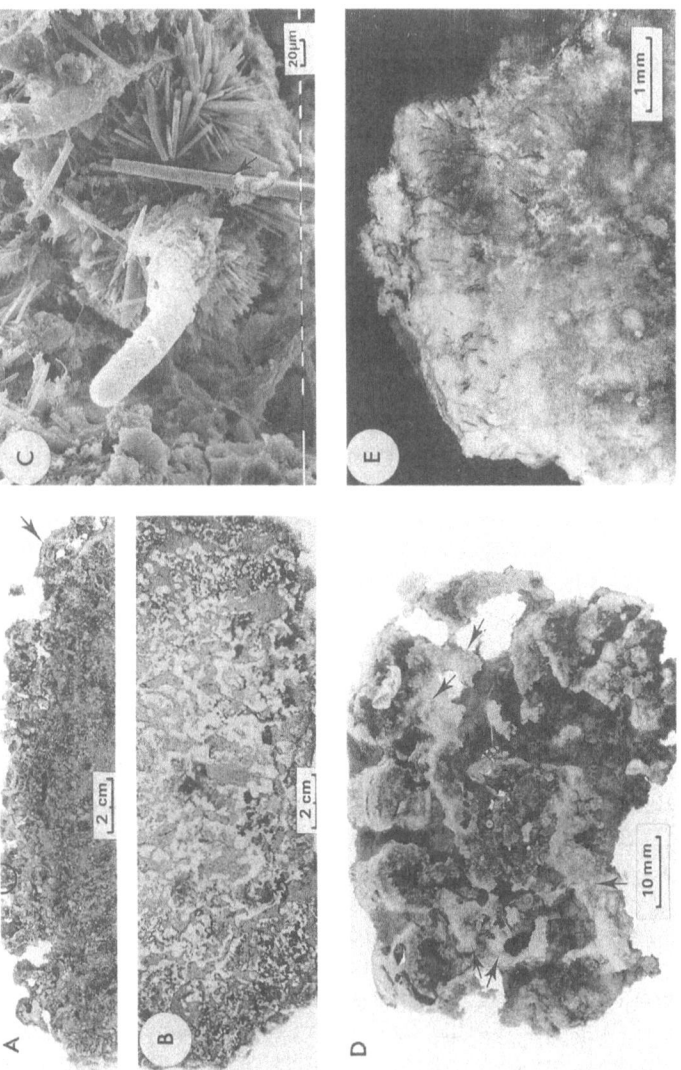

Figure 6: a,b Cross-sections of discoidal thrombolites: domical protuberances and the 'transient' fine laminations (arrowed) on pustular upper surface are evident in 6a, while the framework of aragonitic mesoclots (white) and interframework of sediment-filled fenestrae (grey) are clearly discernible in 6b. In Figure 6a, silica deposition and organic matter reduces the contrast between clots and the open and infilled fenestrae. A developing mesoclot in the circled protuberance is shown, magnified, in Figure 6c; c) Scanning electron micrograph of the surface of a developing mesoclot,showing dense arrays of small aragonite crystals surrounding a protruding filament of *Scytonema*. Needle-like diatom (*Synedra*) is arrowed; d) Cross-section of a small thrombolite showing the pustular and partially laminated upper surface comprising the 'living' BMC, and the lack of laminae where mesoclots have developed from previous surface pustules (arrows); e) View of part of a mesoclot from the upper surface of the thrombolite depicted in Figure 6a, showing *Scytonema* filaments embedded in the aragonitic matrix.

"aragonite lath cement" described by Schroeder (1972) from the algal cup reefs in Bermuda.

Although fine laminae are sometimes conspicuous in the top 2-5 mm BMC layer of a thrombolite (Figs 6a arrowed, 6d), continued precipitation of aragonite and the formation of mesoclots leads to their destruction. Thus microscopic examination of the mesoclots lying just below the living BMC frequently reveals numerous radially-arranged and intact filaments of *Scytonema* embedded in a microcrystalline matrix i.e. the forming mesoclot (Fig. 6e). It is not yet clear, however, whether the continued precipitation of aragonite in the layers immediately below the surface of the Clifton thrombolites is due to *Scytonema* alone, or to bacteria, physico-chemical processes or a combination of all three (see e.g. Monty, 1976; Chafetz and Folk, 1984). Bacteria can produce local areas of CO_2 build-up leading to internal secondary carbonate dissolution and subsequent reprecipitation (Golubic, in Monty, 1976), and thus bacterial action may also account for the more reticulate patterns of the mesoclots that are occasionally discernible in the deeper parts of the framework (Fig. 6b).

The fenestrae comprising the interframework are either open, thereby creating convoluted cavities and tubes, or filled with fecal pellets, quartz sands and unconsolidated carbonate sediments, including the carapaces of ostracods, shells of bivalves (*Arthritica semen*) and gastropods (principally *Coxiella*) (Fig. 7a,b). Open fenestrae occur predominantly near the surface (Fig. 4c, 6a), a feature probably related to the activities of metazoan inhabitants. The pattern of growth exhibited by the Lake Clifton thrombolites strongly indicates that these fenestrae are intrinsic, and that their formation is primarily related to the topography of the surface of the developing microbialite rather than to excavation of the structure by metazoan activity.

Many of the open and filled fenestrae are lined by a dark brown coating up to 0.5 mm thick which covers the surface of the mesoclots (Fig. 7b). This lining is characteristically smooth when wet and cracked when desiccated, and SEM with EDAX and XRD has shown the layer to be composed almost entirely of non-crystalline silica. The origin of this amorphous silica probably results from the dissolution of diatom frustules and subsequent reprecipitation.

The small, low-relief stromatolitic structures from areas along the upper eastern shoreline have a very similar, *Scytonema*-dominated BMC to that of the thrombolites, but the soft internal fabric of these "potential stromatolites" (*sensu* Gerdes and Krumbein, 1984) lacks mesoscopic clots of aragonite. The internal structure exhibits discrete, alternating light and dark laminations 0.2-0.5 mm in width, interspersed with grains of carbonate and quartz sand (Fig. 5d). The absence of the prominent mesoclots which characterise the thrombolites may in part be due to the fact that the upper foreshore region is only inundated by saline lakewaters during winter and early spring, when both salinity and water temperatures are low and day length is short. In other words, such conditions may not be conducive to the precipitation of sufficient quantities of carbonate to form the large clots that replace the fine laminae. Grey and Thorne

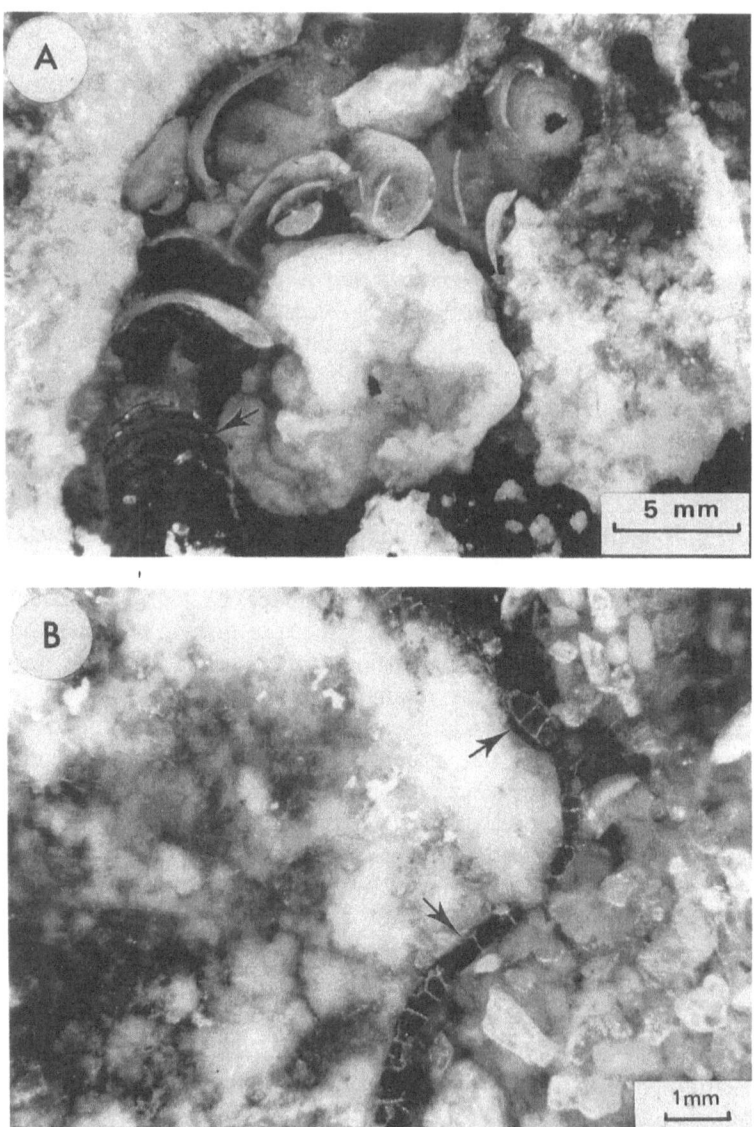

Figure 7: a) Shell material (*Coxiella* and *Arthritica*) infilling fenestrae. An isopod is just visible in lower left-hand corner (arrow); b) Carbonate and quartz grains infilling a fenestra. Dark lining of amorphous silica (arrows) can be seen along edge of mesoclot

(1985) have compared the internal fabric of the Clifton "pincushions" with that of the 2 Ga old *Asperia ashburtonia* from the Duck Creek Dolomite (Wyloo Group) of north-western Australia. These authors attributed the formation of the light and dark laminae of the "pincushions" to cyclical inundation and exposure.

Comparisons between the unlithified, partially lithified and lithified formations along the eastern shore of Lake Clifton provide an almost unbroken 'sequence' in microbialite development. Scytonemacean cyanobacteria characteristically display radial arrays of erect filaments which produce a pustular or "botryoidal" (*sensu* Monty, 1976) surface of mm- to cm-sized domical protuberances (Fig. 6a,d,e). This growth pattern produces a number of active zones surrounded by regions of less relief, many of which are often rapidly covered by sediment. Thus the growth and penecontemporaneous mineralisation of two or more coalescing active pustular surfaces leaves behind a complex clotted structure made up of aragonitic mesoclots and voids, the latter being later infilled principally by lake sediments. Monty (1976, p.231) cites similar modern examples where "the constitutive clots represent calcified algal colonies endowed with a botryoidal surface growth form. As the various algal knobs merge, numerous cavities are created, thereby producing an irregular fenestral fabric". Finally, in those situations in Lake Clifton where insufficient aragonite is precipitated to form the clots, as appears to be the case for the small "pincushion" stromatolites along the upper eastern foreshore, original laminations formed by the growing BMC are not destroyed but remain intact throughout much of the structure.

The above observations strongly indicate that the formation of fenestrae is related far more to the surface topography of developing thrombolites than to the activities of the diverse and abundant metazoan fauna which they harbour. The presence of an amorphous-silica coating in both filled and unfilled fenestrae is consistent with this view. This is not to say that metazoan activity cannot lead to minor modifications of the cavities. However, there are no macroborings of a scale similar to those described by Ekdale, Brown and Feibel (1989) from Pleistocene stromatolites of Kenya. The fauna also contributes to the eventual infilling of fenestrae via transportation of lake sediments and deposition of fecal pellets.

While the clotted texture of the thrombolites precludes the retention of fine laminae, some of the larger domical microbialites exhibit a series of coarse layers between 1 and 5 cm thick (Fig. 8). It is clear from the size of these coarse and often indistinct bands that they are far too large to represent seasonal growth rings. However, while they do not strictly parallel the coalescent rings and cloaks of the tabular and conical thrombolites, it is likely that they also reflect past interruptions to BMC activity and mineralisation.

The internal structure of the Clifton thrombolites parallels that of Cambrian thrombolites, with the various shapes of the mesoclots and the interframework of infilled fenestrae comparable with those found in the Shannon Formation (Kennard and James, 1986; Kennard, this volume), although the size range of the Clifton mesoclots is somewhat smaller. Their size is, however, close to that of the mesoclots of the

24   L. S. MOORE AND R. V. BURNE

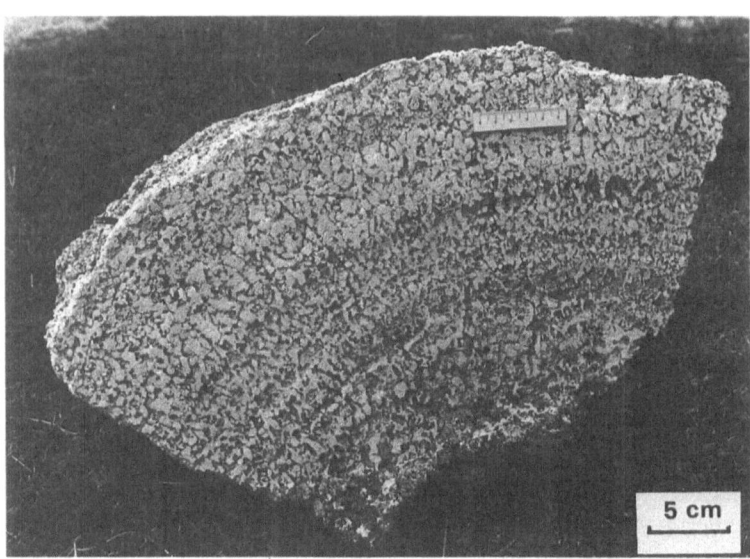

Figure 8 : Cross-section of a large domical thrombolite (50 cm diameter) from the edge of the platform, showing two distinct and four indistinct coarse laminations.

middle-Cambrian thrombolites from the Petit Jardin Formation of Newfoundland (Kennard, Chow and James, 1989).

The Lake Clifton thrombolites develop from the biologically-influenced precipitation of aragonite within a BMC microenvironment dominated by the filamentous *Scytonema*. Kennard and James (1986), on the other hand, suggested that microstructures within the calcitic mesoclots of most Cambrian/Ordovician thrombolites resulted from *in-situ* calcification of coccus-dominated communities.   Apart from the remains of the *Scytonema* filaments, complex microstructures are not discernible in the Lake Clifton mesoclots. Kennard and James (1986) also suggested that "thrombolites *sensu stricto* do not appear to occur in any carbonate environment today", since they believed that the closest analogous modern structures are only "partially clotted", in which "the individual microbial clots are generally poorly-defined". However, the Lake Clifton microbialites clearly exhibit well-defined mesoclots throughout their structure, and a recent comparative study by Kennard (1988) has confirmed that close analogies can be drawn not only between Cambrian thrombolites and those in Lake Clifton, but also with those in Great Salt Lake, Utah (Halley 1976) and Green Lake, New York (Eggleston and Dean, 1976; Thompson, Ferris and Smith, 1990).

The recent study by Thompson *et al.* (1990) of the living thrombolites in Green Lake supports Kennard's (1988) conclusion that the mesoclots of Cambrian thrombolites resulted from mineralisation associated with coccoid microorganisms rather than filamentous types, since the Green Lake structures appear to be produced by the colonial coccoid cyanobacterium, *Synechococcus*. However, while coccoid communities may have dominated the BMCs responsible for both the Cambrian and Green Lake examples and led to complex microstructures, the situation in Lake Clifton demonstrates that not all thrombolites exhibit mesoclots with diverse and complex microstructures, or are necessarily the product of coccus-dominated BMCs.

CONCLUSION

Both modern and Cambrian thrombolites appear to be actively lithifying structures able to coexist with a grazing and burrowing fauna. This feature is not inconsistent with the fact that throughout the Phanerozoic, a suite of microbial structures (stromatolites, thrombolites and endostromatolites [*sensu* Monty, 1984]) continued to make significant contributions to reefs of more mixed assemblages (Monty, 1973, 1984; Pratt, 1982; Kennard and James, 1986; Fagerstrom, 1987; Geldsetzer, James and Tebbutt, 1988). Indeed, even Garrett (1970) noted that "reef stromatolites" which became cemented penecontemporaneously would have been less susceptible to destruction by burrowers. Thus calcifying BMCs were able to persist as 'cryptoflora' in marine reefs (Pratt, 1982), and continued to form significant carbonate formations in certain environments after the Palaeozoic, such as the Upper Permian reefs of the Zechstein basin in central Europe (Paul, 1980).

The modern thrombolites of Lake Clifton and the fossil thrombolites of the Shannon Formation exhibit a complex microbial-metazoan ecosystem. Kennard and James (1986) state that the Cambrian thrombolites "were commonly inhabited by an abundant and diverse skeletal and soft-bodied metazoan fauna which included various trophic groups such as grazers, detritus feeders, suspension feeders and carnivores", as is also the case in Lake Clifton.

Modern microbial mats appear to be particularly susceptible to the pressures of grazing and burrowing, and their development is restricted to 'extreme' environments which preclude disruptive metazoan faunas (Garrett, 1970). An apparent vulnerability to burrowing and grazing has been used to explain the "crisis" in stromatolite diversity and abundance at the end of the Proterozoic (e.g. Garrett, 1970; Awramik, 1971, 1991; Stanley, 1973; Walter and Heys, 1985). It has also been suggested that fossil thrombolites represent stromatolites that were burrowed and bored and, hence, owe their origin to the activity of these animals (Walter and Heys, 1985).

However, the internal structures of stromatolites that have been burrowed and bored are distinct from those of thrombolites (*c.f.* Fig. 10 of Burne and Moore, 1987, with Figs 4-6 of Cao, 1988). Moreover, the pattern of growth exhibited by the Lake Clifton thrombolites strongly indicates that their fenestrae are intrinsic, further supporting the view of Monty (1976) and Kennard and James (1986) that bioturbation need not be the principal cause of fenestrae within fossil thrombolites.

If it is accepted that at least the Palaeozoic thrombolites were able to withstand grazing pressures, their eventual demise in oceanic and littoral environments must then be attributed to factors other than non-competitive exclusion by grazing organisms. Proposed factors have included competition for nutrients, competition for space, and/or changing chemical conditions (Monty, 1973, 1984; Grotzinger, 1990).

Competition for nutrients is unlikely to have been a significant factor in the decline of marine microbialites. The low nutrient demands of microbialite forming BMCs (see e.g. Skyring and Bauld, 1990) strongly support this view. Indeed, low nutrient levels would tend to favour BMCs by inhibiting metaphyte growth. In this context, increasing phosphate levels in Lake Clifton as a result of human activities has been paralleled by increased growth of the green alga, *Cladophora vagabunda* in many parts of the Lake, including the thrombolite reef (Moore and Turner, 1988). The development of *Cladophora* can be seen as a form of spatial competition between this epiphytic macroalga and the BMC forming the thrombolites rather than competition for nutrients.

So far as competition for space is concerned, early thrombolites were able to co-exist with archeocyathans, early corals and stromatoporoids which grew at similar rates (Eicher and McAlester, 1980; Pratt and James, 1982: Pratt, 1984). A significant factor causing the decline of marine thrombolites may have therefore been the evolution of faster-growing, calcifying and reef-building organisms capable of outstripping the 10cm per 100 year growth rates of lithified microbialites (Pratt and James, 1982; Kennard and James, 1986; Fagerstrom, 1987).

Kennard and James (1986) suggested that "changes in the composition of the earth's atmosphere or oceans" may have promoted the penecontemporaneous mineralisation achieved by BMCs associated with Cambrian thrombolites. The rapid lithification and durable nature of the Lake Clifton thrombolites is consistent with the fact that the Cambrian examples were able to diversify and radiate despite the concomitant explosion of metazoan life. However, calcification and rapid mineralisation by the BMC in Lake Clifton is clearly being facilitated by the seepage of carbonate-enriched groundwaters into a saline environment (Moore, 1987; Moore, 1993), and Grotzinger (1990) has suggested that the Late Proterozoic decline of stromatolites was related not to metazoan evolution, but to changes in marine chemistry inhibiting calcification. Thus it remains unclear to what extent the subsequent decline of marine thrombolites during the Ordovician was due to spatial competition from the more rapid eukaryotic calcifiers or to changes in sea-water chemistry affecting cyanobacterial mineralisation.

ACKNOWLEDGMENTS

Gratitude is expressed to R. Folk, J. Marshall and J. Schroeder for discussion of carbonate cements, to K. Grey, R.W. Hilliard, J.M. Kennard, C. Monty, J. Sarfati and P. Thorpe for helpful comment and constructive criticism of the text, and to K. Barrett, J. Mifsud, F. Shilton and P. Williams for assistance with reproduction of the figures. This work forms part of LSM's PhD thesis and research was supported by a Commonwealth Postgraduate Research Award. RVB publishes with the permission of the Director, Australian Geological Survey Organisation.

REFERENCES

Aitken, J.D. (1967) "Classification and environmental significance of cryptalgal limestones and dolomites with illustrations from the Cambrian and Ordovician of southwestern Alberta", J Sedimentary Petrology 37, 1163-1178

Awramik, S.M. (1971) "Precambrian columnar stromatolite diversity: reflection of metazoan appearance", Science 174, 825-827.

Awramik, S.M. (1981) "The pre-Phanerozoic biosphere - three billion years of crises and opportunities", in M.H. Nitecki (ed) Biotic crises in ecological and evolutionary time, Academic Press, New York, pp 83-102.

Awramik, S.M. (1984) "Ancient stromatolites and microbial mats", in Y. Cohen, R.W. Castenholz and H.O. Halvorson(eds), MBL Lectures in biology. Microbial mats: stromatolites, Alan R Liss, New York, pp 1-22.

Awramik, S.M. (1991) "Archaean and Proterozoic stromatolites", in R. Riding (ed), Calcareous Algae and Stromatolites, Springer-Verlag, Berlin, pp 289-304.

Awramik, S.M. and Margulis, L. (1974) "Definition of stromatolite", Stromatolite Newsletter, 2, 5.

Biggs, E.R. (1977) Mandurah (2032 IV). "Urban Geology Map (1:50 000)", Geological Survey of Western Australia, Perth.

Burne, R.V. and Moore, L.S. (1987) "Microbialites: organosedimentary deposits of benthic microbial communities", Palaios 2, 241-254.

Cao, R.J. (1988) "Study on stromatolitic decline event in terminal Precambrian", Acta Palaeontologica Sinica 27, 737-750.

Chafetz, H.S. and Folk, R.L. (1984) "Travertines: depositional morphology and the bacterially constructed constituents", J. Sedimentary Petrology 54, 289-316.

Commander, D.P. (1988) "Geology and hydrogeology of the 'superficial formations' and coastal lakes between the Harvey and Leschenault Inlets (Lake Clifton Project)", Western Australian Geological Survey Professional Papers, Report No 23, 37-50.

Eggleston, J.R. and Dean, W.E. (1976) "Freshwater stromatolitic bioherms in Green Lake, New York", in M.R. Walter (ed) Stromatolites. Developments in Sedimentology, vol 20. Elsevier, Amsterdam, Oxford, New York, pp 479-488.

Eicher, D.L. and McAlester, A.L. (1980) "History of the Earth", Prentice Hall, New Jersey, 413p.

Ekdale, A.A., Brown, F.H. and Feibel, C.S. (1989) "Nonmarine macroborings in Early Pleistocene Algal biolithites (stromatolites) of the Turkana Basin, Northern Kenya", Palaios 4, 389-396.

Fagerstrom, J.A. (1987) "The evolution of reef communities", John Wiley and Son, New York. 600p.

Garrett, P. (1970) "Phanerozoic stromatolites: Non-competitive ecologic restriction by grazing and burrowing animals", Science 169, 171-173.

Geldsetzer, H.H.J., James, N.P. and Tebbutt, G. (1988) "Reefs: Canada and adjacent areas", Canadian Society of Petroleum Geologists, Memoir 13, Calgary, Alberta. 775p.

Gerdes, G. and Krumbein, W.E. (1984) "Animal communities in recent potential stromatolites of hypersaline origin", in Y. Cohen, R.W. Castenholz and H.O. Halvorson(eds), MBL Lectures in biology. Microbial mats: stromatolites, Alan R Liss, New York, pp 59-83

Golubic, S. (1991) "Modern stromatolites: A review", In R. Riding (ed) Calcareous Algae and Stromatolites, Springer-Verlag, Berlin, pp 541-561.

Grey, K. and Thorne, A.M. (1985) "Biostratigraphic significance of stromatolites in upward shallowing sequences of the early Proterozoic Duck Creek Dolomite, Western Australia", Precambrian Research 29, 183-206.

Grotzinger, J.P. (1990) "Geochemical model for Proterozoic stromatolite decline", American Journal of Science 290A, 80-103.

Halley, R.B. (1976) "Textural variation within Great Salt Lake algal mounds" in M.R. Walter (ed) Developments in Sedimentology, vol 20. Stromatolites. Elsevier, Amsterdam, Oxford, New York, pp 435-445.

Hardie, L.A. (1977) "Sedimentation of the modern carbonate tidal flats of northwest Andros Island, Bahamas", John Hopkins University Press, Maryland.

Heckel, P.H. (1974) "Carbonate buildups in the geological record: a review", Society of Economic Paleontologists and Mineralogists, Special Publication 18, 90-154.

Kalkowsky, E. (1908) "Oolith und stromatolith in Norddeutschen Buntsandstein", Deutsche Geologisches Gesellschaft Zeitschrift 60, 68-125.

Kennard, J.M. (1988) "The structure and origin of Cambro-Ordovician thrombolites, Western Newfoundland", PhD Thesis, Department of Earth Sciences, Memorial University of Newfoundland, Canada.

Kennard, J.M. and James, N.P. (1986) "Thrombolites and stromatolites: two distinct types of microbial structures", Palaios 1, 492-503.

Kennard, J.M., Chow, N. and James, N.P. (1989). "Thrombolite-stromatolite bioherm, middle Cambrian, Newfoundland", in H. Geldsetzer, N.P. James and G. Tebbutt (eds), Reefs: Canada and adjacent areas, Canadian Society of Petroleum Geologists, Memoir 13, Calgary, Alberta, pp 151-155.

Logan, B.W., Hoffman, P. and Gebelein, C.D. (1974) "Algal mats, cryptalgal fabrics and structures, Hamelin Pool, Western Australia", American Association Petroleum Geologists Memoir 22, 140-194.

Monty, C.L.V. (1967) "Distribution and structure of Recent stromatolic algal mats, eastern Andros Island, Bahamas", Annals de la Societe Géologie Belgique Bull 90, 55-100.

Monty, C.L.V. (1973) "Precambrian background and Phanerozoic history of stromatolitic communities, an overview", Annals de la Societe Géologie Belgique Bull 96, 585-624.

Monty, C.L.V. (1976) "The origin and development of cryptalgal fabrics", in M.R. Walter (ed), Developments in Sedimentology, vol 20. Stromatolites. Elsevier, Amsterdam, Oxford, New York, pp 193-249.

Monty, C.L.V. and Hardie, L.A. (1976) "The geological significance of the freshwater blue-green algal calcareous marsh", in M.R. Walter (ed), Developments in Sedimentology, vol 20. Stromatolites. Elsevier, Amsterdam, Oxford, New York, pp 447-477.

Monty, C.L.V. (1984) "Stromatolites in Earth history", Terra Cognita 4, 423-430.

Moore, L.S. (1987) "Water chemistry of the coastal saline lakes of the Clifton-Preston lakeland system, south-western Australia, and its influence on stromatolite formation", Australian Journal of Marine and Freshwater Research 38, 647-660.

Moore, L.S. (1988) "Modern thrombolitic microbialites and their geological significance", Terra Cognita 8 (3), 225 (abstract).

Moore, L.S. (1993) "The modern microbialites of Lake Clifton, south-western Australia", PhD Thesis, Department of Microbiology, University of Western Australia.

Moore, L.S. and Turner, J.V. (1988) "Stable isotopic, hydrogeochemical and nutrient aspects of lake-groundwater relations at Lake Clifton", in Proceedings of the Swan Coastal Plain Groundwater Management Conference, Western Australian Water Resources Council, pp 252-282.

Moore, L.S., Knott, B. and Stanley, N.F. (1983) "The stromatolites of Lake Clifton, Western Australia", Search 14, 309-314.

Neil, J. (1984) "Microbiology of the mats and stromatolites of the Clifton-Preston lake complex", Honours Thesis, Department of Microbiology, University of Western Australia.

Paul, J. (1980) "Upper Permian algal stromatolite reefs, Hartz Mountains (F. R. Germany)", Contributions to Sedimentology 9, 253-268.

Playford, P.E. (1980) ."Environmental controls on the morphology of modern stromatolites at Hamelin Pool, Western Australia", Western Australian Geological Survey Annual Report 1979, 73-77.

Playford, P.E. and Cockbain, A.E. (1976) "Modern algal stromatolites at Hamelin Pool, a hypersaline basin in Shark Bay, Western Australia", in M.R. Walter (ed), Developments in Sedimentology, vol 20. Stromatolites. Elsevier, Amsterdam, Oxford, New York, pp 389-411.

Playford, P.E., Cockbain, A.E. and Lowe, G.H. (1976) "Geology of the Perth basin, Western Australia", Western Australian Geological Survey Bulletin 124, 311p.

Pratt, B.R. (1982) "Stromatolite decline - a reconsideration", Geology 10, 521-515.

Pratt, B.R. (1984) "Epiphyton and Renalcis - Diagenetic microfossils from calcification of cocooid blue-green algae", J. Sedimentary Petrology. 54, 948-971.

Pratt, B.R. and James, N.P. (1982) "Cryptalgal-metazoaan bioherms of Early Ordivician age in the St George Group, western Newfoundland", Sedimentology 29, 543-569.

Schroeder, J.H. (1972) "Fabrics and sequences of submarine carbonate cements in Holocene Bermuda cup reefs", Geologische Rundschau 61, 708-730.

Scoffin, T.P. and Stoddart, D.R. (1978) "The nature and significance of microatolls" Philosophical Transactions of the Royal Society of London B 284, 99-122.

Searle, D.J. and Semeniuk, V. (1985) "The natural sectors of the Rottnest Shelf adjoining the Swan Coastal Plain", J Royal Society Western Australia 67, 116-136.

Semeniuk, V. and Johnson, D.P. (1985) "Modern and Pleistocene rocky shore sequences along carbonate coastlines, southwestern Australia", Sedimentary Geology 44, 225-261.

Semeniuk, V. and Meagher,T.D. (1981a) "The geomorphology and surface processes of the Australind-Leschenault Inlet coastal area", J Royal Society Western Australia 64, 33-51.

Semeniuk, V. and Meagher, T.D. (1981b) "Calcrete in Quaternary coastal dunes in southwestern Australia: a capillary rise phenomenon associated with plants", J. Sedimentary Petrology 51, 47-68.

Semeniuk, V. and Semeniuk, C.A. (1991) "Radiocarbon ages of some coastal landforms in the Peel-Harvey estuary, south-western Austarlia", J Royal Society Western Australia 73, 61-71.

Skyring, G.W. and Bauld, J. (1990) "Microbial mats in Australian coastal environments", Advances in Microbial Ecology 11, 461-498.

Sprigg, M. and Bone, Y. (1993). "Bryozoa in Coorong-type lagoons, Southern Australia", Transactions of the. Royal Society of South Australia 117, 87-95.

Stanley, S.M. (1973) "An ecological theory for the sudden origin of multicellular life in the Late Precambrian", Proceedings of the National Academy of Science USA 70, 1486-1489.

Stoddart, D.R. and Scoffin, T.P. (1979) "Microatolls: Review of form, origin and terminology", Atoll Research Bulletin 224, 1-17.

Thompson, J.B., Ferris, F.G. and Smith, D.A. (1990) "Geomicrobiology and sedimentology of the mixolimnion and chemocline in Fayetteville Green Lake, New York", Palaios 5, 52-75.

Walter, M.R. and Heys, G.R. (1985) "Links between the rise of metazoa and the decline of stromatolites", Precambrian Research 29, 149-174.

Walter, M.R., Bauld, J. and Brock, T.D. (1976) "Microbiology and morphogenesis of columnar stromatolites (Conophyton, Vacerrilla) from hot springs in Yellowstone National Park", in M.R. Walter (ed), Developments in Sedimentology, vol 20. Stromatolites. Elsevier, Amsterdam, Oxford, New York, pp 273-310.

Woodroffe, C. and McLean, R. (1990) "Microatolls and recent sea level change on coral atolls", Nature 344, 531-534.

BACTERIALLY CONTROLLED CALCIFICATION OF FRESHWATER SCHIZOTHRIX-STROMATOLITES: AN EXAMPLE FROM THE PIENINY MTS, SOUTHERN POLAND

J. SZULC[1] and B. SMYK[2]
[1] Inst. of Geol. Sci. Jagellonian University, Oleandry Str. 2a, Cracow, Poland
[2] Dept. of Microbiology, Agricult. University, Mickiewicza Str. 24/28, Cracow, Poland

ABSTRACT

The recent *Schizothrix* stromatolites develop within a slope-spring zone in the Pieniny Mts. The stromatolite-forming processes were studied in terms of the seasonally controlled microbial succession within the mat.

The microbial mat comprises abundant algal and bacterial communities including: *Schizothrix calcicola*, *Schizothrix rubra*, diatoms (e.g. *Synedra ulna*, *Gomphonema*, *Achnantes* sp.) as well as chemolithotrophic and chemoorganotrophic, alkalophilic bacteriae (e.g. *Streptomyces*, *Thioploca*, *Seliberia*, *Pseudomonas*, *Bacillus*, *Arthrobacter* sp.).

The complex interrelationships between the microorganisms related to their ecological succession during the year, clearly influence the mineral fabrics of the resulting growth laminae. The spring bloom of both cyanobacteria and diatoms leads to carbonate supersaturation and intense calcite precipitation within the mat. All this results in the dense carbonate fabrics of the spring lamina. The summer declining of the diatoms leads to limitation of the active microbial calcification of the mucus coating the *Schizothrix* trichomes, hence the calcified framework becomes more porous. The autumn diatoms activity and the atrophy of cyanobacteriae result in porous, chaotically arranged carbonate facies.

The study revealed that the major part of the carbonates occurring within the stromatolites, consists of low-Mg calcite rods; mineral replicas of the bacteria cells and colonies. The preserved segmentation of the rods reflects the process of the bacterial fission accompanied by autolytically induced cell calcification.

The polyhedral-like habit of the calcified colonies is thought to be resultant of the competition between the biological growth mechanisms and the physicochemical crystallisation powers. The observed "post mortem" evolution of the calcified colonies into the polyhedral and euhedral calcite grains, supports such inference.

J. Bertrand-Sarfati and C. Monty (eds.), Phanerozoic Stromatolites II, 31–51.
© 1994 *Kluwer Academic Publishers.*

INTRODUCTION

Bacteria as an agent of carbonate precipitation has long been recognized (cf. compilation by Pia, 1934). The calcifying bacteria may themselves build carbonate sediments although they may also collaborate with other organisms (mainly cyanobacteria and algae) enhancing the calcification of microbial mat (Krumbein, 1978).
This paper deals with the recent formation of freshwater cyanobacterial stromatolites where the bacteria seem to have a decisive role to play. The studies included *in vitro* culture of the microorganisms participating in the stromatolite formation and the examination of the carbonates found within microbial mats.

GENERAL SETTING AND HYDROCHEMISTRY

The *Schizothrix* stromatolites described here occur in the Szopczanski Gorge in the Pieniny Mts, Southern Poland (Fig.1). They grow along a fault-spring zone several metres long, on the western slope of the gorge (Fig.2). The bedrock comprises Upper Jurassic, light-coloured limestones with cherts (Birkenmajer, 1979). The site is situated at the altitude of about 650 m. Average daily temperatures range from -6.0 °C in February to +15 °C in July, with annual mean of 6 °C. The annual rainfall amounts to 850 mm (Kostrakiewicz, 1982).

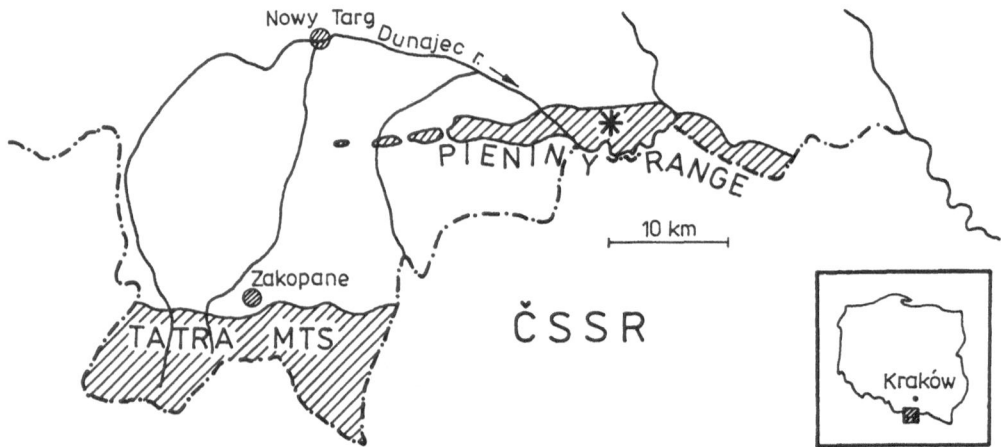

Figure 1 : Location of the study site (marked by asterisk).

The stromatolites occur in discontinuous small patches reaching 50 cm in diameter and up to 8 cm in thickness (Fig.3). A debris fan, up to 5 m high, forms at the base of the slope as a result of peeling off of the stromatolite, following winter frost-effects.
The chemical properties of the spring water were measured at the head part of the spring, above the mat, and in the lower part of the slope, below mat (Fig.2), during the

Figure 2 : View of the study site. Stromatolite cover occurs in the central, dark part of the site. Arrows mark the places of hydrochemical measurements. f - stromatolite-debris fan.

four seasons of 1986. All properties, but the Ca^{2+}-content, were almost constant through the year:
pH ranges between 7.2 and 7.4, CO_2 (free)-contents; 3.5 mg/l and HCO_3^--contents up to 200 mg/l. The Ca^{2+}-contents drops from 180 mg/l in the headwater to 35-60 mg/l in the lower part of the spring, suggesting significant water decalcification within the microbial mat zone. The contents of the other ions (Mg^{2+}, SO_4^{2-}, Fe^{3+} and Cl^-) are negligible.

MATERIALS AND METHODS

Samples were collected seasonally: in January, April, July and September of 1986. In order to examine the processes of carbonate formation and diagenesis, samples were taken from the topmost, living part of the stromatolite as well as from the depths of 3 and 8 cm in the fossil one.

Samples were examined, within 24 hours, using conventional and scanning electron microscopy (JXA 50A and Jeol 72 - electron scanning microscope). Infra-red Fourier's spectrometry was used to determine the mineralogical composition of the stromatolite. In order to trace the effect of early diagenesis on stable isotopes contents, $\delta^{13}C$ and $\delta^{18}O$-compositions were measured in selected samples using a Micromass 602 c Spectrometer. The hydrochemical measurements were made directly in the field using the field laboratory - MP-1 (Markowicz and Pulina, 1979).

The micriobiological examination of 18 samples collected in the field were carried out in the Department of Microbiology, Agricultural University, Cracow. The isolation of chemoautotrophic and chemolithotrophic bacteria capable of dissolution and precipitation of calcium carbonates was carried out on special selective media with the use of standard geomicrobiological techniques (Aaronson, 1970; Boquet *et al.*, 1973; Horikoshi and Akiba, 1981; Silverman and Ehrlich, 1964; Smyk, 1970; Smyk and Drza, 1964; Smyk and Ettlinger, 1963).

The isolation and culture of bacteria active in formation of calcium carbonates were carried out on a modified solid medium, in climatic chamber at 25 °C and at various pH (from 8.0 - 8.8), according to the method by Boquet *et al.*, (1973).

Natural calcium carbonates (calcite and dolomite) and aluminosilicates (olivine, wollastonite, uranotile, orthoclase and biotite) were used in all the microbiological and biogeochemical investigations. The determination of biochemical and biogeochemical abilities of bacteria in dissolution and formation of calcium carbonates was based on standard methods (Gould and Corry, 1980; Krumbein, 1983; Silverman and Ehrlich, 1964; Smyk and Ettlinger, 1963).

The diagnostic studies along with identification of systematic position of the isolated bacteria were based on Bergey's Manual of Determinative Bacteriology (1975).

RESULTS OF MICROBIOLOGICAL STUDIES

The microbiological studies revealed that, in the environments under study, numerous bacteria belonging to genera *Arthrobacter*, *Bacillus*, *Pseudomonas*, *Seliberia*, *Streptomyces* and *Thioploca* occurred (Tab. 1). These were mostly bacteria classified as alkalophilic microorganisms developing in ecological microniches of stromatolite. Owing to specific abilities such as producing organic and mineral acid, chelate compounds, and bacteria siderophores affecting calcium carbonates and other minerals (eg. alumino-silicates) they may finally result in the complex process of calcification (Zajic, 1969; Vaughan and 5). It seems that in the initial phase of the mat calcification, the formation of mineral fabrics has been aided by specialised

Taxonomic determination	Forms of activity
1. *Arthrobacter* sp. - cf. *Arthrobacter globiformis* Conn and Dimnic Aerobic-chemoorganotrophic.	Decomposition of mineral compounds dissolution of calcium carbonates and aluminosilicates, producing of bacterial siderophores.
2. *Bacillus alcalophilus* Vedder. Aerobic or facultative, chemoorganotrophic.	Decomposition (dissolution) of calcium carbonates.
3. *Bacillus psychrophilus* Larkin and Stokes. Aerobic/facultative, chemoorganotroph Catalase usually produced.	Decomposition (dissolution) of calcium carbonates.
4. *Pseudomonas facilis* Davis. Aerobic or facultatively chemolithotrophic.	Decomposition (dissolution) of calcium carbonates.
5. *Pseudomonas flava* Davis. Aerobic or facultative, chemoorganotrophic and facultatively chemolithotrophic.	Decomposition (dissolution) and binding of calcium carbonates.
6. *Seliberia* sp. - cf. *Seliberia stellata* Aristovskaya and Parvinkina. Aerobic, chemoorganotrophic and facultatively chemolithotrophic.	Binding of calcium carbonates.
7. *Streptomyces crystallinus* Tresner, Davies and Backus. Aerobic-chemoorganotrophic.	Binding of calcium carbonates.
8. *Thioploca* sp. ? Facultatively anaerobic-chemolithotrophic.	Binding of calcium carbonates.

Table 1. The bacterial assemblage isolated from the *Schizothrix* stromatolite.

communities of chemoorganotrophic (*Seliberia*) and chemolitotrophic (*Thioploca*) bacteria associated with *Streptomyces crystallinus*. The other bacteria may become involved later, decomposing newly formed carbonate fabrics and intensifying the calcification of the mat.

DESCRIPTION OF THE STROMATOLITE

The stromatolite forms a crust up to 8 cm thick, but usually the thickness does not exceed 3 cm. Morphology of the crust varies from irregular, pustular forms to colloform cover composed of flattened, domal features (Fig.3a). The stromatolite grows perpendicularly to substrate surface and generally follows the substrate disposition and

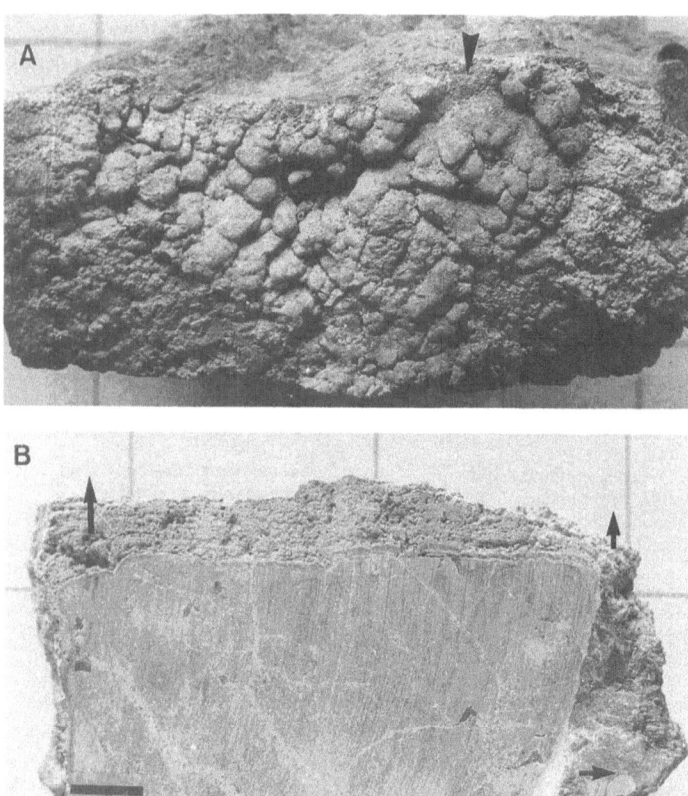

Figure 3 : Morphology and internal structure of stromatolite. a) Plan view of the stromatolite surface. Irregular, pustular morphology visible at sample margins. Flattened domal forms dominate in central part. Note a slight inclination of the shingle-like arranged flat domes reflecting water flow direction (arrow). b) Cross section of the stromatolite crust developed upon the Jurassic bedrocks. Small arrows indicate growth direction of the stromatolite. Large arrow shows the water flow direction. Scale = 3 cm.

micromorphology (Fig.3). The internal fabric consist of mm-thick, slightly undulated layers which in turn, are bipartite and composed of dense, thin (25-250 µm) band and porous, thicker (100-700 µm) one (Fig.4a). The superficial, living part of the stromatolite is coated with mm-thick gelatinous cover encompassing colonies of *Schizothrix calcicola* and *Schizothrix rubra*. Aggregates of very fine-grained carbonate flakes, floating in the mucus, are visible even to the naked eye.

This chapter analyzes anatomy of the living stromatolite mat in terms of the seasonally-controlled microbial succession observed *in statu nascendi*. The last part of the chapter presents the carbonate microfabrics of subjacent ("dead") fossil and subfossil laminae.

Seasonal ecological succession of the microbial communities

Examination of fresh samples revealed a distinctive ecological succession of the stromatolite-forming microbial assemblages. *Schizothrix* and bacteriae are the main, permanent components of the microbial mat, although their growth is greatly reduced during the cold season. Other microorganisms appear and decline seasonally.
In spring, the intense growth of *Schizothrix* colonies is accompanied by bloom of diatoms, e.g. *Synedra ulna* and *Gomphonema angustata*. In summer, the growth of *Schizothrix* remains intense, while diatoms disappear. Autumn and early winter are characterized by substantial reduction of the *Schizothrix* growth and massive development of diatoms, mostly *Achnantes minutissima* and *Achnantes lanceolata*. The microorganisms hibernate during the winter since the water-soaked mat freezes completely.

Seasonal succession of the carbonate microfabrics

1. Spring mat (Fig.4b - d)

The most characteristic feature of the spring fabric is the close packing of the mixed organic and mineral components. Carbonate grains of clotty clusters are bound by cyanobacteria and particularly by the mucus-producing diatoms. The clusters are closely packed, nonetheless the fine, primary ultrastructures - rods, globules, needles are fairly discernable. In some cases, the proceeding early accretion may result in the roughly polyhedral calcite formation.

2. Summer mat (Fig.5)

Despite the relatively great number of *Schizothrix* colonies, the summer mat is more porous than the spring one, because of the withdrawal of diatoms. The mucus is limited to the cyanobacterial and filamentous bacteria envelopes which form a reticulated framework that fix the calcite grains. The grains form coatings growing around the filaments, or randomly arranged grapes adjoining the filaments. The grains commonly comprise clusters of rods up to 3 µm long and 0.1-0.3 µm in diameter. Shape of the rods varies from globular, curved and irregular forms to the most common straight needles and sticks. Some rods show a discrete internal articulation and display various extent of calcification.
The rods either show an irregular arrangement or occur in bundles. The bundles may organize themselves into polyhedra and from these to a rhombohedral habit (Fig.5d). In the larger grains, the baculiform texture may be slightly obliterated.

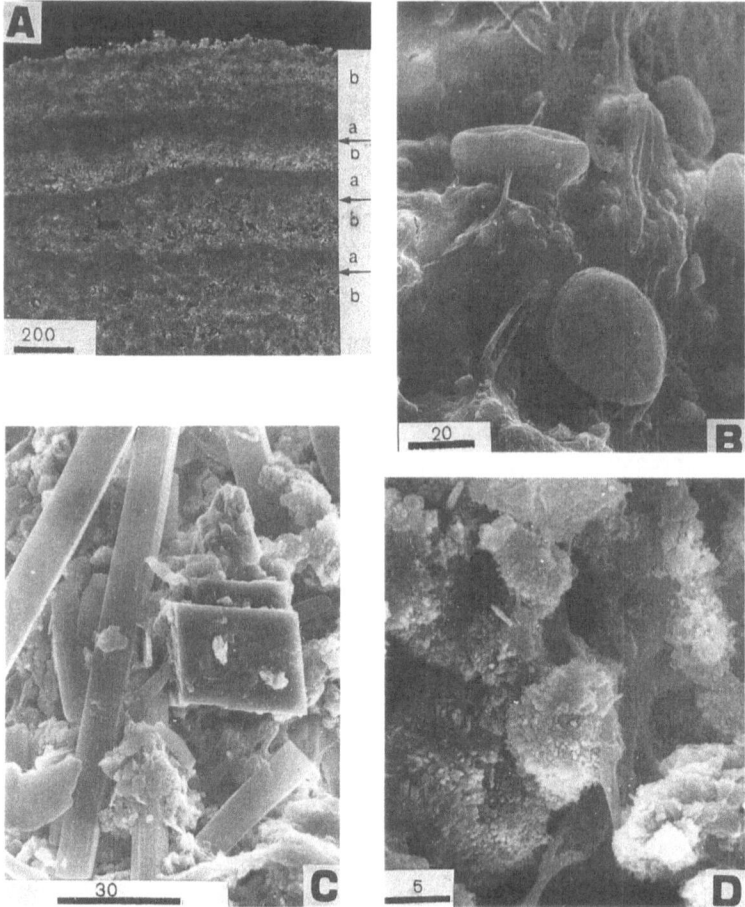

Figure 4 : a) Thin section microphotograph of 3-year's increments of the stromatolite. Dark, dense bands (a) - spring laminae, light, porous bands (b) - summer-autumn laminae. Arrows indicate the boundaries between the annual growth couplets. Origin of a very sharp limit between the spring and the summer-autumn laminae of the middle part of the section is not quite clear, but presumably it may reflect climatic anomalies e.g. extremely dry and/or cold summer. The reduced total thickness of this couplet (half of the adjacent couplets) could also results from the climatic disturbances. Scale bar - 200 μm. b) and c) : Living, spring mat (SEM views). b) *Cocconeis pediculus* Ehr. colony of the superficial mat. Mucus forms a continuous coating, binding detrital particles. c) Details of the *Synedra ulna* colony accompanied by calcite grains. The newly-formed euhedral calcite crystal engulfs the primary granular clumps. Relics of primary ultrastructures are still visible within the crystal. d) Carbonate fabric typical of the spring mat. The mat is made of carbonate clumps composed of minute rods and cocci. Carbonate grains are bound by mucus visible as a film due to the SEM-drying procedure. All scales in μm.

3. Autumn-early winter mat (Fig.6)

This mat is porous owing to the substantial atrophy of cyanobacteria and scarcity of the baculiform calcite bundles. Moreover, part of the calcite grains formed during the warm season, may undergo a slight dissolution (Fig.6c). Silica tests of the autumn-flourished diatoms are dissolved immediately after the death of diatom, and the decomposed frustules are molded with minute bladed calcite (Fig.6d). This indicates a generally alkaline reaction within the autumn mat. On the other hand, a commonplace growth of the blocky polyhedral sparite lacking baculiform ultrastructures, suggests inorganic precipitation of $CaCO_3$ as an important, carbonate-forming mechanism of the autumn lamina.

SUBFOSSIL AND FOSSIL CARBONATE MICROFABRICS

Stromatolite crust is composed of slightly undulated, friable layers which, in turn, consist of alternating dense and porous laminae (Fig.3, 7). Thickness of the dense lamina ranges between 25-250 µm while the porous laminae are thicker and reach up 700 µm. Total thicknes of such bipartite layers reaches up to 1 mm.

The dense laminae (Fig.7, 8) have a closely packed and compact carbonate fabrics. The fabrics include fine, highly aggregated but discernable globules, rods and needles. The messed, ferric hydroxide encrusted colonies of *Galionella* sp. have been also found in this lamina (Fig.8b). Elongate, 2 µm-wide fenestra correspond to the molds of *Schizothrix* filaments. Other kinds of voids are lacking almost completely within the dense lamina.

The porous laminae are not uniform and, as a rule, their lower part consists of densely packed encrustations around tufts of erected *Schizothrix* filaments (Fig.7B, 8C). The fabrics become more loose and chaotically arranged upsection, however the baculiform carbonate ultrastructures are the same as those of the dense laminae described above (Fig.8d).

The comparison of the carbonate fabrics of the living mats with the fossil laminae allows the unequivocal recognition of the dense laminae as spring deposit and the porous ones as summer-autumn increment.

Stromatolite-forming processes. Interpretation and discussion

The described rhythmical alternation of the microfabric arrangement clearly follows the ecological succession of the stromatolite-forming microorganisms. This chapter deals with the possible microbiological interactions within the living mat and the consequential processes of carbonate mineralization.

Figure 5 : Living, summer mat (SEM views). a) The mat is composed of *Schizothrix* trichomes (arrow), accompanied by filamentous bacteria and posssibly hyphae; note the grape-like carbonate clumps. b) Baculiform calcite aggregates encompassing the *Schizothrix* sheath (now dried and collapsed). c) Close-up of the calcite clusters. Granular and elongate, minute bodies (arrows) represent bacteria cells. According to the ED microprobe analysis the light-coloured part of the cells are not completely calcified and may represent the youngest, evolving cells. d) Detail of Fig.5a. Incipient polyhedral habit of some of the clumps is recognizable (arrows). e) Details of the baculiform ultrastructures. The light-coloured rods (arrows) display very low or close to zero contents of Ca^{2+} and are thought to be the evolving cells of the microbes. Many rods show morphology typical of *Bacillus* sp. The faintly preserved "segmentation" of some rods reflects presumably, the successive stages of cell fissions accompanied by autolytically-controlled calcification of the parent cells (small arrows). See also text for further comments. Note the fast coalescence of the subjacent older, "dead" calcified fabrics (c). Scales in μm.

Figure 6 : Autumn-early winter, living mat (SEM view). a) Mat is dominated by filamentous bacteria and diatoms. b) Resulting fabrics become more porous though the baculiform clumps are still present. c) Some of the clumps undergo a slight etching (arrow) owing probably to a localized acidic reaction, due to the diatom physiology (see text). d) *Post mortem* dissolution of the frustules accompanied by molding of the test with minute, bladed calcite (arrow). Scale= μm.

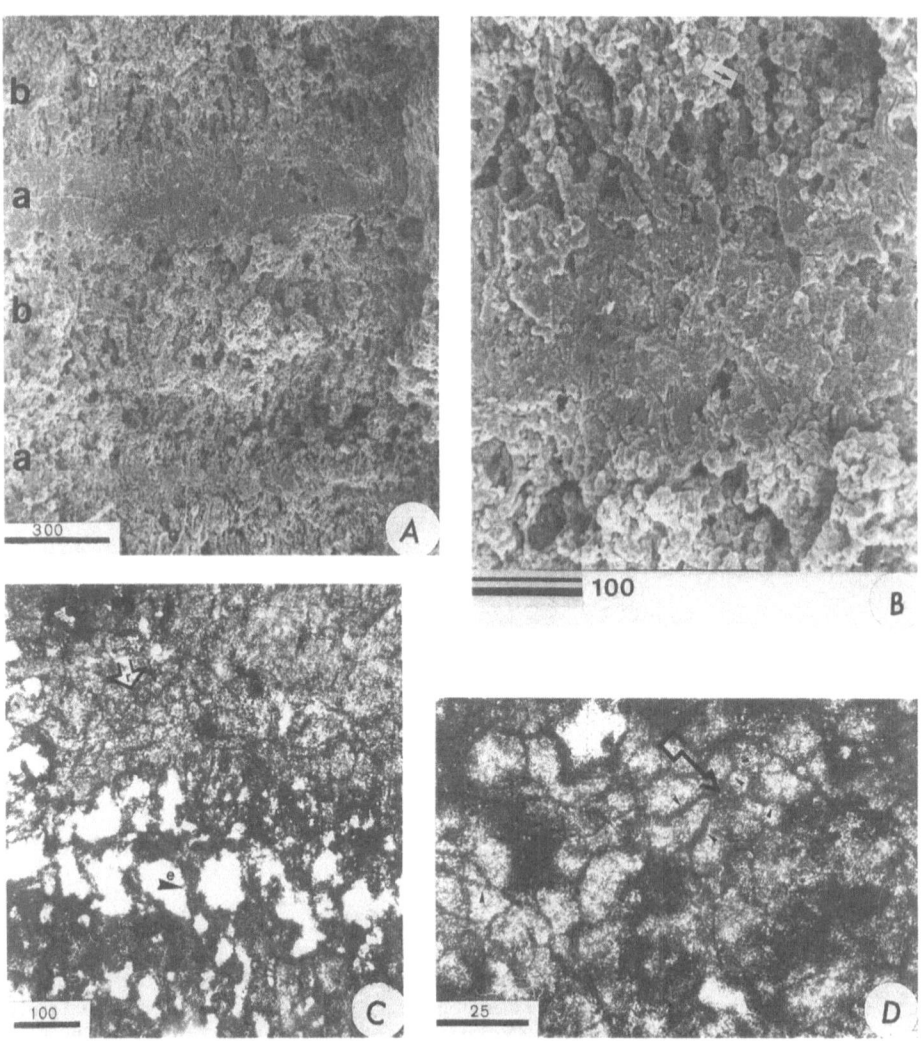

Figure 7: The fossil laminae (photomicrographs). Figures from the sample taken from the depth of 3 cm. a) General view of 2-year growth laminae; a - spring, dense lamina, b - summer-autumn porous lamina. b) Close-up of the transition between the dense and the porous laminae. The carbonate fabrics are concentrated around the erected *Schizothrix* filaments (arrow). c) Transition between cool season porous lamina and spring, dense lamina. Thin section microphotographs. Note the reticulate pattern of the *Schizothrix* filaments in the dense lamina (arrow - r) and erected disposal in the porous one (arrow - e). Note also a partly corrosive boundaries of some of the fenestrae within the porous lamina (arrow -c), reflecting probably the physicochemical degradation of the mat during the cool season. d) Polygonally fitted clumps (small arrows) of the calcite envelopes encompassing the *Schizothrix* filament. Large arrow indicates position of the filament. Transverse thin section microphotographs. All scales in μm.

Calcification of the microbial mat

1. Spring. Very intense spring growth of diatoms and cyanobacteria follows the ice melting. The simultaneous release of nutrients, particularly the nitrates, accumulated during winter, supports the microbial bloom (Alexander, 1971). Diatoms along with cyanobacteria produce slimy mucus that forms a voluminous and continuous film, rich in amino acids, carboxyhydrates, proteins, lipids as well as mono- and polysacharides (Reimann *et al.*, 1966; Nichols, 1973). The mucus may also trap and bind particles of the detrital carbonates and siliciclastic material. The CO_2-uptaken by the photosynthesis and the dark fixation (Krumbein, 1979) lead to carbonate saturation and $CaCO_3$-precipitation inside the mucus film. The film, rich now in both organic and mineral components, may be regarded as a specific fertile medium for the growth of chemoorganotrophic and chemolithotrophic bacteriae (Tab. 1). Bacteria may absorb carbon derived from the organic components of the mat. Moreover, the alkalophilic bacteria, living in the mat, may dissolve the carbonate substrate e.g. detrital carbonates or the earlier-formed authochthonous calcite, and hence utilize the Ca^{2+}-ions. On the other hand, some of the other chemolithotrophic bacteriae (cf. Tab. 1) may simultaneously drive the pH of the microenvironments to the alkaline range, resulting finally in calcite precipitation. Such very complex processes of penecontemporaneous dissolution and (re)precipitation of carbonates, has been clearly shown by experiments of Caudwell (1987). This author has evidenced the intrinsic nature of the dissolution-precipitation cycle resulted from the trophic interactions within the mat; the bacterial utilisation of sugars, comprised within the algal mucus, acidifies the microniche while the subsequent bacterially-controlled decay of the other organic compounds, e.g. peptones, drives the microenvironments toward pH > 8.5.

Since the time of the microbial cycles (e.g. cell binary fission, ecosuccession etc.) is extremely short (several minutes to several tens of hours - Carlile, 1979; Kunicki-Goldfinger, 1982) the two, opposite processes, i.e. carbonate dissolution and (re)precipitation, may be recognized as penecontemporaneous. However, in the general balance, the precipitation prevails over dissolution. Multiplication of these short-term cycles within the prolific gelatinous milieu of the spring mat, result finally in compact and coalescent carbonate lamina.

Summarizing, we wish to stress that the dense and compact carbonate fabric of the spring lamina should be ascribed to the coincident blooms of the *Schizothrix* sp. and diatoms. The voluminous and coherent as well as nutrient-rich mucilage cover, resulting from the bloom, promotes unlimited growth of the mucus-dwelling bacteria and hence enables the overwhelming calcification of the entire spring mat.

2. Summer. The decline of diatoms following the rise in water temperature, results in the limitation of the zone of the active, microbial calcification to the mucus coating the *Schizothrix* trichomes (Fig.8c). The precipitated $CaCO_3$ forms envelopes limited to the filments in the lower part of the porous lamina. In the course of summer, the mat growth

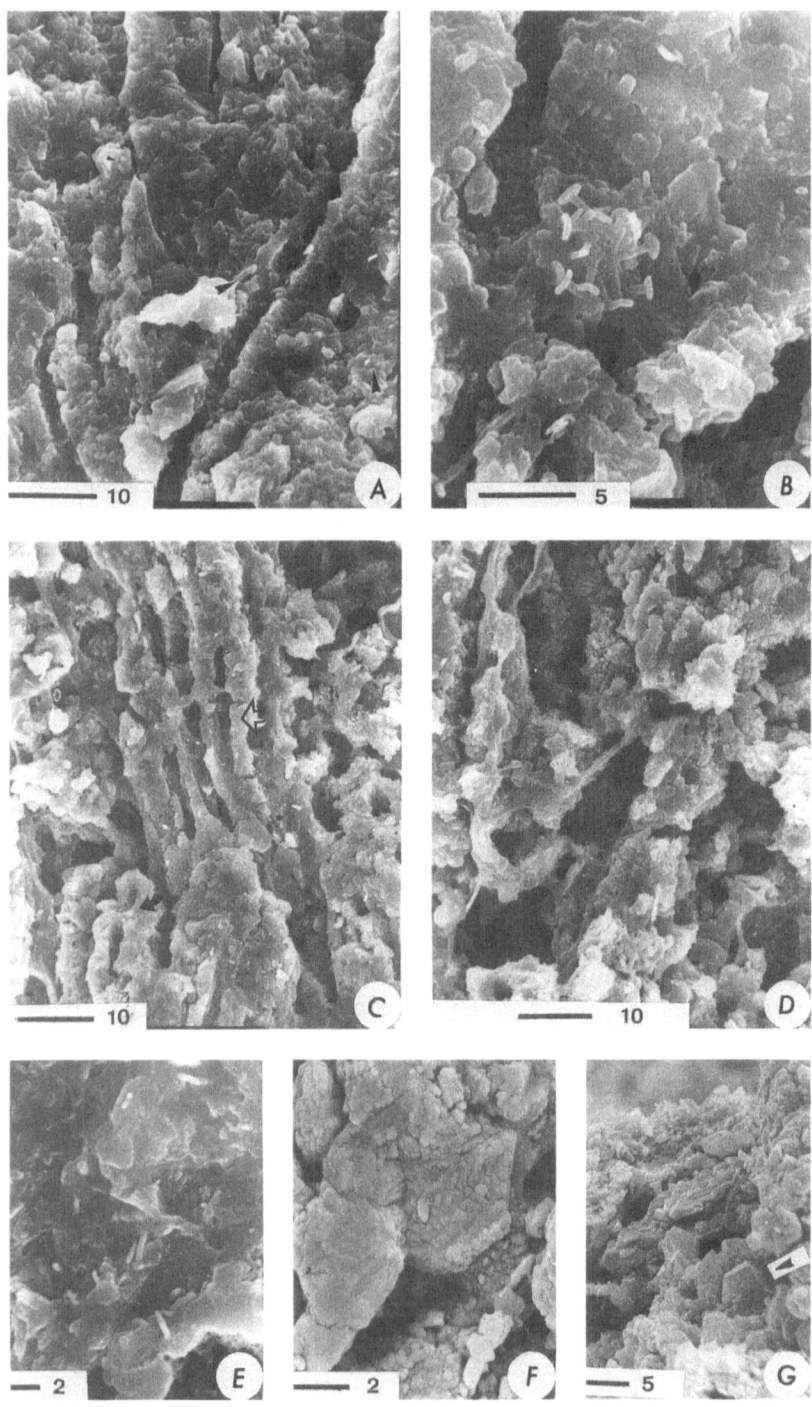

decreases gradually and the calcified envelopes become more loose. The summer microfabric consists of calcite grains similar to those from the spring laminae. The ultrastructure of the grains does not differ from the spring microfabrics and comprises clumps of calcified bacteria rods (Fig.8d). This suggests the same mechanism of the mat calcification.

3. Autumn-early winter. During cold season, the carbonate production diminishes gradually, following the decline of the *Schizothrix* trichomes. Some of the summer fabrics display even traces of weak dissolution (Fig.6c), related presumably to activity of the chemoorganothrophic bacteria and/or to the slight acidic coating of some of the cryophilic diatoms (Lewin, 1961; Szulc, 1984). Nonetheless, as a whole, the autumn mat remains alkaline and its calcification proceeds.

Calcification of bacteria cells

Bacteria as possible carbonate-forming agents have been commonly reported from various environments (Lalou, 1957; Krumbein, 1968; Berner, 1971; Monty, 1976, 1986; Monty and Van Laer, 1984; Adolphe, 1981; Danielli and Edington, 1983; Chafetz, 1985; Chafetz and Folk, 1984; Adolphe et al. 1991). Nevertheless, the analytical studies on the mechanism of the bacterially-controlled calcification are rare. Most of the investigations concern the organic rich environments, particularly anoxic milieu, where carbonate is a by-product of the bacterial metabolism. Krumbein and Cohen (1974, 1977), Krumbein et al. (1977), Danielli and Edington (1983), Deelman (1975), Monty and Van Laer (1984), Thompson and Ferris (1990), described the extracellular carbonate precipitation on the surface of bacteria cells. Assuming the circumcellular calcification of the bacteria body, one should expect its collapse after dissolution of the carbonate shell. Such phenomenon has been observed by Folk (1990) however beside the collapsed cells, he has found protuberant forms surviving the HCl-etching procedure. The full-relief, calcified bacteria cells, were also reported elsewhere

Figure 8 : The fossil lamina (SEM views). a to d - details of the Fig.7. e to g -sample from the depth of 8 cm. a) Dense, spring lamina. Molds of *Schizothrix* filaments running through the coherent carbonate fabrics. Despite the coalescent habit of the grains, the baculiform ultrastructures are recognizable (arrow). Note the lack of diatom remnants. .b) Detail of Fig.7a. Well preserved, calcified bacteria rod and interwoven *Galionella* colony mineralized with ferric hydroxide (central part of the photograph). c) Carbonate fabric of the summer, porous lamina. The summer carbonate grains are confined to the *Schizothrix* filament (arrow). d) Carbonate fabric of the autumn, porous lamina. The fabric is similar to those of the summer lamina, although grains arrangement becomes more chaotic, following the decline of the *Schizothrix* filaments. e) Detail of the dense lamina from the depth of 8 cm. The fabric resembles that of the dense lamina from the depth of 3 cm (cf. Fig.8b). f) and g) Details of the porous lamina (8 cm depth). The roddy ulstructures are obliterated but still recognizable (f). Subhedral to euhedral grains (g - arrows) could result from a transformation of the polyhedral clumps such as these figured on Fig.5d. See text for further comments. All scales in µm.

(Deelman, 1975; Krumbein and Cohen, 1977, Fig.15; Maurin and Noël 1976; Camoin and Maurin, 1988) but the genesis of this phenomenon has not been explained.

The present study revealed that the major part of the carbonates occurring within the *Schizothrix* stromatolite consists of low-Mg rods, making the mineral replicas of bacteria bodies (Fig.5, 6, 8). Since the HCl-etching of the material taken from the living mat did not exhibit any collapsed cells, the entire cell calcification may be inferred. Furthermore, the preserved segmentation or articulation of the rods (Fig.5e) suggests that the process of bacterial fission was accompanied by simultaneous cell calcification. This fact, in turn, suggests that such calcification might be controlled by intrinsic bacterial processes. The microprobe examinations of the fresh samples revealed variable repartition of Ca^{2+} within the rods; some of the rods, or part of them, were still nonmineralized (or partly mineralized) organic cells (Fig.5c, e). This indicates variable and gradual calcification and supports our notion on the intrinsic, physiologically-controlled automineralization of the bacteria cell. We suggest that the entire calcification of the cell may be related to the bacteria reproduction cycle; after the cell fission, the dead parent cells may undergo autolytical degradation by immediate molding with carbonate. Significant amount of Ca^{2+}, as important component of both membrane and internal structure of microbes (Silverman and Ehrlich, 1964) particularly in alkalophilic bacteriae, could facilitate the carbonate mineralization of the entire bacteria cell. It is noteworthy that similar process of the entire calcification of microbial bodies has been recently found and evidenced from cave "moonmilk" speleothems (Gradzicski *et al.*, 1992).

It is difficult to judge, what pattern of cell division geometry: one, two or three-dimensional (sensu Hoffmann, 1964, and Golubic and Barghoorn, 1977), dominates within the studied bacterial communities? The very fast, concomitant calcification of the dividing cells, seems to hinder the unconstrained reproducing growth, hence the resultant colonies pattern may be very haphazardous.

The range and intensity of the bacterial calcification is also noteworthy. The high Ca^{2+}-concentration inside the described microenvironment could act as a toxic agent, even for the alkalophilic bacteria (Kunicki-Goldfinger, 1982). The bacterial anti-toxic reaction depends probably on the mass growth of the cells, Ca^{2+}-molecular binding, and finally, on the $CaCO_3$ precipitation (negative chemotaxis). This process may be, to some extend, compared with phenomenon of the global Ca-detoxification by the mass biogenic carbonate formation as suggested by Kacmierczak *et al.*, (1986).

Diagenesis

The previously mentioned, characteristic orientation of the rod clusters (c.f. Fig.5) may result from the chemotactic development of the microbial colony (Goldacre, 1954). The polyhedra-like habit of the calcified colonies (Fig.5d) is thought to result from the competition between the biological growth processes and the physicochemical crystallisation powers (Fig.9). The observed *post mortem* evolution of the calcified colonies into polyhedral or even euhedral calcite grains (Fig.5d, 8f,g) support such

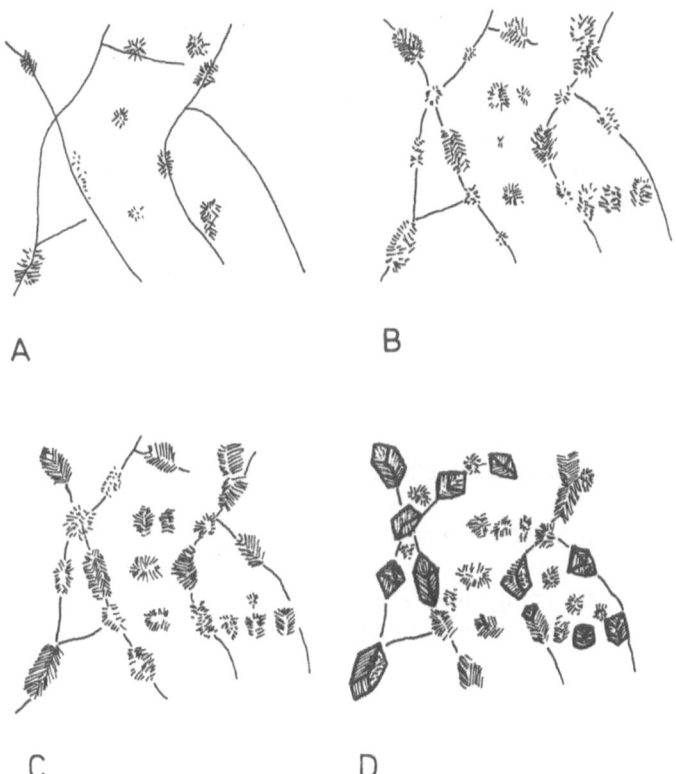

Figure 9 : Scheme of the transformation of polyhedral clumps of rods into subhedral or euhedral calcite crystals. a) initial bacterial colonisation on *Schizothrix* filaments, b) and c) - subsequent stages of colonies growth and intensive calcification, d) construction of carbonate crystals with aggrading neomorphism processes.

suggestion. The comparison of the samples taken from the depth of 3 and 8 cm revealed a progressive aggregation (coalescive neomorphism) of the carbonate microfabrics. Nevertheless, relics of the primary bacteria rods were still well-marked (Fig.8e, f).

In order to examine the possible $\delta^{13}C$ and $\delta^{18}O$-compositional changes following the diagenesis, the sample of the living carbonate mat as well as samples from the depths of 3 and 8 cm were measured. The results (in ‰vs PDB) for these three samples are respectively:

| $\delta^{13}C$ | - | -7.4 | -7.1 | -7.1 |
| $\delta^{18}O$ | - | -9.5 | -8.9 | -9.2 |

They suggest that the described, early diagenetical changes have not resulted in considerable compositional changes of the stable isotopes; accordingly the material could be used as a reliable tool for environmental reconstruction.

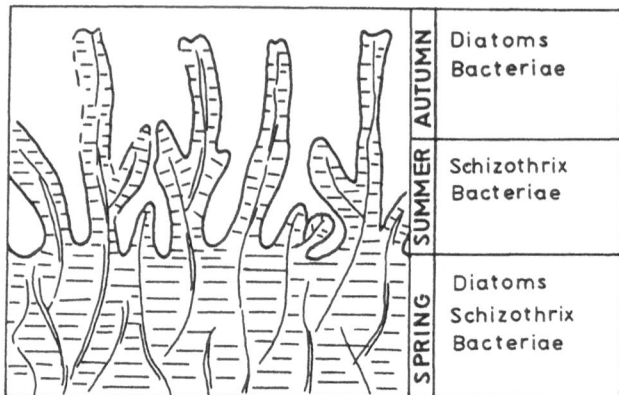

Figure 10 : Scheme diagram of the seasonal succession of the stromatolite-forming microorganisms and the resultant carbonate fabric (hatched area).

Unlike the gradual and moderate diagenetical changes of the carbonates, the siliceous frustules subject a very instant *post mortem* dissolution, thus the stromatolite increments older than one year, are completely lacking diatom remnants. As presented here, the diatoms play a crucial role in the mineralisation of the microbial mat, therefore a complete dissolution of the diatom tests within the aged laminae seems to be the most impressive feature of the early diagenesis. In addition, this feature emphasizes a general need of a very careful and critical interpretation of the growth mechanisms of the fossil, spongiostromata stromatolites.

CONCLUSIONS

The formation of *Schizothrix*-stromatolite is a process involving activity of complex communities of microorganisms: diatoms, cyanobacteria and bacteria. Their interrelationships vary throughout the year and accordingly affect the internal features (fabrics and microstructures) of the stromatolite. Cyanobacteria filaments make a scaffolding network colonized by the mostly sessile diatoms. The latter may bind the filamentous framework with their, massively produced, mucilagenous gel. The ensuing mat forms an accessible, fertile medium of bacteria expansion. The bacteria seem to play a decisive role in the calcification of the microbial mat. They may actively precipitate $CaCO_3$ as a result of CO_2 assimilation or biochemical process of Ca^{2+}-binding. The bacteria are also deeply involved in the decomposition of the organic-mineral mixture which makes stromatolite mat.

The seasonally-controlled changes within the mat-forming microbial community, result finally in the rhythmical changes of the carbonate fabrics (Fig.10) and then, in the seasonal lamination of the stromatolites.

ACKNOWLEDGEMENTS

We thank K. Wasylik and T. Mrozicska for the diatoms and cyanobacteriae identification, J. Siemicska and G. Haczewski for the valuable suggestions. We are specially indebted to J. Sarfati and C. L. V. Monty for their helpful comments.

REFERENCES

Aaronson, S. (1970) Experimental microbial ecology, Academic Press, New York.

Adolphe, J. P. (1981) "Exemples de contribution microorganiques dans les constructions carbonatees continentales" Actes du Coll. de l'AGF, Formation carbonatees externes, tufs et travertins, Paris, 15-30.

Adolphe, J. -P., Choppy J., Choppy, B., Loubiere, J. -F., Paradas, J., and Soleilhavoup, F. (1991) "Biologie et concretionnement: un exemple, les baguettes du gours", Carstologia, 18, 49-55.

Alexander, M. (1971) Microbial ecology, Wiley and Sons, New York.

Bergey's Manual of Determinative Bacteriology. Eight Edition, (1975) The Wiliams and Wilkins Co. Baltimore.

Berner, R. A. (1971) "Bacterial processes effecting the precipitation of calcium carbonate in sediments", in O.P. Bricker (ed.), Carbonate cements, The Hopkins Press, London, pp. 247-251.

Birkenmajer, K. (1979) Przewodnik geologiczny po pienicskim pasie skackowym, WG, Warsaw.

Boquet, E., Boronat, A., and Ramos-Cormenzana A. (1973) "Production of calcite (calcium carbonate) by soil bacteria is a general phenomenon", Nature, 246, 621-623.

Camoin, G. and Maurin A. -F. (1988) "Roles de micro-organismes (bacteries cyanobacteries) dans la genese des "Mud Mounds". Exemples du Turonien des Jebels Bireno et Mrhila (Tunisie)", C. R. Acad. Sci. Paris, T 307 sII, 401-407.

Carlile, M. . (1979) "Bacterial, Fungal and Slime Mould Colonies", in G. Larwood and B.R. Rosen (eds.), Biology and Systematics of Colonial Organisms, Academic Press, London, 3-27.

Caudwell, Ch. (1987) "Etude experimentale de la formation de micrite et de sparite dans les stromatolites d'eau douce a *Rivularia*", Bull. Soc. geol. France, 8, 299-306.

Chafetz, H. S. (1985) "Marine peloids: a product of bacterially induced precipitation of calcite", J. Sediment. Petrol., 56, 812-817.

Chafetz, H. S. and Folk, R. L. (1984) "Travertines: depositional morphology and bacterially constructed constituents", J. Sediment. Petrol., 54, 289-316.

Danielli H. M. and Edington, M. A. (1983) "Bacterial calcification in limestone caves", Geomicrobiol., 3, 1-16.

Deelman, J. C. (1975) "Two mechanisms of microbial carbonate precipitation", Die Naturwissenschaften, 62, 484-485.

Folk, R. L. (1990) "Bacterial bodies and carbonate fabrics: Recent to Triassic", Abstracts of the Carbonate Microfabrics Symposium and Workshop, College Station, Texas, p. 43.

Goldacre, R. J. (1954) "Crystalline bacterial arrays and specific long range forces", Nature, 174, 732-734.

Golubic, S. and Barghoorn, G. S. (1977) "Interpretation of microbial fossils with special reference to the Precambrian", in E. Flügel (ed.), Fossil algae, Springer, Berlin, pp. 1-14.

Gould, G. W. and Corry, J. E. L. (1980) Microbial growth and survival in extremes of environment, Academic Press, London.

Gradzicski, M., Szulc, J., Dziadzio, P., Rocniak, R., and Smyk, B. (1992) "Moonmilk -origin, microbiology, diagenesis - selected examples from Southern Poland", in Abstracts of 13th IAS Reg. Meet., Jena, 1992, pp. 51-52.

Hoffmann, H. (1964) "Morphogenesis of bacterial aggregations", A. R. Microbiol., 18, 111-130.

Horikoshi, K. and Akiba, T. (1981) Alkalophilic microorganisms a new microbial world, Springer, Berlin.

Kacmierczak, J., Degens, E. T., and Ittekot, V. (1986) "Cellular response to Ca^{++} stress and its geological implications", Acta Paleontol. Pol., 30, 115-126.

Kostrakiewicz, L. (1982) "Klimat", in K. Zarzycki (ed.), Przyroda Pienin w obliczu zmian, PWN, Warsaw.

Krumbein, W. E. (1968) "Geomicrobiology and geochemistry of the "Nari-Lime Crust" (Israel)", in G. Müller and G.M. Friedman (eds.), Recent developments in carbonate sedimentology in central Europe, Springer, Berlin.

Krumbein, W. E. (1978) "Algal mats and their lithification", in W. E.Krumbein (ed.), Environmental biogeochemistry and geomicrobiology, Ann. Arbor Science Publishers Inc Ann. Arbor, pp. 209-227.

Krumbein, W. E. (1979) "Calcification by bacteria and algae", in P.A. Trudinger and D.J. Swain (eds.), Biogeochemical cycling of the mineral-forming elements, Elsevier, New York, pp. 27-68.

Krumbein, W. E. (1983) Microbial geochemistry, Blackwell Sc. Pub. Ltd., Oxford.

Krumbein, W. E. and Cohen, Y. (1974) "Biogene, klastische und evaporatische Sedimentation in einem mesothermen, monomiktischen, ufernahen See (Golf von Alaba)", Geol. Rundsch., 63, 1035-1065.

Krumbein, W. E. and Cohen, Y. (1977) "Primary production, mat formation and lithification: contribution of oxygenic and facultative anoxygenic cyanobacteria", in E. Flügel (ed.), Fossil algae, Springer, Berlin, pp. 37-56.

Krumbein, W. E., Cohen, Y., and Shilo, M. (1977) "Solar Lake (Sinai). 4. Stromatolitic cyanobacterial mats", Limnol. Oceanogr., 22, 635-656.

Kunicki-Goldfinger, W. J. H. (1982) _ycie bakterii, PWN, Warsaw.

Lalou, C. (1957) "Studies on bacterial precipitation of carbonates in sea water", J. Sediment. Petrol., 27, 190-195.

Lewin, J. C. (1961) "The dissolution of silica from diatom walls", Geochim. Cosmochim. Acta, 21, 182-198.

Markowicz, M. and Pulina, M. (1979) "Ilozciowa pócmikroanaliza chemiczna wód w obszarach krasu wcglanowego. Pr. Nauk. UC, 289, 1-67.

Maurin, A. F. and Noël, D. (1977) "A possible bacterial origin for Famennian micrites", in E. Flügel (ed.), Fossil algae, Springer, Berlin, pp. 136-142.

Monty, C. L. V. (1976) "The origin and development of cryptalgal fabrics", in M.R. Walter (ed.), Stromatolites, Elsevier, New York, pp. 193-249.

Monty, C. L. V. (1986) "Microbial dolomites", in 12th Intern. Sediment. Congress, Abstracts, Canberra, p. 215.

Monty, C. L. V. and Laer, P. (1984) "Experimental radial calcite ooids of microbial origin and fossil counterparts", in 5th European Reg. Meeting of IAS, Abstracts, pp. 296-297.

Nichols, B. W. (1973) "Lipid composition and metabolism", in N. Carr and B. Whitton (eds.), The biology of blue-green algae, Blackwell, Oxford, pp. 144-161.

Pia, J. (1934) "Die Kalkbildung durch Pflanzen", Beih. Bot. Zentralbl., A 52, 1-72.

Reimann, B. E. F., Lewin, . C., and Volcani, B. E. (1966) "Studies on the biochemistry and fine structure of silica shell formation in diatom. II. The structure of the cell wall of *Navicula pelliculosa* (Breb) Hilse", J. Phycol., 2, 74-84.

Silverman, M. P. and Ehrlich, H. L. (1964) "Microbial formation and degradation of minerals", Adv. Appl. Microb., 6, 153-206.

Smyk, B. (1970) "The microbial degradation of silicates and aluminium silicates", Post. Microb., 9, 121-125.

Smyk, B. and Ettlinger, L. (1963) "Recherches sur quelques especes *d'Arthrobacter* fixatrices d'azote isolees des roches karstique alpines", Ann. Inst. Pasteur, 105, 341-348.

Smyk, B. and Drzac, M. (1964) "Untersuchungen über den Einfluss von Microorganismen auf das Phänomen der Karstbildung", Erdkunde, 18, 102-113.

Szulc, J. (1984) "Sedimentation of the Quaternary travertines from Southern Poland", Ph.D. Thesis, Pol. Acad. Sci, Kraków.

Thompson, J. B. and Ferris, F. G. (1990) "Cyanobacterial precipitation of gypsum, calcite and magnesite from natural alkaline lake water", Geology, 18, 995-998.
Vaughan, D. and Malcolm. R. F. (1985) Soil organic matter and biological activity, Martinus Nijhoff Publ., Amsterdam.
Zajic, J. E. (1969) Microbial biogeochemistry, Academic Press, New York.

STROMATOLITIC MATS IN ANTARCTIC LAKES

R. A. WHARTON, JR.
Desert Research Institute ; P.O. Box 60220 ; Reno, Nevada 89506 USA

ABSTRACT

Stromatolitic microbial mats composed primarily of bacteria, cyanobacteria, eukaryotic algae are found in the cold, dimly-lit, perennially ice-covered antarctic lakes of southern Victoria Land, Antarctica (77°32-43'S, 161° 33' -163°7' E).   The morphology of a particular mat results from a combination of biological, geochemical, and sedimentological processes, some of which may be unique to ice-covered lakes. Prostrate, lift-off, columnar, and pinnacle mats are trapping, binding and/or precipitating carbonates and various other minerals forming organosedimentary structures.   The ice-covered lakes of Antarctica may serve as an important model for understanding the formation of stromatolites in cold environments. Studies of antarctic stromatolitic mats enhance our understanding of the range of environmental conditions capable of supporting stromatolite formation, particularly cold facies including those present during the Precambrian.

INTRODUCTION

Microbial mats are a common feature of the perennially ice-covered lakes in southern Victoria Land, Antarctica (Simmons et al., 1979; Wharton et al., 1983; Parker and Wharton, 1985).   The mats are trapping and binding sediment, precipitating minerals, and remaining undisturbed on the lake bottom forming laminated, organosedimentary structures. These mats have been the focus of several studies which have recognized that the antarctic microbial mats are stromatolitic (Parker et al., 1981; Wharton et al., 1982, 1983; Love et al., 1983). The discovery of modern stromatolitic microbial mats in the cold, dimly lit, lakes of Antarctica may provide a new approach for understanding the distant past.   As pointed out by Walter and Bauld (1983), there has been a misconception that Precambrian stromatolites and their associated carbonates, sulphates and chloride evaporites were formed in warm climates. It is well known that several periods of major glaciation occurred during the Precambrian (Frakes, 1979; Anderson, 1983, Walter and Bauld, 1983). Based upon studies in the cold antarctic lakes, it is apparent that stromatolites, carbonates, and various evaporites ·may be components of

J. Bertrand-Sarfati and C. Monty (eds.), Phanerozoic Stromatolites II, 53–70.

frigid lake facies, and do not necessarily reflect warm climates. Studies of antarctic microbial mats and their development into stromatolites may play an important role in the re-interpretation of their occurrence in Precambrian polar environments. This chapter synthesizes what is currently known about the antarctic stromatolitic microbial mats.

ENVIRONMENTAL SETTING

Southern Victoria Land, Antarctica (77°32-43'S, 161° 33' -163°7' E) contains several relatively ice-free valleys often referred to as the McMurdo Dry Valleys (Fig.1, 2). The

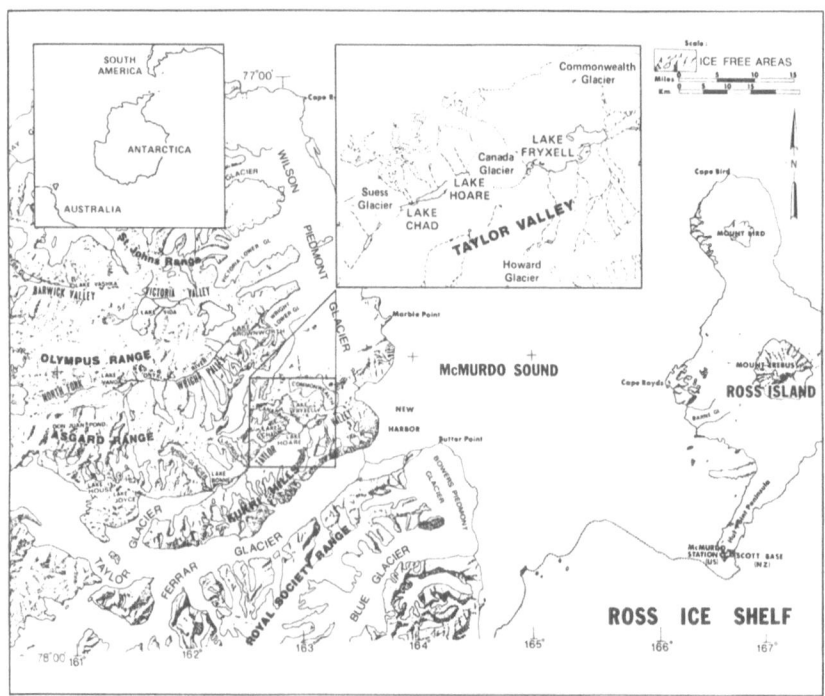

Figure 1 : a) Map of Antarctica showing the McMurdo Dry Valleys of southern Victoria Land.

climate in this region of Antarctica is characterized by a mean annual temperature of ~ -20°C and an annual precipitation of < 10 cm (Clow et al., 1988). The winter climate is controlled by the wind with cold calm conditions being repeatedly interrupted by strong foehn winds originating on the Polar Plateau. Air temperatures rapidly increase by 20 - 30°C (e.g. from -40°C to -10°C) during these episodes. The Summer climate is dominated by radiative heating and the associated instability of the atmosphere boundary layer within the low albedo dry valleys. Temperatures often approach 0°C in December and January and occasionally exceed this value during sporadic calm periods.

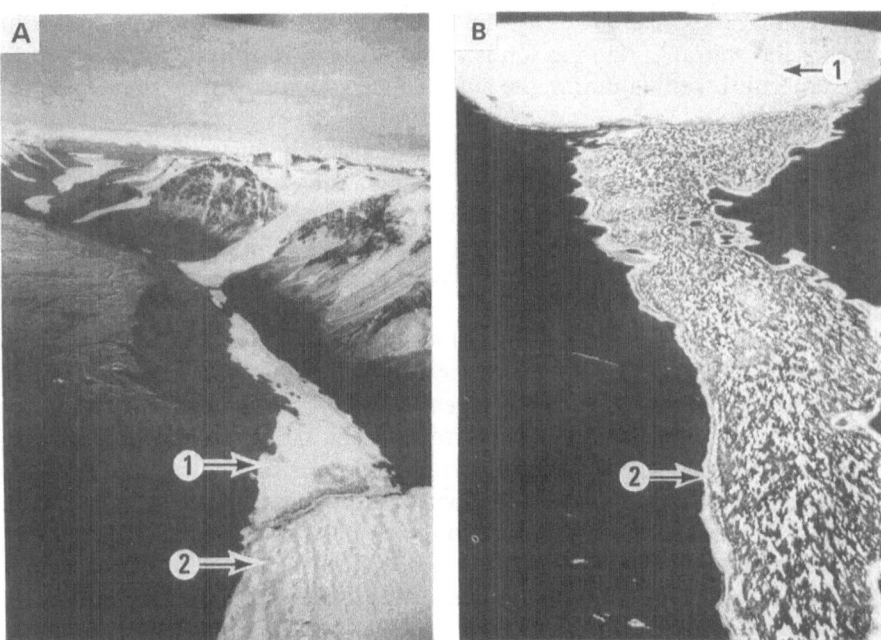

Figure 2 : a) : Aerial photograph looking west up Taylor Valley. Canada Glacier (2) forms a dam for perennially ice-covered Lake Hoare (1) which is 1 km wide and 4 km long. b): Aerial photograph looking east down Taylor Valley. Canada Glacier (1) forms a dam for perennially ice-covered Lake Hoare (2) which has abundant sediment load on the surface of its ice cover.

The biological systems in the McMurdo Dry Valleys are relatively simple. For example, there are no higher plants or vertebrates and very few insects. The dry soils and seasonal streams contain isolated populations of mosses, lichens, algae, fungi, bacteria, protozoans, tardigrades, and nematodes (Vincent, 1988). The floors of the valleys contain closed-basin lakes in which planktonic and benthic microorganisms live. The brief 2-3 months austral summer of continuous solar radiation allows up to 10 weeks of glacial meltstream flow to the lakes. This meltwater input is balanced by losses of water from the surface of the ice cover by ablation (McKay et al., 1985; Wharton et al., 1993). An important feature possessed by all the lakes is the insulation of their water columns with perennial ice covers of 3-6 m in thickness. These perennial ice covers protect the lakes from wind effects and eliminate turbulence or mixing, creating a condition favorable for considerable physical, chemical, and biological stratification (Parker et al., 1982a). Another important effect of the ice cover is the filtering of photosynthetically active radiation (PAR) before it reaches the water column

(Palmisano and Simmons, 1987; Wharton et al., 1989). The light regime in this region of Antarctica is characterized by approximately four months each of sunlight, twilight, and darkness. The effect of this light regime is manifested in the biological activity of the microorganisms. Instead of having many light/dark diurnal cycles as occurs in more temperate regions, the antarctic lakes are characterized by having one long diurnal cycle: during the austral Spring the sun is rising; the sun is up continuously during the austral Summer; it is setting during the austral Fall; and total darkness prevails during the austral Winter.

The perennial ice covers also contribute to the constant temperature profiles of these lakes. The majority of the lakes have water temperatures <10°C (Parker et al., 1982a). Lake Vanda is unusual with the temperature increasing with depth to 25°C at the bottom (70 m). Seaburg et al. (1981) measured the temperature-growth responses of algal isolates from the antarctic lakes and found that the isolates were well-adapted to growing at lower temperatures.

The relatively complete seal of the perennial ice covers on these lakes is reflected in the dissolved gases (Mikell et al., 1984; Wharton et al., 1986, 1987, 1989; Craig et al., 1992). The shallower depths of these lakes have about 4 times the normal oxygen equilibrium concentration, grading downward to anaerobic conditions in the poorly lit bottom waters. Pronounced chemical stratification is exhibited in conductivity, especially for lakes Bonney and Vanda which have hypersaline deep waters. By contrast, Lake Hoare is uniformly fresh throughout its water column. Numerous other variables (e.g. alkalinity, various ions) show pronounced chemical stratification in these lakes (Parker et al., 1981, 1982a). Furthermore, biological stratification is clearly evident in the cell numbers and species distribution of planktonic algae, bacteria and yeasts within the water columns of these lakes.

SEDIMENTATION IN ICE-COVERED LAKES

The physical environment for sedimentation in the antarctic lakes is unusual because of the presence of the perennial ice cover. The ice catches and traps wind blown sediment and provides a surface for the movement (by saltation, rolling and drift on the ice) of larger sediment particles into the middle of a lake. Although several of the lakes contain large boulders, most of the mass (>95%) of the ice cover sediment burden is in the form of sand-sized and finer particles. Wharton et al. (1989) estimate that the average sediment loading is between 0.2 and 2.0 g cm^{-2}.

Sediment cores from the bottom of the antarctic lakes contain alternating layers of organic and inorganic material. Until recently, the source of the inorganic material was unclear. Glacial meltstreams which feed the lakes have relatively low velocities and flow for only a few weeks each austral summer. These streams do deposit sediment locally and mostly in deltas < 100 m from shore (Nedell et al., 1987). How then do coarse grained materials (e.g. sand particles) reach the bottom in the center of an ice-covered lake? Love et al. (1983) first suggested that sediment was deposited through the ice cover. Their suggestion was based upon observations from Lake Vanda, which in the late 1970's and early 1980's had a relatively thin ice cover (~3 m thick) and lacked abundant sediment deposits on the lake's ice cover surface. Also, cracks were present in Lake Vanda's ice cover which allowed gas bubbles produced by SCUBA divers to escape from the water below into the atmosphere. From these observations, it was inferred that any sediment which was deposited on the ice cover surface could eventually make its way through the ice cover and enter the water column below.

Nedell et al. (1987) provided the first indirect evidence that sediment was able to pass through the ice cover. They compared the mineralogy and grain size distribution of sediment obtained from the ice cover surface, the bottom of the lake and the meltstream. Their results showed a high degree of similarity between the ice cover surface sediment and sediment collected from the lake bottom. They suggested that sediment enters the lake through cracks or gas bubble channels in the ice cover. Further evidence for the passage of sediment through the ice cover came from sediment trap data collected at Lake Hoare (Wharton et al., 1989; Squyres et al., 1991). While coarse sand-sized sediment were collected in all the traps, the most interesting observation was that the relative amounts of sediment within each set of traps often differed significantly. In addition, small mounds and ridges of sediment have been observed on the lake bottom. The observation that different amounts of sediment were being deposited within close proximity on the lake bottom supports the hypothesis that sediment enters the lake through the ice cover at distinct locations via cracks and/or gas bubble channels in the ice.

One important aspect of the above deposition scenario is its dependence upon a relatively thin ice cover. It appears that the development of cracks in the ice cover, which link the atmosphere with the water column and allow sediment to pass through the ice, develop only when the ice cover thins to ~3 m (Andersen et al., 1992; Wharton et al., 1989, 1993). Because most of the sediment entering the lake must penetrate the ice cover, changes in the thickness and character of the ice cover could control the rate of sediment accumulation. Alternating layers of terrigenous and biogenic sediment on the lake bottom show a record of past changes in the ice cover and may provide a sensitive record of external controls, such as regional climate change.

MICROBIAL MATS

Microbial mats (Fig.3) composed primarily of cyanobacteria, heterotrophic bacteria, and eukaryotic algae occur throughout much of the benthic regions of Lakes Bonney, Chad, Hoare, Fryxell, Joyce, and Vanda (Wharton et al., 1983; Mikell et al., 1984; Parker and Wharton, 1985; Vincent, 1988). The filamentous cyanobacterium *Phormidium frigidum* Fritsch (Fig.4) forms the matrix of all mat types thus far studied. This species is ubiquitous in southern Victoria Land, occurring in soils, lakes, glacial meltstreams, and cryoconite holes (Parker et al., 1981; Simmons et al., 1981; Wharton et al., 1981). *Lyngbya martensiana* Menegh. is another filamentous cyanobacterium often found in the benthic mats of Lakes Fryxell and Vanda (Allnutt et al., 1981, Love et al., 1983). Diatoms comprise the largest number of algal species in the mats. Other

Figure 3 : a) Pinnacle mat in Lake Vanda, Wright Valley, Antarctica. b) Gas-charged (numerous gas bubbles are observed in the mat matrix) lift-off mat in Lake Hoare, Taylor Valley, Antarctica. c) : Indurate columnar calcite structure in Lake Fryxell, Taylor Valley, Antarctica. d) Columnar and prostrate mat in Lake Fryxell, Taylor Valley, Antarctica.

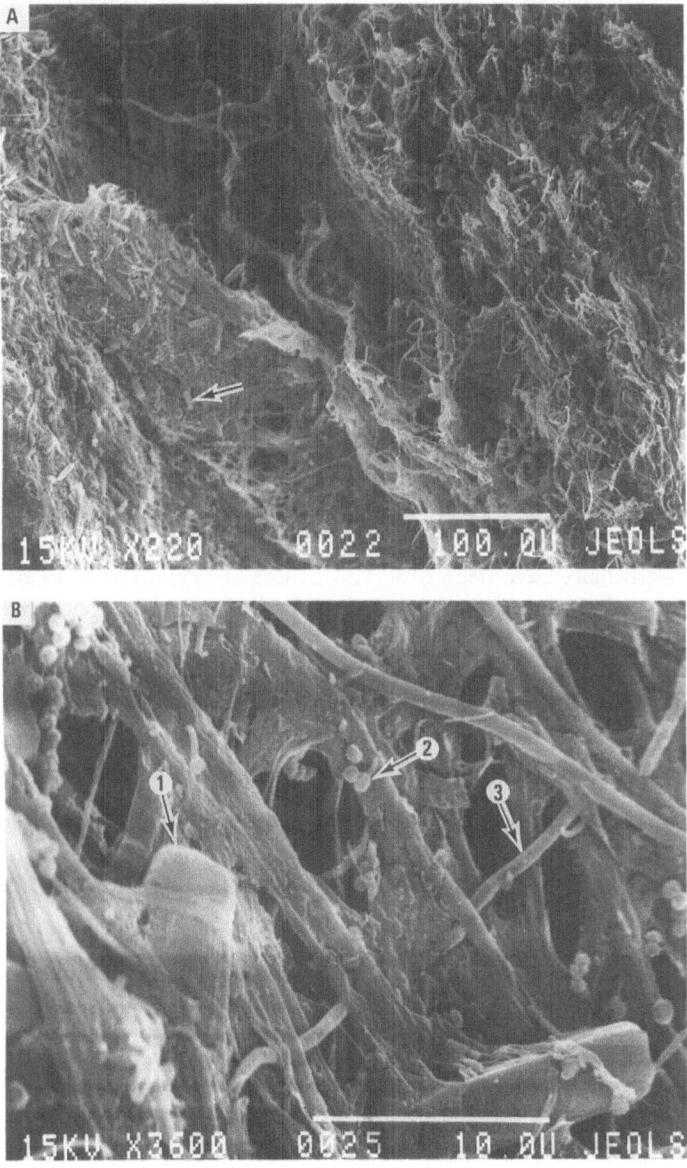

Figure 4 : a) : SEM cross section of prostrate microbial mat collected from Lake Bonney, Taylor Valley, Antarctica. The interwoven filaments of cyanobacteria are readily observed. The arrow points to one of the may pennate diatoms in the mat. b) : Enlargement of same mat shown in A. Diatom frustules (1), heterotrophic bacteria (2), and filamentous cyanobacteria (3) are the main biological components of microbial mats in antarctic dry valley lakes.

microorganisms observed in the benthic mats include fungi (Baublis et al., 1991) and protozoans (Cathey et al., 1981).

While there are differences in relative abundances of the species which make up the microbial mats (for species list see Wharton et al., 1983; Parker and Wharton, 1985), the most interesting differences concern the distribution of mat morphologies within a lake and between the lakes. The four major categories of microbial mat morphologies that have been observed are: prostrate, lift-off, columnar, and pinnacle morphologies.

Prostrate Mats

Prostrate mats (Fig.5) are observed in all of the dry valley lakes thus far studied. These mats occur under either aerobic or anaerobic conditions. The upper surfaces of aerobic prostrate mats are smooth or flocculose in texture, and dark green, pink, purple, or brown in color. Lakes Fryxell, Hoare, Chad, Bonney, and Joyce have extensive development of this mat type. The actively growing surficial mat in aerobic prostrate mats is generally < 5 mm thick. Calcite crystals are abundant in this mat type. Aerobic prostrate mats typically contain the same species composition as found in lift-off, columnar and pinnacle mats.

Anaerobic prostrate mats have been observed in Lakes Fryxell and Hoare. These mats are black, coarse and flocculose with a distinct hydrogen sulfide odor. While calcite is not observed, abundant granules of iron, manganese and sulfur are present. Organisms observed in anaerobic mat include the bacteria *Leptothrix* sp., *Achroonema* sp., *Planctomyas* sp., *Thiothrix* sp., *Clostridium* sp., and several other as yet unidentified species. The anaerobic, photosynthetic green sulfur bacterium, *Chloroflexus* sp., is common in Lake Hoare anaerobic mats. This species has previously been reported from thermal springs and temperate lakes. Its discovery in an antarctic lake may represent the first report of this species from a perennially cold environment. *Chlorella* sp. and empty diatom frustules are also commonly observed. It is interesting to speculate on the origin of the empty diatom frustules. They are pennate forms which represent the same species as those observed living in aerobic mats. It appears that in recent years the anaerobic zone in Lake Hoare has been rising in step with a rising lake level (Chinn, 1993; Wharton et al., 1992, 1993). As a result, regions which were formerly aerobic and therefore contained aerobic prostrate mat (including diatoms) have subsequently turned anaerobic.

Alternating laminae of organic and sediment material (Fig.5) have been observed below the aerobic and anaerobic prostrate mats studied thus far (Wharton et al., 1983; Squyres et al., 1991). While the total thickness of these layers has not been determined, they are at least 2-3 m in thickness at Lakes Fryxell and Hoare (Love et al., 1982). The number of laminae varies but 300-400 alternating layers in a 0.5 m section of sediment core are not uncommon. As shown in Figure 4, the distribution of laminae within the sediment profile also varies with sediment depth. Squyres et al. (1991) provided a detailed analysis of nine cores of aerobic prostrate mat and underlying sediment taken from a 3 m grid in Lake Hoare. The cores contained alternating layers of fine-to-coarse sand and

Figure 5 : a) : Longitudinal section of box cores obtained from an area of aerobic prostrate mat in Lake Hoare, Taylor Valley, Antarctica. Arrow 1 indicates the actual top of the core (i.e. the sediment-water interface. Arrows 2 and 3 show organic-rich and sediment-rich layers, respectively. b) : Benthic cores obtained from an area of aerobic prostrate mat in Lake Hoare, Taylor Valley, Antarctica. Arrow 1 and 2 in B point to sediment-rich and organic-rich layers (~100 organic-rich laminations), respectively. c) : Benthic cores obtained from area of anaerobic prostrate mat in Lake Hoare. Arrow points to organic-rich layers.

organic-rich layers. Cyanobacterial filament sheaths and rhombohedral calcite crystals are found only within the organic-rich laminae. Diatom frustules are observed in both organic- and sediment-rich laminae; however, the organic-rich layers usually contained at least an order of magnitude more frustules than the sediment-rich layers.

Lift-off Mats

Lift-off mats have been observed in every lake studied thus far, except for lake Vanda (Wharton et al., 1983; Parker and Wharton, 1985). Lift-off mat results from an interesting combination of physical and biological processes (see below). In lakes Hoare, Bonney, and Joyce gas-charged pieces of prostrate mat "lift-off" the bottom of the lake (Fig.3b) forming vertically-oriented sheets and columnar structures. The upper layer of actively growing mat is <5 mm thick, smooth or flocculose in texture, and dark green or purple in color. The sheets and columnar structures are between 2 cm to 1.5 m high. The microorganisms observed in this mat type are similar to those found in prostrate mats (namely cyanobacteria, heterotrophic bacteria, and eukaryotic algae). Calcite crystals are abundant within the lift-off mat. The sediment profile beneath the lift-off mats is also similar to that observed for prostrate mat as described in the section above. That is, the 2-3 m sediment profile consists of alternating laminae of organic-rich and sediment-rich material of varying thickness. Layers of $CaCO_3$ are common in the sediment profile beneath lift-off mat.

Columnar Mats

In lake Fryxell, mat overlying calcite structures (Fig.3c,d) is observed in the aerobic benthos at water depths between 6 and 9 m (Wharton et al., 1982). The calcite structures are indurate, unbranched, and oriented either vertically or horizontally. The vertically-oriented, columnar structures (1-5 cm diam, 1-10 cm tall) are usually hollow, terete and covered by 0.2-1.0 cm thick layer of microbial mat. Horizontally-oriented calcite plates ~5 mm thick also occur beneath actively growing mat in between the columnar structures. The top layer of mat consists primarily of *P. frigidum*, pennate diatoms, and heterotrophic bacteria. The sediments beneath the calcite structures contain alternating bands of organic- and sediment-rich layers to a depth of ~1 m (Lawrence and Hendy, 1985). The sediments contain five distinct units, three of which are calcareous. The uppermost unit (0-47 cm) contains the columnar and plate-like calcite structures mentioned above. The second unit (47-58 cm) is a varve-like aragonite deposit dated at ~10,000 years B.P. The next unit (58-64 cm) is transitional between the aragonite layers and a still lower unit (64-87 cm) of calcareous mud of mixed aragonite-calcite mineralogy. This later unit is dated at ~20,000 years B.P. The lowermost unit (87-107 cm) is primarily fine sands deposited >20,000 years B.P. Wharton et al. (1982) identified five organic-rich layers, isolated from 7-32 cm in the sediment that had an average organic matter content of ~9%. Microscopic examination of both the columnar and horizontal calcite structures revealed diatom frustules and filamentous material (1-2 mm diam) resembling sheaths of *P. frigidum*.

Pinnacle Mats

Pinnacle mats have thus far been reported from lakes Hoare, Bonney, and Vanda (Wharton et al., 1983). The greatest abundance of this mat type is in lake Vanda (Fig.6a) where it has been observed to a depth of 30 m (Love et al., 1983). The pinnacles are generally 2-5 cm high and 3-5 cm wide at their bases and spaced 6-12 cm apart. In cross section, the lake Vanda pinnacles consist of about four sets of translucent bands of light brown to tan layers 1-2 mm thick alternating with greenish purple layers < 1 mm thick (Fig.6a). Sand and calcite crystals are common in these layers (Fig.6b). The pinnacles are relatively solid structures consisting of superimposed conical mat layers with no gas-charged voids or hollow central areas as observed in lift-off or columnar mats. The pinnacles retain their structure when removed from the water and dried. Microscopic examination revealed that the light-brown to tan layers were primarily *P. frigidum* and diatoms, and the greenish-purple layers consisted mostly of *L. martensiana*, *P. frigidum* , and diatoms. Sand grains are sparsely dispersed through the living portion of the mat. Abundant small (~300 mm diam) euhedral to subhedral calcite crystals are embedded within the *P. frigidum* layers.
Alternating laminae of organic- and sediment-rich layers are found below pinnacle mats. The organic-rich layers are from 0.5-1.0 cm thick and contain abundant calcite crystals. These layers also contain empty diatom frustules representing the same taxa found living in the upper pinnacle mat layers. The sediment-rich layers are similar (i.e. fine- to coarse-grained sands) to those found below prostrate and columnar mats.
An interesting biological feature of the pinnacle mat in lake Vanda is the presence of *Bryum* cf. *algens* Card (Kaspar et al., 1982). To date, mosses have not been observed in mats from any of the other dry valley lakes. The presence of *B. algens* in lake Vanda is probably related to the relatively higher irradiance available in the lake compared to other lakes in the region It does not appear that this moss is an essential component in pinnacle mat formation as many of the pinnacles lack the moss.

FORMATION OF MODERN STROMATOLITES

The microbial mats in the perennially ice-covered antarctic lakes are currently trapping and binding sediment, and precipitating various minerals (Parker et al., 1981; Wharton et al., 1983). These stromatolitic mats are morphologically either prostrate, lift-off, columnar, or pinnacle shaped with alternating organic-and sediment-rich layers. The morphology of a microbial mat and subsequent modern stromatolitic structure is dependent upon the microorganisms involved and local environmental conditions (Golubic, 1976; Walter, 1977).
Prostrate mats form by the gliding of filaments of cyanobacteria over the lake bottom forming a cohesive tissue-like fabric. Local environmental conditions determine

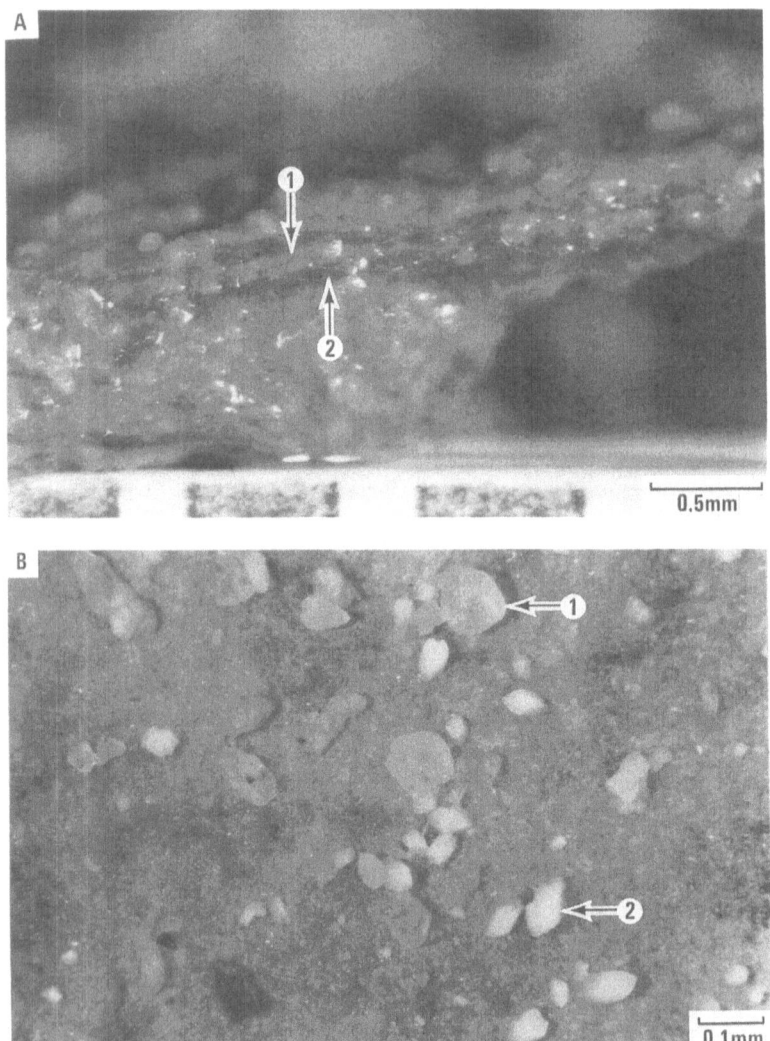

Figure 6 : a) Section through a single pinnacle of mat collect in lake Vanda, Wright Valley, Antarctica. Arrow 1 points to a calcite-rich layer in the mat, while arrow 2 points to a cyanobacterial-rich layer. b) Silicate minerals (1) and calcite crystals (2) isolated from lake Vanda's pinnacle mat.

whether a mat remains prostrate or develops into lift-off or columnar mat. As mentioned previously, the relatively complete seal of the perennial ice cover results in elevated atmospheric gas levels within the shallower parts of the lake. Lift-off mat is produced when the pressure of dissolved gases inside the prostrate microbial mat exceeds the local hydrostatic pressure at that depth and bubbles form inside the mat causing it to "lift-off" the lake bottom. Wharton et al. (1986, 1987) have developed a quantitative model which predicts the maximum depth of lift-off mat in lake Hoare. Lift-off mats do not form below 10 m in lake Hoare because increase hydrostatic pressure prohibits bubble formation. In relatively deeper areas (>10 m) of the lakes less light and an undersaturation of the water with atmospheric gases results in a stable prostrate mat.

In some cases, lift-off mats tear completely loose from the lake bottom and float to the undersurface of the ice cover. Once at the bottom of the ice cover, these pieces of microbial mat freeze into the ice cover and through ablation of surface ice eventually make their way to the top of the ice cover. As much of the mat is still viable after passage through the ice, the escape mechanism described above is important in the distribution of microbes between the lakes and other environments in the region (Parker et al., 1982b).

Some of the lift-off mats do remain in place forming stable structures. Calcite crystals are usually observed in the microbial layers of the lift-off mats and may provide a stabilizing effect. This is particularly obvious in the columnar calcite structures which are covered with growing mat in lake Fryxell. Gas bubbles develop inside the mat matrix and cause the mat to lift-off the lake bottom as described above. However, in lake Fryxell, much of this gas-charged mat is stabilized by precipitated calcite and does not tear loose from the substrate, as it often does in lakes Bonney and Hoare. Lake Fryxell has a much higher alkalinity and therefore favors $CaCO_3$ precipitation compared with some of the other less alkaline lakes.

The hollow nature of the lake Fryxell structures suggests that calcite precipitation follows the orientation of the growing mat. There is also the possibility that ground water inflows into the lake may play some role in the formation of these structures (Wharton et al., 1989). However, the influence of groundwater on microbial mats and stromatolite formation in the antarctic lakes has yet to be investigated.

The antarctic lake pinnacle mat forms by a process similar to that previously described for this type of mat in hot springs (Walter et al., 1976; Walter, 1977; Brock, 1978) and is suggested for the formation of Triassic algal tuft structures (Mayall and Wright, 1981). Gliding filaments become entangled forming knots on the apices or upper surfaces of pieces of sand or gravel. Vertical protrusions or tufts begin forming at the knots by movement of filaments upwards over one another towards the light. This reorientation initiates positive phototaxis and additional upward gliding results (Brock, 1978). The lake bottom is now dotted with vertical protuberances. Increasing numbers of gliding filaments are diverted up the sides of the tufts, which become enlarged and pointed. As suggested by Brock (1978), the three essential processes in pinnacle mat formation are gliding, phototaxis, and cohesion. The pinnacles accrete upwards at rates determined by the sedimentation rate, and by the supply of nutrients and light. Variations in these factors produce intermittent accretion, and therefore lamination.

These modern stromatolitic mats are considered analogs of *Conophyton*, a columnar stromatolite very abundant in Precambrian rock (Walter, 1977; Love et al., 1983).

Sharp-edged rhombohedrons and aggregates of calcite of non-detrital origin have been observed in all shallow water (< 12 m) mat types thus far observed (Wharton et al., 1983). Calcite precipitation and crystal growth occurs within the mat during the growth process. The shallow waters (<12 m water depth) of lakes Fryxell, Joyce, Hoare, and Vanda are all supersaturated with respect to calcite (Green et al., 1988). Calcite precipitation is in part the result of the photoautotrophs removing CO_2 from the interstitial water. As these molecules are removed, the pH in the microenvironment near the cells increases and $CaCO_3$ precipitates (Wetzel, 1960; Golubic, 1973). In lake Fryxell, carbonate precipitation also occurs in the water column as a result of planktonic CO_2 fixation and falls from suspension to the lake bed where it is trapped and stabilized within the growing microbial mat (Lawrence and Hendy, 1985).

Superimposed over the biological and geochemical processes is the unusual sedimentation process occurring in the antarctic lakes (Squyres et al., 1991; Andersen et al., 1993). Based upon the observation of microbial remains (i.e., sheath material of cyanobacteria and diatom frustules) and pigment evidence, the lower organic-rich layers in the antarctic lake sediments represent relic microbial mat layers similar to those presently forming at the sediment-water interface (Simmons et al., 1983). Sediment which is deposited through the perennial ice cover or carried in glacial meltwater will settle to the lake bottom and bury portions of the surficial microbial mat (Fig.7a). Depending upon the amount and rate of deposition, two subsequent outcomes are possible. One large or several relatively rapid, small depositional events will probably not allow the buried microorganisms to grow up through the sediment back to the sediment water interface. Cells for recolonization of these areas will probably be from adjacent, unburied areas of mat. Following small or relatively slow depositional events, microorganisms would probably be capable of growing up through the sediment to the sediment-water interface. In either case, once established at the sediment-water interface the microbial mat will stabilize the sediment surface and form a relatively thick mat until the next depositional event occurs in that specific area. Hence, this activity provides a mechanism by which alternating laminae of inorganic and organic material, that is organosedimentary structures (modern stromatolites) develop.

CONCLUSIONS

Stromatolitic microbial mats composed primarily of bacteria, cyanobacteria, and eukaryotic algae are found in cold, dimly-lit, perennially ice-covered antarctic lakes. Four mat morphologies have been reported: prostrate, lift-off, columnar, and pinnacle. The morphology of a particular mat results from a combination of biological, geochemical, and sedimentological processes, some of which may be unique to the

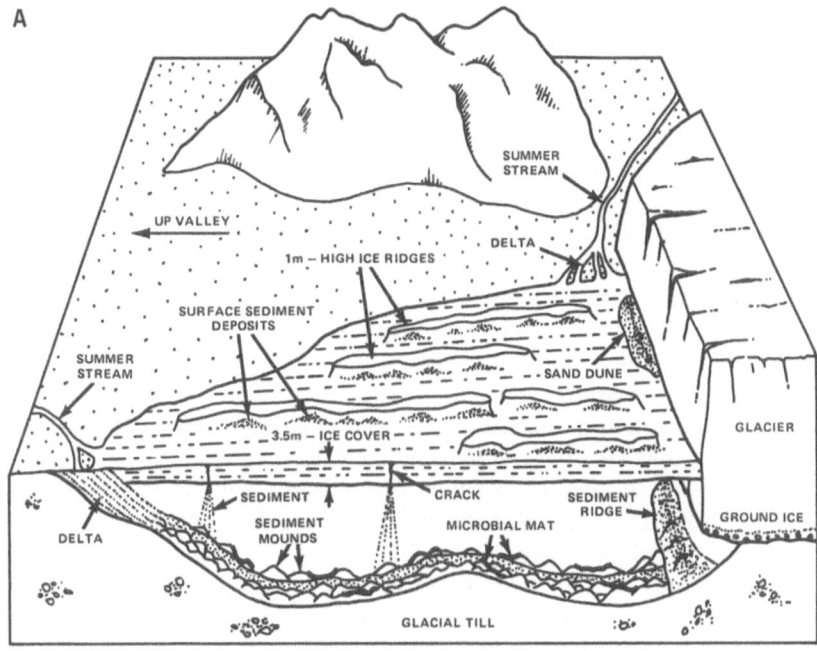

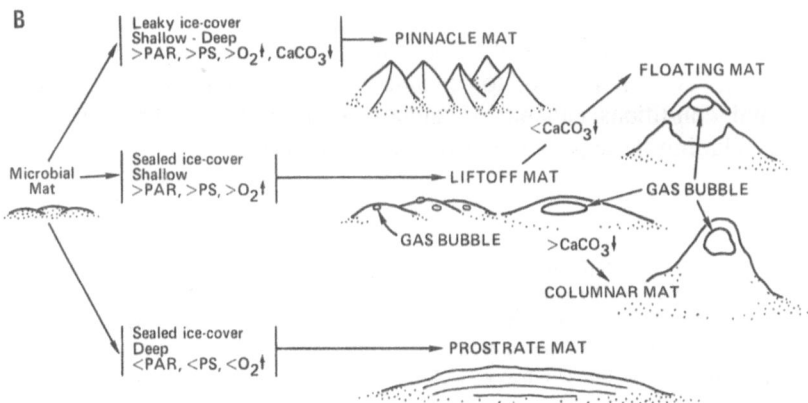

Figure 7 : a) : Diagram of lake Hoare, Taylor Valley, Antarctica showing sediment deposition through cracks in the ice cover and the formation of unique depositional features on the lake bottom. b) : Flow diagram showing key environmental features influencing the morphology of developing microbial mat in ice-covered antarctic lakes. PAR = photosynthetically active radiation (400-700 nm), PS = photosynthetic rate.

ice-covered antarctic lakes. The mats are trapping, binding and/or precipitating carbonates and various other minerals forming organosedimentary structures. Using the definition of Awramik et al. (1976), these structures can be classified as modern stromatolites.

Several biological and physico-chemical features (Fig.7b) of the antarctic lakes allow for the formation and preservation (at least on the order of thousands of years) of the stromatolitic structures (Parker et al., 1981; Lawrence and Hendy, 1985; 1989). Many of these features are directly dependent upon the perennial ice cover and its control on the lake environment (Wharton et al., 1993). These include: (1) varying light conditions which play a key role in determining the morphology of the stromatolitic microbial mats, (2) relatively high levels of dissolved atmospheric gases in certain areas of a lake that result in gas bubble formation within the mats, and subsequent lift-off of the mat from the lake bottom, (3) calcite precipitation in aerobic areas and manganese, iron, and sulfur deposition in anaerobic areas, (4) minimal mixing of lake water and a lack of burrowing and browsing organisms such as cladocerans, copepods, annelids, gastropods, insect larvae, fish leaving the mat and sediment undisturbed (Cathey et al., 1981). The antarctic stromatolitic mats and their associated calcareous and evaporitic sediments reflect a climatological and biogeochemical history of ~20,000 years (Lawrence and Hendy 1985, 1989; Clayton-Greene et al. 1988). The prospects for preservation of these modern stromatolites on time scales of 10^3 years is good given the cold, arid conditions in the McMurdo Dry Valleys. However, determining the potential for long-term preservation and lithification into stromatolitic rocks is difficult at best given the vagaries of climate and geology over time scales of 10^6 years.

The ice-covered lakes of Antarctica may serve as an important model for understanding the formation of stromatolites in cold environments. Additional studies of antarctic stromatolitic mats could help to enhance our understanding of the range of environmental conditions capable of supporting stromatolite formation, particularly cold facies including those present during the Precambrian.

ACKNOWLEDGEMENTS

This research was supported by the National Science Foundation's Office of Polar Programs (DPP-8416340 and OPP-9211773) and the National Aeronautics Space Administration's Exobiology Program (NCA2-2, 1R675-402, NAGW-1947). This chapter is dedicated to Bruce C. Parker who introduced me to the wonders of the antarctic microbial mats. Excellent support and assistance in Antarctica was provided by the US Navy, Holmes and Narver, and ITT's Antarctic Services. I am particularly thankful to my many colleagues for help in the field and laboratory. I am also thankful to P. Hirsch for the identification of bacteria in the lake Hoare anaerobic mat. Helpful reviews were provided by Rocco Mancinelli, Chris McKay, George Simmons, and three other anonymous reviewers. I am grateful for the persistence and patience of C.L.V. Monty and J. Sarfati, the editors of this volume.

REFERENCES

Allnutt, F.T.C., Parker, B.C., Seaburg, K.G., and Simmons, G.M. (1981) "*Insitu* nitrogen (C_2H_2)-fixation in lakes of Southern Victoria Land, Antarctica", Hydrobiol Bull 51: 99-109.

Andersen, D.W, Wharton R.A., Squyres, S.W. (1993) "Terrigenous clastic sedimentation in antarctic dry valley lakes", In: W. Green and E.I. Friedmann (eds.), Physical and biogeochemical processes in antarctic lakes. Antarct Res Ser, Amer Geophys Union, Washington DC, Vol. 59, pp. 71-82.

Anderson, J. (1983) "Spatial and temporal distribution of glacial marine sediment", In: B.F. Molnia (ed.) Global marine sedimentation. Plenum Press, New York, pp. 1-99.

Awramik, S.M., Margulis, L., and Barghoorn, E.S. (1976) "Evolutionary processes in the formation of stromatolites", In: M.R. Walter (ed.), Stromatolites. Elsevier, Amsterdam, pp. 149-162.

Baublis, J.A., Wharton, R.A., Volz, P.A. (1991) "Diversity of micro-fungi isolated in an antarctic dry valley", J. Basic Microbiol 31: 10-20.

Brock, T.D. (1978) "Thermophilic microorganisms and life at high temperatures", Springer-Verlag, New York, pp. 1-468.

Cathey, D.D., Simmons, G.M., Parker, B.C., Yongue, W.H. (1981) "The microfauna of algal mats and artificial substrates in southern Victoria Land lakes of Antarctica", Hydrobiol 85: 3-15.

Chinn, T.J. (1993) "Physical hydrology of dry valley lakes", In: W. Green and E.I. Friedmann (eds.), Physical and biogeochemical processes in antarctic lakes. Antarct Res Ser, Amer Geophys Union, Washington DC, Vol. 59, pp. 1-52.

Clayton-Greene, J.M., Hendy, C.H., Hogg, A.G. (1988) "Chronology of a Wisconsin age proglacial lake in the Miers Valley, Antarctica", New Zealand J. Geol Geophys 31: 353-361.

Clow, G.D., McKay, C.P., Wharton, R.A., Simmons, G.M. (1988) "Climatological observations and sublimation rates at Lake Hoarse, Antarctica", J. Climate 1: 715-728

Craig, H., Wharton, R.A., McKay, C.P. (1992) "Oxygen supersaturation in ice-covered antarctic lakes: biological versus physical contributions", Science 255: 318-321.

Frakes, L.A. (1979) "Climates through geologic time", Elsevier, New York, pp. 1-310.

Garrett, P. (1970) "Phanerozoic stromatolites: noncompetitive ecologic restriction by grazing and burrowing animals", Science 169: 171-173.

Golubic, S. (1976) "Organisms that build stromatolites", In: M.R. Walter (ed.), Stromatolites. Elsevier, Amsterdam, pp. 113-126.

Golubic, S. (1973) "The relationship between blue-green algae and carbonate deposits", In: N.G. Carr and B.A. Whitton (eds.) The biology of blue-green algae. Univ of Calif Press, Berkeley, pp. 434-472.

Green, W.J., Angle, M.P., Chave, K.E. (1988) "The geochemistry of Antarctic streams and their role in the evolution of four lakes in the McMurdo Dry Valleys", Geochim Cosmochim Acta 52: 1265-1274.

Kaspar, M., Simmons, G.M., Parker, B.C., Seaburg, K.G., Wharton, R.A. (1982) and Louis-Smith, R.I. "A species of Bryum Hedw. from Lake Vanda, Antarctica", Bryologist 85: 424-430.

Lawrence, M.J.F., Hendy, C.H. (1985) "Water column and sediment characteristics of Lake Fryxell, Taylor Valley, Antarctica", New Zealand J. Geol Geophys 28: 543-552.

Lawrence, M.J.F., Hendy, C.H. (1989) "Carbonate deposition and Ross Sea ice advance, Fryxell basin, Taylor Valley, Antarctica", New Zealand J. Geol Geophys 32: 267-277.

Love, F.G., Simmons, G.M., Parker, B.C., Wharton, R.A., Seaburg, K.G. (1983) "Modern conophyton-like microbial mats discovered in Lake Vanda, Antarctica", Geomicrobiol J. 3: 33-48.

Love, F.G., Simmons, G.M., Wharton. R.A., Parker, B.C. (1982) "Methods for melting dive holes and vibracoring beneath ice", J. Sed Petrol 52: 644-647.

Mayall, M.J., Wright, V.P. (1981) "Algal tuft structures in stromatolites from the Upper Triassic of Southwest England", Palaeont 24: 655-660.

McKay, C.P., Clow, G.A., Wharton, R.A., Squyres, S.W. (1985) "Thickness of ice on perennially frozen lakes", Nature 313: 561-562.

Mikell, A.T., Parker, B.C., Simmons, G.M. (1984) "Response of an antarctic lake heterotrophic community to high dissolved oxygen", Appl Environ Microbiol 47: 1062-1066.

Nedell, S.S., Andersen, D.W., Squyres, S.W., Love, F.G. (1987) "Sedimentation in ice-covered Lake Hoare, Antarctica", Sedimentology 34: 1093-1106.

Palmisano, A.C., Simmons, G.M. (1987) "Spectral downwelling irradiance in an Antarctic lake", Polar Biol 7: 145-151.

Parker, B.C., Seaburg, K.G., Cathey, D.D., Allnutt, F.C.T. (1982a) "Comparative ecology of plankton communities in seven antarctic oasis lakes", J. Plankton Res 4: 271-286.

Parker, B.C., Simmons, G.M., Love, F.G., Wharton, R.A., Seaburg, K.G. (1981) "Modern stromatolites in antarctic dry valley lakes", BioScience 31: 656-661.

Parker, B.C., Simmons, G.M., Wharton, R.A., Seaburg, K.G., Love, F.G. (1982b) "Removal of organic and inorganic matter from antarctic lakes by aerial escape of bluegreen algal mats", J. Phycol 18: 72-78.

Parker, B.C., Wharton, R.A. (1985) "Physiological ecology of bluegreen algal mats (modern stromatolites) in antarctic oasis lakes", Arch Hydrobiol Alg Stud 38/39: 331-348.

Seaburg, K.G., Parker, B.C., Wharton, R.A., Simmons, G.M. (1981) "Temperature growth responses of algal isolates from antarctic oasis lakes", J. Phycol 17: 353-360.

Simmons, G.M., Parker, B.C., Allnutt, F.T.C., Brown, D., Cathey, D.D., Seaburg, K.G. (1979) "Ecological comparisons of oasis lakes and soils", Antarct J. US 24: 181-183.

Simmons, G.M., Parker, B.C., Wharton, R.A., Love, F.G., Seaburg, K.G. (1981) "Physiological adaptations of biota in antarctic oasis lakes", Antarct J. US 26: 173-174.

Simmons, G.M., Wharton, R.A., Parker, B.C., Andersen, D.T. (1983) "Preliminary observations on chlorophyll-a and ATP concentrations in antarctic and temperate lake sediments", Microbial Ecol 9: 123-135.

Squyres, S.W., Andersen, D.W., Nedell, S.S., Wharton, R.A. (1991) "Lake Hoare, Antarctica: sedimentation through a thick perennial ice cover", Sedimentology 38: 363-379.

Vincent, W.F. (1988) "Microbial ecosystems of Antarctica", Cambridge Univ Press, pp. 1-304.

Walter, M.R., Bauld, J., Brock, T.D. (1976) "Microbiology and morphogenesis of columnar stromatolites (*Conophyton, Vacerrilla*) from hot springs in Yellowstone National Park", In: M.R. Walter (ed.) Stromatolites. Elsevier, Amsterdam, pp. 273-310.

Walter, M.R. (1977) "Interpreting stromatolites", Amer Sci 65: 563-571.

Walter, M.R., Bauld, J. (1983) "The association of sulphate evaporites, stromatolitic carbonates and glacial sediments: examples from the proterozoic of Australia and the Cainozoic of Antarctica", Precambrian Res 21: 129-148.

Wetzel, R.G. (1960) "Marl encrustation on hydrophytes in several Michigan lakes", Oikos 11: 223-236.

Wharton, R.A., McKay, C.P., Clow, G.D. (1993) "Perennial ice covers and their influence on antarctic lake ecosystems", In: W. Green and E.I. Friedmann (eds.) Physical and biogeochemical processes in antarctic lakes. Antarct Res Ser, Amer Geophys Union, Washington DC, Vol. 59, pp. 53-70.

Wharton, R.A., McKay, C.P., Clow, G.D., Andersen, D.T., Simmons, G.M., Love, F.G. (1992) "Changes in ice cover thickness and lake level of Lake Hoare, Antarctica: implications for local climatic change", J. Geophys Res 97: 3503-3513.

Wharton, R.A., McKay, C.P., Mancinelli, R.L., Simmons, G.M. (1987) "Perennial N_2 supersaturation in an antarctic lake", Nature 325: 343-345.

Wharton, R.A., McKay, C.P., Simmons, G.M., Parker, B.C. (1986) "Oxygen budget of a perennially ice-covered antarctic Lake", Limnol Oceanogr 31: 437-443.

Wharton, R.A., Parker, B.C., Simmons, G.M. (1983) "Distribution, species composition and morphology of algal mats in antarctic dry valley lakes", Phycologia 22: 355-365.

Wharton, R.A., Parker, B.C., Simmons, G.M., Seaburg, K.G., Love, F.G. (1982) "Biogenic calcite structures forming in Lake Fryxell, Antarctica", Nature 295: 403-405.

Wharton, R.A., Simmons, G.M., McKay, C.P. (1989) "Perennially ice-covered Lake Hoare, Antarctica: physical environment, biology, and sedimentation", Hydrobiol 172: 305-320.

Wharton, R.A., Vinyard, W.C., Parker, B.C., Simmons, G.M., Seaburg, K.G. (1981) "Algae in cryoconite holes on Canada Glacier, southern Victoria Land, Antarctica", Phycologia 20: 208-211.

RECENT FRESH-WATER LACUSTRINE STROMATOLITES, STROMATOLITIC MATS AND ONCOIDS FROM NORTHEASTERN MEXICO

B.M. WINSBOROUGH[1], J-S. SEELER[2], S. GOLUBIC[3], R.L. FOLK[4], B. MAGUIRE JR.[5]

1 Winsborough Consulting, 5701 Bull Creek Road, Austin, Texas, U.S.A. 78756
2 Department of Internal Medicine, University of Texas Southwest Medical Center, Dallas, Texas, U.S.A. 75235-8594
3 Department of Biology, Boston University, Boston, Mass., U.S.A. 02215
4 Department of Geology, The University of Texas, Austin, Texas, U.S.A. 78712
5 Department of Zoology The University of Texas, Austin, Texas, U.S.A. 78712

ABSTRACT

Stromatolites, stromatolitic mats and oncoids are forming in two adjacent lakes (El Mojarral East and West) located in the desert basin of Cuatro Ciénegas, Coahuila, Mexico. Characteristic of the stromatolites and stromatolitic mats is a surface pattern of offset, horizontally oriented, outwardly projecting ledges produced by the calcifying cyanobacterium *Homoeothrix balearica,* in conjunction with the uncalcified cyanobacterium *Schizothrix lacustris.* In areas of low light and hence reduced carbonate accretion, endolithic cyanobacteria obscure the boundary between newly precipitated carbonate above and bedrock below. Fish and a diverse and abundant benthic animal population are associated with these biogenic structures. Fish alter the lamination pattern through grazing activities ; the microbenthos have little effect on the integrity of the lamination. Diatoms are well preserved in the stromatolites and oncoids under certain conditions and dissolved in other situations. Quartz crystals are present in inverse proportion to the quality of diatom preservation.

INTRODUCTION

Stromatolites and related microbial accretions constitute a major part of the littoral lithofacies in spring-fed lakes and streams of the Cuatro Ciénegas Basin, Coahuila, Mexico. Their habitats range from deeply shaded settings near the bottom of lakes and streams where the current is rapid, to shallow, well-illuminated less energetic current conditions. Several different and distinct stromatolite morphologies are presently forming (Winsborough and Seeler, 1986, Winsborough and Golubic 1987, Winsborough,

J. Bertrand-Sarfati and C. Monty (eds.), Phanerozoic Stromatolites II, 71–100.
© 1994 *Kluwer Academic Publishers.*

1990). In El Mojarral, the lake system described here, stromatolites and stromatolitic mats develop with a characteristic lobate ridged growth pattern, produced primarily by the activity of a distinct and vertically differentiated microbial community. Oncoids, built by the same microbial community as the stromatolites, but lacking well-developed ridges also form.

SETTING

The Cuatro Ciénegas Basin is a small desert valley of about 1200 km² enclosed by mountains of the Sierra Madre Oriental of Coahuila, Northeastern Mexico. The northern tip of Sierra San Marcos, protruding from the south, almost bisects the basin (Fig.1).

Figure 1 : Location map of the Cuatro Ciénegas Basin, Coahuila, Mexico, showing El Mojarral lakes.

This mountain is a faulted anticline composed of Cretaceous shallow water marine limestone, dolomite, and gypsum. The steep to vertical canyon slopes of Sierra San Marcos are rimmed by alluvial fans, which give way to a flat desert floor (elevation 740m). The mean annual air temperature is 23°C (Morafka, 1977) ; annual rainfall is less than 200 mm, over half of that occuring in September (Garcia et al, 1975). This arid climate with highly seasonal rainfall provides the setting for a flat basin filled with a facies mosaic of evaporitic minerals.

Springs and seeps (reaching a density of 12-15 per km²) emerge along the distal margin of the alluvial fans. Most of the springs are warm (up to 35°C), but springs as cool as

19°C are also found. Some of the springs have nearly constant water temperatures throughout the year, whereas others vary between 19 and 35°C. The springs range in size from small shallow marsh seeps to deep spring-fed lakes and streams. Some springs are permanent, some dry out gradually, and some reverse their flow and then drain the lake they had been feeding. The spring-fed lakes and streams at Cuatro Ciénegas, in which stromatolites and oncoids are forming, are all very similar in their water chemistry. The cations are dominated by calcium, sodium, and magnesium while the most abundant anions are sulfates, followed in much less abundance by carbonates and chlorides. What is distinctive about these waters is that, for water with so little chloride, they have unusually high levels of sulfate relative to alkalinity. The water in which these stromatolites form is different in chemistry from stromatolite locations in other parts of the world because it is fresher than that found in a playa or brackish setting and has a greater amount of total dissolved solids than most fresh-water lake and stream examples (Winsborough, 1990).

The El Mojarral system is typical of several Cuatro Ciénegas spring-lake systems as is the growth pattern of their stromatolites and related microbial accretions. Laguna El Mojarral West (westernmost lake) appears to be the primary spring source for it and the adjacent lake, El Mojarral East. The lakes are about 200 m apart, connected by a shallow surface stream (and possibly by subsurface conduits). Extensive marshlands to the west contribute to surface and probably subsurface inflow. El Mojarral West is a roughly constant level, spring-fed lake about 13 m wide, 48 m long, and 1 to 5 m deep. Water temperature varies only about 7 degrees annually, between 28° and 35°C. Physical and chemical characteristics of El Mojarral West are summarized on Table 1. The lake appears to have been partly formed by the collapse of sections of roof

Mg	109.4	Zn	0.0047	Ca	372.7
Cd	0.0004	Na	142.5	Pb	not detected
K	8.0	Ni	0.0026	SO_4	1373.7
Co	0.0001	Cl	102.8	Fe	0.0040
Alk	164.8	Mn	0.00005	Sr	12.7
Cu	0.0033	Si	18.3	TDS	2283
F	2.7	pH	7.1	NO_3	6.8
Cond.	2800 μmhos	PO_4	0.003	Elev.	720 m
Size	14x48 m	Depth	0.5-6 m	Water T	33.3°C
Current	variable				

Table 1 : Chemical and physical characteristics of Laguna El Mojarral (West), Cuatro Ciénegas, Coahuila, Mexico. Water samples collected July 1983. Analyses by R. Murnane, Princeton University. Data in mg/l unless otherwise specified.

covering a subterranean watercourse, thus it is technically a sinkhole. Water inflow is in the north end of the sinkhole near the bottom, at a depth of about 4.5 m. It comes through a horizontal cave tunnel, which is passable for about 3 m. Another tunnel in the sinkhole bottom, at the opposite end, carries water out, presumably to El Mojarral East. In contrast to the sinkhole, El Mojarral East is about 50 m wide, 200 m long and 0.5 to 2 m deep. It has a firm, indurated carbonate bottom, in places largely devoid of loose sediment. Water inflow is primarily the surface stream at the west end, from El Mojarral West. The lake has an outflow channel on the east side that flows into the Rio Mesquites, a stream a few hundred meters to the east of El Mojarral. The *Phragmites* marsh associated with El Mojarral contributes so much particulate organic matter that, in still, marginal areas the sediment that accumulates is largely dark, organic, anoxic, copropelic mud (gyttja). Soft, white carbonate mud, of an undetermined thickness, covers parts of the sinkhole and lake bottoms. On these unconsolidated muds there is no development of microbial mats or stromatolites. Aquatic vegetation associated with the shallow margins of these lakes includes the sedge *Eleocharis*, the macroscopic alga *Chara*, the bladderwort *Utricularia* and the water lotus *Nymphaea*. Stromatolites were found only in El Mojarral West, oncoids only in El Mojarral East, and stromatolitic mats in both lakes.

METHODS

Some of the living surface mat from the various biogenic structures was removed from the lake, and, while keeping it submerged in lake water, examined and dissected in the field to obtain qualitative information about community structure. Preliminary examination of fresh material was carried out using dissecting microscopes and compound microscopes equiped with brightfield and phase contrast optics (the lights were powered by a portable generator). Material was routinely fixed in 3% formaldehyde or gluteraldehyde in environmental water. Cyanobacteria, diatoms, protozoa, and micrometazoa were later cultured in the lab for further study. Microscopy of field collected material was performed in the laboratory either as whole mounts or on gently decalcified samples (3% HCl) and preserved as semi-permanent slides mounted in glycerol.

DESCRIPTION OF MICROBIAL DEPOSITS

Stromatolites and stromatolitic mats

Along the shallow margin of the sinkhole, stems and roots of sedges, rushes and grasses frequently become cemented into a porous, calcareous travertine sometimes called "tufa". Eroded pebbles of this travertine are eventually distributed in the lake by water currents. The surface of this material as well as snails, discarded cans and bottles, and fragments of dead vegetation are colonized by a specific set of microorganisms responsible, under favorable conditions, for formation of stromatolites oncoids and stromatolitic mats.

Stromatolites are distributed on a shallow, broad, marginal shelf area that covers part of one end of the sink, and on the sides and bottom of the sink where the current is sufficient to carry away the fine silts, clays, and organic debris, leaving a firm substrate. The sediment covering most of the shelf consists of a poorly-sorted, sand-sized mixture of living and dead snails (mostly *Mexipyrgus churinceanus* Taylor with some *Euboria bella* Conrad and *Mexistobia manantiali* Hershler) and small carbonate pebbles. In addition to isolated individual stromatolites, stromatolitic mats up to 1.5 cm thick, build up on exposed bedrock and any other solid substrate both in El Mojarral West and El Mojarral East.

Macroscopic features of stromatolites and stromatolitic mats

The stromatolites are irregularly rounded in shape. Some are short and others are taller than they are wide. Typical examples reach up to about 20 cm in diameter, and are usually a bit wider than tall. Figure 2 shows the outline, internal and external appearance of a typical stromatolite. Stromatolitic mats up to 1.5 cm thick build up on

Figure 2 : Internal view of typical stromatolite from El Mojarral West, showing ridged appearance of surface, and laminated interior. Scale bar = 2 cm.

exposed bedrock and any other solid substrate. Both the free-standing stromatolites and the stromatolitic mats lithify rapidly, often requiring a hammer and chisel to collect the entire thickness of the living part of the mat. In the case of both biogenic structures the surface is contoured with a constant, regular, repeating growth pattern as seen in Figures 2 and 3. Small, horizontally projecting ridges resembling tiny ledges or terraces are distributed regularly over most or all exposed surfaces. The ledges are lobate and project outward from the stromatolite and stromatolitic mat surface over a distance of about 2 mm. The ledges have an upper surface that is variably convex, leading to a horizontally convex outer rim. The lower side of the projection is variable in shape and sometimes exhibits partial "peeling" (see Monty, 1972) which lifts the projection slightly, shearing it away at the base. Over the surface of the stromatolite or stromatolitic mat these ridges are usually arranged in an offset pattern somewhat similar to that of roof shingles. Well developed ridges are remarkably regular in size and convexity (Fig.3a). When growing vertically, they are radial-campanulate in shape (Fig.3b). In older deposits the ridges fuse horizontally, and form long anastomosing "tuning fork" patterns (Fig.3c).

A general view of the internal pattern of the stromatolite laminae can be seen in Figure 2. Cross-sections through a stromatolite mat, cut parallel to growth (Fig.4a,b) shows, in greater detail, the typical lamination pattern present on the stromatolites and stromatolitic mats. The pattern consists of a series of light-colored, palisade-like layers of porous carbonate, alternating with dark green (or pink-purple) unencrusted organic-rich layers. The thickness of the laminae is not the same everywhere ; it varies with the nature and growth rates of the individual organisms. Within the major calcite-rich layers there are sometimes up to about 6 thin carbonate-rich laminae approximately 0.2 mm apart separating sequential organic-rich laminae.

Community architecture of stromatolites and stromatolitic mats from well-lit habitats

By community architecture we mean the lamination and three-dimensional arrangement of the microbial contents. The surface layer of the stromatolites and mats found in well-lit areas is built by photosynthetic cyanobacteria and algae. Differentiation of this layer through time leads to alternation between zones dominated by the rivulariacean *Homoeothrix balearica* Bornet and Flahault and zones consisting of *Schizothrix lacustris* A.Braun; another *Schizothrix* sp. characterized by a slightly thicker trichome and a thinner less diffluent sheath has been recognized but not identified (Fig.5 a,b). Below this come pink-purple layers consisting primarily of empty *Schizothrix* sheaths surounded and/or inhabited by unidentified flexuous filamentous purple photosynthetic bacteria (Fig.5c), and occasional clusters of *Chromatium* sp. (not shown). In places this pink-purple layer extends almost to the surface of the mat, indicating an elevation of the redoxcline. The mat thus shows a classical vertical zonation with a surficial layer of

Figure 3 : a) b) ; c)   El Mojarral Stromatolites developed around substrates projecting in
shallow (0.5 m) areas note well developed ridge patterns. a) & b) general view of mature
stromatolites with elongate, sometimes anastomosing ridges. a) & b) scale bars in cm and mm.
c) overgrowth of mushroom or campanulate shaped cushions built by *Homoeothrix balearica*
around a dead plant stem. This structure will eventually topple over as the stem inside rots away
then it will become a potential nucleus for future stromatolite development. Scale bar is in cm.

oxygenic photosynthesizers (cyanobacteria, diatoms and other algae) overlying a
reduced layer of anoxygenic photosynthesizers and dissimilatory sulfate reducers. A
high rate of sulfate reduction is indicated by the odor of these sediments.

Under high light conditions the brownish appearance of the mat surface is due to the
sheath pigmentation of *H. balearica* filaments, while under low light conditions (such
as near the underwater cave mouth and on the underside of stromatolites) these
filaments take on a dark purple color due to increased accumulation of phycoerythrin in

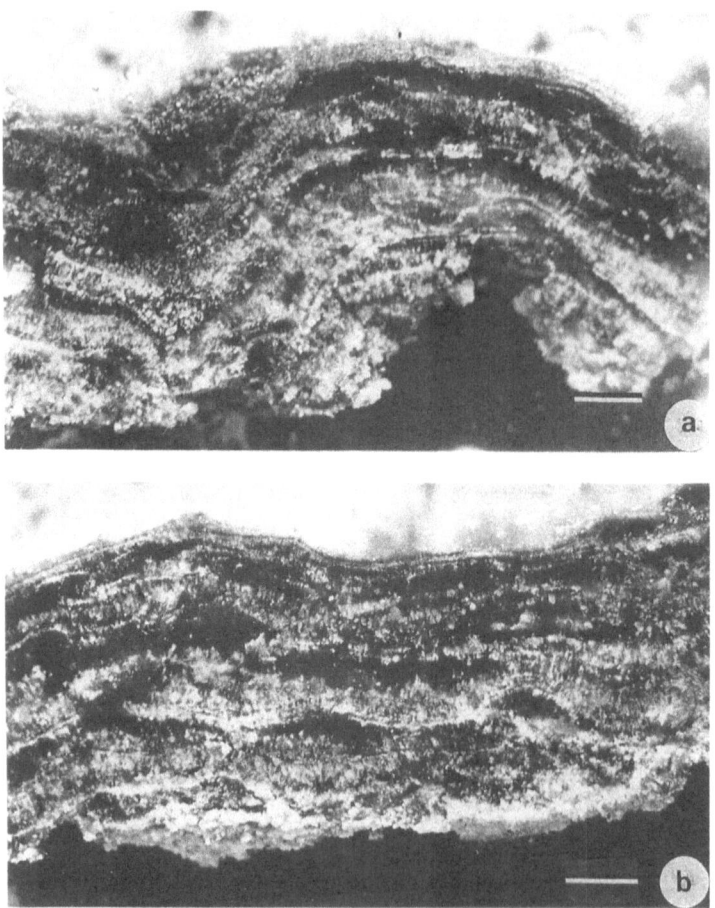

Figure 4 : a) ; b)   Sections through the actively growing stromatolitic mat showing the
alternation of light colored carbonate-rich and dark colored organic-rich laminae comprising a
vertical sequence from bedrock to present growth surface. Scale bar = 1 mm.

the cells and reduced sheath pigmentation. Packed among the upright cyanobacterial filaments and often attached to them is a thick gel-bound zone composed primarily of diatoms, testate and other protozoans, annelids, ostracods, harpacticoid copepods, tiny snails, insect larvae, calcite crystals, and fecal pellets (primarily from snails and fish). Toward the base of this layer, calcite crystals and fecal pellets become more densely packed. Occasionally, colonies of *Gongrosira calcifera* Krieger are found beneath the uppermost layer, sometimes attached to filaments of *H. balearica*. Diatoms such as *Gomphonema intricatum* var. *vibrio* (Ehr.) Cl. (also called *G. angustum* Agardh), *Cocconeis placentula* Ehr. and *Achnanthes* spp. are epiphytic on the *Homoeothrix* filaments, whereas others, especially *Cymbella* spp. and *Anomoeoneis vitrea* (Grun.) Ross form dense clusters among the pockets and crevices on the irregular mat surface.

The diatom population in a well developed stromatolite mat dominated by *Homoeothrix balearica* consists of about 5 very commonly occurring species and about 30 less common taxa. The most common species are as follows. 1. *Denticula kuetzingii* Grun., 2. *Cymbella cesatii* (Rabh.) Grun. ex A.S., 3. *Denticula elegans* Kütz., 4. *Achanthes affinis* Grun., and 5. *Gomphonema intricatum* var. *vibrio*. There is a spatial component to the distribution of different diatom species as there is to the cyanobacteria. *D. kuetzingii* is sometimes motile, attaches by a mucilage pad, or forms gelatinous masses which are distributed throughout the mat. *C. cesatii* may attach by a short stalk, but is also able to crawl quite rapidly, and is commonly seen among the loose carbonate particles as well as associated with colonies of the protozoan *Ophrydium* cf *versatile* (O.F.M.). *D. elegans* often forms dense mucilaginous colonies between the filamentous cyanobacteria, whereas *A. affinis* lives adpressed to the sediment. *G. intricatum* var. *vibrio* attaches to other plants with long branching stalks, allowing access to the uppermost parts of the mat where nutrients and light may be more available. These species are all alkaliphilous forms, characteristic of temperate to warm spring-fed streams and ponds with high conductivity water, and prefer or are indifferent to moderate water current. Small diatoms, such as *Achnanthes affinis*, *Anomoeoneis vitrea*, *Cocconeis placentula*, and *Amphora* spp., and the much larger *Epithemia* Ehr Kütz. are to be found living in crevices of the mat or as early colonizers of freshly deposited calcite layers. *Achnanthes* and *Cocconeis* produce a mucilage pad that leaves a thick bas-relief impression of the cell on the carbonate.

Community architecture of stromatolitic mats from low-light habitats

The general community architecture described above for the well-lit sinkhole stromatolites applies also for the stromatolitic mats lining the deeply-shaded vertical ledge leading to the mouth of the inflow cave. Indeed, there is a continuity of the community architectures from the stromatolites and multilayered stromatolitic mats to these thin surface communities. Thus, with increasing depth (i.e. reduced light) the thickness of the surface encrustation diminishes to about 0.7 mm, and a different suite of cyanobacteria is found in the sub-surface layer. *Homoeothrix* and *Schizothrix* still predominate in the surface layer, at times foundtogether with patches of *Lyngbya*

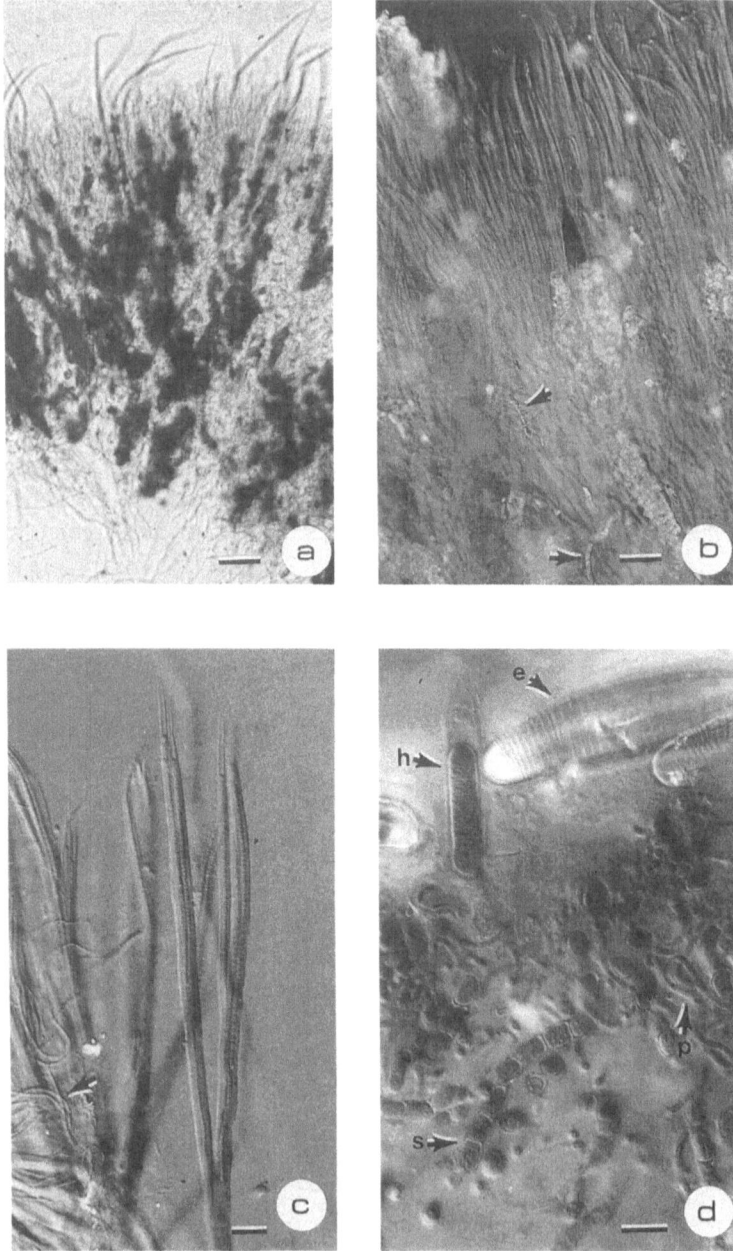

kuetzingii Schmidle, *Gongrosira* and the rhodophyte *Adouinella sp.* However, at the base of the filaments a diverse community of the coccoid chroococcalean and pleurocapsalean cyanobacteria takes over. Examples include *Pleurocapsa minor* Hansg. em. Geitler, *Cyanostylon microcystoides* Geitler, and various members of the genera *Chlorogloea, Aphanocapsa, Synechococcus, Synechocystis,* and *Microcystis.* Exact determination of these small-celled taxa (1-5 μm diam. cocci) was impossible due to the mixed nature of these populations and each taxon's patchy distribution. Primary cultures on agar media routinely yielded up to a dozen different forms as judged by cell size and colony morphology, but in no case was it possible to correlate cultured cells definitively with the feral material from which it was inoculated. Such low light communities, whose pigmentation is dominated by phycoerythrin, have been termed "red-colored deep water biocoenoses" ("rotbunte Tiefenbiocoenose", see Kann and Sauer, 1982), and are characteristic of low-light habitats of clear, hard-water lakes.

In places where the low light level permits only patchy and reduced growth of the filamentous forms (*Homoeothrix, Schizothrix, Gongrosira, Adouinella* and *Lyngbya*), a 0.1-0.2 mm deep zone of endolithic cyanobacteria penetrates the bedrock below the coccoid forms described above (Fig.5d). It is dominated by *Plectonema terebrans* Bornet & Flahault, *Iyengariella endolithica* Seeler & Golubic (Seeler and Golubic 1991) and another undescribed stigonematalean endolith (Seeler and Golubic unpublished data). These changes in community architecture along the light gradient in El Mojarral West are summarized diagrammatically in Figure 6.

Microstructure and calcification processes of stromatolites and stromatolitic mats

Within the actively growing part of the mat of both stromatolites and stromatolitic mats (up to 10 mm or more in thickness) calcite is precipitated in the vertical-radial masses of *Homoeothrix* filaments to form calcified laminae. The stromatolitic mats consist of an alternation of carbonate-rich and organic-rich laminae. These laminae show a palisade-like pattern made of tubular or fused-tubular carbonate encrustations (Fig.5a) made of rounded to loaf- shaped crystals about 10 μm wide (Fig.7b). Successive layers of calcified *Homoeothrix*-dominated laminae are usually interrupted or separated by thin,

Figure 5 : a); b) ; c) ; d) a) Micrograph of the top layer of El Mojarral stromatolite mat showing dark calcitic encrustations surrounding tapering filaments of *Homoeothrix baleairca* among *Schizothrix lacustris.* Scale= 50 μm ; b) Detail showing encrusted *H. balearica* filament on lower right, dark area represents filament emerging from encrustaion. Note thin monocrystalline calcitic tubes precipitated around *Schizothrix* sp. filaments (arrows). Scale= 20 μm ; c) *Schizothrix lacustris* exhibiting typical branching of filament. Note also filamentous purple bacteria among *Schizothrix* sheaths (arrow). Scale= 10 μm. d. Resin- embedded and decalcified thin section through cyanobacterial and diatom community from mouth of the inflow cave. Note *Eunotia* cf. *maior* (e) and *H. balearica* (short filament, h) on the surface and endolithic undescribed stigonematalean cyanobacterium (s) and *Plectonema terebrans* (p) below. Scale= 10 μm. (a) = plane transmitted light ; b), c), d) = Nomarski differential interference contrast.)

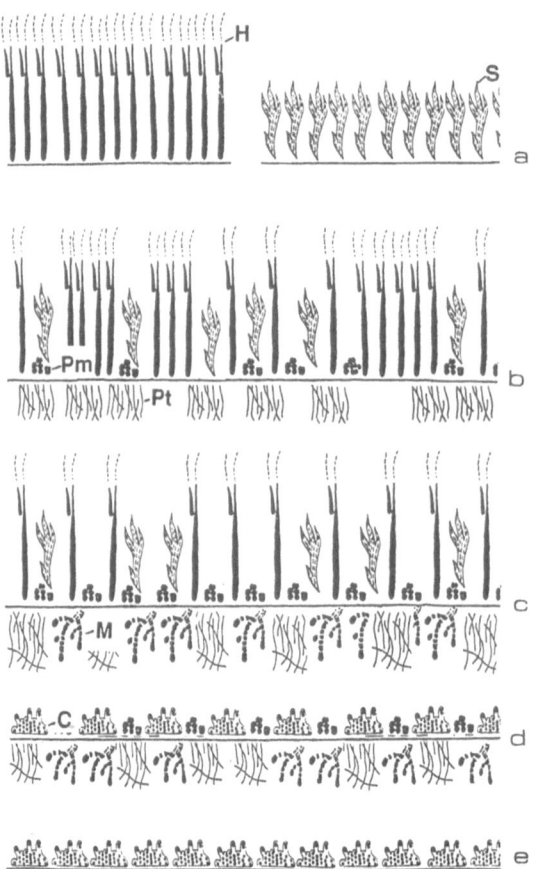

Figure 6 : a) ; b) ; c) ; d) ; e) Profile diagram showing cross-sections of cyanobacterial communities encountered down the depth and light gradient of El Mojarral West illustrating transitions from the high-light surface community a) to the shaded low- light community at depth e). The horizontal line in each cross-section separates the epilith zone (above) from the endolith zone (below). Only representative taxa are included. H = *Homoeothrix balearica*, S = *Schizothrix lacustris*, Pm = *Pleurocapsa minor*, Pt = *Plectonema terebrans*, M = *Mastigocladus*-like and *Iyengariella endolithica* stigonematalean endoliths, C = *Chlorogloea* sp. and other coccoid epiliths. a) Multi- layered stromatolitic mats consist of alternating *Homoeothrix* and *Schizothrix*-dominated laminae, diagrammed here as only a single layer. b) Single epilithic layer from intermediate depth with endolith zone consisting exlcusively of *Plectonema terebrans*. c) More shaded than community in "b)" with the addition of stigonematalean endoliths and exhibiting sparser epilithic growth. d) Deeply shaded mat in which filamentous epiliths have been replaced by coccoid forms. e) In the lowest light regions of the inflow cave only a thin epilithic surface layer of coccoid cyanobacteria (e.g. *Chlorogloea* sp.) remains.

relatively fine laminae of precipitated material, consisting of cloudy, tightly packed masses of calcite crystals ; these sometimes surround *Homoeothrix* filaments, extending through from the layer beneath. The organic-rich laminae are dominated by the two *Schizothrix* species reported above. Occasionally monocrystalline calcite encrustations occur as thin tubes surrounding the sheath of the unidentified *Schizothrix* species (Fig.5b). The surface layers of *Homoeothrix balearica* and *Schizothrix lacustris* are thickest on and beneath the crests of the ridges such as illustrated on Fig.4. Carbonate crystals also nucleate upon diatom stalks and the surface of gelatinous spherical envelopes produced by diatoms. In the protected environment of the gel produced by cyanobacteria and diatoms, clusters of sparry calcite crystals with a radial fan-like growth habit develop. A comparable example of *in situ* mineralization in the shape of spherulitic fans takes place in the mucilage surrounding *Scytonema*, in the supratidal freshwater cyanobacterial mats illustrated by Monty and Hardie (1976, p. 454, Fig.4h). Calcite is also precipitated in long, plate-like crystal masses around the filaments of *Gongrosira*, producing hard patches of calcite among the more loosely distributed grains surrounding the cyanobacteria. The crystal morphology associated with *Gongrosira* is the same as that described and illustrated in Golubic and Fischer, (1975). Micrite and micro-sparitic calcite cement (Fig.8), fills some of the cavities as lithification proceeds. These low-magnesian calcite crystals appear as deformed "gothic arch" and "edge-guttered" calcite (Folk, Chafetz, and Tiezzi, 1985) perhaps due to the poisoning effect of the sulfate ion on the calcite crystal lattice. Crystal size of cement ranges from 2 to >30 µm.

This generally loose fabric changes gradually downward, toward the interior of the mat, to more rigidly packed sheaths of *Homoeothrix* surrounded at first by loose anhedral, then tighter interlocking coarser crystals (see Fig.10, next section). This internal pattern of change in the crystallographic habit represents gradual crystal growth concomitant with bacterial degradation of the organic components of the mat. Some filaments in the interior of the mat may become coated by closely packed fine-grained calcite crystals (micrite). This collection of carbonate fabrics may represent various stages in the syndepositional and early diagenetic processes that take place during growth of the surface mat. The moldic porosity may be altered to a denser fabric as the holes are filled with cement.

In some cases the decomposed remains of *H. balearica* and *S. lacustris* produce fenestrae, (cf. Monty, 1976 and Monty and Mas, 1981). Particularly in the low-light stromatolitic mats, such interstices may be secondarily filled with growth of *Plectonema terebrans*, other cyanobacteria of the LPP type (*Lyngbya-Plectonema-Phormidium*, see Rippka *et al*, 1979), the chroococcalean *Cyanostylon microcystoides*, or they may remain open and eventually become lined with sparry calcite cement.

Whereas the stromatolite microfabric is largely the consequence of *in situ* calcite precipitation, there is evidence of detritus incorporation into the structures. Sediment consisting primarily of snails, shell fragments, fecal pellets, micrite and sparite grains is episodically trapped and bound into the surface community. Fish activity, for example, often stirs up clouds of fine sediment, which settles on the nearby stromatolites.

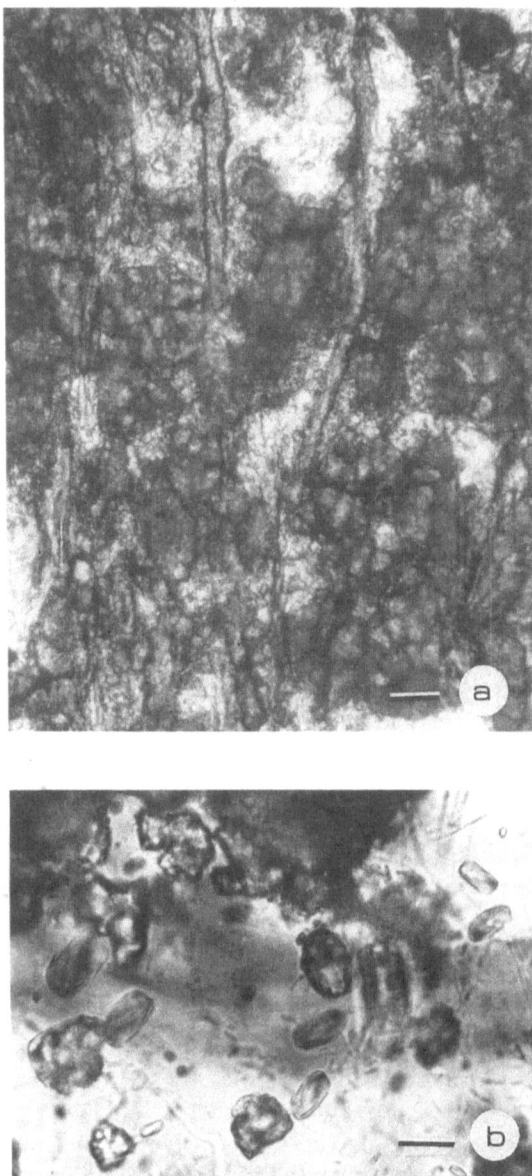

Figure 7 : a) ; b) Micritic and rounded spar crystals photographed *in situ* from the surface of an actively growing stromatolite mat ; a) Crystals attached to the surface mucilage of *Homoeothrix balearica* filaments ; b) Crystals associated with diatom mucilage. Scale bars = 10 μm.

Oncoids

Macroscopic features of oncoids

Spherical to subspherical oncoids (Fig.9a) are found in scattered areas along the margin of El Mojarral East in water depths of 0.5 to 2 m, where the bottom is hard, and the current is sufficient to remove material smaller than a pebble. The average diameter of the oncoids observed is about 10 to 20 cm. The most concentrated deposit occurs along the initial 25-50 m or so of a broad outflow stream that leads to the Rio Mesquites. The flat floor of the stream is almost covered with a layer of oncoids. The oncoids show a well-laminated cortex composed of alternating layers of dense, light colored carbonate-rich laminae and porous, darker, organic-rich laminae (Fig.9b).This lamination pattern is essentially indistinguishable from that of the stromatolites described above. The nucleus may be a calcite pebble, snail shell, or piece of organic debris. The material that forms the nuclei of the oncoids is variable in porosity and microstructure. Some pebbles consist of an unlaminated snail-fragment biosparite beneath the laminated cortex. Others are composed of fragments of marsh travertine or calcified *Chara*.
The oncoid surface is usually slightly wavy or lumpy, with round holes 1-2 mm wide distributed somewhat unevenly over the surface (Fig.9a). These holes are lined with living microorganisms. The holes are the external expressions of irregularly shaped channels and chambers oriented roughly perpendicular to the growing surface ; they extend at varying distances, often more than 1 cm into the interior of the oncoid, and frequently reach or even penetrate the nucleus. These channels appear to be maintained by the activities of the asssociated benthos.

Community architecture of oncoids

The organisms responsible for construction of the oncoids are *Schizothrix* and *Homoeothrix*, producing a rhythmic depositional pattern similar to that of the stromatolites. The surface of the oncoids lacks the ledges, and has an overall lumpy appearance produced by tufts of *Homoeothrix*. The cyanobacteria, diatoms and other eukaryotic algae are the same species as those found on the stromatolites. Channels and interior chambers in the oncoid cortex are lined with microbial growth, and the underlying carbonate is bright green for a thickness of almost a millimeter due to the extensive activities of endolithic cyanobacteria (principally *P. terebrans* and other LPP forms). These cavities in the stromatolites show decreased *Homoeothrix* growth. Detrital material, particularly fecal pellets, partially fills some holes.

Microstructure and calcification processes of oncoids

Oncoid laminae consist of alternating dense carbonate-rich, and porous organic-rich

Figure 8 : SEM view of cavity-lining micritic and microsparitic cement in cortex of stromatolite. Crystal irregularities such as ragged faces, bezeled or guttered edges and corners, pits in the center of faces and arched curves are common. Scale bar = 5 μm.

layers as do the stromatolites, but the laminae are generally somewhat thinner, tighter and more compact. They also do not exhibit secondary internal *Schizothrix* growth to the same degree as do the stromatolites. In thin section *Homoeothrix* filaments are clearly evident in growth position, to a distance of more than a cm into the cortex of oncoids (Fig.10a). Micritic calcite indicates what was formerly the surface of the cyanobacterial filament as seen in Fig.10b. The sheath cavity has been filled with micrite. Some organic matter (possibly last remains of the original sheath) appears as thin black linings (arrows, Fig.10b). In addition to the micrite described above, interlocking fringes of isopachous sparry calcite crystals fill voids between *Homoeothrix* filaments in Fig.10b. Carbonate particles (including snail shells) smaller than 5 mm diameter are usually bored by *Plectonema terebrans* and other LPP forms (*sensu* Rippka *et al*, 1979) to such an extent as to constitute a cohesive organic matrix even when all carbonate is dissolved away with acid. Other cyanobacterial endoliths contributing to carbonate boring in oncoids include the undescribed stigonematalean depicted in Fig.5d and at least three new species of *Hyella* (J.-S. Seeler and S. Golubic, unpublished).

Figure 9 : a) ; b)   a) External view of oncoids. Note undulating surface pattern and scattered holes leading to internal chambers ; b) Oncoids sawn open to illustrate laminated cortex, and unlaminated nuclei. Scale bars in cm.

Imperfectly formed bipyramidal quartz crystals (Fig.11) are found in oncoids as well as occasionally in stromatolites. These crystals range in size from about 4.5 x 8.2 µm to 20.0x62.5 µm. An authochthonous origin for the quartz crystals is supported by the presence of colorless filaments trapped in some of them (Fig.11, lower left crystal, with arrows). In the residues of porous internal structures of stromatolites with alternating thin, dense, and thick, sparsely calcified laminae, there are large numbers of well preserved diatoms and fewer quartz crystals. In thinly-laminated oncoids and stromatolitic mats from the deeply shaded areas, there are few diatoms preserved in the insoluble residue and numerous quartz crystals. These preliminary observations show an inverse relationship between the number of diatoms preserved in a structure and the number of quartz crystals in the residue. This suggests that diatom opal, which resembles silica gel (Krumbein and Werner, 1983), is being dissolved to feed the growth of quartz crystals precipitating locally in the mat.

Grazers and predators associated with stromatolites, stromatolitic mats and oncoids

There is a varied and numerous benthic fauna associated with these biogenic structures, relying on them for shelter and food. In a typical mat from a well-lit area of the sinkhole the benthos usually includes amoebae, flagellate protozoans, euglenas, ciliates, rotifers, neorhabdocoels, gastrotrichs, nemerteans, nematodes, harpacticoid copepods, flatworms, chironomid larvae, small snails, ostracods, water mites and amphipods. Nematodes are very common in the interstices of the mat, and because of their size and numbers appear to disturb the softer laminae as they move through them. The thin, deeply shaded stromatolitic mats do not exhibit significant benthic fauna. In addition to the invertebrates, several species of fish graze on the stromatolites and related structures, leaving groups of parallel, tooth-produced grooves on the stromatolite surface sometimes removing entire patches of mat growth, exposing the greenish-white calcite interior.

DISCUSSION

Abundance of stromatolites, stromatolitic mats, and oncoids

Stromatolites are distributed in El Mojarral West in water from 0.5 to 4.5 m depth. Their distribution is limited primarily by the availability of an appropriate substrate and adequate current which prevents burial by detrital sediment. Their abundance has diminished over the years due to vandalism, and excessive siltation. The stromatolitic mats are extensive in distribution in both El Mojarral East and West and relatively undisturbed. From experiments using artificial substrates, the growth rate of stromatolites and stromatolitic mats is estimated to be on the order of several mm or more per year. The distribution of oncoids is very current dependent and large numbers are concentrated only in the outflow area of El Mojarral East.

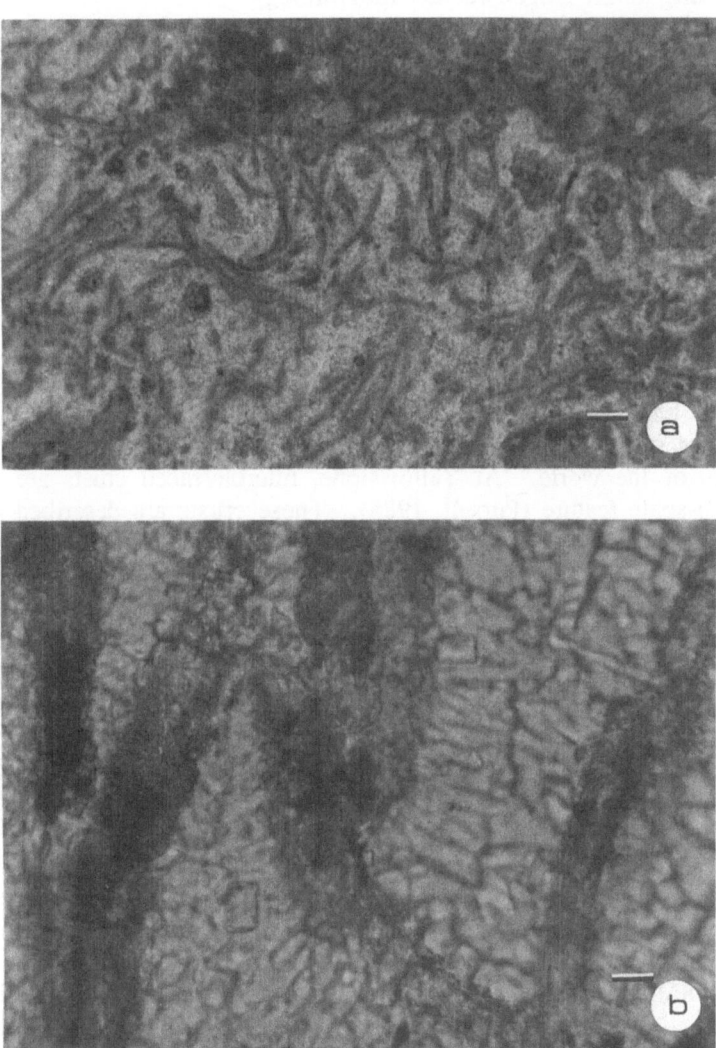

Figure 10 : a) Thin section 1 cm into interior of oncoid cortex showing *Homoeothrix* filaments in growth position. Scale= 50 μm ; b) Closer view showing micritic calcite at position of former filament surface. Note interlocking spar-calcite crystals between filaments. Scale= 10μm.

Factors controlling lobate ridged surface morphology

The structure of the El Mojarral stromatolites and stromatolitic mats is primarily the consequence of *in situ* precipitation of carbonate, rather than trapping and binding of detrital sediment particles by its constituent microbiota. Both biological and physical factors play a role in the morphology of the ridged pattern of stromatolites and stromatolitic mats. The offset pattern of small lobate ridges or micro-terraces characteristic of El Mojarral stromatolites and stromatolitic mats appears to be produced by the regularly spaced, upright intermingled, closely bound growth habit of *Homoeothrix* combined with a steady water current which prevents the accumulation of detritus. The algae on the surface appear to grow in such a way as to maximize light and nutrient absorption ; the result is enhanced growth on the outer and upper surfaces which leads to the formation of the ledges. Each such ledge forms a "canopy" which shades the area directly around and beneath it.

Similar patterns of ridges or terraces are observed in modern non-marine stromatolites in various parts of the world. At Yellowstone, microterraced crusts are the most ubiquitous small-scale feature (Pursell, 1985). These crusts are described as quasi-horizontal microterraces with a raised rim (0.5-1.33 cm high) and a small irregular "basin" or depression (1.3-2.5 cm wide) behind the smooth microcrystalline calcium carbonate rim. Pursell (1985) found that cuts perpendicular to the microterraced surface show a composition of dendritic crystals oriented with their main axis perpendicular to the substrate. Similar structures called anastomosing ridges, with a "torn-tissue" texture, have been described from steeply sloping rimstone faces in a stream in Dyfed, South Wales (Braithwaite, 1979). Gomes (1985) described tabular "crinkled stromatolites" from Wondergat Sinkhole in the Western Transvaal of South Africa, the pattern of which, from illustration and description, appears to be similar in size and shape to that at El Mojarral.

Another term that may refer to the same type of pattern is "ridge and furrow structure" R. Riding, pers. comm. 1986). Monty (1972) discusses and illustrates lineations called ridges on pinnacle-bearing stromatolites subjected to water currents driven by persistent trade winds, at Andros Island in the Bahamas. Another interpretation of ridges, suggested by Casanova (1986) who studied similar Plio-Pleistocene lacustrine stromatolites in the Gregory Rift Valley of Tanzania and Kenya, is that they represent a regular succession of ripples or ridges induced by the development of a microbial community which is at its climax condition. His illustrations (Casanova, 1986, plates 50-54) resemble the most mature examples at El Mojarral (Fig.3b), where the ledges have anastomosed into continuous ridges. The structure of the El Mojarral mats has not changed significantly in the decade we have been observing them, which may suggest some degree of long term stability. Dixit (1984) refers to Pleistocene lacustrine ridged oncoids from Lake Manyara in Tanzania as having "peaked ridges formed by laminae having a cuspate profile". His conclusion regarding the formation of the ridges was that the "laminae of these oncolites are clearly not formed by binding of carbonate detritus by the mucilaginous filaments of

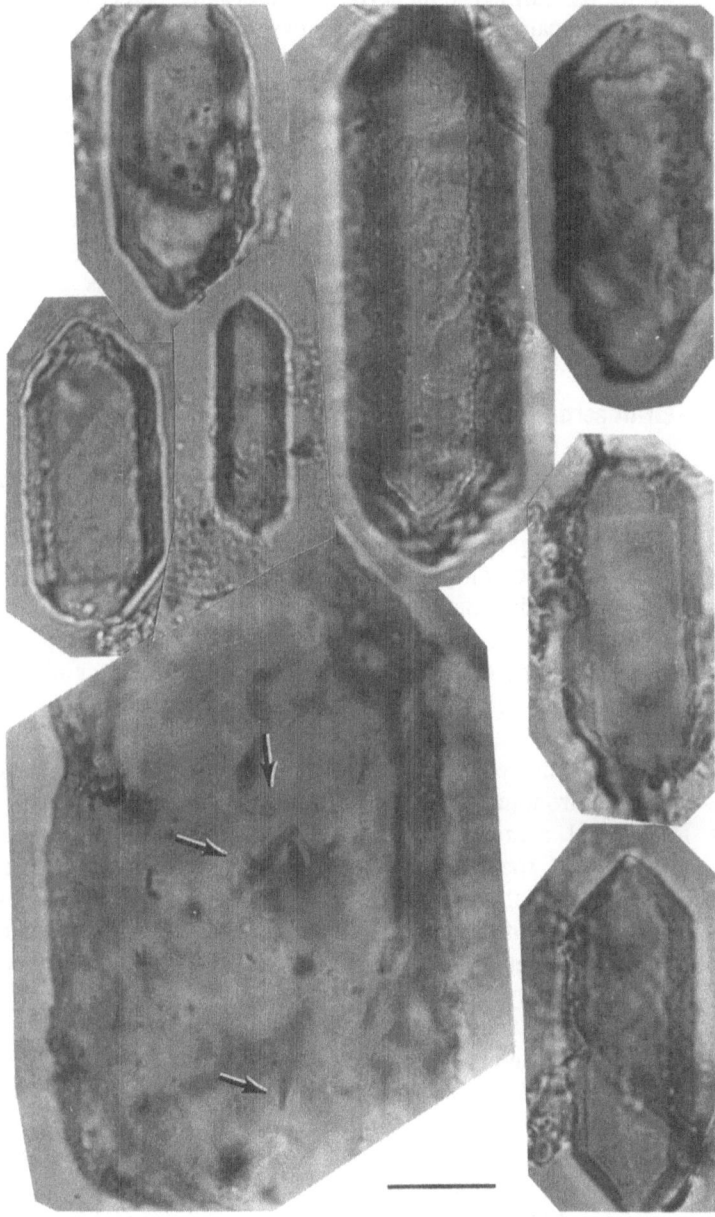

Figure 11 : Quartz crystals liberated by acid treatment of lake stromatolite mats. Note filaments in crystal on lower left (see arrows). Scale bar = 10 μm.

algae...they appear to be formed either by the inorganic precipitation of calcium bicarbonate...or by biological precipitation because of the photosynthetic activity of algae".

Although conventional wisdom holds that oncoids are associated with agitated conditions providing regular or constant movement of the structures, oncoid movement such as rolling or flipping has not yet been observed at El Mojarral. Large oncoids do sometimes develop a rudimentary ridged pattern and some with a ridged pattern exhibit sharply juxtaposed growth orientations. The ledges always form close to horizontal in each of the sequential growth periods, suggesting that movement of at least the larger oncoids is episodic in nature rather than continuous or regular. This same observation was made by Jones and Wilkinson (1978) who recognized that their large oncoids were overturned infrequently.

Lamination in well-lit settings

In order to frame the description of the mat architecture into a temporal perspective it should be noted that the reported distribution or dominance of the microbiotic components is not static. Population dynamics is, as discussed by Monty (1973), a basic parameter controlling not only the formation of laminated fabric but also the development of particular biodiagenetic features throughout the mat. He distinguishes 2 types of lamination patterns in microbial mats. The first is *historical lamination* resulting from the succession of the alternating microbial populations at the surface of the mat following diurnal, periodic, seasonal, or accidental changes in the environmental parameters. For example, Monty (1976) reports three different types of laminated fabrics built by the same two cyanobacteria, each heavily calcified, or not, depending on slight environmental changes in the same setting. This lamination process leads to vertical accretion of mat. The second type of lamination is *instantaneous or biological stratification* resulting from the in-depth distribution of successive specialized microbial populations occupying specific zones and niches along steep physico-chemical gradients ; this "metabolic" stratification, superimposed on the historic lamination, contributes to the diagenesis of the mat and moves upwards following their accretion (Monty, 1976).

The alternation of uncalcified, *Schizothrix* dominated laminae with calcified laminae of *Homoeothrix* may thus be considered as an example of "instantaneous" lamination. The local coalescence of *Homoeothrix* filaments at the mat surface results in the formation of sculptured ridges and provides the microhabitat necessary for the growth of *Schizothrix* below. *Homoeothrix* is found in conditions ranging from high light near the water surface to deeply shaded areas near the inflow cave. Hence it appears that *Homoeothrix* is adapted to life in a broad range of light conditions. Under conditions of high light, *Homoeothrix* exhibits the characteristic dark brown sheath pigmentation, while at low light intensities, colorless sheath and increased phycoerythrin pigment production in the cells enhances its ability for efficient light capture. Further, laboratory cultures of this organism have confirmed its ability to undergo complementary chromatic adaptation (*sensu* Tandeau de Marsac, 1977), a low light physiological response to changes in the spectral quality of light (Seeler, unpublished results). *Schizothrix*, on the other hand, appears only in the sub-surface layer of the mat, where it is shaded by *Homoeothrix*, and thus restricted to a much narrower photic environment than *Homoeothrix*. It therefore appears likely that the alternating lamination of *Homoeothrix* and *Schizothrix* is the consequence of the different ecological responses of these organisms to light.

Calcification process in laminations

The El Mojarral stromatolites and stromatolitic mats exhibit species-specificity with respect to in situ carbonate precipitation. Carbonate precipitation at El Mojarral is strongly light-dependent, therefore suggesting that it is driven by photosynthesis. Carbonate precipitation by *Homoeothrix* (and not *Schizothrix*) may be related to the difference in their sheath thicknesses, affecting CO_2 diffusion rates to the photosynthetic trichomes. Hence the thicker trichomes (and relatively thinner sheaths) of *Homoeothrix* cause more severe local CO_2 depletion leading to $CaCO_3$ precipitation. The calcite layers separating *Homoeothrix* laminae are due perhaps to faster accumulation of carbonate on the mat surface during the dry season when the water is most saturated with carbonate ; possibly coincident with the end of the growth cycle of *Homoeothrix*. Pentecost (1987) observed a seasonal pattern of growth in *Rivularia haematites* (D.C.)Ag. colonies in North Yorkshire, where two different processes of calcification may be operating : when cyanobacterial growth was slow, a dense annual winter zone of calcite precipitation (attributed to nucleation at the sheath surface and possibly from calcite seeds trapped from the surrounding water) accumulated on the surface of the colonies, with much the same appearance as the calcite rinds at El Mojarral. Additionally, he describes a set of narrow calcified summer bands associated with zones of active *Rivularia* growth and cell division, induced by photosynthesis. Similar sets of thin bands occur between the rinds at El Mojarral. The organization of the lamination in El Mojarral stromatolites, consisting of loose laminae 1-2 mm thick, separated by calcite-rich rinds, resembles the architecture of the lamination of Type 2 oncoids described from Lower Cretaceous (Wealdian) deposits of the Province of Valencia, Eastern Spain (Monty and Mas, 1981). These fossils were built by the superposition of spongious loose laminae produced by populations of a carbonate-coated filamentous cyanobacterium, separated by micritic rinds marking the end of a growth phase.

In another nonmarine cool-water travertine setting, in the Arbuckle Mountains of Oklahoma, a seasonal pattern of alternating sparry, "bushy" layers (spring-summer) and darker micritic layers (fall-winter) was interpreted by Love (1985). This seasonal origin of the laminae was later confirmed by measuring the $\delta^{18}O$ and $\delta^{13}C$ signatures of seasonally alternating sparry and micritic laminae in the same travertine stromatolites (Chafetz et al., 1991). These authors found significant variations in the stable isotope compositions, particularly the $\delta^{18}O$ values between sparry and micritic laminae, with $\delta^{18}O$ values higher in the micritic laminae, and $\delta^{13}C$ values higher in the sparry laminae.

An example of succession involving the vertical growth of long thick cyanobacterial filaments through a felt of thinner ones has been documented by Kann (1941) for lacustrine cyanobacterial crusts built by *Calothrix* and *Dichothrix*. Oncoids that occur in Lake Constance are constructed by two cyanobacteria, *Schizothrix*, forming dense layers, and *Phormidium*, *Calothrix*, and/or *Dichothrix* forming spongy layers (Schafer and Stapf, 1978). The sequential layering of *Homoeothrix* produces a vertically-differentiated palisade-type structure which is much like that described by Monty (1972, 1976) for stromatolites in pools of the Fresh Creek area on Andros Island, where long vertical filaments of *Scytonema*, calcified below their growing tips, produce a vertically-oriented feature, and by Monty (1976) for stromatolites from Shark Bay where rapid cementation and incomplete oxidation produce a similar fabric.

Oncoids with a nuclear composition and cortex lamination pattern resembling those of El Mojarral have been reported by Jones and Wilkinson (1978) from marl lakes in Michigan. Stream oncoids with similar calcite grain morphology (but lacking the regular alternation of dense and porous layers) constructed by *Gongrosira*, *Homoeothrix* and *Phormidium* were described by Roddy (1915) and later by Golubic and Fischer (1975) from Little Connestoga Creek in Pennsylvania. Minckley (1963) described a calcareous encrusting community from Doe Run (a stream in Kentucky) consisting of *Gongrosira*, *Phormidium*, *Schizothrix* and diatoms. His oncoids grew around the shells of *Goniobasis*. El Mojarral coated pebbles with snail nuclei, that develop into irregularly rounded oncoids, are also similar to those described by Weiss (1970) for oncoids forming around *Goniobasis* snails. An example of fossil lacustrine oncoids that resemble the type of laminated oncoids forming at El Mojarral comes from the Paleogene nonmarine algal deposits of the Ebro Basin in Northeastern Spain (Anadón and Zamarreño, 1981). These oncoids contain radially growing calcified filaments periodically pervaded by concentrically disposed micrite films or disrupted by the invasion of chironomid larvae. Chironomid larvae are sometimes abundant in the interstices of the El Mojarral structures as well.

Impact of fish and other grazers

The assumption has been made that metazoan evolution has caused a decrease in stromatolite distribution in marine strata beginning with the Precambrian/Cambrian boundary, (Awramik, 1971, 1981, 1984). Several authors have even suggested that recent potential stromatolites cannot develop in the presence of grazing and burrowing animals (Garrett, 1970 a,b, Gerdes and Krumbein, 1984). This hypothesis however has been criticized by Monty (1972), Winsborough (1990) and Golubic (1991). The presence of a diverse animal community associated with living, actively growing stromatolites at Cuatro Ciénegas shows that at least some aquatic communities, which appear to be as complex as average fresh-water communities, do not prevent stromatolite or oncoid formation and growth. The vertical channels that are present in oncoids penetrate deep into the structures and allow the circulation of carbonate-rich water to reach the interior, increasing the likelihood of rapid internal cementation.

Movement of benthic animals, particularly amphipods, in these channels enhances water circulation thus contributing in a positive way to stromatolite diagenesis. Fish are responsible for the most extensive damage to stromatolites and stromatolitic mats, (and oncoids to a lesser degree) as they sometimes clear areas a cm or more in diameter thus disrupting the continuity of laminae. An uneven growth pattern produced by fish grazing has been documented by in in-situ colonization experiments, where fish grazing appears to be responsible for removing patches of the mat. Pieces of mat up to a cm in diameter have been removed from the artificial substrate by predation. It is not uncommon that one side of a structure (that which is on the down-current side) will be cleaned of much surface mat growth, while the other side is relatively undisturbed.

Relationship of diatoms and quartz in stromatolites and oncoids

With regard to the disappearance of diatoms, and concomitant appearance of quartz crystals in the interior of some stromatolites and oncoids, the following mechanism is suggested. Dissolution of diatoms in the stromatolite mat may be taking place in the ecological zone where dissimilatory sulfate reduction by bacteria such as *Desulfovibrio* causes local chemical characteristics to be significantly altered. This would be deep enough into the mat that many of the diatoms present are likely to be dead. After death of the algal cell, the protecting organic coating is degraded exposing unprotected diatom frustules to dissolution ; further, removal of metallic complexes may also contribute to these processes (Lewin, 1961). Birnbaum and Wireman (1984) suggest that metabolically-mediated pH changes by *Desulfovibrio* may significantly influence decomposition of certain silicate minerals due to hydrolysis reactions, and later provide the microenvironment for the nucleation and precipitation of diagenetic silica from a saturated solution. These authors discuss a mechanism by which soluble silica in the form of monosilicic acid, H_4SiO_4, dissociates to H_3SiO_4 - at pH values above 9.7. The initial increase in pH leading to silica dissolution is attributed to the release of ammonia during proteolysis ; *Desulfovibrio* then releases CO_2 and sulfide ions, lowering pH and while other cyanobacteria and diatoms, through the production of copious amounts of mucilage, contribute to carbonate precipitation as well as sediment trapping and binding. This microbial community is also responsible for the shaping and microstructure of the stromatolite's lamination. In low-light areas, endolithic cyanobacteria may contribute significantly to net carbonate removal. Such microhabitats therefore require detailed study of the respective roles (ie. carbonate precipitating or dissolving) of the cyanobacteria enmeshed within the carbonate matrix.
It is noteworthy that in such microhabitats carbonate precipitation (principally by *Homoeothrix*) and carbonate dissolution (by the endoliths below) occur simultaneously, and in close spatial proximity, thus obscuring the boundary between newly accreted carbonate and the bedrock below (cf. Fig.5d, 6b,c). Culture studies have confirmed the carbonate penetrating ability of stigonematalean endoliths (Seeler and Golubic, unpublished).

Characteristic of stromatolitic structures at El Mojarral and certain physico-chemically similar lakes in the Cuatro Ciénegas basin is the development of a remarkably constant surface growth pattern of horizontally oriented lobate ridges or ledges. These ridges are linear areas of enhanced cyanobacterial growth and carbonate precipitation. They result from the horizontal coalescence of individual cushions of *Schizothrix lacustris*, capped by the growth and carbonate precipitation of *Homoeothrix balearica*. These two dominant filamentous cyanobacteria exhibit species-specific differences in biogenic calcium carbonate precipitation. Similar terraced or ridged patterns have been described in the literature, from modern and fossil nonmarine stromatolite deposits in various parts of the world.

The stromatolites, mats, and oncoids of El Mojarral display similar internal microstructure and lamination. Differences appear to depend primarily on composition of the biological communities, available light, the kind and amount of grazing, and the amount of internal cyanobacterial and mineralogical growth and diagenesis. Typical oncoids and stromatolites which form in full sun and moderate current, particularly within about 2 meters depth, show a repeated alternation of gelatinous cyanobacteria-rich laminae and dense carbonate-rich ones. In some habitats, the original cyanobacterial or algal material is trapped as inclusions in the lithified structures, in others the mold of the organic material has been filled by calcite.

The surface relief of the structures may also be affected by feeding activities of fish and crustaceans. To a much lesser degree, locomotion and feeding activities of smaller benthic animals such as flat worms, nematodes and dipterans can disrupt the fabric of internal laminae. Some of these animals appear to be important in the maintenance of open vertical and horizontal channels irregularly distributed within the living mat. Our observations suggest that both fish and microbenthos may play important roles in controlling the growth and population dynamics of the algal/cyanobacterial portion of the stromatolite community by affecting the rate and pattern of stromatolite accretion. Fish, in particular, through their feeding activities, disrupt the surface relief, influencing the ultimate shape of the structures. Besides the direct effects of the interactions between the grazers and the stromatolites, predators in this community increasing carbonate alkalinity in the localized environment surrounding the cells, thereby causing precipitation of silica. Stoessell (1992) outlines the overall anaerobic aqueous sulfate reaction of carbohydrates, presumably controlled by *Desulfovibrio*, in a discussion of the pathways for sulfate-reduced mixing-zone pore-water fluids, under various pore-water geochemical regimes.

An example of the precipitation of silica in oncoids was reported by Schafer and Stapf (1978). They observed the precipitation of quartz in the interior of oncoids forming in Lake Constance and attributed the process to the production of ammonia. They suggest that the silica was dissolved from the abundant diatom tests embedded in the algal fabric, as diatom tests could only be detected within the outer layer of the oncoids. Diatom dissolution and the precipitation of silicate minerals were also linked in a study by Stoffers and Holdship (1975) who found that diatom-poor intervals in a core from Lake Manyara contain abundant quantities of analcine and erionite, and suggest that these silicates may arise from the dissolution of biogenic silica. Holdship (1976) interpreted the diatom-poor periods as representing a more concentrated hydrochemical environment than the diatom-bearing intervals. Barker (1992) experimentally looked at differential diatom dissolution of Lake Manyara sediments and concluded that the almost complete dissolution of diatoms represents the interaction of specific hydrochemical conditions, with high alkalinity accompanied by undersaturation with respect to silica.

The dissolution of diatoms may be the reason why no diatoms were found in otherwise well preserved oncoids and stromatolites of the Green River Formation (Winsborough, unpublished). The associated biota described by Bradley (1929, 1964) suggests that the habitat was amenable to diatom growth, yet diatoms are notably absent. No diatoms were found associated with the stromatolites of the Pliocene Ridge Route Formation either (Link et al., 1978), although the fauna and flora reported are generally associated with diatoms. In contrast, dense accumulations of well-preserved diatoms are observed in some fossil stromatolites. For example, diatoms are easily recognizable in material from the 6300-8900 yr. old sabkha of Chemchane (Mauritania) (Casanova pers. comm. 1989). The presence or absence of diatoms in fossil biogenic deposits may thus reflect the nature of the biological and chemical conditions present during early diagenesis.

CONCLUSIONS

Microbial structures, which include oncoids, stromatolites and stromatolitic mats are forming in the warm, spring-fed sulfate-rich lakes and sinkholes of the Cuatro Ciénegas Basin. They are built by a diverse community of microorganisms, principally cyanobacteria, associated with carbonate precipitation. The stromatolite morphologies presently forming in the El Mojarral Lakes, and the nature of the microbial populations of which they are composed, reflect physical settings that differ primarily in the intensity of light, depth, and possibly current. The community structure of the living surface mat includes a diverse assemblage of both prokaryotes and eukaryotes. *Schizothrix* and *Homoeothrix* are responsible for the overall shape of the structures, may influence stromatolites indirectly, perhaps beneficially (see Sterner, 1986) by controlling grazers. Early induration of stromatolitic structures by *in situ* precipitation of calcium carbonate may be responsible for their preservation in spite of the presence of diverse populations of grazers. In the case of oncoids, actively swimming benthos, such as amphipods, that inhabit the open channels within the oncoids, may actually contribute to rapid cementation by increasing water circulation to the interior of the structures. From our observations of living material, and the lack of benthos- free stromatolites with which to compare, we have no evidence that the smaller animals disrupt stromatolite growth and morphology significantly.

Although never dominant in biomass, the diatoms are by far the most diverse class of stromatophilic eukaryotes at El Mojarral, and are sometimes well preserved in the lithified stromatolites. Many diatom species have autecological preferences or requirements regarding salinity, temperature, pH, nutrient and oxygen concentration, light intensity, water current and substrate. Careful taxonomic and ecological documentation of the subtle patterns and relationships between modern stromatolite diatoms and their biotic and abiotic environments is a prerequisite to paleoecological analysis. There is evidence suggesting *in situ* precipitation of quartz crystals following dissolution of diatom tests (under certain conditions) during mat growth or early diagenesis. Diagenesis allowing, however, it may even be possible to use the autecological characteristics of diatoms to make interpretations about the specific ecological setting of fossil Cenozoic diatom-bearing stromatolites.

ACKNOWLEDGMENTS

This manuscript is part of the dissertation research of B. Winsborough and J.-S. Seeler. B. Winsborough is responsible for its preparation and overall content. B. Maguire identified the protozoa and micro-metazoans and S. Golubic and J. Seeler identified and cultured the cyanobacteria and clarified their role in mat community structure. R. Folk examined the thin sections and provided much input on carbonate mineralogy. All authors participated in field studies at Cuatro Ciénegas, including the in situ microscopic examination of living material. We would like to acknowledge the assistance of Srs. Jose "Pepe" Lugo G., Manuel Gonzalez R., and Jose Castaneda M. of Cuatro Ciénegas. G. Johnson provided SCUBA diving assistance ; J.- S. Seeler and B. Maxim field and technical support ; and P. Winsborough managed overall field logistics. R. Stallard and R. Murnane made the water chemistry collections and R. Murnane did the analyses. Discussions by B.W. with R. Riding, J. Casanova, and C. Caran have been helpful in understanding various aspects of stromatolite biogenesis. The Department of Geology, The University of Texas at Austin, provided SEM and EDEX facilities. Geochemical analyses and X-rays were done at the Department of Geological and Geophysical Sciences, Princeton University. The critical reviews of C.L. Monty and J. Bertrand-Sarfati greatly improved the manuscript.

REFERENCES

Anadón, P. and Zamarreño, I. (1981) "Paleogene nonmarine algal deposits of the Ebro Basin, Northeastern Spain", in C.L.V. Monty (ed.), Phanerozoic Stromatolites, Case Histories, Springer, Berlin Heidelberg New York London Paris Tokyo Hong Kong Barcelona, pp. 140-154.
Awramik, S.M. (1971) "Precambrian columnar stromatolite diversity : reflection of metazoan appearance", Science, 174, 825-827.
Awramik, S.M. (1981) "The Pre-phanerozoic biosphere - three billion years of crises and opportunities", in M.H. Nitecki (ed.), Biotic Crises in Ecological and Evolutionary Time, Academic Press, New York, pp. 83-102.
Awramik, S.M. (1984) "Ancient stromatolites and microbial mats", in Y. Cohen, R.W. Castenholz, H.O. Halvorson (eds.), Microbial Mats : Stromatolites, Alan R. Liss, New York, pp. 59-83.
Barker, P. (1992) "Differential diatom dissolution in Late Quaternary sediments from Lake Manyara, Tanzania : an experimental approach", J. Paleolimnology 7, 235-251.
Birnbaum, J. and Wireman, J.W. (1984) "Bacterial sulfate reduction and pH : Implications for early diagenesis", Chem. Geol. 43, 143-149.
Bradley, W.H. (1929) "Algal reefs and oolites of the Green River Formation", U.S. Geol. Survey Prof. Paper 154-G, Washington, pp. 203-223.
Bradley, W.H. (1964) "Geology of Green River Formation and associated Eocene rocks in southwestern Wyoming and adjacent parts of Colorado and Utah", U.S. Geol. Survey prof. paper 496-A, Washington, pp. 1-86.
Braithwaite, C.J.R. (1979) "Crystal textures of recent fluvial pisolites and laminated crystalline crusts In Dyfed, South Wales", J. Sediment Petrol. 49, 181-194.

Casanova, J. (1986) Les stromatolites continentaux : paleoécologie, paleohydrologie, paleoclimatologie. Application au Rift Gregory. Thesis, Université d'Aix- Marseille II.

Chafetz, H.S., Utech, N.M. and Fitzmaurice, S.P. (1991) Differences in the $\delta^{18}O$ and $\delta^{13}C$ signatures of seasonal laminae comprising travertine stromatolites", J. Sediment. Petrol. 61, 1015-1028.

Dixit, P.C. (1984) "Pleistocene lacustrine ridged oncolites From the Lake Manyara area, Tanzania, East Africa", Sedimentary Geology 39, 53-62.

Folk, R.L., Chafetz, H.S. and Tiezzi, P.A. (1985) "Bizarre forms of depositional and diagenetic calcite in hot-spring travertines, central Italy". in N. Schneidermann, and P.M. Harris (eds.) Carbonate Sediments, S.E.P.M. Spec. Pub. 36, 349-369.

Garcia, E., Vidal, R., Tamayo, L.M., Reyna, T., Sanchez, R., Soto, M. and Soto, E. (1975) Precipitation y probabilidad de la lluvia en la Republica Mexicana y su evaluation. Comision de Estudios del Territorio Nacional, Instituto de Geografia, UNAM, Mexico.

Garrett, P. (1970a) "Deposit feeders limit development of stromatolites", Am. Assoc. Petrol. Geol. Bull. 54, 848.

Garrett, P. (1970b) "Phanerozoic stromatolites : noncompetitive ecologic restriction by grazing and burrowing animals", Science 169, 171-173.

Gerdes, G., and Krumbein, W.E. (1984) "Animal communities in recent potential stromatolites of hypersaline origin", in Y. Cohen, R.W. Castenholz, H.O. Halvorson (eds.), Microbial Mats : Stromatolites, Alan R. Liss, New York, pp. 59-83.

Golubic, S. (1991) "Modern stromatolites : a review", in R. Riding (ed.), Calcareous Algae and Stromatolites, Springer, Berlin Heidelberg New York London Paris Tokyo Hong Kong Barcelona, pp. 541-561.

Golubic, S., and Fischer, A.G. (1975) "Ecology of calcareous nodules forming in Little Conestoga Creek near Lancaster, Pennsylvania", Verh. Int. Verein. Limnol. 19, 2315-2323.

Gomes, N.A. (1985) "Modern stromatolites in a karst structure from the Malmani Subgroup, Transvaal Sequence, South Africa", Trans. Geol. Soc. S. Afr. 88, 1-9.

Holdship, S.A. (1976) The Palaeolimnology of Lake Manyara, Tanzania : a diatom analysis of a 56 meter sediment core, PhD Dissertation, Duke Univ.

Jones, F.G., and Wilkenson, B.H. (1978) "Structure and growth of lacustrine pisoliths from Recent Michigan marl lakes", J. Sediment. Petrol. 48, 1103-1110.

Kann, E. (1941) "Krustensteine in Seen", Arch. Hydrobiol. 37, 495-503.

Kann, E., and Sauer, F. (1982) Die "Rotbunte Tiefenbiocoenose" (Neue Beobachtungen in osterreichischen Seen und eine zusammenfassende Darstellung), Arch. Hydrobiol. 95, 181-195.

Krumbein, W.E. and Werner, D. (1983) "The microbial silica cycle", in W.E. Krumbein (ed.), Microbial Geochemistry, Blackwell, Oxford, pp. 125-157.

Lewin, J.C. (1961) "The Dissolution of Silica From Diatom Walls", Geochimica et Cosmochimica Acta, 21, 182-198.

Link, M.H., Osborne, R.H. and Awramik, S.M. (1978) "Lacustrine stromatolites and associated sediments of the Pliocene Ridge Route Formation, Ridge Basin, California", J. Sediment. Petrol. 48, 143-158.

Minckley, W.L. (1963) "The ecology of a spring stream- Doe Run, Meade County, Kentucky", Wildlife Monographs 11, 5- 124.

Monty, C.L.V. (1972) "Recent algal stromatolitic deposits, Andros Island, Bahamas, Preliminary Report", Geol. Rundschau 61, 742-783.

Monty, C.L.V. (1973) "Remarques sur la nature, la morphologie et la distribution spatiale des stromatolithes", Sciences de la Terre, Nancy, 8, 189-212.

Monty, C.L.V. (1976) "The origin and development of cryptalgal fabrics", in M.R. Walter (ed,), Stromatolites, Elsevier, Amsterdam pp. 193-249.

Monty, C.L.V. and Hardie, L.A. (1976) "The geological significance of the freshwater blue-green algal calcareous marsh", in M.R. Walter (ed.), Stromatolites, Elsevier, Amsterdam, pp. 454.

Monty, C.L.V. and Mas, J.R. (1981) "Lower Cretaceous (Wealdian) blue-green algal deposits of the province of Valencia, eastern Spain", in C.L.V. Monty (ed.), Phanerozoic Stromatolites. Springer, Berlin Heidelberg New York, pp. 85- 120.

Morafka, D.J. (1977) A Biogeographical Analysis of the Chihuahuan Desert Through Its Herpetofauna. W. Junk, the Hague.

Pentecost, A. (1987) "Growth and calcification of the freshwater cyanobacterium *Rivularia haematites*", Proc. R. Soc. London Ser. B 232, 125-136.

Pursell, V.J. (1985) The petrology and diagenesis of Pleistocene and recent travertines from Gardiner, Montana, and Yellowstone National Park, Wyoming. Thesis, Univ of Texas, Austin.

Rippka, R., Deruelles, J., Waterbury, J.B., Herdman, M., and Stanier, R.Y. (1979) "Generic assignments, strain histories and properties of pure cultures of cyanobacteria", J. Gen. Microbiol. 111, 1-61.

Roddy, H.J. (1915) "Concretions in streams formed by the agency of blue-green algae and related plants", Amer. Philos. Soc. Proc., 54, 246-258.

Schafer, A. and Stapf, K.R. (1978) "Permian Saar-Nahe Basin and Recent Lake Constance (Germany) : two environments of lacustrine algal carbonates", Spec. Pub. Int. Assoc. Sedimentol. 2, 83-107.

Seeler, J-S. and Golubic, S. (1991) "*Iyengariella endolithica* sp. nov. a carbonate boring stigonematalean cyanobacterium from a warm spring-fed lake : nature to culture", Arch. F. Hydrobiologie Algological Studies 65, 399-410.

Sterner, R.W. (1986) "Herbivores' direct and indirect effects on algal populations", Science, 231, 605-607.

Stoessell, R.K. (1992) "Effects of sulfate reduction on $CaCO_3$ dissolution and precipitation in mixing zone fluids", J. Sediment. Petrol. 62, 873-880.

Stoffers, P. and Holdship, S.A. (1975) "Diagenesis of sediments in an alkaline lake : Lake Manyara, Tanzania". Proc. IX th Int. Cong. Sedimentology, Nice.

Tandeau de Marsac, N. (1977) "Occurrence and nature of chromatic adaptation in cyanobacteria", J. Bacteriol. 130, 82-91.

Weiss, M.P. (1970) "Oncolites forming on snails", J. Sediment. Petrol. 44, 765-769.

Winsborough, B.M. (1990) Some ecological aspects of modern fresh-water stromatolites in lakes and streams of the Cuatro Ciénegas Basin, Coahuila, Mexico. Dissertation, Univ of Texas, Austin.

Winsborough, B.M. and Seeler, J-S. (1986) "The relationship of diatom epiflora to the growth of limnic stromatolites and microbial mats", in M. Ricard (ed.,) Proceedings of the 7th international symposium on living and fossil diatoms, Paris, 1984, pp 395-407.

Winsborough, B.M. and Golubic, S. (1987) "The role of diatoms in stromatolite growth : two examples from modern freshwater settings", J. Phycol. 23, 194-201.

PERITIDAL POTENTIAL STROMATOLITES – A SYNOPSIS

G. GERDES[1] and W.E. KRUMBEIN[2]
1   Carl von Ossietzky University of Oldenburg, Institute for Chemistry and Biology of the Marine Environment/Marine Laboratory, Schleusenstraße 16, D-26382 Wilhelmshaven, Germany
2   Carl von Ossietzky University of Oldenburg, Institute for Chemistry and Biology of the Marine Environment, P.O.Box 2503, D-26111 Oldenburg, Germany

ABSTRACT

The term peritidal refers to areas "within and slightly outside the influence of tides". The diversity and distribution of potential stromatolites growing in these areas is discussed. The chapter deals with Recent examples to better understand the available ecospace of procaryotes that produce biolaminations to considerable thickness. Two main variations are considered: the macro- to mesotidal coast controlled by a severe tidal rhythm, and the microtidal coast, usually more protected against ocean dynamics and controlled by climate factors. In macrotidal to mesotidal areas of temperate climate, microbial mats usually develop first at the intertidal/supratidal boundary. Their distribution seems to be controlled by effects of tidal flushing which runs twice a day, erodes sediment and supplies the intertidal bioturbating fauna with plenty food and oxygen. At microtidal coast which are more frequent in lower latitudes, effects of tides are less intense, climatic factors getting more important for benthic systems. Schizohaline, hypersaline or even freshwater conditions prevail. Distribution of microbial mats towards the intertidal takes place. Likewise, structural diversification of microbial mats increases. Smooth, tufted, pinnacle- and pincushion-like, pustular and blistered surface structures indicate low energy conditions. Microbial bioherms may basically indicate the increase of water depth. Three basic mechanisms producing biolaminated buildups are discussed: (a) the grain-supported type resulting from the movement of motile microbes corresponding to repeated sedimentation of lower rate, (b) buildups according to filaments erected above the mat surface causing baffling of currents and settling of sediment grains, (c) mat-by-mat overgrowth without sedimentation resulting from a competitive growth between populations of mat-forming microbes in response to fluctuations of local ecological conditions. Grain-supported growth bedding may indicate settings controlled by tidal activity, while biolaminations without sedimentation refer to settings in which almost non-deposition prevails (e.g. microtidal

J. Bertrand-Sarfati and C. Monty (eds.), Phanerozoic Stromatolites II, 101–129.
© 1994 *Kluwer Academic Publishers.*

lagoons and anchialine pools). Several sedimentological features are discussed which in association with biolaminated deposits contribute to peritidal characteristics (e.g. open space structures, non-skeletal carbonate particles forming within microbial mats, burrows, physical deformation structures, peritidal organic matter and metal accumulation).

INTRODUCTION

Peritidal habitat variability

The term "potential stromatolite" (Krumbein 1983) is used to indicate modern biolaminated deposits whose fate of fossilization is unknown even in case of syngenetic cementation. The compilation of data from modern peritidal settings (for definitions see Table 1) reveals that the constitutive microbes obviously show an "anthipathy to norms". Rather than being restricted to the intertidal, low-energy,

References	Definitions
Folk (1973)	Sediments deposited within and slightly outside the range of tides; it includes supratidal, intertidal and shallow sub-tidal areas
Friedman and Sanders (1978)	Sea-marginal areas subject to tidal fluctuations
Bates and Jackson eds (1987), Glossary of Geology	Depositional environments in a zone from somewhat above highest storm or spring tides to somewhat below lowest tides; a broader term than "intertidal"

Table 1 : Definitions of the term peritidal

hypersalinity or unbioturbated sediments, potential stromatolites tend to occupy the most varied peritidal environments that include supratidal marshes and sabkhas, intertidal settings, subtidal channels and lagoons (Table 2). They appear in low, moderate and episodic high energy environments under hypersaline, normal saline, brakish and freshwater conditions as well as in coexistence with burrowing fauna. In summary, stromatolite-building microbes find varied ecospaces available within and slightly outside the range of tides.

Approach to variability

Peritidal conditions responsible for habitat change need to be selected along bathymetric and geographic scales. According to the bathymetric scale, the peritidal includes shallow subtidal, intertidal and supratidal flats. The shallow subtidal is characterized by tidal current paths, sand bars and shoals and typically displays abrupt variations in sediment texture and composition (Frey and Howard, 1986). The intertidal extends from the low water line at spring tides (LLWS) to the mean high water level (MHW). High bioturbation rates in intertidal flats are indirectly the result of tidal flushing since transport of food and oxygen is plenty and normal salinity values commonly reinstated twice a day (Fig. 1). The supratidal extends from MHW to somewhat above mean high water at spring tides (MHWS), high lying flats being affected by marine waters mainly during storm tides.

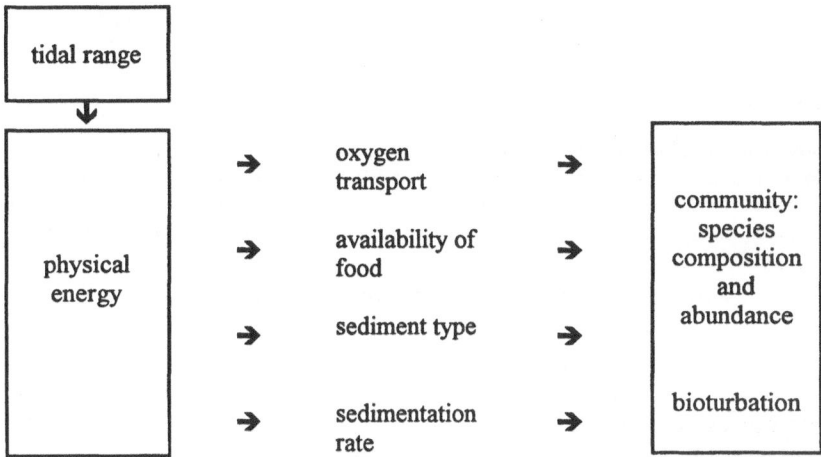

Figure 1: Ecologically important physical parameters of peritidal flats controlled by higher tidal range (also controlling coastal morphology, see Fig. 2).

Analyzing peritidal habitat diversity, the aspect of geographic variation of coastlines is of interest (Davies, 1980). Hayes (1979) basically differentiated between three geomorphic types corresponding with tidal range (Fig. 2): Macrotidal coasts (tidal range increasing above 4 m at spring tide), microtidal coasts (tidal range less than 2 m), and intermediated mesotidal coasts. On macro- and mesotidal coasts, large flats between intertidal levels exist, strongly affected by the hydraulic regime of the tidal prism which enters and leaves the flats in semidiurnal rhythm (Figs. 2a and b). Coasts with low tidal ranges are associated with long and linear barriers which protect shallow back-barrier lagoons of reduced tidal activity (Fig. 2c). The ecological differentiation between intertidal and shallow subtidal parts of these back-barrier lagoons is less clear, since ponded shallow water also prevails between intertidal levels.

Although neither a tidal type nor a tidal range is significantly related to climatic zones, coasts with low tidal ranges and barrier-protected lagoon complexes are more frequent in low latitudes (Davies, 1980). The protected inshore flats are affected by high annual temperatures, aridity or humidity that control salinity conditions of remaining seawater.

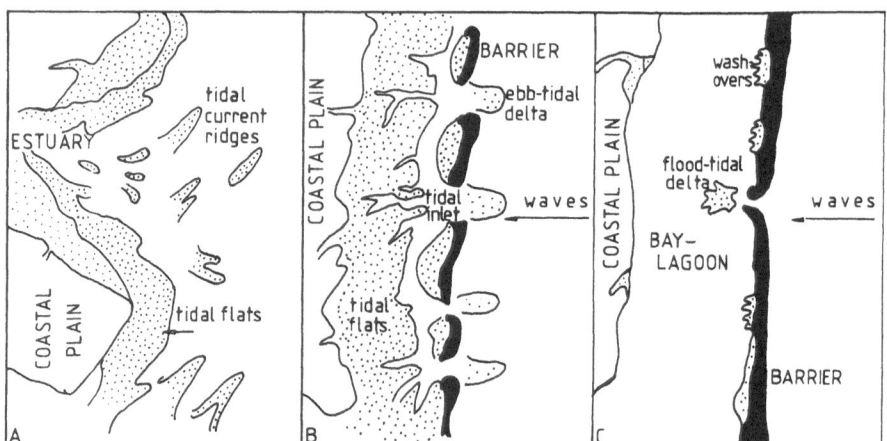

Figure 2: Morphological summary of coastal variations; a) macrotidal coast, lack of barrier islands, wide open tidal flats; b) mesotidal coast, barrier islands broken by frequent tidal inlets, extensive back-barrier tidal flats between intertidal zone; c) microtidal coast, elongated barrier islands, rare tidal inlets, inshore lagoon and storm washovers (modified after M. O. Hayes, 1979).

Schizohaline conditions (strongly fluctuating but never normal marine salinities; Folk and Siedlecka, 1974) and hypersalinity make survival of marine fauna increasingly difficult. Mat-constructing microbes, on the other hand, remain unharmed by the changing conditions. Thus, from a global viewpoint, microbial mats seem to be less abundant on macro-and mesotidal flats subject to the short-term periodicity of tidal flushing, but widespread in climate-controlled peritidal lagoons.

Special cases of peritidal lagoons are anchialine pools. The term (Por, 1985) is used to describe depressions close to the seashore which are only connected subterraneously with the open sea. Water levels shift slightly in tidal rhythm, but at a reduced level and with a certain time delay compared with the open marine areas. There is also no connection with any permanent flow of freshwater on their landward side. These pools have an active and permanent subterranean outflow. Of ecological importance is the fact that environmental conditions, such as salinity signatures, remain seasonally predictable (Gerdes et al., 1985c). Anchialine pools are well-known sites of luxuriant microbial mats (Gavish et al., 1985). Due to the secular stability of environmental conditions in annual cycles, finely laminated mats grow to extraordinary thickness. Man-made anchialine pools are saltworks where engineering designs set limits to tide-dominated processes while favouring prolonged weather-dominated periods. These are important models for understanding peritidal stromatolite diversification and biogeochemical cycles (Javor, 1989, Reineck et al., 1990, Cornée et al., 1992).

The following overview is based on an environmental grid of coastal variations overlying the peritidal bathymetric view (Fig. 3). Environmental factors at macro- and

Location	Peritidal zones	Mat-colonized areas	Organism dominance	References (selected)
Spencer Gulf, South Australia	Intertidal/-supratidal boundary	semi-closed back-barrier tidal flats; mesotidal? temperate, semiarid	*Microcoleus, Lyngbya*	Burne & Coldwell 1982; Bauld 1984
Hamelin Pool, Shark Bay, W. Australia	Intertidal, shallow subtidal	Topographic lows, water retention; subtidal slope wave-dominated; microtidal, subtropical arid	*Microcoleus, Lyngbya, Entophysalis*	Hoffman 1976; Playford & Cockbain 1976; Hagan & Logan 1975; Bauld 1984
Boca Jewfish, Bonaire Netherl. Antilles	Intertidal	Spit complex wave-dominated, protected inshore, microtidal, semiarid	(no taxonomic details)	Pratt 1979
Trucial coast, Persian Gulf	Supratidal, intertidal, subtidal	Supratidal sabkha, intertidal water-logged depressions, meso- to microtidal, subtropical arid	*Microcoleus, Oscillatoria, Entophysalis*	Evans 1970; Golubic 1976; Shinn 1983
Gavish Sabkha, Aqaba Gulf, Sinai	Intertidal, supratidal	Hypersaline anchialine pool, sabkha formation, microtidal lagoon, subtropical arid	*Microcoleus, Oscillatoria, Pleurocapsale-ans, others*	Krumbein et al. 1979; Gavish et al. 1985; Gerdes & Krumbein 1987
Solar Lake, Aqaba Gulf, Sinai		Hypersaline anchialine pool, microtidal lagoon, subtropical arid	*Microcoleus, Oscillatoria, others*	Krumbein et al. 1977; Cohen 1984; Gerdes & Krumbein 1987; Javor 1989
Bay St. Jean, Mauritania, W-Africa	Intertidal/-supratidal belt	Quartz-sandy tidal flats drowned dune valley, microtidal, arid	*Microcoleus, Oscillatoria,*	Schwarz 1975
Laguna Mormona, Baja Cal., Mexico	Intertidal	Back-barrier ponded water, microtidal, anchialine, semiarid	*Lyngbya, Microcoleus, Entophysalis*	Horodyski 1977; Horodyski & von der Haar 1975; Stolz 1984, 1990
Andros Island, Bahamas	Intertidal, supratidal	Back-barrier lagoon, supratidal marshes, channel margins, microtidal?, subtropical humid	*Schizotrix, Scytonema*	Golubic 1973; Monty & Hardie 1976
Stocking Island, Bahamas	Intertidal	Reef complex, normal salinity, microtidal	*Schizothrix,* various eukaryotes	Reid & Browne 1991
Exuma Cays, Bahamas	Subtidal	Inter island channels, tidal currents, ooid sand	Chlorophyte *Schizothrix* sp., benthic diatoms	Dill et al. 1986; Shinn 1987; Riding et al. 1991a, b
Chetumal Bay, Belize, Cent.Amer.	Subtidal-lower intertidal	Brackish coastline south of mouth of Rio Hondo, microtidal, tropical	*Scytonema, Schizothrix, Phormidium*	Rasmussen et al. 1993
Plum Island, Massachusetts	Intertidal to supratidal	Metahaline pools, back-barrier siliciclastic tidal flats, mesotidal? temperate humid	*Microcoleus, Lyngbya*	Cameron et al. 1985
Mellum Isl. southern North Sea	Intertidal to supratidal belt	Back-barrier siliciclastic tidal flats, meso-macro-tidal, temperate humid	*Microcoleus, Oscillatoria, Merismopedia*	Gerdes et al. 1985a; Stal et al. 1985

Table 2. Selected data from Recent localities of microbial mat growth

mesotidal flats, microtidal lagoons, and anchialine pools show, in this order, decreasing influence of tides (Fig. 1) and increasing weather signature. Against this background, the distribution, diversity or poverty of potential stromatolites will be discussed. References will concentrate upon details from modern environments (Table 2).

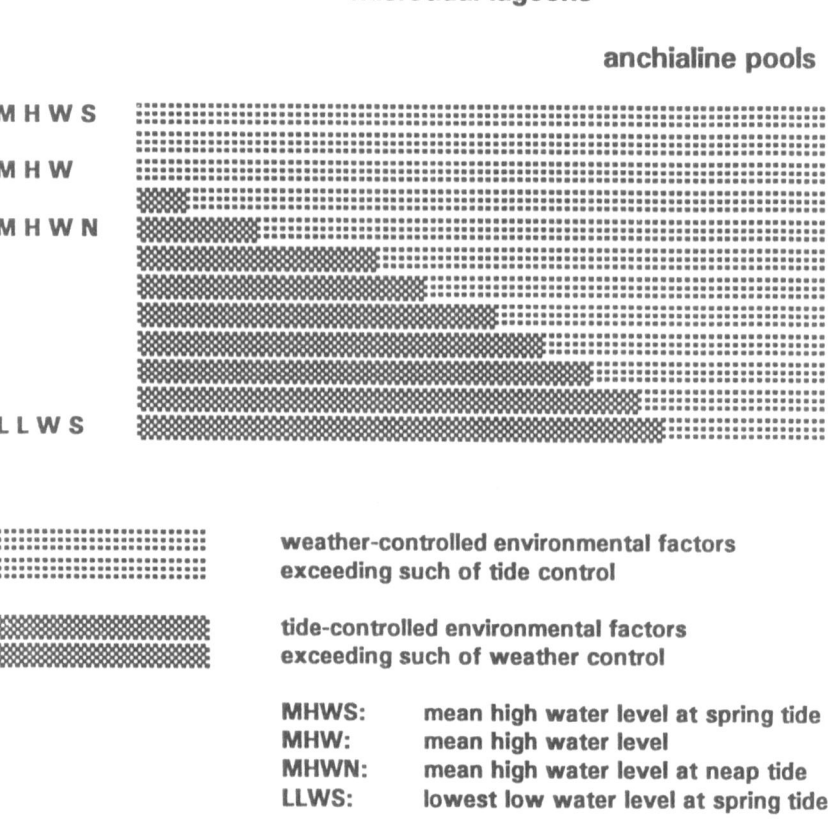

Figure 3: Environmental grid of coastal variations (such as illustrated in Figure 2) overlying the peritidal bathymetric view. At macrotidal to mesotidal coasts of temperate climate, microbial mats usually develop first at the intertidal/supratidal boundary (MHWN to MHWS), since the intertidal is controlled by effects of tidal flushing (running twice a day, eroding sediment and supplying intertidal bioturbating fauna with plenty food and oxygen; compare Figure 1). At microtial coasts which are more frequent in lower latitudes, effects of tides are less intense, climatic effects (e.g. schizohaline, hypersaline or even freshwater conditions) getting more important for benthic systems. In anchialine pools, tidal effects are almost excluded.

FRAMEWORK BUILDERS OF PERITIDAL MICROBIAL MATS

Monty (1976) emphasized hierarchical ranks between microbial mats (termed "cryptalgal fabrics") and biostromes/bioherms. The latter describe the overall morphology of depositional units built by and composed of various mats, while mats represent individual laminae. Kalkowsky (1908) used the term "stromatoid" to define individual stromatolitic laminae. A microbial mat may be seen as the modern analogue of a stromatoid.

In shallow marine and peritidal settings, microbial mats are primarily built by cyanobacteria. Usually, one or few species tend to aggregate in the form of "mats", i. e. coherent coatings on sediment and rock. The term "microbial mat" (Brock, 1976, Krumbein, 1986) was created to emphasize multiple microbial communities commonly succeeding once the mat is established by initial framework builders.

In spite of different climates, saline signatures or degrees of wetting, striking similarities exist between framework species (Tables 2 and 3). Filamentous cyanobacteria with the genera *Microcoleus*, *Oscillatoria* and *Lyngbya* are reported from various peritidal settings, in monospecific and composite mats. The cosmopolitan cyanobacterium *Microcoleus chthonoplastes* (Fig. 4) is known for its motility, production of cohesive fabric and a pronounced tolerance for salinity fluctuations and microaerobic conditions (Golubic, 1976). Mats dominated by *Microcoleus chthonoplastes* occur in the most varied peritidal environments (Table 3). The similarities in the type of constitutive microbiota support statements that mat-forming microbes are non-reliable organisms and do not constitute proof of any growth environment of stromatolites (Friedman and Sanders, 1978).

Dominant genera	Localities										
	GS	SL	TC	SB	SG	LM	MA	AI	CB	PI	ME
Microcoleus	x	x	x		x	x	x			x	x
Oscillatoria	x	x	x			x	x			x	x
Lyngbya	x	x	x	x	x	x				x	x
Scytonema				x				x	x		
Pleurocapsa	x										
Entophysalis	x		x	x		x				x	
Synechococ.	x	x		x							x
Gloeocapsa	x	x									x
Climate	SA	SA	SA	SA	SA	SA	SA	SH	T	TH	TH
Salinity	hy	hy	hy	hy	hy	hy	hy	fr	b	n/h	n/b

Table 3. Cyanobacteria (main-structuring taxa) of microbial mats from various peritidal settings
Abbreviations (for references see Table 2):
Localities: GS: Gavish Sabkha, SL: Solar Lake, TC: Trucial Coast, SB: Shark Bay, SG: Spencer Gulf, LM: Laguna Mormona, MA: Mauritania/Africa, AI: Andros Island, CB: Chetumal Bay, Belize; PI: Plum Island, ME: Mellum Island
Salinity conditions: hy: hypersaline, fr: freshwater marsh, n/h: normal marine to slightly hypersaline, n/b: normal marine to brackish, b: brackish
Climate: SA: Subtropical arid, SH: Subtropical humid, T: Tropical, TH: Temperate humid

In response to differences in local ecological conditions, however, or the presence of constitutive genera in different portions according to the site, mats of different microstructures will form. If constitutive species are drastically different like *Scytonema* sp. which form mats on the high marsh of Andros Island, and at the brackish coastline of Chetumal Bay, Central America (Rasmussen et al.,1993), significant internal microstructures evolve. *Scytonema* is a filamentous freshwater species. Its growth proceeds in a combined prostrate/reticulate pattern. Lamination results in alternating phases of horizontally growing filaments with phases characterized by vertical bundles (Monty, 1976). In *Scytonema* mats, the flabellate growth of the reticulate phase is well discernible and distinguishable from euryhaline species such as *Microcoleus chthonoplastes*.

Much of the literature on modern peritidal mats notes the abundance of algae in peritidal microbial mats. Studying mat composition of Bahamian subtidal stromatolites, Riding et al. (1991a) described filamentous chlorophytes associated with cyanobacteria, producing a dense felt of erected filaments up to 1 cm above the mat surface. The erected filaments cause baffling of sediment grains held in suspension by tidal currents. Various investigations of microbial mats note also the occurrence of diatoms. Ehrlich and Dor (1985) considered their role as contributors to the total biomass of microbial mats. Diatom frustules may provide solids for microbial biofilm attachment (Fig. 5b), although Krumbein (1978) stated that due to bacterial decay processes, silica frustules of diatoms dissolve completely within stromatolitic mats.

STRUCTURAL DIVERSITY OF SURFACE MATS

Smooth mats

Mats formed due to a dominant presence of filamentous cyanobacteria, e. g. *Microcoleus* sp., are usually smooth when associated with low-rate sedimentation (Fig. 4a, b). Sediment trapping and binding by the sticky mat surface is a common phenomenon on tidal flats. Sand flats biostabilized by this mat withstand largely erosive forces (Heinzelmann and Wallisch ,1991).

Reid and Browne (1991) also reported smooth mats from the intertidal of Stocking Island, Bahamas, formed by *Schizothrix* sp.

Tufted, pinnacle- and pincushion-like mats

Tufted mats frequently occur a few centimeters to a few decimeters below water level. In the Laguna Figueroa, several square meters to several thousand square meters in area are flooded in this way. Millimeter-sized tufts projecting through the sediment-water interface consist of vertical filament bundles of cyanobacteria, often of *Lyngbya aestuarii* (Horodyski, 1977).

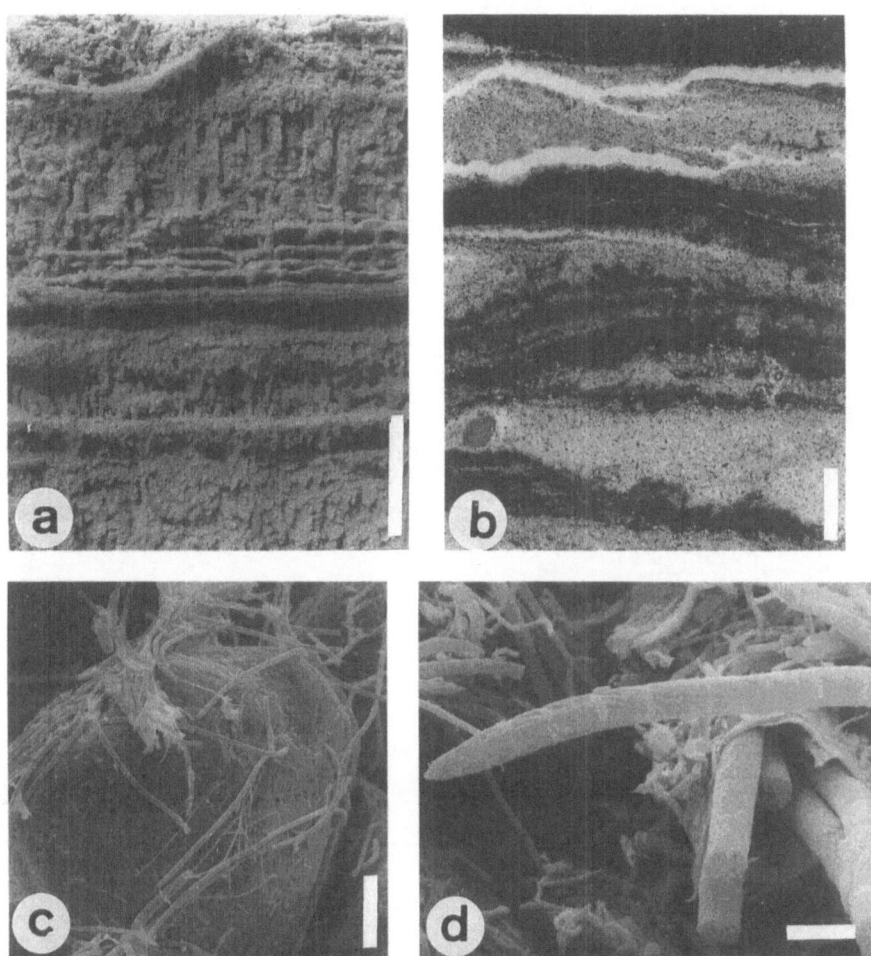

Figure 4: Smooth *Microcoleus* mats from Mellum Island, North Sea; a) sediment core, containing mats between quartz-sandy sediments; vertical burrows of marine polychaetes and amphipodes visible; ripple on top of the core coated by a mat (scale is 5 cm); b) thin section from core in a); mats appear as dark organic layers between light quartz sand (scale is 1 cm); c) *Oscillatoria* sp. and *Microcoleus* sp. colonizing a quartz grain (scale is 30 µm); d) bundled filaments of *Microcoleus* sp. enclosed by polysaccharid sheath and other LPP-forms isolated from a microbial mat (scale is 5 µm).

Pinnacle-like structures similar to tufted mats are recorded from the Solar Lake (Krumbein et al., 1977) and saltworks (pers. observation). Pinnacles are 1 to 5 mm high and composed of a mixed framework of *Microcoleus* bundles and diatoms, embedded in extracellular polymeric substances (EPS) of bacterial and diatom origin (Fig. 5a, b).

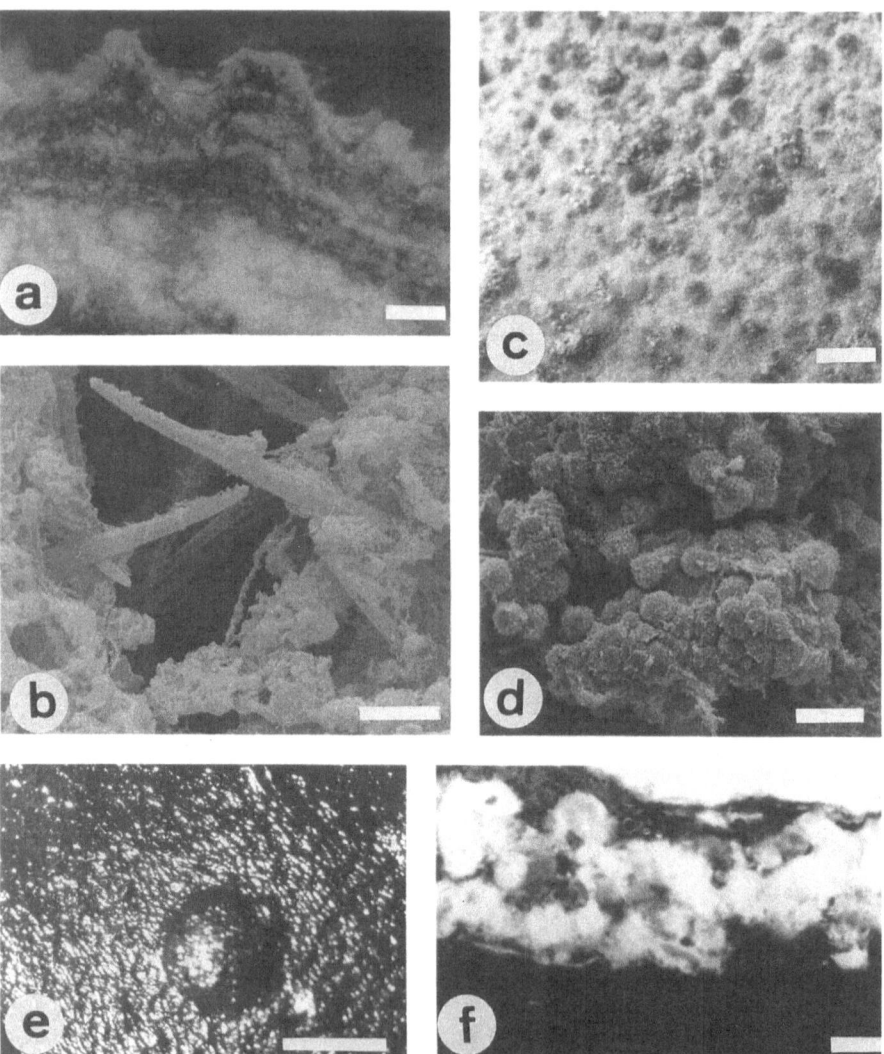

Figure 5: Structural diversity of surface mats; a) Pinnacle mat from Bretagne saltworks formed by *Microcoleus* sp. and diatoms (scale is 5 mm); b) Pinnacle interior showing frustules of *Nitzschia* sp. (scale is 20 µm); c) pustular mat from Gavish Sabkha showing knobby surface structures due to Pleurocapsaleans (scale is 1 cm); d) organisms that produce knobby structures of pustular mats as shown in Fig. 5d (scale is 5 µm); e) gas accumulation beneath smooth lab-cultured mat causing blistering (scale is 5 mm); f) vertical section through surface mat (Gavish Sabkha), blistered surface encrusted by gypsum crystals (scale is 1 mm).

Diatoms predominantly belong to *Nitzschia*. Pinnacles contribute to the characteristic wavy appearance of biolaminations in microbial build-ups and stromatolites. The underlying biological reason is not yet quite clear. *Nitzschia* often colonizes the outermost parts of mats floating in the water, possibly to maintain a certain distance from toxic substances inside the mat (e.g. H_2S). On the other hand, cyanobacteria and other organisms in the mats attach themselves to solid substrates, even to diatom frustules. The final effect may be a competitive growth between benthic diatoms and cyanobacteria which may be responsible for pinnacle formation (Fig. 5a).

Similar are also *Scytonema* "pincushions", recorded from the Andros freshwater marsh (Golubic, 1973; Monty and Hardie, 1976). *Scytonema* sp. imparts a characteristic porous fabric to pinnacles, due to the accretion of high-Mg calcite around the sheaths. Such fabrics are reminiscent of fossil Orthonella (Monty 1981).

Pustular mats

This type is known from various settings (e. g. intertidal belt of Shark Bay, Gavish Sabkha lagoon). It is formed by coccoid cyanobacteria of the Pleurocapsalean or Chroococcalean type (e.g. *Entophysalis* sp.) Thick, concentrically lamellated sheaths encasing individual cells contribute to the pustular surface patterns. Pustular, mammillate (Golubic, 1973) and cinder mats (Kendall and Skipwith, 1969) describe the same morphological type (Fig. 5c). The group of colonial coccoid cyanobacteria mentioned (Fig. 5d) is known to switch from the pustular mat also to nodular growth. In that way, nodules of several centimeters in diameter can develop, lying unattached on a mat or sediment surface (Gerdes et al., 1985b). Golubic (1973) mentioned also this type at the dams and floors of tidal channels of the Persian Gulf.

Blister mats

This type is reported mainly from intertidal/supratidal transitions. Blisters are small elevations of surface mats usually with a hollow center (Fig. 5c). According to Golubic (1973), blistering is caused by aerobic decomposition of organic matter beneath flat (smooth) mats. North Sea supratidal flats are covered in places with small dome-like elevations, the hollow spaces filled with methane. The gas migrates from buried organic deposits toward the surface. Here, the coherent fibrillar mat inhibits the escape of the gas into the air or water. On arid coastlines, blisters are commonly encrusted by gypsum crystals (Fig. 5f).

Environmental significance of mat types

The environmental significance of mat types described above may be interpreted as follows: *Smooth mats* (Fig. 4) are common in tide-dominated settings of low to moderate energy. On macro- and mesotidal coasts, their occurrence is restricted to subaerially exposed flats. In areas of low hydrologic energy, they tend to spread into

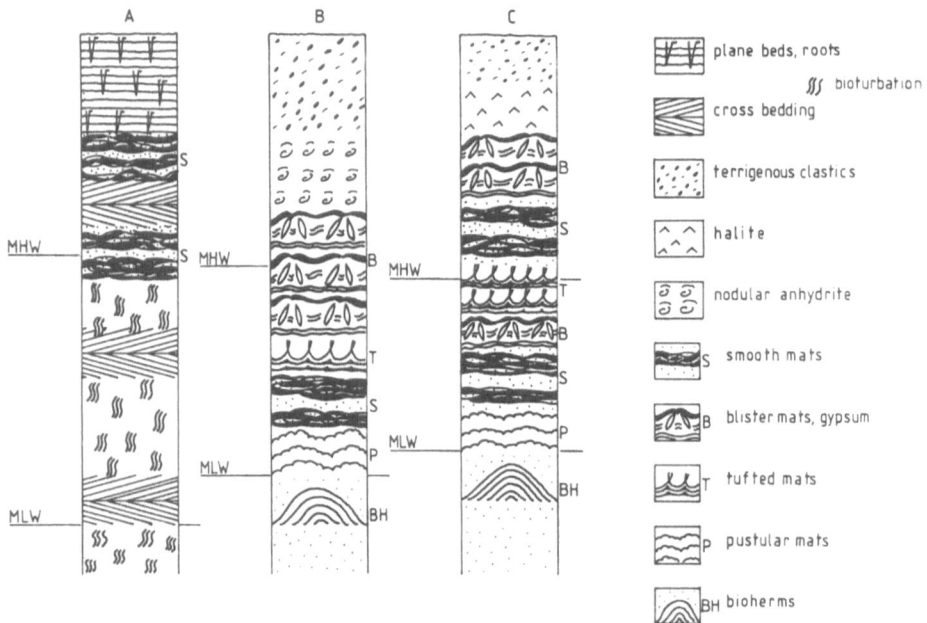

Figure 6: Idealized stratigraphic columns illustrating occurrence of microbial mats and potential stromatolites at different coastal types; a) tidal flats close to macrotidal range (details from Mellum Island, North Sea); note smooth, grain-supported mats at the intertidal-supratidal boundary zone; b) tidal flats close to microtidal range (details from Abu Dhabi/Persian Gulf; c) microtidal lagoon (details from Shark Bay, Western Australia). Note diversification of mat types and spreading into the intertidal and subtidal zones in b) and c).

topographically lower-lying areas. Golubic (1973) mentioned their occurrence in subtidal channels. *Tufted*, *pinnacle* and *pustular mats* (Fig. 5a-d) are common in low-energy microtidal lagoons but scarce on tidal flats. The reason may be that sedimentation occurs on tidal flats at a higher rate, and shallow water ponds between intertidal levels hardly occur. Pustular mats indicate schizohaline conditions. This type is restricted by increasing hypersaline values. *Blister mats* (Fig. 5e), dependent on gas formation in the sediment, generally may not be restricted to a certain peritidal zone. Their gypsum encrustation (Fig. 5f) may indicate mainly conditions of exposure.

Fig. 6 summarizes the diversification of mat types and their spreading capacity into peritidal subzones.

MODES OF GROWTH BEDDING

Grain-supported vertical accretion

This type occurs on siliciclastic tidal flats of the North Sea (Fig. 4a and b, 7a). Grains deposited by wind or water on top of the mats, stimulate the microbes to move upwards. *Microcoleus* sp. is known to produce a multitude of new mat generations with the aid of low-rate sedimentation to which hormogones from the buried mat below respond (Gerdes and Krumbein, 1987). Under prolonged periods of exposure at supratidal level, air-borne deposition seems to be the main process which accelerates vertical accretion, provided that moisture is sufficient. The fine- to medium-grained quartz-sand facilitates the capillary movement of groundwater (Hoffmann, 1942).

Sediment trapping and binding by the sticky mat surface is a common phenomenon in peritidal environments, resulting in contrasting sediment-rich/organic-rich laminae. Loose, unstabilized sand deposited on the mat surface is frequently rippled by currents or waves and subsequently overgrown by mats. Mats replicating ripple morphology are thus not infrequent in peritidal areas (Fig. 4a).

Microbial mats on siliciclastic shorelines have been increasingly discovered in recent years. High latitude quartz-rich mats have been described from European tidal flats, e. g. from the Wash, England (Evans, 1965), Normandy (Le Gall and Larsonneur, 1972), and the southern North Sea coast (Reineck, 1979; Gerdes et al., 1985a; Stal et al., 1985). Cameron et al. (1985) reported occurrences from the east coast of the United States. A number of reports note the presence of siliciclastic microbial mats at lower latitudes (e. g. Schwarz et al., 1975 for the west coast of Africa, Mauritania; Frey and Basan, 1978 for Georgia).

Grain-supported growth bedding has also been noted in the subtidal potential stromatolites of Shark Bay (Logan et al., 1964; Playford and Cockbain, 1976) and Bahamian channels (Dill et al., 1986, Shinn, 1987, Riding et al., 1991a, b). Awramik and Riding (1988), and Riding et al. (1991a) emphasized a poorer lamination in Shark Bay and Bahamian potential stromatolites than is usually the case in prokaryotic buildups. The authors assume that this may be due to the abundance of algae which contribute substantially to the entrapment and accretion of sediment grains.

According to Monty (1976), physical discontinuities ("microdiastems") play a decisive role in the buildup of distinctly laminated stromatolitic fabrics. We made the observation that at least four days of non-deposition are needed to re-establish well-defined individual laminae (Fig. 7a). In the Bahamian channels (Dill et al., 1986), rhythmic sedimentation occurs due to tidal currents sweeping in and out of the channels three hours out of every six (Riding et al., 1991a).

On North Sea tidal flats (Fig. 2a and b, 6a), microbial mats experience flooding and related sediment transport in the spring tide cycle (on a yearly average, about 50% of high tides reach or cross the areas of microbial mat growth; Gerdes et al., 1985a). Siliciclastic examples lack syngenetic cementation prevailing in potential stromatolites of subtropical and tropical regions, thus spreading of siliciclastic buildups are rare into intertidal and subtidal zones. This again underlines the importance of climate for peritidal stromatolite distribution.

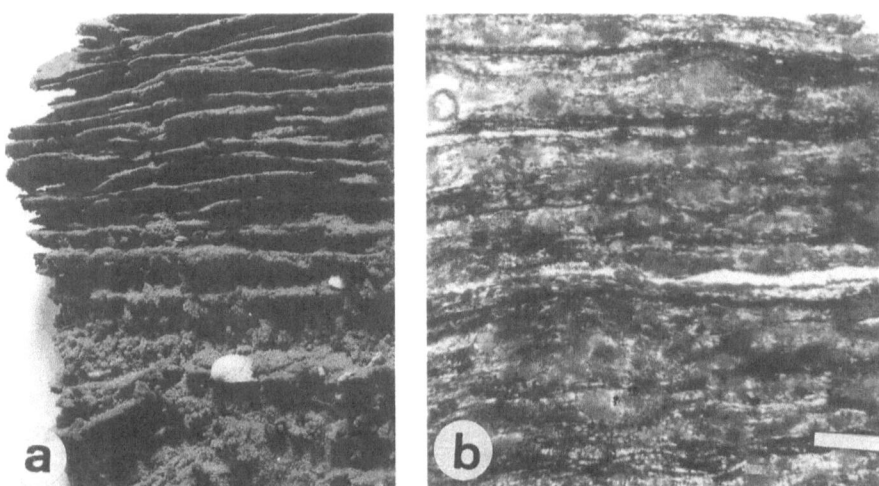

Figure 7: Types of vertical accretion; a) Subaerial rise of biolaminated quartz sand (experiment). Undisturbed sediment wit a mat on top was cored, and twelfth instances of oversedimentation carried out in the lab, simulating low-rate deposition. Relief cast preparation showing generations of microbial mats that originated from one and the same preexisting mat (scale is 5 cm); b) vertical section through a biostromate buildup from Solar Lake showing dark (*Microcoleus*-dominated) and light laminae (gel-supported due to dominance of coccoid cyanobacteria and diatoms); light laminae containing carbonate biogenically formed in situ (scale is 1 cm).

Mat-by-mat overgrowth without aid of sedimentation

This type of growth bedding (Fig. 7b) was mainly observed in anchialine pools (Gavish et al., 1985, Gerdes et al., 1985b, Gerdes and Krumbein, 1987, Gerdes et al., 1991). Sites of mat growth are characterized by a combination of shallow water, very low energy level and hypersalinity. Filaments of *Microcoleus chthonoplastes* and other LPP-forms (Fig. 4d) were found to create biolaminites in association with slime-secreting coccoid cyanobacteria and diatoms. Whereas burial by sedimentation is rare, self-burial proceeds via mat-by-mat overgrowth. The process of self-burial is triggered by temporal changes in ecologic conditions (temperature, light intensity, salinity). In response to fluctuations of these environmental variables, microbes overgrow other surface mats in order to gain the most favorable position within the environmental gradient. Recurrence is most likely in a seasonal pattern. The chemocline shifting in day-night rhythm across the mat/water interface may also account for the overgrowth phenomenon (Monty, 1976). In the Solar Lake and Gavish Sabkha, coccoid cyanobacteria and diatoms, embedded in large amounts of slime, overgrow *Microcoleus* mats in summer when the water table is low. With rising water level in fall, the filamentous cyanobacteria again overgrow the summer generation. In the Solar Lake, mat-by-mat overgrowth without the aid of sedimentation has created a varvite-like sequence of alternating dark (*Microcoleus*-dominated) and light layers (dominated by coccoid cyanobacteria and diatoms). The vertical sequence already extends over 1.20 m.

Such a vertical extension may be an exceptional case in modern peritidal environments. Even back-barrier lagoons are usually more frequently interrupted by storm washovers and sheetflood-derived sedimentation which repeatedly disturbs vertical mat-by-mat accretion. Thus, even in closed coastal settings, beds of biogenic varvites usually only reach a few decimeters although a total thickness of more than 1 m was also observed in a sediment core of 3 m length taken in the Gavish Sabkha. In summary, environments allowing for the development of growth bedding by self-burial are characterized by ponded water of almost zero energy. The type is excluded at tide-dominated and subaerially exposed flats.

MICROBIAL BUILDUPS - OUTER MORPHOLOGY

Two types of stromatolitic morphotypes may be roughly differentiated: bedding plane-oriented (stratiform) sets of laminae, and those which swell into three dimensional mound- or dome-like forms. In this chapter, the terms "biostrome" and "bioherm" are used to differentiate between these main types. Distinctly bedded, widely extensive, blanket-like biolaminated buildups are termed "biostromes", and nodular, biscuit-like, dome-like or columnar stromatolites are referred to as "bioherms".

Distributional aspects of microbial biostromes

Biostromes commonly develop under low energy conditions (Fig. 7). Thus, by far the greatest proportion of literature relates to settings protected by bars in combination with a low tidal range (Fig. 6b and c) excluding short-term tidal periodicity and sedimentation (a selection is given in Table 2). Since weather-dominated processes take over, temperature and salinity deviations from marine values can develop in the ponded seawater. Mat-constructing microbes remain almost unharmed by these changing environmental conditions and grow to luxuriant thickness (see for example productivity measurements in hypersaline waters; Bauld, 1984, and reviews about halophily and halotolerance in cyanobacteria; Golubic, 1980). The diversity of microhabitats (persisting thin water sheets, permanently water-filled ponds which experience only seasonal changes between flooding and exposure, considerably raises biological and sedimentological diversity of microbial growth patterns (see e.g. smooth, tufted, pinnacle-like surface mats etc.).

Microbial bioherms

Microbial bioherms are accumulations of more or less distinct biolaminations that swell into mound, dome, or columnar form. Eggleston and Dean (1976) used this term for dome-like Recent stromatolites, although it may include series of head-like to palisade-like buildups (see also the classification scheme of basic geometric forms, Logan et al. 1964).

Besides spectacular findings of columnar potential stromatolites from Shark Bay (Hagan and Logan, 1975) and Bahamas (Dill et al., 1986; Riding et al., 1991a, b), subtidal microbial buildups have been reported from various places, such as the Great Bahama Bank and Bermuda Islands. According to Golubic (1973), current velocity is the main determinant of subtidal morphologies (biostromes developing at a low current velocity, while stronger currents modulate such types as domes or columns).

Summarizing mechanisms to shape bioherms, the following criteria were suggested: (i) particle transport held in suspension by subtidal currents, (ii) rhythmic changes of transport e.g. due to tidal slack causing swash and backwash of suspended sediments which become bound on crests and limbs of microbial bioherms, (iii) specific capacity of component organisms to trap and bind the grains from suspension, (iv) presence of organic matter and catalyzing effects of internal biofilms, (v) fast synsedimentary cementation.

The vertical extent of bioherms may be primarily controlled by water depth. Monty (1979) reported small domes from the Australian Great Barrier Reef achieving only 2 - 4 cm in diameter. The author observed that the combination of low water depth and high energy modulated the size of the domes and concluded: "They are easily washed away by the waves, once they reach a given size and oppose too great a resistance to the currents".

A corresponding environmental requisite of all types of bioherm growth (including nodules, i.e. unattached subspherical bodies) seems to be sufficient water cover, either permanently as in the subtidal, or at least periodically as in the lower intertidal (see Fig. 6b and c).

Growth success of bioherms in turbulent water

As already mentioned, a fast synsedimentary cementation seems to be a prerequisite for microbial bioherms which grow in tide-controlled settings. Pratt (1979) observed early carbonate cementation in nodular structures which occur in the wave-dominated lower intertidal zone of the Boca Jewfish spit complex. The author suggested that cementation within these structures could result from the interplay between species composition, internal architecture, and rapidity of drainage during intertidal exposure. Other authors also emphasize the role of rapid changes of concentrations and reaction velocities in the microenvironment of pore spaces. However, the presence of organic matter and katalyzing effects of internal biofilms should also be stressed since they seriously alter the geochemical milieu (Javor, 1989). Basically physical forces, e. g. capillary action, evaporative or tidal pumping, adhesive forces, the formation of water films in meniscus patterns etc., all this is controlled by the microbially trimmed internal geometry and more than that, by physiological processes which constantly convert organic material to inorganic, gaseous, dissolved, or solid phases (Krumbein, 1986).

To return to the initial question, rapid syngenetic cementation of microbial buildups could be a way to withstand erosive forces in peritidal high energy environments. One should be cautious, however, about making an all too generalised high energy interpretation based on cemented microbial bioherms for two reasons: (a) carbonate also precipitate in microbial biostromes indicating low energy; (b) non-agitated intertidal ponds of lagoons also bring about microbial bioherms (see following review on the local coexistence of bioherms and biostromes on Andros Island).

Coexistence of bioherms and biostromes

The east and west coast of Andros Island (Monty, 1965, 1967, 1972, 1976) shows a mutual distribution of microbial biostromes and bioherms:

--   In the supratidal marsh, small bioherms form in depressions, surrounded by pinnacle mats forming the biostromate type. Bioherms and biostromes are formed by *Scytonema myochrous* and *Schizothrix* sp. (Monty, 1972).

--   At the lower marsh, these species also form up to 5 cm high tufts which precipitate $CaCO_3$ (Monty and Hardie, 1976).

--   In intertidal ponds, the biostromate tufted type prevails, formed by *Scytonema crustaceum* and *Microcoleus chthonoplastes*.

--   On the shallow intertidal, small bioherms and smooth mats are formed by *Dichothrix* sp. and *Rivularia* sp. (no biolaminations).

--   The permanently flooded ponds and the low intertidal are characterized by smooth biostromate structures and small columns (5 - 6 cm high 3 - 4 cm wide). Both biostromes and columns are built by *Phormidium hendersonii* (*Schizothrix calcicola* in Monty, 1965, 1967, 1972).

Bioherms in this low energy region are clearly controlled by water cover. An estimated minimum water level may be slightly higher than the buildup crest.

Rasmussen et al. (1993) observed a similar mutual occurrence of smooth biostromate and small columnar structures formed by dominance of *Scytonema* spp. at the brackish coastline of Chetumal Bay, Belize.

Reid and Browne (1991) observed the occurrence of dome-shaped heads and tabular forms in the lower intertidal of a fringing reef complex at Stocking Island, Bahamas. Energy conditions here are moderate, since waves break seaward on a subtidal reef.

ASSOCIATED FEATURES

Features associated with potential stromatolites may help in peritidal reconstruction (Table 4; see also Pratt and James, 1986; Shinn, 1983, 1986).

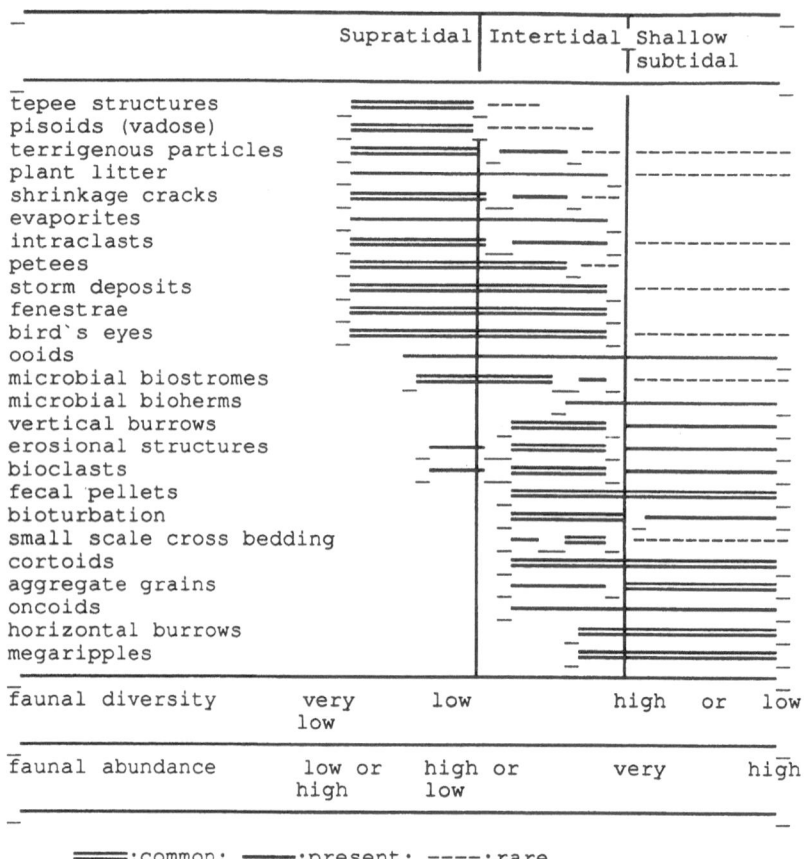

	Supratidal	Intertidal	Shallow subtidal
tepee structures			
pisoids (vadose)			
terrigenous particles			
plant litter			
shrinkage cracks			
evaporites			
intraclasts			
petees			
storm deposits			
fenestrae			
bird`s eyes			
ooids			
microbial biostromes			
microbial bioherms			
vertical burrows			
erosional structures			
bioclasts			
fecal pellets			
bioturbation			
small scale cross bedding			
cortoids			
aggregate grains			
oncoids			
horizontal burrows			
megaripples			

| faunal diversity | very low | low | high or low |
| faunal abundance | low or high | high or low | very   high |

=====:common; ——:present; ----:rare

Table 4 : Range of Recent peritidal characteristic carbonate evaporite shorelines (after Flügel, 1982).

Open space structures

Fenestrae (Tebutt et al., 1965) are typical of microbial buildups including oval, elongated, irregular shapes (Monty, 1976). Such interspaces develop in the strongly entangled
microbial meshwork, especially if physical sedimentation is lacking and completely
biogenic laminated fabrics build up (Fig. 7b). Genetically, they may be related to
biochemical processes occurring in microbial mats, including decay, gas and liquid
formation. Also bioturbation may play a role.

Non-skeletal carbonate particles

In subtropical lagoons, authigenic carbonates generate within the mats (Friedman, 1978; Dahanayake et al., 1985; Gerdes and Krumbein, 1987; Gerdes et al., 1991). The formation of these particles is obviously climate-controlled. There is no *in-situ* production of carbonates in biolaminations from temperate regions, not even in hypersaline settings (e. g. salterns of Bretagne, France). The particles grow predominantly within hydroplastic, gel-supported laminae (Fig. 8). Bacterial cells embedded in gel probably act as nuclei which catalyze or limit diffusional processes (Fig. 8d).

Syngenetic growth of biolaminated particles within slowly growing buildups occurs in coexistance with the carbonaceous smooth paste of decaying cyanobacteria. Dahanayake et al. (1985) interpreted the genesis of well laminated biostromate buildups and oncoids from Jurassic Minette in such a milieu. Particle shapes are largely determined by the kind and shape of the nucleus. Larger oncoids forming on gastropod shells were described by Gerdes and Krumbein (1987) and Weiss (1969). Monty and Mas (1981) described nuclei formed by knobby microstromatolites; Dahanayake et al. (1985) mentioned microooncoids forming around microbial colonies and biogenic gas and liquid bubbles which become trapped within gel-supported hypersaline microbial mats.

Different particle types are usually associated within one and the same biolaminate including oncoids, ooids, peloids (Fig. 8), finally characteristic grape-shaped arrangements resembling Kalkowsky`s "ooid bags". Such a "bad sorting" may be an useful criteria of *in-situ* formation. Indications of *in-situ* formation are also "sedimentary eye structures" (Dahanayake et al., 1985), imparted by the the growing particle to the surrounding soft organic matrix (Fig. 8a).

Burrowing and grazing

Shinn (1972) described the unique association between living cyanobacterial buildups and burrowing worms on intertidal flats of the Persian Gulf (West coast of Quatar). The structures appear as small (about 10 cm high) mounds. Internally they show convex-upward laminations produced by the combined growth of cyanobacteria and excretion of worm pellets on the surface of the mounds.

Gerdes and Krumbein (1987) described growth success of microbial mats associated with burrowing polychaetes and amphipods at the intertidal/supratidal boundary of tidal flats (Mellum Island, North Sea coast). No destruction of laminations was visible here either. The examples show that in the presence of burrowing, accretion of microbial mats may prevail over destruction (Fig. 4a).

The case of destruction prevailing over accretion is reported by Schwarz et al. (1975). These authors studied the impact of burrowing fiddler crabs and grazing fishes on microbial mats on intertidal flats within the Bay of Saint-Jean (Mauritania, West Africa). Also Monty (1967) and Gerdes et al. (1985a) described fish grazing on intertidal mats, although this did not lead to the total mat destruction.

Survival of mats is said to depend mainly on the decrease of faunal diversity and abundance. However, a diversity decrease can also select for opportunistic species

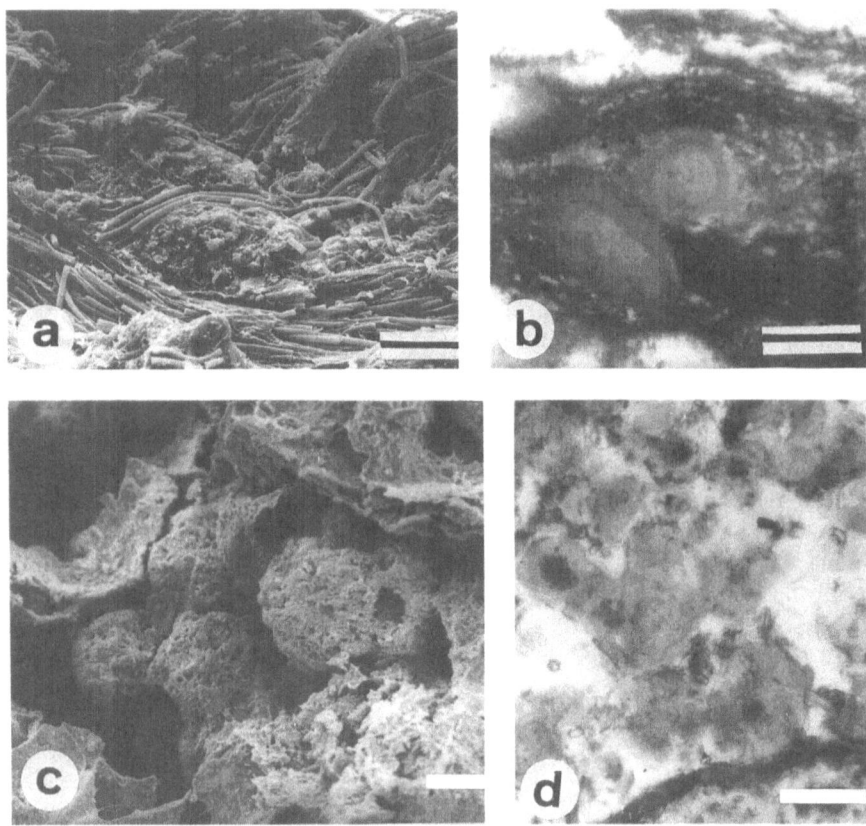

Figure 8: Carbonate precipitation in microbial mats; a) carbonate particle growing between *Microcoleus* filaments (scale is 30 μm; Gavish Sabkha); b) thin section of laminated carbonate particles in *Microcoleus* mats from Gavish Sabkha (scale is 30 μm); c) aggregated carbonate particles between filamentous mats from Lanzarote saltworks (scale is 100 μm); d) thin section of carbonate particles from Gavish Sabkha showing concentrically oriented filaments embedded, and bacterial colonies forming the dark organic nuclei (scale is 100 μm).

which reproduce to extraordinary abundance. Examples were found on Mellum Island (North Sea): Here, burrowing marine polychaetes and amphipods considerably increase their relative abundance across the MHW-level where they meet the mats (Fig. 4a).
The fact that grazing fauna is abundant where mats are scarce, however, could mean that ecospace available for cyanobacteria and the grazing eucaryotes is juxtaposed (see also Monty's discussion of this problem, 1979).

Surface deformation structures

The gel-sticky, fibrillar substrates formed by microbes interfere with weather-or tide-controlled physical processes which affect the sedimentary surfaces. Thus, peritidal potential stromatolites show a variety of surface deformation structures, including erosion pockets with ripples (Reineck, 1979), shrinkage cracks (Shinn ,1983), dome-like upheaval and folding (Reineck et al.,1990). Comparative studies on shrinkage cracks in mineral clay and microbial mats have shown that clay produces a clear discrete lineation, whereas in microbial mats, crack margins are altered into obscure patterns (Fig. 9a and b).

In sabkhas and evaporation swamps, upfolding of surfaces due to crystallization pressure is common (Kendall and Warren, 1987). Reineck et al. (1990) differentiated between folds and buckles in evaporite crusts laking microbial mats (tepees; Adams and Franzel, 1950) and those colonized by microbial mats. Mat growth on evaporite crusts takes place where moisture is sufficiently supplied to the surface so that populations of mat- and gel-forming microbes can thrive. Usually, this takes place in lower sabkha subzones. Reineck et al. (1990) and Gerdes et al. (in press) also described buckles and folds derived from subsurficial gas pressure. In this case, the surface arrangement of buckles and folds shows a somewhat chaotic arrangement (Fig. 9c) which is easy to differentiate from orderly polygonal patterns shaped by processes of lateral expansion and upfolding (Fig. 9d). Multiple overgrowth of gas domes after re-wetting as indicated in Fig. 10 lead to cabbage head structures, displaying a mode of bioherm formation under low energy conditions.

Peritidal organic matter production

Since their evolution in Phanerozoic times, aquatic vascular plants share growing space with microbial mats. Marsh grasses are able to compete successfully with cyanobacterial mats (Howarth and Marino, 1984). The authors assumed that prior to the evolution of aquatic vascular plants, microbial mats may have been much more widespread between intertidal levels. The Andros Island freshwater marsh allows for a local coexistence of microbial mats and marsh plants. The coexistence is made possible by a seasonal change between dryness and rainfall, and between different tidal ranges. Dryness reduces the abundance of sawgrass; a low tidal range reduces the abundance of aquatic plants adapted to salt water (mangroves form only stunt stands in the marsh). The microbial mats, on the other hand, remain relatively unharmed by dryness and thrive in freshwater and saltwater. During dryness as well as during low tidal range, they grow to luxuriant thickness because space competition with marsh plants is reduced.

Sabkhas are counterparts of marshes in hot arid regions. Here, the plant life is severely restricted, and roots are scarce. Nevertheless, coastal sabkhas in arid and semiarid climate zones are recent models for hydrocarbon accumulation in evaporite salts, due to the intimate association between extensive evaporite precipitation and microbial mats (Warren, 1986, Busson, 1988, Javor, 1989).

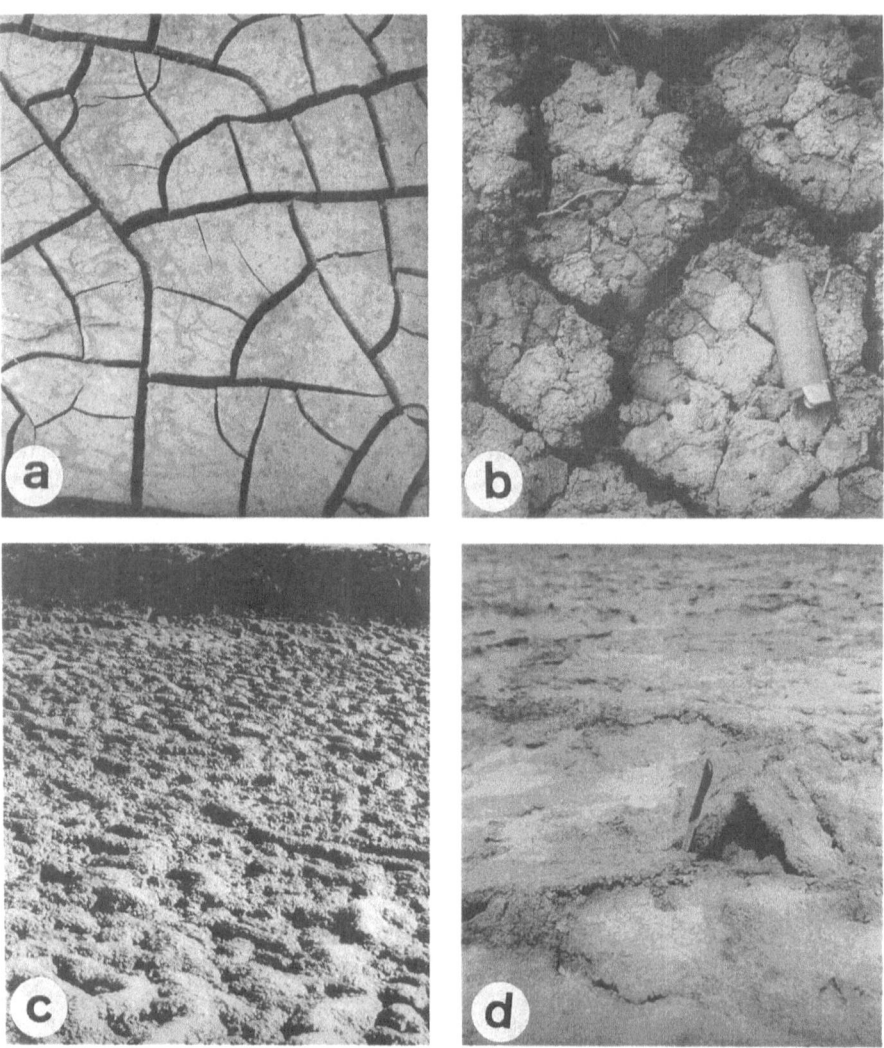

Figure 9: Surface deformation structures; a) desiccation cracks in mineral clay; b) desiccation cracks in microbial mats; c) doming of microbial mats due to gas pressure and subsequent gypsum encrustation; note coexistence of domes and folds (arrow); d): tepee structures and polygons.

Peritidal metal accumulation

It is evident that climate conditions are important peritidal factors. In particular the climate extremes of humidity and aridity are easily discernible, as they control vegetation, weathering and run-off, salinity and metal mineralisation. A proposed model for metal mineralisation involves (i) biomass accumulation by microbial mats to considerable thickness, (ii) burial of organic matter, (iii) development of bacterial sulfate reduction, and (iv) fixation of metals as a constituent of sulfide from metal-enriched groundwater which passes through the buried organic matter (Trudinger and Williams, 1982).
Salt swamp situations with prolific mats (the example of Gavish Sabkha) have been common in Earth`s history (Dunlop et al., 1978; Dean and Anderson, 1978; Sarg, 1981). Characteristic examples are the Messinian evaporite depositions of the Mediterranean and their association with stromatolites (Decima et al., 1988, Rouchy, 1988, Vai and Ricci Lucchi, 1977).

Figure 10: Doming of a surface mat took place due to the accumulation of subsurficial gas; subsequent mat-by-mat overgrowth triggered by a repeated change between desiccation and wetting, typically producing cabbage head-like structure (displaying microbial bioherm development under low energy conditions (Lanzarote saltworks).

Seen against the background of Phanerozoic seas, the present oceans and their margins may be unusual in many respects, being well oxygenated, poor in shelf seas and peritidal shallows. An anchialine setting like the Gavish Sabkha lagoon may represent a recent model for formerly probably more extended peritidal areas, in which procaryotes produced biogenic matter to considerable thickness, sulfide preservation was possible, and gypsum and polyhalite accumulation started with decreased bioactivity, indicated by changes from biogenic layers to evaporites. In such settings, laminated dolomite in association with crinkled microbial mats may indicate lower supratidal sediments attended with magnesium-rich waters by active tidal or evaporative pumping (Purser, 1985; Carballo et al., 1987; Gasiewicz et al.,1987).

SUMMARY

This chapter describes patterns of distribution, diversity and form of potential stromatolites in the peritidal zone. Usually, a bathymetric scale divides the peritidal into subtidal, intertidal and supratidal subzones. This linear approach alone, however, could hardly justify the great variability of peritidal potential stromatolites. Thus, the approach of the present chapter is to emphasize the chance combination of weather- and tide-dominated processes which is characteristic of the transitional zone between land and sea. The degree of intensity and persistence of the one or other set of processes decisively determines biotopic conditions and biofacies characteristics.

The view of coastal variation reveals an environmental differentiation between macro- and mesotidal flats, microtidal lagoons and anchialine lagoons (with only a subterranean connection to the open sea). In the given order of environments, effects of tides decrease, and the importance of weather signature increases.

Organic, often peat-like, smooth horizons between layers of well-rounded and sorted detrital clastics may indicate lower supratidal conditions of temperate macrotidal to mesotidal coasts. Irregular biolaminated units with rare clastics but nests of carbonate particles (see Fig. 8) may characterize mats in hypersaline water of topographic lows where water retention is prolonged and kinetic energy is weak (e. g. intertidal ponded water of lagoons).

The distribution of microbial mats between intertidal levels is particularly abundant in low latitudes where protected microtidal lagoons occur more frequently and provide varied ecospace (persisting thin water sheets, permanently water-filled ponds, flats subject to daily flooding and those of seasonal exchange between flooding and exposure). The diversity of microhabitats corresponds with the increasing sedimentological diversity of mats.

Peritidal processes controlling the form of microbial mats are: (i) sedimentation (smooth mats, lack of tufted forms); (ii) shallow ponded water (tufted or pinnacle-like mats); (iii) gas and gypsum formation in surface mats (blistering); (iv) species dominance (pustular mats by dominance of *Entophysalis* sp. and Pleurocapsaleans); smooth mats by dominance of *Microcoleus chthonoplastes*; (v) increased water depth (growth of larger bioherms); (vi) rapid syngenetic cementation (allows for growth of microbial bioherms in agitated water). Vertical accretion proceeds with and without sedimentation. Biolaminated growth without sedimentation indicates long-term protected conditions. Sedimentation usually accelerates the vertical growth of microbial buildups. Peritidal biostromate types produce a variety of biogenic particles and open space structures and are associated with various peritidal lithotope characteristics (plant litter, roots, shrinkage cracks etc.).

On subaerially exposed peritidal flats, microbial buildups are common in the form of biostromes. Rare, if ever, do stromatolitic bioherms occur which need deeper water for their development, at least periodically as in the lower intertidal. Here, synsedimentary cementation seems to be necessary due to current-and wave-related erosion.

REFERENCES

Adams, J.E. and Frenzel, H.N. (1950) "Capitan barrier reef, Texas and New Mexico", J. Geol. 58, 289-312.

Awramik, S.M. and Riding, R. (1988) "Role of algal eukaryotes in subtidal columnar stromatolite formation", Proceedings, Nat. Acad. Sci., USA 85, 1327-29.

Bates, R.L. and Jackson, J.A. (ed.) (1987) "Glossary of geology Am. Geol. Inst., Falls Church Virg, 751 pp.

Bauld, J. (1984) "Microbial mat in marginal marine environments; Shark Bay, Western Australia and Spencer Gulf, South Australia", in Y. Cohen, R.W. Castenholz and H.O. Halvorson (eds.) "Microbial Mats: Stromatolites", Alan Liss Publ, New York, pp 39-58.

Bernier, P., Gaillard, C., Gall, J.C., Barale, G., Bourseau, J.P., Buffetaut, E. and Wenz, S. (1991) "Morphogenetic impact of microbial mats on surface structures of Kimmeridgian micritic limestones (Cerin, France)" Sedimentology 38, 127-136.

Brock, T.D. (1976) "Biological techniques for the study of microbial mats and living stromatolites", in M.R. Walter (ed.), Developments in sedimentology 20, "Stromatolites", Elsevier, Amsterdam, pp 21-30.

Burne, R.V. and Colwell, J.B. (1982) "Temperate carbonate sediments of Northern Spencer Gulf, South Australia: A high salinity "foramol" province", Sedimentology 29, 223-238.

Busson, G. (1988) "Evaporites et Hydrocarbures", Sciences de la Terre 55, 1-139.

Cameron, B., Cameron, D. and Jones, J.R. (1985) "Modern algal mats in intertidal and supratidal quartz sands, Northeastern Massachusetts, USA", in H.A. Curran (ed.), SEPM Spec Publ No 35, "Biogenic structures: Their use in interpreting depositional environments", SEPM, Tulsa Oklahoma, pp. 211-223.

Carballo, J.D., Land, L.S. and Miser, D.E. (1987) "Holocene dolomitization of supratidal sediments by active tidal pumping, Sugarloaf Key, Florida", J. Sed. Petrol. 57, 153-165.

Cohen, Y. (1984) "The Solar Lake cyanobacterial mats: strategies of photosynthetic life under sulfide", in Y. Cohen, R.W. Castenholz and H.O. Halvorson (eds.) "Microbial Mats: Stromatolites", Alan Liss Publ, New York, pp 133-148.

Cornee, A, Dickman, M. and Busson G. (1992) "Laminated cyanobacterial mats in sediments of solar salt works: some sedimentological implications", Sedimentology 39, 599-612.

Dade, W.B., Davis, J.D., Nichols, P.D., Nowell, A.R.M., Thistle, D., Trexler, M. & White, D.C. (1990) "Effects of bacterial exopolymer adhesion on the entrainment of sand", Geomicrobiol. J. 8, 1-16.

Dahanayake, K., Gerdes, G., Krumbein, W.E. (1985) "Stromatolites, oncolites and oolites biogenically formed in situ", Naturwissenschaften 72, 513-518.

Davies, J.L. (1980) "Geographic variation of coastal development", in K.M. Clayton (ed.), Geomorphology text 4, Longman Group Ltd, London.

Dean, W.E. and Anderson, R.Y. (1978) "Salinity cycles: evidence for subaqueous deposition of Castile Formation and lower part of Salado Formation, Delaware Basin, Texas and New Mexico", N. Mex. Bur. Mines Mineral. Resoure Circ. 159, 15-20.

Decima, A., McKenzie, J.A., Schreiber, B.C. (1988) "The origin of "evaporative" limestones: An example from the Messinian of Sicily (Italy)", J. Sed. Pet. 58, 256-272.

Dill, R.F., Shinn, E.A., Jones, A.T., Kelly, K. and Steinen, R.P. (1986) "Giant subtidal stromatolites forming in normal salinity waters", Nature 324, 55-58.

Dunlop, J.S.R., Muir, M.D., Milne, V.A. and Groves, D.I. (1978) "A new microfossil assemblage from the Archaean of western Australia", Nature 274, 676-678.

Eggleston, J.R. and Dean, W.E. (1976) "Freshwater stromatolitic bioherms in Green Lake, New York", in M.R. Walter (ed.) Developments in sedimentology 20, "Stromatolites", Elsevier, Amsterdam, pp. 479-488.

Ehrlich, A. and Dor, I. (1985) "Photosynthetic microorganisms of the Gavish Sabkha", in G.M. Friedman and W.E. Krumbein (eds.) "Hypersaline Ecosystems The Gavish Sabkha", Ecological studies 53, Springer, Berlin, pp. 296-321.

Evans, G. (1965) "Intertidal flat sediments and their environments of deposition in the Wash", Q. J. Geol. Soc. London 121, 209-245.

Evans, G. (1970) "Coastal and nearshore sedimentation: A comparison of clastic and carbonate deposition", Proc. Geol. Assoc. 81, 493-508.

Flügel, E. (1982) "Microfacies analysis of limestones", Springer, Berlin.

Folk R.L. (1973) "Evidence for peritidal deposition of Devonian Caballos Novaculite, Marathon Basin, Texas", AAPG Bull. 57, 702-725.

Folk, R.L. and Siedlecka, A. (1974) "The `schizohaline` environment: its sedimentary and diagenetic fabrics as exemplified by Late Paleozoic rocks of Bear Island, Svalbard", Sed. Geol. 11, 1-15.

Frey, R.W. and Basan, P.B. (1978) "Coastal salt marshes", in R.A. Davies (ed.) "Coastal sedimentary environments", Springer, New York, pp. 101-169.

Frey, R.W., Howard, J.D. (1986) "Mesotidal estuarine sequences: A perspective from the Georgia Bight", J. Sed. Petrol. 56, 911-924.

Friedman, G.M. (1978) "Solar Lake: A sea-marginal pond of the Red Sea (Gulf of Aqaba or Elat) in which algal mats generate carbonate particles and laminites", in W.E. Krumbein (ed.), "Environmental Biogeochemistry and Geomicrobiology. The Aquatic Environment" 1, Ann Arbor Sci. Publ. Inc., Michigan, pp. 227-235.

Friedman, G.M. and Sanders, J.E. (1978) "Principles of sedimentology", J. Wiley Sons, New York.

Gasiewicz, A., Gerdes, G., Krumbein, W.E. (1987) "The peritidal sabkha type stromatolites of the Platy Dolomite (CaCO$_3$) of the Leba elevation (Northern Poland)", in T.M. Peryt (ed.) "The Zechstein facies in Europe", Lecture Notes in Earth Science 10, Springer, Berlin, 253-272.

Gavish, E., Krumbein, W.E. and Halevy, J. (1985) "Geomorphology, mineralogy and groundwater geochemistry as factors of the hydrodynamic system of the Gavish Sabkha", in G.M. Friedman and W.E. Krumbein (eds.) "Hypersaline Ecosystems The Gavish Sabkha", Ecological studies 53, Springer, Berlin, pp.. 186-217.

Gerdes, G. and Krumbein, W.E. (1987) "Biolaminated deposits", in S. Bhattacharji, G.M. Friedman, H.J. Neugebauer and A. Seilacher (eds.), Lecture Notes in Earth Sciences 9, Springer, Berlin.

Gerdes, G., Krumbein, W.E. and Reineck, H.E. (1985a) "The depositional record of sandy, versicolored tidal flats (Mellum Island, southern North Sea)", J. Sedim. Petrol. 55, 265-278.

Gerdes, G., Krumbein, W.E. and Holtkamp E.M. (1985b) "Salinity and water activity related zonation of microbial communities and potential stromatolites of the Gavish Sabkha", in G.M. Friedman and W.E. Krumbein (eds.) "Hypersaline ecosystems The Gavish Sabkha", Ecological studies 53, Springer, Berlin, pp. 238-266.

Gerdes, G., Spira, Y. and Dimentman, C. (1985c) "The fauna of the Gavish Sabkha and the Solar Lake - a comparative study", in G.M. Friedman and W.E. Krumbein (eds.) "Hypersaline ecosystems The Gavish Sabkha", Ecological studies 53, Springer, Berlin, pp. 322-345.

Gerdes, G., Krumbein, W.E. and Reineck, H.E. (1991) "Biolaminations -ecological versus depositional dynamics" in G. Einsele, W. Ricken, and A. Seilacher (eds.) "Cycles and events in stratigraphy", Berlin (Springer), 593-607.

Gerdes, G., Claes, M., Dunajtschik, K., Riege, H., Krumbein, W.E. and Reineck, H.E. (1993) "Contribution of microbial mats to sedimentary surface structures", Facies, in press.

Giani, D., Seeler, J., Giani, L. and Krumbein, W.E. (1989) "Microbial mats and physicochemistry in a saltern in the Bretagne (France) and in a laboratory scale saltern model", FEMS Microbiology Ecol. 62, 151-162.

Golubic, S. (1973) "The relationship between blue-green algae and carbonate deposits", in N.Carr and B.A. Whitton (eds.) "The biology of blue-green algae", Blackwell, Oxford, pp. 434-472.

Golubic, S. (1976) "Taxonomy of extant stromatolite-building cyanophytes", in Y. Cohen, R.W. Castenholz and H.O. Halvorson (eds.) "Microbial mats: Stromatolites", MBL Lectures in Biol., Alan Liss, New York, pp.. 245-263.

Golubic, S. (1980) "Halophily and halotolerance in cyanophytes", Origin of Life 10, pp.. 169-183.

Hagan, G.M., Logan, B.W. (1975) "Prograding tidal-flat sequences: Hutchinson Embayment, Shark Bay, Western Australia", in R.N. Ginsburg (ed.) "Tidal deposits", Springer, Berlin, pp. 215-222.

Hayes, M.O. (1979) "Barrier island morphology as a function of tidal and wave regime", in S.P. Leatherman (ed.) "Barrier Islands", Academic Press, New York, pp. 1-27.

Heinzelmann, C. and Wallisch, S. (1991) "Benthic settlement and bed erosion. A review", J. Hydraulic Res. 29, 355-373.

Hoffman, P. (1976) "Stromatolite morphogenesis in Shark Bay, Western Australia", in M.R. Walter (ed.) "Stromatolites", Developments in sedimentology 20, Elsevier, Amsterdam. pp. 262-271.

Hoffmann, C. (1942) "Beiträge zur Vegetation des Farbstreifen-Sandwattes", Kieler Meeresforsch. 4, 85-108.

Horodyski, R.J. (1977) "*Lyngbya* mats at Laguna Mormona, Baja California, Mexico: Comparison with Proterozoic stromatolites", J. Sed. Petrol. 47, 1305-1320.

Horodyski, R.J. and VonderHaar, S.P. (1975) "Recent calcareous stromatolites from Laguna Mormona (Baja California) Mexico", J. Sed. Petrol. 45, 894-906.

Howarth, R.W. and Marino, R. (1984) "Sulfate reduction in salt marshes, with some comparisons to sulfate reduction in microbial mats", in Y. Cohen, R. W. Castenholz and H.O. Halvorson (eds.) "Microbial mats: Stromatolites". MBL Lectures in Biol, Liss New York, pp. 245-263.

Javor, B. (1989) "Hypersaline environments", Springer, Berlin.

Kalkowsky, E. (1908) "Oolith und Stromatolith im norddeutschen Buntsandstein", Z. Deutsch. Geol. Gesellsch. 60, 68-125.

Kendall, C.G. and Skipwith, P.A. (1969) "Holocene shallow water carbonate and evaporite sediments of Khor als Bazam, Abu Dhabi, Southwest Persian Gulf", Bull. Amer. Ass. Petrol. Geol. 53, 841-869.

Kendall, C.G.St. and Warren, J.K. (1987) "A review of the origin and setting of tepees and their associated fabrics", Sedimentology 34, 1007-1027.

Krumbein, W.E. (1978) "Algal mats and their lithification", in W.E. Krumbein (ed.) "Environmental biogeochemistry and geomicrobiology" 1, Ann Arbor Sci. Publ. Inc., Michigan, pp. 209-225.

Krumbein, W.E. (1983) "Stromatolites - The challenge of a term in space and time", Precambrian Research 20, 493-531.

Krumbein, W.E. (1986) "Biotransfer of minerals by microbes and microbial mats", in B.S.C. Leadbeater and R. Riding (eds.) "Biomineralization in lower plants and animals". Oxford Univ. Press, Oxford, pp. 55-72.

Krumbein, W.E., Cohen Y. and Shilo, M. (1977) "Solar Lake (Sinai) 4. Stromatolitic cyanobacterial mats", Limnology and Oceanography 22, 635-656.

Krumbein, W.E., Buchholz, H., Franke, P., Giani, D., Giele, C. and Wonneberger, K. (1979) "O_2 and H_2S coexistence in stromatolites. A model for the origin of mineralogical lamination in stromatolites and banded iron formations", Naturwissenschaften 66, 381-389.

Le Gall J. and Larsonneur (1972) "Sequences et environments sedimentaires dans la Baie des Veys (Manche)", Rev. Geogr. Geol. Dyn. 14, 189-204.

Logan, B.W., Rezak, R. and Ginsburg, R.N. (1964) "Classification and environmental significance of algal stromatolites", J. Geol. 72, 68-83.

Monty, Cl. (1965) "Recent algal stromatolites in the windward lagoon, Andros Island, Bahamas", Ann. Soc. Geol. Belg. 88, 269-276.

Monty Cl. (1967) "Distribution and structure of Recent stromatolitic algal mats eastern Andros Island, Bahamas", Ann. Soc. Geol. Belg. 90, 55-100.

Monty, Cl. (1972) "Recent algal stromatolitic deposits, Andros Island, Bahamas, Preliminary report", Geol. Rundschau 61, 742-783.

Monty, Cl. (1976) "The origin and development of cryptalgal fabrics", in M.R. Walter (ed.) "Stromatolites", Developments in sedimentology 20, Elsevier, Amsterdam, pp. 193-249.

Monty, Cl. (1979) "Scientific reports of the Belgian expedition on the Australian Great Barrier Reefs, 1967, Sedimentology 2, monospecific stromatolites from the Great Barrier Reef tract and their paleontological significance", Ann. Soc. Geol. Belgique 101, 163-171.

Monty, Cl. and Hardie, L.A. (1976) "The geological significance of the freshwater blue-green algal calcareous marsh", in M.R. Walter (ed.) "Stromatolites", Developments in sedimentology 20, Elsevier, Amsterdam, pp. 447-477.

Monty, Cl. and Mas, J.R. (1981) "Lower Cretaceous (Wealdiaan) blue green algal deposits of the province of Valencia, Eastern Spain", in Cl. Monty (ed.) "Phanerozoic stromatolites", Springer, Berlin, pp. 85-120.

Playford, P.E. and Cockbain, A.E. (1976) "Modern algal stromatolites at Hamelin Pool, a hypersaline barred basin in Shark Bay, Western Australia", in M.R. Walter (ed.) "Stromatolites", Developments in sedimentology 20, Elsevier, Amsterdam, pp. 389-411.

Por, F.D. (1985) "Anchialine pools - comparative hydrobiology", in G.M. Friedman and W.E. Krumbein (eds.) "Hypersaline ecosystems The Gavish Sabkha", Ecological studies 53, Springer, Berlin, pp. 136-144.

Pratt, B.R. (1979) "Early cementation and lithification in intertidal cryptalgal structures, Boca Jewfish, Bonaire, Netherlands Antilles", J. Sed. Petrol. 49, 371-386.

Pratt, B.R. and James, N.P. (1986) "Carbonate islands in epeiric seas", Sedimentology, 313-343.

Purser, B.H. (1985) "Coastal evaporite systems", in G.M. Friedman and W.E. Krumbein (eds.) "Hypersaline ecosystems The Gavish Sabkha", Ecological studies 53, Springer, Berlin, pp. 72-102.

Rasmussen, K.A., Macintyre, I.G. and Prufert, L. (1993) "Modern stromatolite reefs fringing a brackish coastline, Chetumal Bay, Belize", Geology 21, 199-202.

Reid, R.P. and Browne, K.M. (1991) "Intertidal stromatolites in a fringing Holocene reef complex, Bahamas", Geology 19, 15-18.

Reineck, H.E. (1979) "Rezente und fossile Algenmatten und Wurzelhorizonte", Natur u. Museum 109, 290-296.

Reineck, H.E., Gerdes, G., Claes, M., Dunajtschik, K., Riege, H. and Krumbein, W.E. (1990) "Microbial modification of sedimentary surface structures", in D. Heling, P. Rothe, U. Förstner and P. Stoffers (eds.) "Sediments and environmental geochemistry", Springer, Berlin, 254-276.

Riding, R., Awramik, S.M., Winsborough, B.M., Griffin, K.M. And Dill, R.F. (1991a) "Bahamian giant stromatolites: microbial composition of surface mats", Geol. Mag. 128, 227-234.

Riding, R., Braga, J.C. and Martin, J.M. (1991b) "Oolite stromatolites and thrombolites, Miocene, Spain: analogues of Recent giant Bahamian examples", Sedimentary Geol. 71, 121-127.

Rouchy, J.M. (1988) "Relations évaporites-hydrocarbures: l'association laminites-récifs-évaporites dans le Messinien de Méditerranée et ses enseignements". in G. Busson (ed.) "Evaporites et Hydrocarbures", Sciences de la Terre 55, pp. 43-70.

Sarg, J.F. (1981) "Petrology of the carbonate evaporite facies transition of the Seven Rivers Formation (Guadalupian, Permian) southeast New Mexico", J. Sed. Pet. 51, 73-75.

Schwarz, H.U., Einsele, G. and Herm, D. (1975) "Quartz-sandy, grazing-contoured stromatolites from coastal embayments of Mauritania, West Africa", Sedimentology 22, 539-561.

Shinn, E.A. (1972) "Worm and algal-built columnar stromatolites in the Persian Gulf", J. Sed. Petrol. 42, 837-840.

Shinn, E.A. (1983) "Tidal flat environment", AAPG Mem 33, 172-210.

Shinn, E.A. (1986) "Modern carbonate tidal flats: Their diagnostic features", in L.A. Hardie and E.A. Shinn (eds.) "Carbonate depositional environments, modern and ancient", Colorado School of Mines Quarterly 81, 7-35.

Shinn, E.A. (1987) "Sand castles from the past: Bahamian stromatolites discovered", Sea Frontiers 33, 334-343.

Stal, L.J., Gemerden, H. and Krumbein, W.E. (1985) "Structure and development of a benthic marine microbial mat", FEMS Microbiol. Ecol. 31, 111-125.

Stolz, J.F. (1984) "Fine structure of the stratified microbial community at Laguna Figueroa, Baja California, Mexico: II transmission electron microscopy as a diagnostic tool in studying microbial communities in situ", in Y. Cohen R.W. Castenholz and H.O. Halvorson (eds.) "Microbial mats: Stromatolites", Alan Liss Publ, New York, pp. 23-38.

Stolz, J.F. (1990) "Distribution of phototrophic microbes in the flat-laminated microbial mat at Laguna Figueroa, Baja California, Mexico", Biosystems 23, 345-357.

Tebutt, G.E., Conley, C.D. and Boyd, D.W. (1965) "Lithogenesis of a distinctive carbonate rock fabric", Wyoming Univ. Contr. Geol. 4, 1-13

Tribovillard, N.P., Gorin, G.E., Belin, S., Hopfgartner, G. and Pichon, R. (1992) "Organic-rich biolaminated facies from a Kimmeridgian lagoonal environment in the French Southern Jura mountains - A way of estimating accumulation rate variations", Palaeogeography, Palaeoclimatology, Palaeoecology 99, 163-177.

Trudinger, P.A. and Williams, N. (1982) "Stratified sulfide deposition in modern and ancient environments", in Holland H.D. and M. Schidlowski (ed.) "Mineral deposits and the evolution of the biosphere", Dahlem Konferenzen, Springer Berlin, 177-198.

Vai, G.B. and Ricci Lucchi, F.R. (1977) "Algal crusts, autochthonous and clastic gypsum in a cannibalistic evaporite basin: a case history from the Messinian of Northern Apennines", Sedimentology 24, 211-244.

Warren, J.K. (1986) "Shallow-water evaporitic environments and their source rock potential", J. Sed. Pet. 56, 442-454.

Weiss, M.P. (1969) "Oncolites, paleoecology, and Laramide Tectonics, Central Utah", AAPG Bull 53, 1105-1120.

STROMATOLITE AND SERPULID BIOHERMS IN A HOLOCENE RESTRICTED LAGOON (SABKHA EL MELAH, SOUTHEASTERN TUNISIA)

E. DAVAUD[1], A. STRASSER[2] and Y. JEDOUI[3]

1 Department of Geology, 13 rue des Maraichers, 1211 Geneva, Switzerland
2 Department of Geology, 1700 Pérolles, Fribourg, Switzerland
3 E.N.I.S., route de Soukra, 3038 Sfax, Tunisia

ABSTRACT

The Sebkha el Melah, located near the Libyan-Tunisian border, is fringed by a relict stromatolite-serpulid belt. These bioherms developed 5500 years ago during the Flandrian sea-level highstand when the area of the present-day sebkha was covered by a wide restricted lagoon. Relict lagoonal beaches and spits, mainly composed of cerebroid ooids and cerithid gastropods, are flanked by mushroom-shaped stromatolites reaching up to 50 cm in height. These stromatolites grew on top of small serpulid bioherms as well as on top of beachrock and beachrock blocks in a shallow subtidal pre-evaporitic realm.

The primary porosity of these bioherms is obliterated by botryoidal and spherulitic aragonite, probably of microbial origin. This sub-Recent association of stromatolites, serpulid bioherms and cerebroid ooids provides a useful additional model for the interpretation of upper Paleozoic deposits, where similar facies have been described.

INTRODUCTION

The Sabkha El Melah is a large evaporitic basin located on the southeastern coast of Tunisia, about 10 kilometers south of the city of Zarzis (Fig.1). The sabkha formed during the last 5000 years in a depression cut in late Tertiary (Pontian) and early Quaternary (Villafranchian) continental deposits. Along the coast it is bordered by Tyrrhenian marine deposits (well cemented coastal dunes and beach ridges).

During the Flandrian sealevel highstand (+ 2 m), about 5500 years ago, the area of the present-day sabkha was occupied by a restricted lagoon connecting with the Mediterranean sea through a narrow channel (Fig.2a). The facies deposited along the borders of this lagoon are still observable all around the sabkha. They consist

J. Bertrand-Sarfati and C. Monty (eds.), Phanerozoic Stromatolites II, 131–151.

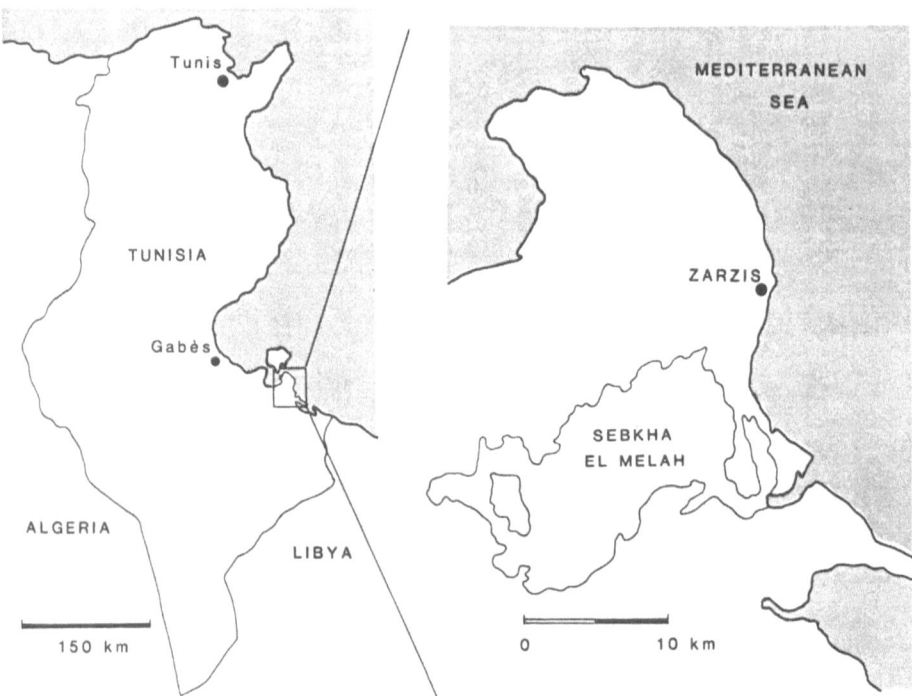

Figure. 1 : Locality map

mainly of oolitic/bioclastic sands and of a well developed stromatolite/serpulid belt.
The present paper describes this sub-Recent co-occurrence of stromatolites and
serpulids. Special attention is given to the synsedimentary cements which, in most
outcrops, are beautifully preserved .

GEOLOGICAL SETTING

The Sabkha el Melah has been studied by Perthuisot and Floridia (1973) and Perthuisot
(1975). The presence of stromatolite buildups was mentioned for the first time by
Perthuisot (1975). The pre-evaporitic lagoonal sediments deposited during the Flandrian
sealevel highstand reach up to 5 meters in thickness and have been observed in
numerous cores below thick and complex evaporitic deposits (Perthuisot, 1975). The
distribution of lithologies of the lagoonal sediments was controlled mainly by
bathymetry and by local inputs of terrigenous material (alluvial fans and fan deltas ;
Fig.2a,b). The deepest parts of the lagoon were characterized by organic-rich clays with
hypermagnesian carbonates. This suggests that only the upper levels of the water were
renewed by wind and tidal currents. Perthuisot et al. (1972) observed that the carbonate
muds surrounding the euxinic sediments contain various taxa of macro- and microfauna
very similar to those of the present-day coast.

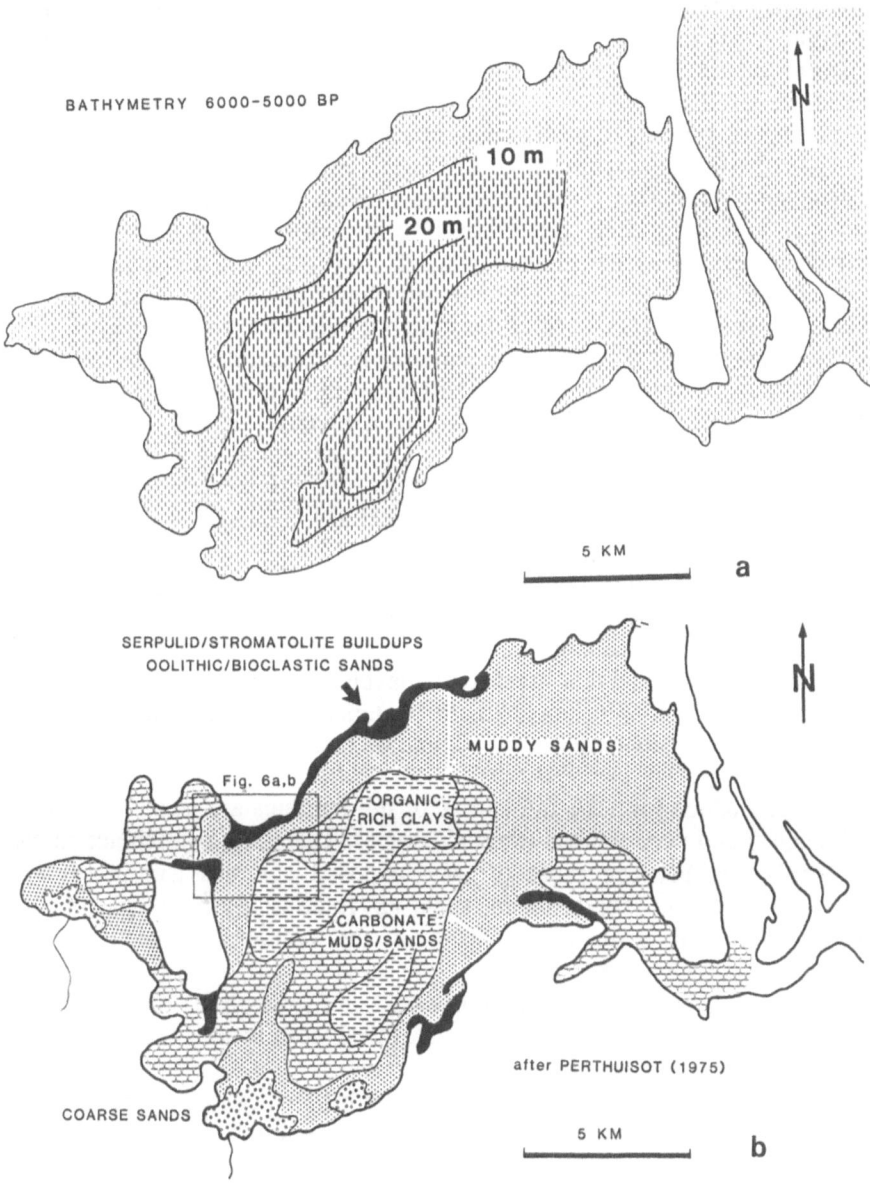

Figure 2 : a) Paleobathymetry during the Flandrian sealevel highstand (after Perthuisot, 1975); b) Schematic distribution of lagoonal sediments deposited during the Flandrian sealevel highstand (after Perthuisot, 1975). The serpulid/stromatolite belt has been marked only where observed in outcrop. However, the buildups bordered probably the entire Flandrian lagoon.

The Flandrian coastal sediments occur all around the sabkha at an elevation corresponding to the last sealevel highstand (+ 2 m, Perthuisot et al. , 1972). Geological investigations are facilitated by numerous trenches dug by local workers looking for loose sand. The sediment is composed mainly of oolitic/bioclastic sands in which cerithid gastropods prevail. It locally contains abundant cerebroid ooids (Davaud et al., 1988, 1990). Sediment accumulated on beach ridges often shows evidence of synsedimentary cementation (beachrock). Most inlets (Fig.2b,6a) have been isolated from the open lagoon by sand spits prograding from northeast to southwest under the action of waves driven by the prevailing wind. The sediment deposited behind these spits is made up of black laminated mud and abundant monospecific cerithid shells (*Potamides conicus*). This is characteristic for a pre-evaporitic environment.

The beach and spit ridges are commonly bordered by stromatolite and serpulid buildups which formed a more or less continuous belt around the Flandrian lagoon. The width of the belt varies from a few tenths of meters to a hundred meters, depending on the subtidal topography.

SERPULID AND STROMATOLITE BIOHERMS

The outer margin of the bioherm belt can be observed in many places where erosion exposed it at the surface of the present-day sabkha (Fig.3a,4a). There, dense serpulid buildups reach up to forty centimeters in height and form elongated ridges parallel to the ancient coastline. Behind this "reef front" developed scattered ellipsoidal colonies. Their diameters vary from 2 meters to 20 centimeters and tend to decrease landward. In both cases the serpulid bioherms were anchored on stable substrates made of beachrock blocks or coarse gravel.

The worm tubes which form the framework of the buildups are small-sized (about 0.3 mm in diameter) and very sinuous (*Spirorbis* type). There is no difference in shape or spatial organisation between tubes of wave-exposed colonies and those of wave-protected ones. Benthic foraminifera such as rotalids (*Ammonia*) and miliolids (*Quinqueloculina*) are frequently trapped in the serpulid framework, and thin stromatolitic layers may occur.

The bioherms are cemented mainly by botryoidal and spherulitic aragonite (Fig. 10a,b and 12d) filling most of the inter- and intraskeletal pore spaces. The origin of these cements will be discussed later.

Close to the Flandrian shoreline, the serpulid bioherms are covered by a thin and porous stromatolitic crust. In some places, the water was deep enough to allow development of domal stromatolites on top of serpulid colonies, coating them completely (Fig.4c). These stromatolites reach up to 30 centimeters in height and half a meter in diameter. They frequently display loaf-like shapes, with the long axis arranged perpendicular to the Flandrian wave front.

When water depth was reduced, the stromatolites tended to develop a pseudo-columnar morphology and grew horizontally. This led to mushroom shaped colonies (Fig. 3b and 4 b,d) and further to coalescent colonies which invaded the upper shoreface.

Figure 3 : a) General view of the stromatolite belt cropping out at the surface on the NW border of the present-day sabkha. The open lagoon was towards the right of the picture. Scattered stromatolite heads (SH) can be observed in front of the stromatolite pavement (SP, light-colored elongated zone) which developed on beachrock (B). The infilling sediment is mainly silt deposited during flash floods. b) Trenches revealing mushroom-shaped, large coalescent stromatolite heads.The basement of the stromatolites consists of serpulid bioherms strongly cemented by botryoidal aragonite. The picture was taken from the Flandrian beach, towards the centre of the lagoon.

Figure 4 : a) Elongated serpulid buildups (SE). The wave- exposed margin is higher and better cemented than the protected one. These buildups commonly are coalescent and can form a continuous ridge. The open lagoon was towards the left of the picture (white arrow). b) Trench showing mushroom-shaped stromatolite (ST) which grow on top of a serpulid buildup (SE). c) Well developed domal stromatolite reaching up to 30 centimeters in height. The infilling sediment consists of cerithid sand. The open lagoon was towards the front (white arrow). d) Trench showing pseudo-columnar stromatolite (ST) anchored on two small serpulid buildups (SE). The open lagoon was located towards the left (arrow).

The foreshore itself is very well exposed in numerous trenches on the northwestern border of the sabkha. The oolitic/bioclastic sands show a typical low-angle planar stratification dipping lagoonward (Fig.5). An early phase of cementation often led to beachrock formation. The abundance of beach-conglomerates composed of large imbricated slabs indicates that cementation was poorly developed and could not prevent undermining by waves. The beachrocks are commonly overlain by a continuous wedge-

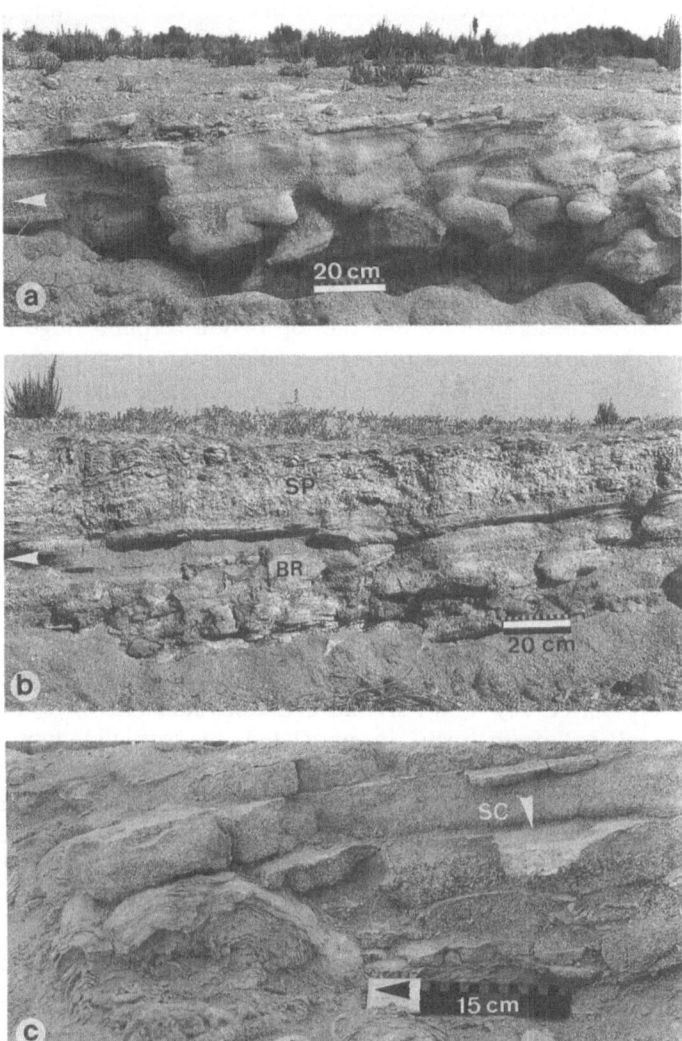

Figure 5 : Trenches showing the internal structure of the Flandrian beaches. The lagoon was towards the left. a) Low-angle seaward-dipping laminated sands. Large beachrock slabs, produced by wave undermining, have been incorporated in the lower beach facies. b) Wedge-shaped stromatolite pavement (SP) established on beachrock(BR). c) Beach facies prograding over domal stromatolites. Early marine cementation led to beachrock formation. The upper surfaces of the beachrock pavement are coated by a stromatolitic film (SC). In some places, these films evolved into dense stromatolitic crusts reaching up to 3 centimeters in thickness (Fig.10 c).

shaped stromatolite pavement made of pseudo-columnar colonies (Fig.3a and 5b). Its thickness reaches 30 centimeters on the exposed margin; its width varies from one to ten meters, depending on the beach slope.

In some areas, where sediment accumulation was important (ooid-producing areas), stromatolite pavement and domal stromatolites have been buried by the prograding beach ridge (Fig.5c). The upper surface of beachrock blocks is commonly coated by a very dense laminated crust, the thickness of which may reach up to 30 millimeters (Fig.10c). All these types of stromatolite still contain serpulids. However, these are encrusting and do generally not contribute to bioherm construction.

Figures 6 a and b show a cross section through an alluvial fan flanked by beach levees and serpulid/stromatolite bioherms observed on the northwestern border of the sabkha . Topographical measurements indicate that the top of the serpulid ridge lies 30 to 40 centimeters below the top of the stromatolite pavement.

Based on observations all around the Sabkha El Melah, the following spatial distribution becomes apparent, going from shoreface to foreshore (Fig.7) :

     1. subtidal serpulid ridge;

     2. scattered serpulid bioherms, the sizes of which decrease landward;

     3. domal, loaf-shaped stromatolites coating the top of small serpulid colonies;

     4. pseudo-columnar, mushroom-shaped to coalescent stromatolites capping
     domal stromatolites or directly serpulid bioherms;

     5. wedge-shaped stromatolite pavement growing on beachrock;

     6. stromatolite-encrusted beachrock blocks.

Few paleoecological data are available on serpulids bioherms. In Recent stromatolites of Shark Bay, Hoffman (1976) and Playford and Cockbain (1976) observed that the surface of subtidal colloform mats is populated by serpulid worms. Personal observations in Shark Bay suggest that serpulids are scattered and never form a rigid monospecific framework such as it occurs along the borders of the Sabkha El Melah.

Daley (1972) mentions that serpulids tolerate wide ranges of salinity, and Heckel (1974) suggests that "serpulids, which do live as minor encrusters on marine buildups, form welded-frame reefs in restricted marine environments that are unfavorable to most other reef builders".

This assessment is supported by the best-known modern serpulid reef occurring in a South-Texas lagoon which is generally hypersaline, but becomes nearly freshwater after seasonal rainfalls (Andrews, 1964). The climatic conditions there are similar to those which prevailed 5500 years ago in southeastern Tunisia (Perthuisot, 1974; Paskoff and Sanlaville, 1983). Small sub-Recent to Recent serpulid bioherms also occur in the shallow euryhaline lagoon of Bahiret el Biban (close to Sabkha el Melah). There, they grow seawards of mushroom-shaped colonies of the red algae *Neogoniolithon* (Thornton et al., 1978).

In the Flandrian lagoon, the evolution from a euryhaline to hypersaline coastal environment is supported by the following field observations :

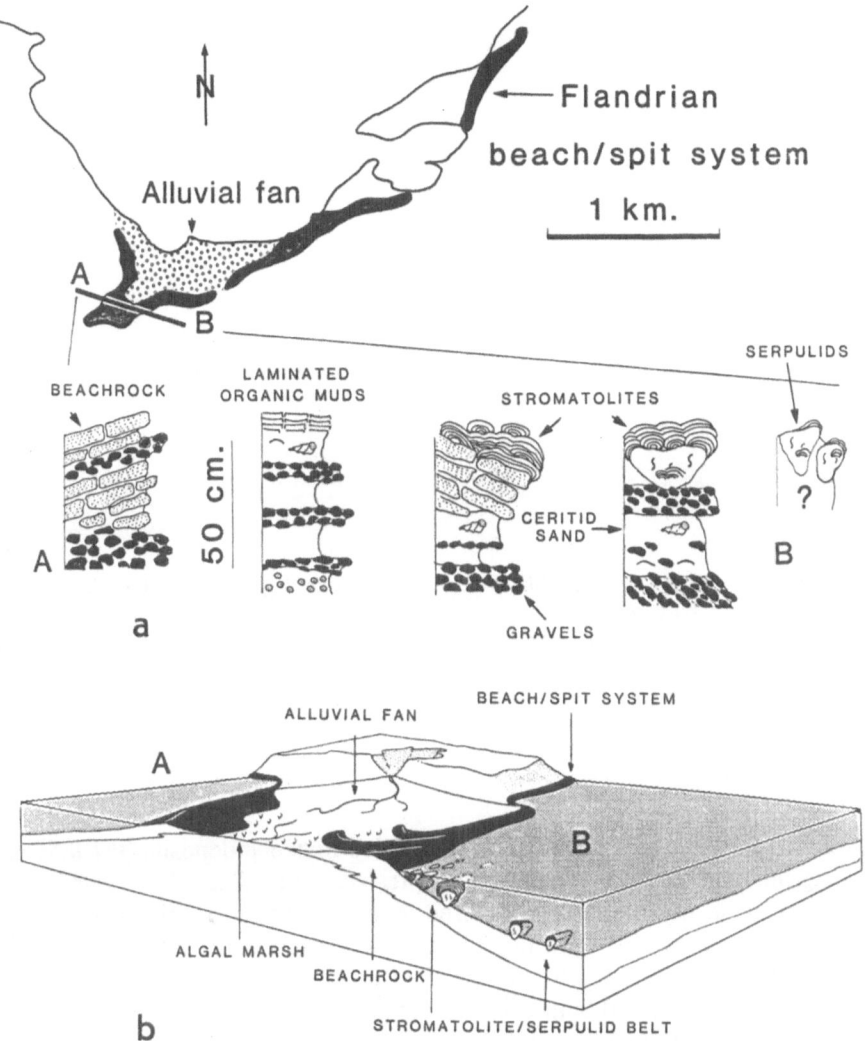

Figure 6 : a) Stratigraphic profiles observed in trenches along the northwestern border of the sabkha (location on Fig. 2 b). An alluvial fan is flanked by beach and spit levees (black areas). On the exposed margin (southeast), the shoreface is lined by a stromatolite pavement, by mushroom-shaped stromatolites capping serpulid bioherms (second section from right), and by serpulid bioherms. Coarse alluvial gravels interfinger with nearshore lagoonal deposits. Behind levees, laminated organic muds were deposited in a supratidal marsh (second section from left). b) Hypothetical landscape in the same area during the Flandrian.

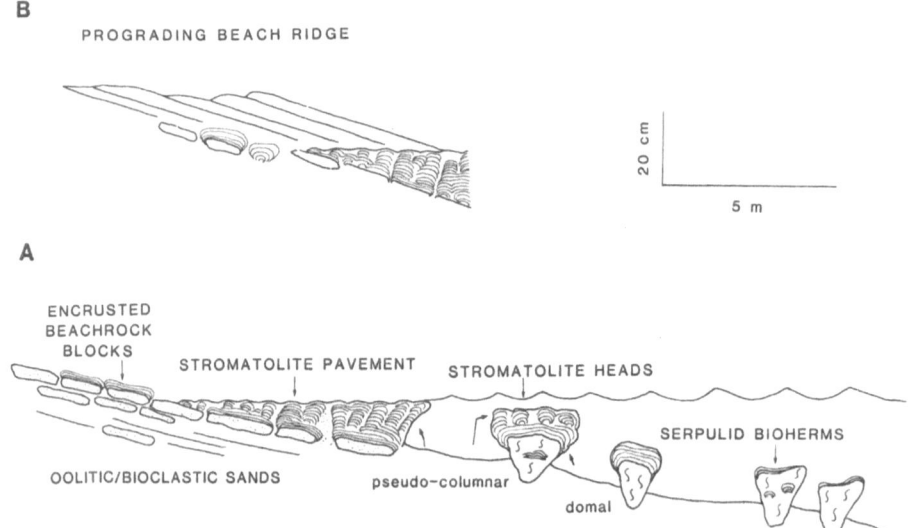

Figure 7 : a) Synthetic profile through a Flandrian lagoonal coastline. The stromatolites developed on beachrock slabs and on serpulid bioherms. Several evidences (see text) indicate that the stromatolites progressively replaced the serpulids when euryhaline conditions evolved into hypersaline ones. b) Where sedimentation was very active (ooid-producing areas), the stromatolites were buried under prograding beach ridges.

1. the presence of cerithid gastropods (*Potamides conicus*, known to be able to support drastic salinity changes; Plaziat pers. comm.) in the stromatolite/serpulid buildups, and their abundance around the buildups;

2. the presence of neolithic tools embedded in beach deposits (Perthuisot, 1975), suggesting human colonisation and thus the presence of available freshwater;

3. the rapid evolution from a closed lagoon to an evaporitic basin. ^{14}C -dating obtained by Perthuisot (1974) on hypermagnesian carbonate mud gives an age of 5330 $\pm$ 170 B.P. These muds, sampled in the most restricted area of the Flandrian lagoon, overlie stromatolite/serpulid buildups. The youngest marine shells are 5160 $\pm$ 170 years old and have been collected near the Flandrian channel.

4. the evaporitic infilling of the lagoon proceded from hypermagnesian carbonates to gypsum and halite, thus monitoring the progressive closure of the basin (Perthuisot, 1975).

The fact that large stromatolites commonly developed on serpulid frameworks (Fig.7) implies therefore that they grew on dead serpulid colonies, after euryhaline conditions had evolved into hypersaline ones (only encrusted worm tubes survived on the surface and in cavities of the stromatolites). Differential cementation between serpulid and stromatolite buildups supports this hypothesis (see next chapter). The assumed sequence of events is schematized in figure 8.

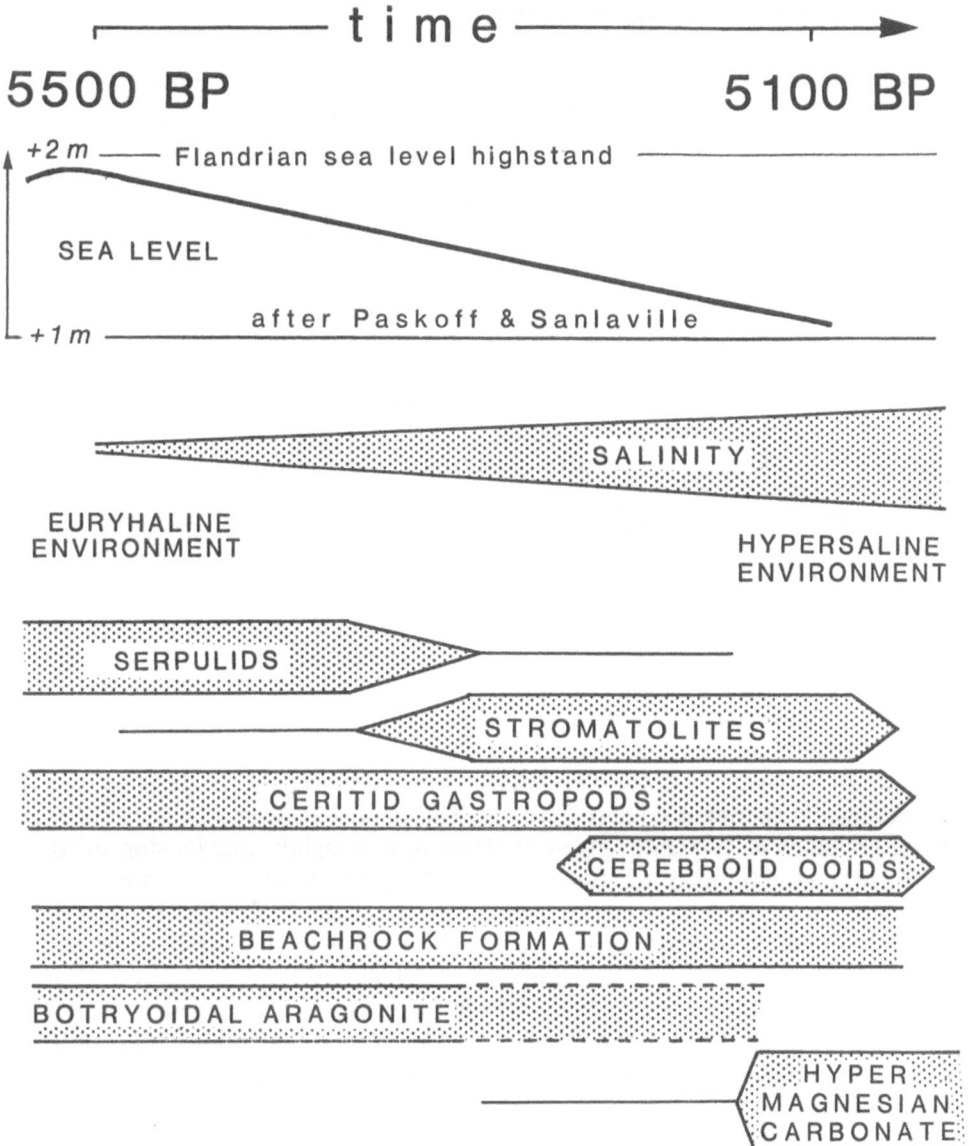

Figure 8 : Sea level and salinity changes from 5500 BP to 5100 BP in the lagoon, and their consequences on fauna, allochems and mineralogy. Sea level curve based on data from Paskoff & Sanlaville (1983) and Perthuisot (1975).

STROMATOLITE MICROSTRUCTURES

The domal stromatolites (Fig.4c) show a laminoid fenestral fabric similar to that observed in subtidal stromatolite heads in Shark Bay (illustrated in Monty, 1976). The framework is made of a micritic reticular net defining irregular fenestrae. The small-sized fenestrae are often invaded by fan-shaped aragonite, whereas large horizontal vugs remain uncemented (Fig.9b and c). The reticular net is bordered by better-defined horizontal micritic films. This pattern could result from ecotonal communities of cyanobacteria (Monty, 1976), in which each taxon imprints its own fabric. Scattered microspherulites (see below) can be observed in the micritic layers.

The pseudo-columnar stromatolites (Fig.4d) occurring on top of the mushroom-shaped buildups and the stromatolite pavements display a radial fenestral fabric. This radial fabric is often interrupted by small-sized elongated fenestrae, which are concentrically arranged. The framework is made of chains of aragonite microspherulites (Fig.9d,e), the mean diameter of which is 10 microns. Microspherulites are known to form on the inner surface of " hollow shells of organic matter" probably within or beneath microbial films (Monty, 1976, fig. 27,29), in decaying organic matter (Krumbein et al., 1977), inside coccoid algal cells (Buchbinder, 1981), or around heterotrophic marine bacteria (Krumbein, 1974).

The radial microspherulitic framework could have been formed in mucilages filling intersticial vugs between bundles of microbial filaments (as shown by Monty and Hardie, 1976, fig.4h), which later decayed and correspond to the radial fenestrae. The radial microspherulitic framework could also result from precipitation which took place directly around microbial filaments as observed by Monty and Hardie (1976, fig.10a,2a,c) in *Scytonema* mats from a coastal freshwater marsh. In this case, the radial fenestrae would represent primary porosity. The microspherulites are made of radially arranged needles of aragonite (Fig.9e,f). They form a tightly interlocking microfabric similar to that described by Buchbinder (1981) in stromatolites from the Lisan Formation.

An interesting feature can be observed at high magnification (Fig.9f): some aragonite needles show an axial canal the diameter of which reaches 0.5 micron. This may indicate that the crystals grew around organic filaments as illustrated by Monty (1981, pl. 1h) and Monty and Mas (1981, fig.18b). It demonstrates the intimate relationship between carbonate precipitation and microbial activity. The pseudo-columnar stromatolites are poorly cemented ; associated fan-shaped aragonite has never been observed.

The dense stromatolitic layers encrusting the upper surface of some beachrock blocks in the upper intertidal zone show a cyclically laminated fabric (Fig.10c-f). It consists of laminae of fibrous, radial to fan-shaped aragonite crystals separated from each other by a continuous micritic film (Fig.10d). Periodically, crystal length is reduced, the radial aragonitic layers become discontinuous, and micritic films are condensed. This suggests that precipitation was controlled by cyclic factors (Chafetz et al., 1991): seasonal changes in light, temperature and salinity, but possibly also by periodic domination of

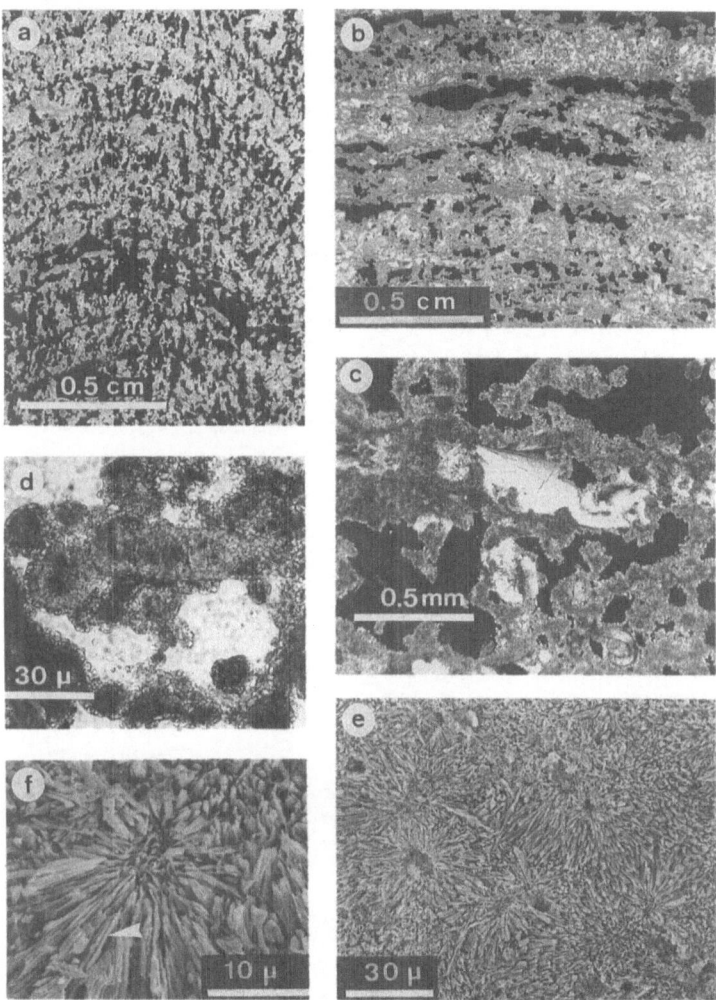

Figure 9 : a-d : Thin sections in normal and polarized transmitted light; e-f : SEM on etched thin sections. a) Microfabric of pseudo-columnar stromatolites .Note the high fenestral porosity. b,c) Laminoid fenestral fabric in domal stromatolites. The small-sized fenestrae are often invaded by fan-shaped aragonite, whereas large horizontal vugs remain uncemented. d) clotted fabric in pseudo-columnar stromatolites consisting of chains of aragonitic microspherulites. The black micritic aspect of the core is probably due to the fact that aragonite needles are cut perpendicularly (see Fig. 9 e,f). e) Aragonite microspherulites composing the framework of pseudo-columnar stromatolites. f) Close-up view of the same area demonstrating the absence of a micritic nucleus in the center of microspherulites. Many aragonite needles reveal an axial canal (white arrow) which could derive from decayed microbial filaments.

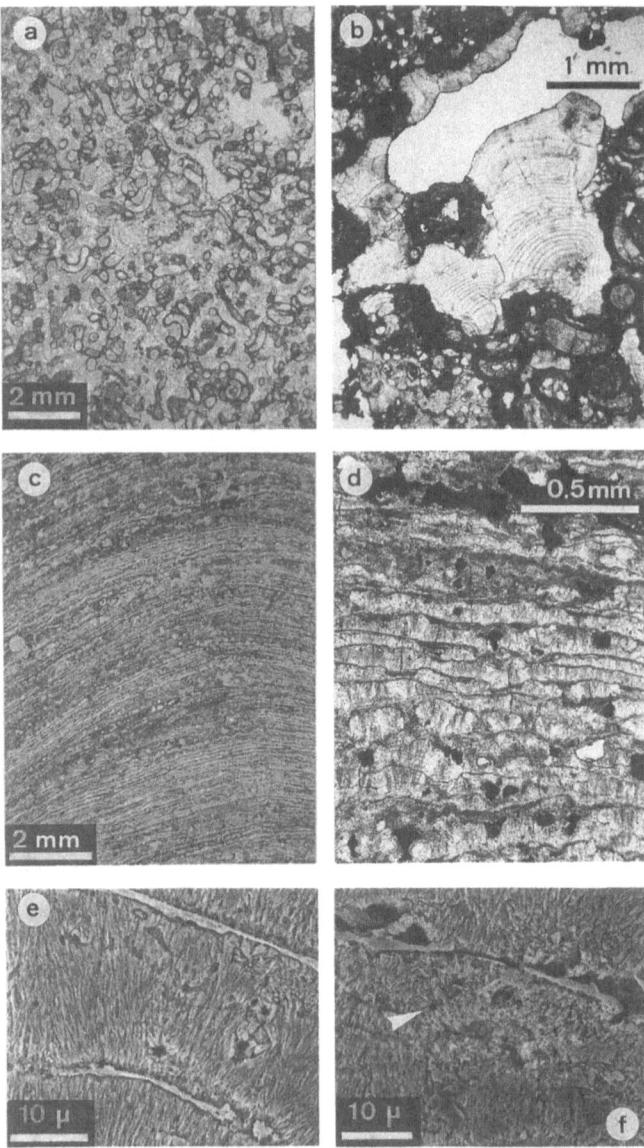

Figure 10 : a-d : thin sections in normal transmitted light ; e-f : SEM on etched thin sections. a,b) Microfabric of a serpulid bioherm. The pore spaces are completely filled with fan-shaped and botryoidal aragonite. c,d) Dense laminated fabric of stromatolitic crusts which developed on beachrock slabs. The light layers are composed of radial to fan-shaped aragonite. Note the well developed cyclicity on Fig. c (see text for comments). e,f) Close-up views of radial aragonite and cryptocrystalline layers. These are often surrounded by very small spherulites (white arrow).

microorganisms triggered by ecological stresses (Monty, 1976). SEM observations show that micritic films are cryptocrystalline (Fig.10e,f). They could result from chemical or biochemical alteration of aragonitic layers.

CEMENTATION

Serpulid bioherms are strongly cemented by zoned botryoidal and spherulitic aragonite which invaded intra- and interskeletal pore-space (Fig.10a,b and 12a-d). Similar submarine cements have been described from the outer face of modern coralgal reefs (Aissaoui et al., 1986; Ginsburg and James, 1976) and are inferred from ancient reef limestones (Davies, 1977; Aissaoui, 1986).

In our case, several evidences indicate that cementation started very early and took place in normal to euryhaline seawater. The fan-shaped aragonite commonly encloses benthic foraminifera (*Ammonia, Quinqueloculina*, Fig.11a-e). They have been trapped at the surface of cements and acted as an obstacle to further crystal growth. This is demonstrated on figures 11c-e illustrating how, toward the center of a pore, aragonite needles are deflected and crystal zoning is refracted.

The *Ammonia* often are trapped with the ombilical face lying on a growth surface of the botryoidal aragonite. This, together with the fact that they are lining the cemented pore walls, suggests that they were trapped in life position attached to the cement surface. This could indicate that they were grazing organic matter coating the cement surface; the same organic matter may have stimulated crystal growth (Mitterer and Cunningham, 1985).

Although no direct evidences such as trapped filaments have been observed in these cements, organic participation in the precipitation of botryoidal aragonite is suggested by additional observations : the well developed zoning of fan arrays is due to periodic interruptions in crystal growth (Fig.12c) and could be explained by alternating organic-stimulated precipitation and re-colonisation by microbes. Numerous micritic drop-shaped clots are included in the botryoidal cements (Fig.11b,c) and seem to locally inhibit aragonite precipitation. SEM examination shows that they tend to encroach on aragonite fibers when crystal growth is interrupted (Fig.11f,g). This suggests a competition for space and demonstrates that these micritic clots are not inert particles but behave as active zones which interact with the surrounding zones of crystal growth. Similar micritic clots have been observed by Cross and Klosterman (1981) in Lower Permian botryoidal cements. The fan-shaped arrays often originate from subspherical micritic cores (Fig.11b) which also form the center of large spherulites (Fig.12d). The spherulites develop in cavities and are commonly not in contact with the walls. This floating position indicates that they grew in or at the expense of microbial mucilages as proposed by Monty (1982). The synsedimentary origin of macrospherulites is supported by the fact that they are frequently found reworked as nuclei in ooids (Fig.12e).

The close association between botryoidal cements and macrospherulites has been reported from the Lower Permian in bryozoan reefs from Ellesmere Island (Davies, 1977) and in stromatolitic/phylloid algal reefs from New-Mexico (Cross and

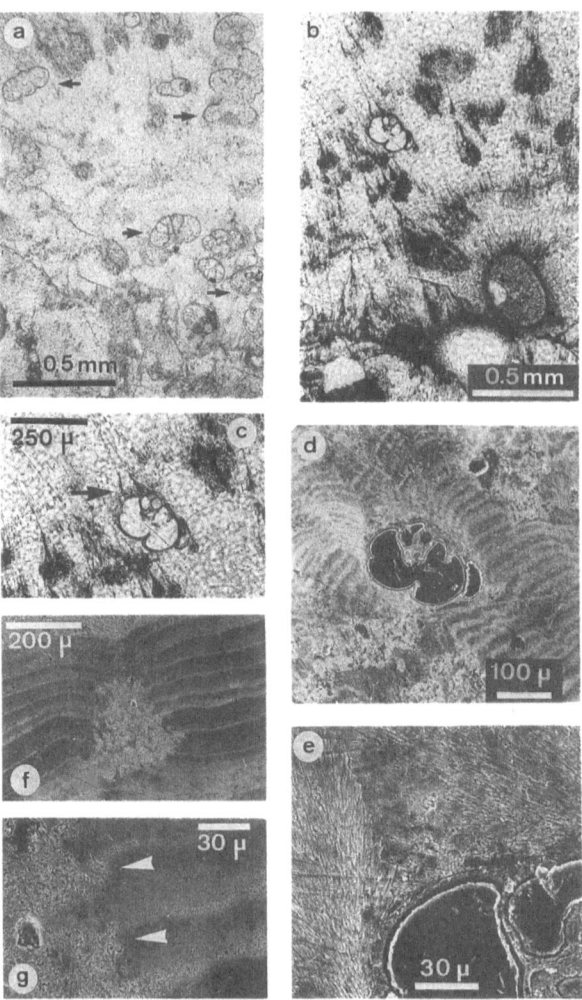

Figure 11 : a-c : Thin section in normal transmitted light ; d-g : SEM on the same etched thin section. a) Trapped benthic foraminifera (*Ammonia*) in botryoidal aragonite. The crystals grew from the lower right corner (pore wall) towards the top of the picture. The *Ammonia* often lie on their ombilical faces (black arrows). b,c,d,e) The trapped foraminifera acted as obstacle for further crystal growth. This is demonstrated by the fact that aragonite needles are deflected towards the centre of the pore (upper left, c,e) and that crystal zoning is refracted (e). Figures c,d and e are close-up views of Fig. b. f,g) SEM close-up views of micritic drop-shaped clots (Fig. 11 b) showing that these areas locally inhibit crystal growth (zoned areas) and tend to encroach on cement surfaces when crystal growth is periodically interrupted (white arrows).

Figure 12 : a-e : Thin sections in normal transmitted light ; b,c : SEM on etched thin section. a,b,c) Botryoidal aragonite showing well-developed zoning. SEM examination of the same area reveals that zoning is due to periodical interruptions of crystal growth (b,c). d,e) In serpulid bioherms, large aragonitic spherulites are associated with botryoidal aragonite. Drop-shaped micritic clots and trapped benthic foraminifera are common. These "cements" have been found reworked as nuclei of ooids (e).

Klosterman, 1981). When comparing figures 10 a and b with figures 9 a and b, it appears clearly that void-filling cements are well developed in serpulid bioherms, scarce in the basal (commonly domal) part of stromatolites, and absent in the pseudo-columnar stromatolites (unless the microspherulitic pattern is interpreted as cement formed in intersticial mucilages, as mentioned above). This differential cementation originated partly from differences in primary porosities : in stromatolites, the high fenestral porosity is secondary in origin and results from decay of mucilages or bundles of algal filaments. The primary porosity and permeability was probably much lower and intersticial waters much more confined in stromatolites than in serpulid bioherms.

A time control on cementation may also be considered : the serpulid bioherms are strongly cemented because they were much longer exposed to circulating seawater than the stromatolite heads and pavements which developed later (as proven by their stratigraphic position), and which emerged earlier when sealevel dropped (Fig.7,8).

OLDER EXAMPLES

Few similar serpulid/stromatolite co-occurrences have been described from the fossil record. Garwood (1931) mentioned the presence of algal stromatolites encrusted with spirorbids in the Lower Carboniferous of northwestern England. In the same area and the same formations, Leeder (1973, 1975) observed lens-shaped serpulid bioherms reaching in average a few decimeters in height. In the Liddel Formation (Tournaisian), stromatolites and serpulids intergrow to form composite bioherms or biostromes. More frequently pluricentimetric serpulid clumps form the initial substrate for stromatolite growth. The stromatolites develop domal and pseudo-columnar structures and often contain "adnate spirorbids". They are surrounded by oolitic/bioclastic sands and overlain by algal mats with gypsum pseudomorphs. Such sequences have been rightly interpreted as shoreline deposits. Although the sizes of stromatolitic buildups differ from those observed in the Flandrian stromatolites, the vertical sequence recorded in the Lower Carboniferous (Leeder, 1975, fig. 2 a) shows many similarities with the facies evolution described in this paper.

More recently, Wright and Mayall (1981) described from the Upper Triassic in Southwest Britain stromatolites showing a " conspicuous alternation of flat or domal laminated algal units with generally thicker layers characterized by the development of dendritic, arborescent structures". These arborescent structures, reaching a few centimeters in height, are composed of micritic laminations which coat tubes of serpulid worms (*Microtubus*). This association has been related to commensalism. The alternation of stromatolitic layers with serpulid/algal dendritic layers has been interpreted as the consequence of environmental changes.

Other serpulid/stromatolite co-occurrences have been reported, but they differ from the Flandrian model. Peryt (1974) reports that spirorbids occur in great numbers inside Lower Triassic stromatolites from Poland and contribute to reinforcing the stromatolite framework. In the Lower Permian of New-Mexico, Toomey and Cys (1977) observed well-developed algal stromatolites containing abundant remains of spirorbid worms which are "enmeshed within the laminae". The bioherms reach up to 55 cm in height and 130 cm in diameter. They are composed of "digitate colonies giving a knobby appearance" (pseudo-columnar) ? and grow on stromatolite fragments and oncoids.

As the bioherms are overlain by continental sandy shales and contain a very poor faunal assemblage, they have been interpreted as a marginal marine/brackish-water deposit.

It should be noted that all these occurrences are limited to the Carboniferous and the Triassic. Whether their absence in other time intervals results from lack of documentation or whether it is biologically or environmentally controlled, remains an open question.

CONCLUSIONS

The observations on the coastal facies in a Flandrian restricted lagoon evolving rapidly into an evaporitic basin provide a useful additional model for understanding older deposits. The following points should be lined out:

1.dropping sealevel and subsequent increasing salinity led to development of high salinity-tolerant stromatolites on top of euryhaline serpulid bioherms;

2.the fact that serpulid bioherms are better cemented than the overlying stromatolites is due to their longer exposure to circulating seawater and to higher primary porosities;

3.botryoidal and macrospherulitic cements (and their neomorphic counterparts), which have in many cases been interpreted as purely inorganic precipitations, should be reconsidered from a microbiological point of vue (as suggested by Monty, 1982, and Van Laer and Monty, 1984);

4.foraminifera trapped in botryoidal aragonite cements suggest rapid crystal growth and the presence of organic mucus lining the pore walls.

5.the surprising convergence between Lower Carboniferous and Late Quaternary serpulid/stromatolite bioherms could be related to environmental similarities

AKNOWLEGMENTS

We are very much indebted to Cl. Monty and J. Sarfati for reviewing the manuscript and for their helpful suggestions. We also thank T. Lajmi and M. Gribi from the Geological Survey of Tunisia for providing facilities in the field. This work was supported by FNSRS grants 2.897.083 and 2000.5322.

REFERENCES

Aissaoui D.M. (1986) Diagenèse carbonatée en domaine récifal. Thesis no 3248, Univ. Paris Sud.
Aissaoui D.M. , Buigues D. , Purser B.H. (1986) "Model of reef diagenesis, Muroroa atoll, French Polynesia ". In Schroeder J.H. & Purser B.H. (eds), Reef diagenesis, Springer Verlag, Berlin, pp. 27-52.
Andrews P.B. (1964) "Serpulid reefs, Baffin bay, southeast Texas." In Depositional environments, south-central Texas coast, Gulf Coast Assoc.Geol. Socs. Fieldtrip Guidebook 1964 Ann.Mtg.
Buchbinder B. (1981) "Morphology, microfabric and origin of stromatolites of the Pleistocene precursor of the Dead Sea, Israel". In Monty C.L.V. (ed). Phanerozoic stromatolites, Springer, pp 181-196

Chafetz H.S., Utech N.M. & Fitzmaurice S.P. (1991) "Differences in the $d^{18}O$ and $d^{13}C$ signature of seasonal laminae comprising travertine stromatolites." J. Sediment. Petrology, 61/6, 1015-1028.

Cross T.A. , Klosterman M.J. (1981) "Primary submarine cements and neomorphic spar in a stromatolitic-bound phylloid algal bioherm, Laborcita formation (Wolfcampian), Sacramento Mountains, New Mexico, USA". In Monty C.L.V. (ed) Phanerozoic stromatolites, Springer , pp 60-73.

Daley B. (1972) "Macroinverterbrate assemblages from Bembridge Marls (Oligocene) of the Isle of Wright, England, and their environmental significance" Paleogeogr., Paleocl., Paleoecology 11, 11-32

Davaud E., Strasser A. & Jedoui Y. (1988) "Cerebroid and spiny ooids from a Holocene restricted lagoon (Sabkha el Melah, Southeastern Tunisia)". Abstracts 9th IAS Regional Meeting of Sedimentology, Leuven 46-47

Davaud E., Strasser A. & Jedoui Y. (1990) "Spiny ooids: early subaerial deformation as opposed to late burial compaction." Geology 18, 816-819.

Davies G.R. (1977) "Former magnesian calcite and aragonite submarine cements in upper Paleozoic reefs of the Canadian Artic : a summary." Geology 5, 11-15.

Garwood E.J. (1931) "The Tuedian beds of Northern Cumberland and Roxburghshire east of Liddel water." Quat. Jour. Geol. Soc. London 87, 87-159.

Ginsburg R.N., James N.P. (1976) "Submarine botryoidal aragonite in Holocene reef limestones, Belize", Geology 4, 431-436.

Heckel P.H. (1974) "Carbonate buildups in the geological record : a review." In Laporte L.F. (ed), Reefs in time and space, Soc. Econom.Paleonto.Miner., sp publ. 18, pp .90-154.

Hoffman P. (1976) "Stromatolite morphogenesis in Shark Bay, Western Australia", in Walter M.R. (ed) Stromatolites, Developments in Sedimentology 20 Elsevier, Amsterdam, pp 261-271.

Krumbein W.E. (1974) "On the precipitation of aragonite on the surface of marine bacteria", Naturwissenschaften 61, 167.

Krumbein W.E., Cohen Y., Shilo M. (1977) "Solar Lake (Sinai) .Stromatolite cyanobacterial mats", Limnology and Oceanography 22, pp. 635-656.

Leeder M.R. (1973) "Lower Carboniferous serpulid patch reefs, bioherms and biostromes", Nature 242, pp 41-42.

Leeder M.R. (1975) "Lower Border Group (Tournaisian) stromatolites from the Northumberland basin", Scott. J. Geol. II, pp. 207-226.

Mitterer R.M., Cunningham R. (1985) "The interaction of natural organic matter with grain surfaces : implications for calcium carbonate precipitation", in Schneidermann N. & Harris P.M. (eds), Carbonate cements ; Spec. Publ. Soc. Econ. Paleont. Mineral 36, pp .17-31.

Monty C.L.V. (1976) "The origin and development of cryptalgal fabrics", in Walter M.R. (ed), Stromatolites , Developments in Sedimentology 20 , Elsevier, Amsterdam, pp .193-249.

Monty C.L.V., Hardie L.A. (1976) "The geological significance of the freshwater blue-green algal calcareous marsh", in Walter M.R. (ed), Stromatolites, Developments in Sedimentology 20, Elsevier, Amsterdam, pp .447-477.

Monty C.L.V. (1981) "Observations pétrographiques et chimiques sur l'éodiagenèse de carbonates du précontinent calvais (Corse)", Bull. Soc. royale Sci. Liège 11-12, pp .470-482.

Monty C.L.V., Mas J.R. (1981) "Lower Cretaceous (Wealdian) blue-green algal deposits of the province of Valencia, Eastern Spain", in Monty C.L.V. (ed), Phanerozoic Stromatolites, Springer, Berlin, pp 85-120.

Monty C.L.V. (1982) "Microbial spars", I.A.S. Abstracts Int. Sedimentol.Congress,1982, Hamilton pp 26.

Paskoff R., Sanlaville P. (1983) "Les côtes de Tunisie. Variations du niveau marin depuis le Tyrrhénien", Coll. Maison de l'Orient méditerranéen,Lyon 14, 1- 92.

Perthuisot J.P. (1974) "Les dépôts salins de la Sebkha El Melah de Zarzis : conditions et modalités de la sédimentation évaporitique", Rev. Geogr. phys. Geol. dynam. 16/2, 177-187.

Perthuisot J.P. (1975) "La Sebkha El Melah de Zarzis: genèse et évolution d'un bassin paralique", Trav. Lab. géol. 9, Ecole Norm. Sup., Paris,

Perthuisot J.P., Floridia S. , Jauzein A. (1972) "Un modèle récent de bassin côtier à sédimentation saline : la Sebkha El Melah (Zarzis, Tunisie)", Rev. Geogr. phys. Geol. dynam. 14/1, 67-83.

Perthuisot J.P., Floridia S. (1973) "Carte géologique de la Sabkha El Melah et de ses bordures", Trav. Lab. géol. 8, Ecole Norm. Sup. Paris.

Peryt T.M. (1974) "Spirorbid-algal stromatolites", Nature 249, 239-240.

Playford P.E., Cockbain A.E. (1976) "Modern algal stromatolites at Hamelin Pool, a hypersaline barred basin in Shark Bay, Western Australia", in Walter M.R. (ed), Stromatolites, Developments in Sedimentology 20, Elsevier ,Amsterdam, pp 389-412.

Thornton S.E., Pilkey O.H. , Lynts G.W. (1978) "A lagoonal crustose coralline algal micro-ridge : Bahiret el Bibane, Tunisia", J. Sediment. Petrol. 48, 743-750.

Toomey D.F., Cys J.M. (1977) "Sprirorbid/algal stromatolites, a probable marginal marine occurence from the Lower Permian of New Mexico, USA", Neues Jahrbuch für Geologie und Paläontologie, Monathefte 6, 331-342.

Van Lear P., Monty C.L.V. (1984) "The cementation of mud mound cavities by microbial spars", Abstracts 5th IAS Regional Meeting of Sedimentology, Marseille, pp 446-447.

Wright V.P., Mayall M. (1981) "Organism-sediment interactions in stromatolites: an example from the Upper Triassic of South West Britain", in Monty C.L.V. (ed), Phanerozoic stromatolites. Springer Verlag, Berlin, pp 75-84.

PART II

Cenozoic Stromatolites in Lakes and Marine Stromatolitic Phosphorites

MICROSTRUCTURES IN TERTIARY NONMARINE STROMATOLITES (FRANCE). COMPARISON WITH PROTEROZOIC

J. BERTRAND-SARFATI *, P. FREYTET** and J.C. PLAZIAT ***
* Institut des Sciences de l' Evolution, URA 327, CNRS,
 Sciences et Techniques du Languedoc,
34095 Montpellier Cedex 5. France
** 41 rue des Vaux mourants,
91370 Verrières-le-buisson, France
*** Laboratoire de Pétrologie sédimentaire et Paléontologie, URA CNRS,723, Université de Paris-Sud, 91405 Orsay Cedex. France

ABSTRACT

Late Cretaceous to Early Miocene series of Southern France and northeastern Spain display an almost continuous record of nonmarine stromatolites. Well preserved organic remains, mainly microbial or algal filaments, are tentatively used as a base for the description of stromatolite microstructures of fluviatile, lacustrine and brackish origins. Four terms can be described as follows :
a) microstructures with well preserved filaments are subdivided according to filament size and disposition within the lamina,
b) microstructures with dubious filaments,
c) zoned structures composed of fibro-radial sparite,
d) associated non-microbial biogenic microstructures.
Recent petrological works and biological studies on modern fluviatile and lacustrine stromatolites afforded terms of comparisons whereas paleoenvironmental considerations suggest comments on salinity requirements for several of the microstructures.
This descriptive approach should allow for the addition of new categories according to similar criterion : nature of the organic remains, size and habit of filaments and spheres. It furthermore facilitates comparisons with some Proterozoic stromatolite microstructures.

INTRODUCTION

Stromatolites have been firstly described and the name forged from nonmarine environments by Kalkovski (1908). However the discovery of living marine

155

J. Bertrand-Sarfati and C. Monty (eds.), Phanerozoic Stromatolites II, 155–191.

stromatolites has emphasized their role in understanding ancient stromatolites. Nonmarine stromatolites have been subsequently neglected and their informations regarding laminae building not used. The discovery of assumed lacustrine setting in the Proterozoic implies a renewed interest for the study of nonmarine recent stromatolites.

Nonmarine stromatolites have been recorded from numerous formations of Late Cretaceous to Early Miocene age in France and northern Spain (Fig.1). Recent studies of these fossil stromatolites have been devoted to microstructural features, laminations, crystal fabric and biologic remains, algal or microbial filaments (1935a-c, Freytet and Plaziat, 1965, 1972, 1982 ; Bertrand-Sarfati et al., 1966 ; Plaziat, 1966, 1974 ; Donsimoni and Giot, 1977 ; Anadon and Zamarreno, 1981 ; Monty and Mas, 1981 ; Casanova, 1981a, 1986 ; Riding, 1983 ; Schneider et al., 1983 ; Casanova and Nury, 1989 ; Freytet and Verrechia, 1989 ; Leinfelder and Hartkopf-Fröder, 1990 ; etc..) improved by studies on living stromatolites (Wallner, 1934 ; Pentecost, 1978, 1987, 1990). Moreover we now dispose of data about the processes of microstructure fabric in several modern Cyanobacteria, especially in *Rivularia* (Caudwell, 1987 ; Pentecost, 1987). One of the authors (P.F.) presently attempts to correlate the microbial content, obtained by crushing and then decalcifying the samples with the microstructure types, using thin sections and SEM studies, in living stromatolites from rivers in the Paris Basin (Freytet, 1989 a,b), Burgundy and other areas in France (Freytet, 1992 ; Freytet and Plet, 1991) and foreign regions (Morocco, unpublished data).

In continental settings, stromatolites are related to specific environments : active fluviatile channels, abandoned channels, lakes, hot and cold springs in karstic environments. Some estuarine and lagoonal environments include oncolites and stromatolites which may have been confused with freshwater stromatolites.

However we suggest that the gross morphology and the microstructure of continental stromatolites (i.e. flat-laminated versus columnar or branched) of Mesozoic to Cenozoic age may be definitely influenced by environmental factors. Complex microbial biocenoses, including several Algae beside the cyanobacteria are today the active agent of the modern stromatolite lamina building. In ancient constructions, Paleozoic and essentially Proterozoic, the biocenoses seem to have been more simple and stable, in environments themselves much more stable through time, giving rise to constant morphologies and structures with a valuable chronologic significance. The ecologic complexity and unstability of more recent nonmarine stromatolites favour an environmental control on their microstructures and morphologies.

According to the obvious morphogenetic influence of biologic components in Modern stromatolites, an attempt to describe nonmarine stromatolite microstructures, based on the generally good preservation of some of the organic remains is proposed here. The importance of microstructure in stromatolite classification is discussed.

As these microstructures display striking similarities with some Proterozoic stromatolite microstructures, comparisons are attempted and a biologic origin may be inferred for these ancient microstructures.

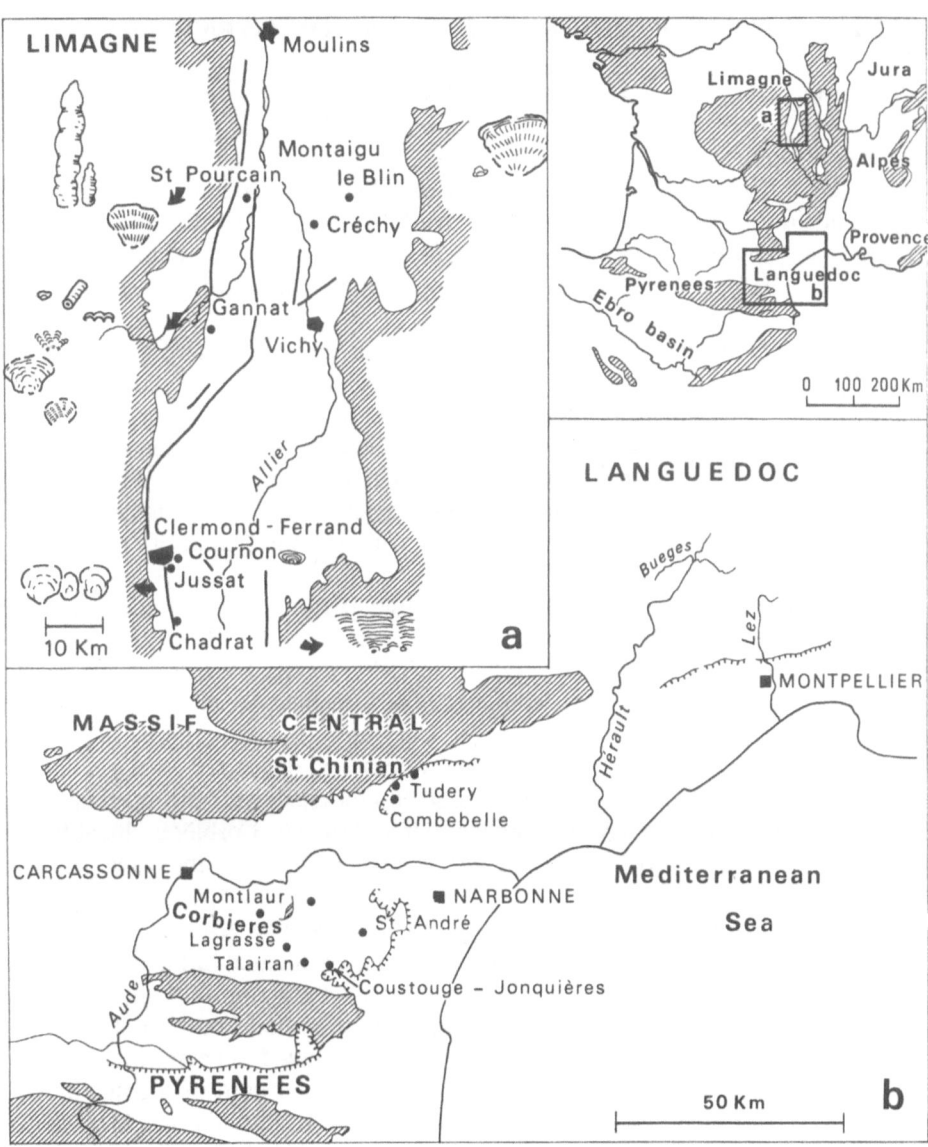

Figure 1 : a) Schematic map of the Limagne grabben, showing the main areas of stromatolite occurences, quoted in the text and their gross morphologies. Solid lines are faults. b) Simplified geographical map of Southern France, showing the main stromatolitic areas mentionned in the text or in illustrations : Languedoc-Corbières : St André (Campanian-Maastrichtian) ; Fontcouverte (Lower Sparnacian) ; Lagrasse (Ilerdian) ; Jonquières, Montlaur, Talairan, Coustouge (Lutetian) ; Languedoc-StChinian : Combebelle, Tudery (Sparnacian) ; Languedoc-Montpellier (Middle Eocene). Provence : Phox-Amphoux (Maastrichtian) ; Provence-Manosque basin : Revest-St-Martin (Oligocene). Spanish Pyrenées : Figueras (Maastrichtian), Berga (Thanetian and Early Ilerdian).

STROMATOLITE MORPHOLOGIES IN RELATION TO ENVIRONMENTS

Freshwater environments in which stromatolites are growing, offer a large diversity of specific settings. Nevertheless morphological or microstructural types cannot be related to one specific environment.

In active *fluviatile channels*, the sandy and conglomeratic sediments indicate a high energy deposition. These environnements have been particularly illustrated in the Late Cretaceous to Middle Eocene of the Languedoc area. The autochtonous encrusted malaco-fauna (*Unio, Theodoxus, Melanopsis*) demonstrates the freshwater conditions (Plaziat, 1966 ; Freytet and Plaziat, 1982).

The *abandoned channels* constitute ephemeral ponds or lakes. There, sedimentation is muddy, more frequently carbonated. The cylindrical coating around roots, twigs and branches is the more frequent stromatolite morphology, while coatings of pieces of bark or leaves affected a concavo-convex shape. Oncoids well rounded or irregular are developed on sand grains gravels or shells. Planar encrustations are less frequent in fossil stromatolite record than in Modern waterways. Large discrete bioherms (0,5 to 2 m high) locally suggest more permanent waters with a limited input of sediment (Early Eocene, Languedoc).

Disruption of levees (crevasses splay) allowed transportation of fluviatile oncoids and associated channel sediments onto the *flood plain* where they are included in mudstones, subsequently affected by pedogenesis (hydromorphic ferruginous mottling). A similar reworking of the channel sediments is documented by an oncolite lense in the Early Eocene marine-shore deposits of the Corbières (Plaziat, 1966). In a lagoonal setting, evidenced by abundant fossils, a freshwater flood deposited locally coated *Unio* and terrestrial snails. This intrusion must not be confused with the brackish oncoids development as demonstrated in another Early Eocene Corbières locality where estuarine malaco-fauna (*Cyrena, Melanatria*) has been coated by the same salt-tolerant stromatolite community.

The *larger lakes* with stable shorelines offer a wide range of sizes and morphologies of the stromatolite buildups. The Oligo-Miocene lakes located along the graben. of Limagne (French Massif Central) exhibit numerous bioherms with different shapes (Fig.2). They differ with respect to the nature of the shoreline : along the deeply sloping shores, fluviatile channels provide oncolites (coated vegetal debris or gravels) or sand-sized grains. Bedded oncolitic sands have been confused with ooids because of their small size (for example, see Jung, 1958, fig.40). The problem has been discussed by Strasser (1986). Where the lake shore dips gently, dessicated mud flats are encrusted, grading from flat domal coatings to centimetric columnar bioherms (20-40 cm high and 10-20 cm in diameter) and pedonculate mushroom-like bioherms (10-15 cm). In deeper parts of the same lake, long-living globular to cushion-like bioherms about 0,5 to 2 m in diameter develop in columnar buildup, several meters high, isolated or closely packed. These bioherms comprise thousands of encrusted tubes of caddies-fly larvae forming their basal and central parts (Fig.3b,c). These larger bioherms are interpreted as initiated around tufts of subaquatic reeds or rushes providing the necessary substrate

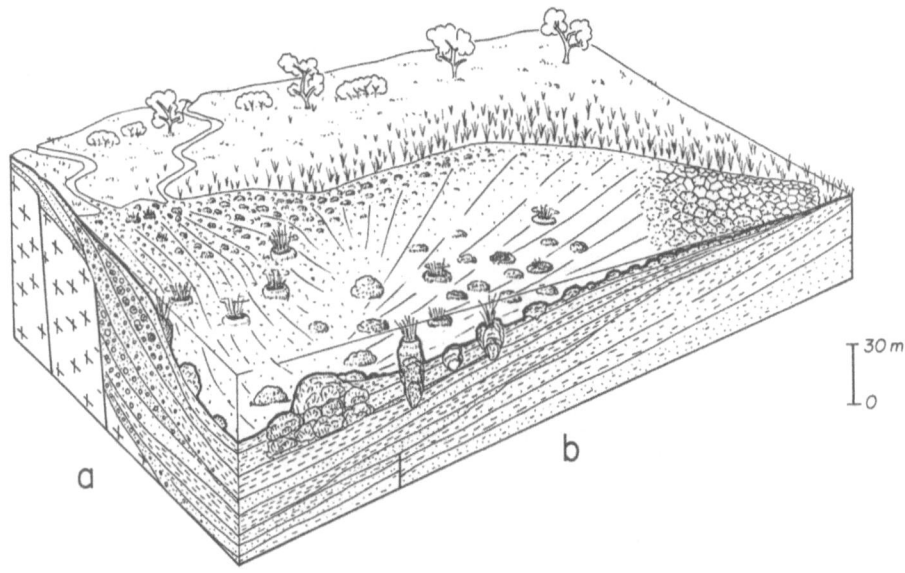

Figure 2 : Bloc diagramme of a lacustrine shoreline in the Limagne graben illustrating the shapes and positions of the different types of stromatolites. a) deeply sloping margin : oncolites and decimetric stromatolite heads are localised in fluviatile channels and along the shoreline. Oolites are interbedded. b) gently dipping lake edge : mud-flats with encrustations, club-shaped and columnar stromatolites. In deeper parts, globular or columnar edifices comprising a core of coated indusiae of caddies-fly, gathered around subaqueous bushes of reeds or other plants and are coated by thick pseudocolumnar laminae. In the deepest parts, complex edifices are due to superimposition of stromatolitic buildups (lesser amount of indusiae).

for the insect nymphosis (Bertrand-Sarfati et al., 1966, 1990 ; Freytet and Plaziat, 1982). Each larval tube is coated. The yearly accumulation of fallen empty indusiae at the base of the perennial plant cluster is coated by a cylindrical flat-laminated stromatolite forming a thick external envelope and resulting in the construction of discrete high erect columns or juxtaposed globular buildups ("indusial limestones" of Hugueney et al., 1990). Generally the vertical development of these complex columns (up to 15 m) comprises alternation of coated accumulations of caddies-fly larvae and stromatolite buildups indicating variations in the shallow water environment (Fig.3a,b). Stromatolites also comprise tiny columnar fascicles embedded in ostracods lacustrine carbonates, bushes or columns of erect large filaments with a thin coating of micrite and thick pseudocolumnar laminae. They probably grow in more deeper waters. Flat biostromes, several decimeters thick are recorded exceptionally (Limagne-Chadrat).

Figure 3 : Large buildups in permanent lakes; a & c, Créchy ; b, St Pourçain. a) surface of a small columnar buildup showing the effect of compaction on the encasing sediments. Scale= 10cm b) two juxtaposed buildups (a third one appears on the right); the buildup on the left, exibits its outer surface ; the central transverse section displays coated and aggregated indusiae with diverse orientations, stacked in the core of the buildup . The outer surface is coated by stromatolitic laminae thining downward. The very base, uncoated, indicates the initial position of the tufts of reeds stuffed with the fallen larvae previously sticked on them. Scale= 10 cm. c) huge columnar buildup (Limagne-Créchy) with different phases of growth, the upper one is disymmetrical with a synoptic relief of 4 metres. Indusiae are found in the basal part, while the upper one is entirely stromatolitic. The diametre of the different parts also changes, the intermediate one displays only the smooth surface of the outer part of the column. Scale= 1,5 m.

BIOLOGICAL SIGNIFICANCE OF STROMATOLITE LAMINAE

Modern freshwater stromatolite laminae reflect the striking high diversity of their biologic components. The stromatolitic laminae represent more or less complex alternating biocenoses comprising a few species prone to fossilization and many others, especially inhabitants of the mat, which are promised to decay : Bacteria, Algae, Fungi and small invertebrates.

Decandolle (1806) first drew attention on calcite precipitation by *Rivularia haematites* when he described this species (in Golubic, 1976a). Tilden (1897) has also shown calcite crystals growing around *Chaetophora calcarea* filaments. When Pia (1934) recorded 150 years of works on continental "Kalkbildung durch Pflanzen" he gave a list of 70 bacteria, 79 cyanophyceae, 5 diatoms, 19 chlorophyceae or xanthophycea, 2 rhodophytes, 31 hepatica and mosses.

Is microbial activity the major agent of calcification ? Lowenstam (1986) and Lowenstam and Wiener (1989) distinguish between "organic matrix mediated" biomineralization controled by a framework constructed organically from "biologically induced" mineralization with no biological control. However a large spectrum of possibilities may be found with different degrees of biological versus environmental control (Golubic, 1990). In the literature, opinions vary, many authors supporting the biologically induced hypothesis (Schneider and Schröder, 1980 ; Adolphe, 1981 ; Chafetz and Folk, 1984, Castagnier et al. 1989, Dahanayake et al., 1985 ; Thompson et al., 1990) while others considered it as less effective (Pentecost, 1990 ; Pentecost and Riding, 1986 ; Pentecost and Terry, 1988 ; Golubic, 1976a) limiting the number of active freshwater algae to about fifteen genera : Cyanophytes, *Rivularia, Lyngbya, Phormidium* and *Schizothrix* especially, Xanthophyceae, *Vaucheria germinata*, a Chetophoral Chlorophyceae *Gongrosira incrustans* and a Desmidial Chlorophyceae *Oocardium stratum*. Wallner (1934) the first, referred to the same organisms and illustrated the different modes of crystallization relating to the diverse species.

Sometimes the morphology of stromatolitic lamination may be confused with that of the living bio-substrate of the filaments : mosses, algae, insect indusiae (caddies-fly larval tubes) or larval tubes of Chironomids (mosquitos). In modern travertines, the local abundance of Chironomid tubes led Langeron (1902) to separate three distinct types of tufas according to their non microbial content : "compact tufas" made by "algae" alone, "cavernous tufas" with "algae" and twigs and "porous tufas" with "algae" and mosquitos larval tubes.

The fact that ancient stromatolites were built by plurispecific biocenoses has been already suggested (Freytet and Plaziat, 1965, 1982 ; Bertrand-Sarfati et al., 1966). It is easily demonstrated in Recent constructions. In the small oncoids of a slow stream from the Paris basin, Bourrelly (in Adolphe and Rofes, 1975) identified *Stigonema sp.*, *Gongrosira sp.* and *Lyngbya calcarea*. In freshwater marshes of the Bahamas, Monty (1965,a) described microbial-laminated sediments built by *Scytonema* and to a lesser extent *Schizothrix*. Freshwater streams from Belgium, as well as the Seine river display incrustations comprising *Schizothrix* and *Phormidium* (Monty, 1976 ; and Monty and

Mas, 1981 ; Freytet, 1989, a,b). Many brooks in Burgundy display coated gravels and oncoids induced by an association of *Phormidium incrustatum*, *Schizothrix fasciculata* and *S. calcicola* with 10 to 20% of *Gongrosira incrustans*. *Oocardium* are strictly limited to incrustations on mosses. *Vaucheria* in fragile bunches of incrustated filaments and *Batrachospermum* present only scattered crystals on their surfaces (Freytet, 1989b ; Freytet and Plet, 1991). Constructions dominated by *Rivularia* have been described from Provence (Casanova, 1981 ; Casanova and Laffont, 1985), England (Pentecost, 1978, 1985, 1987) and in Austria (Kann, 1978), usually associated to *Phormidium*. The oncoids of the Constance lake reflect a complex association, essentially *Phormidium* and *Schizothrix* with rare *Rivularia* in coralloid colonies (Schaffer and Stapf, 1978). Golubic and Fisher (1975) recorded also oncolites built by *Gongrosira*, in North America fluvial environments.

Within the laminae, of nonmarine stromatolites diverse filaments remnants can be seen. The major difficulties in the interpretation of the filaments diversity originate in :

1- the biologic variability of the filaments diameter, sometimes due to changes in the number of trichomes in the sheath.

2- the diverse sizes and forms of original calcite crystals, precipitated around the filaments or sheaths.

3 - the diagenetic evolution, micritisation and crystal enlargement.

1- the precise filament diameter is not necessarily a reliable factor but the average size range appears to be generally significant. Dangeard (1938) and Gauthier-Lièvre (1955), summarizing published data, gave a size range of 50 to 132 μm for *Vaucheria geminata*. On the other hand, *Microcoleus* filaments contain from 5 to more than 20 trichomes, their diameter becoming three times bigger.

2 - In modern stromatolites some of the filament-crystal relationships have been clearly established as follows :

- in *Rivularia haematites* and a few other species, biologically modified sparitic crystals (50-100 μm) appear and include rapidly several filaments. Later on the coalescent crystals form a sparitic layer with a more or less palissadic fabric (Caudwell, 1983, 1987 ; Freytet and Plet, 1991 ; Pentecost, 1990)

- in *Gongrosira calcitans*, sparite crystals are immediately in contact giving rise to a massive sparitic construction (Golubic and Fisher, 1975).

- in the isolated cells of *Oocardium stratum* stuck to the substrate by a mucous string, the calcified base grows as a branching monocrystal (Wallner, 1935a, p.1). In older constructions we can observe juxtaposed but discrete branched colonies (Freytet and Plet, 1991) or sparitic massive layers (Wallner, 1935a ; Golubic and Marcenko, 1965).

- in *Phormidium incrustatum* the large trichomes are freely moving within the sheath and calcification develops around the sheath (Golubic, 1976 a ; Casanova, 1986 ; Freytet and Plet 1991). *Phormidium* is usually associated with a dense *Schizothrix* mat (Monty and Mas, 1981, fig. 12-16 ; Schaffer and Stapf, 1978, fig.15,22 ; Freytet and Plet, 1991).

The first calcite crystals may be perfectly rhomboedral, growing afterwards more irregularly and finaly coalescing as a continous coating of the filaments. Examples are reported of crystal growing on the mucous pedoncle of Diatoms (Wallner, 1935b ; Emeis et al. 1987) and on the filaments of *Chaetophora calcarea* (Tilden, 1897), *Batrachospermum* (Pia, 1934, Freytet and Plet, 1991) and *Vaucheria*. Microcrystals also develop on mosses, chlorophyte algae and subaquatic phanerogam leaves giving a rigid coating of physico-chemical origin. Conversely, early biogenic processes may change the initial crystallinity as Caudwell (1987) demonstrated with episodic micritisation of the *Rivularia biasolettiana* sparite crystals by bacteria.

3- Transposition of these observations concerning the Tertiary stromatolite calcification processes is uncertain. It is worthwile to note that the role played by the environment on crystal fabrics is obvious, but it remains difficult or impossible to correlate crystal habit with primary environmental factors (limpidity of water, temperature, ionic concentration etc..).

DESCRIPTION OF NONMARINE STROMATOLITE MICROSTRUCTURES

Each morphotype of stromatolite commonly contains more than one microstructure, and each microstructural type is present in more than one stromatolite irrespective of the environmental setting. This is the reason why we have attempted to dissociate microstructural description from gross morphology.

This microstructural classification is based on the biologic component microbial filaments : Group A, when microbial remains are preserved, laminae fabric when microbial remains are poorly preserved ; Group B and C when they are unpersistent ; no cocoid cyanobacteria have been recognized in our material. We will add the description of some of the most striking non-microbial or non-algal remains in Group D, characterized by sparitic spheres (insect eggs or fungal spores) and hemitubular tubes of Chironomids mosquitos larvae.

Groupe A - Microstructure with well preserved filaments

This type has been subdivided according to filament apparent diameter. Three size classes A-I, A-II, A-III, are recognized in our Late Cretaceous-Tertiary material. These classes may represent different modes of preservation and/or reflect different species. Each class is subsequently subdivided with respect to filament orientations and/or fabric of laminae. A rapid comparison with modern cyanobacteria building stromatolites is added.

Burial of filaments and subsequent diagenesis tend to decrease the filament size (Golubic and Barghoorn, 1977). Only spores retain their initial diameter (Freytet and Moissenet, 1983). Thus, a great variability is expected according to conditions of fossilization. Preservation of the filaments may be favoured in a micritic layer.

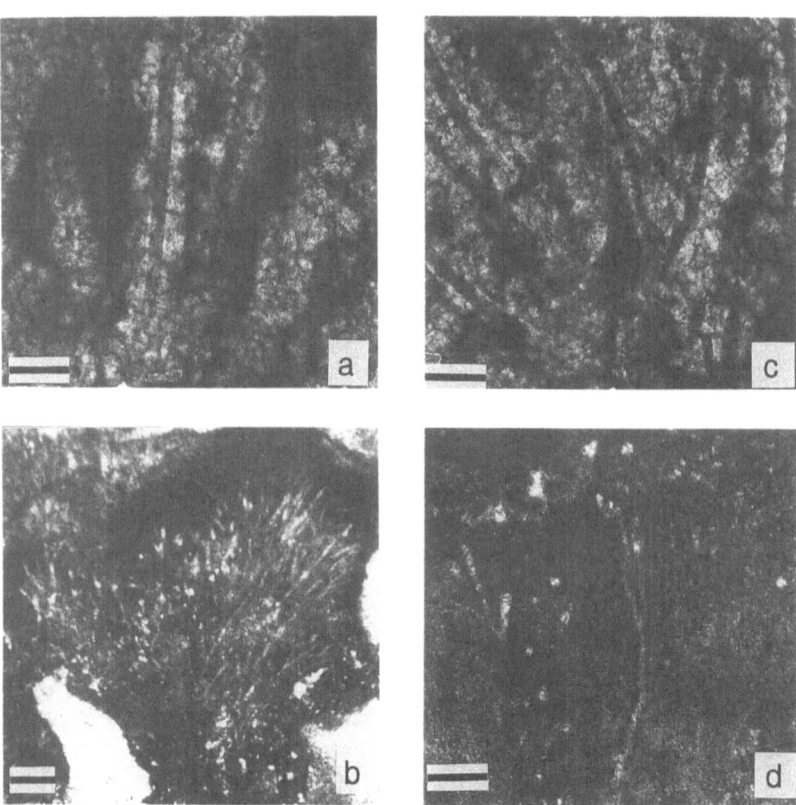

Figure 4 : Well preserved intermediate size filaments : a) and c) Type 1 : filament (dark) encased in a mosaic of calcite crystals, isolated or grouped in bundles (Limagne-Gannat). They form patchy masses in discontinuous laminae in the central part of large bioherms, 1-2 m diameter, deeper part of the lake. Scale bar= 80 μm. b) and d) Type 2 : radiating bushes of filaments in micritic matrix. Filaments are impregnated by sparry calcite with a micritic outline (Corbières-Talairan, bioherms or oncolites along a lake shore). Scale bar= 500 μm.

Aragonitic concretions of the marine *Rivularia* with their micritic grains and needles (0.2 µm diameter) give a valuable model for micritic fossilization (Golubic and Campbell, 1981). On the other hand, radial sparite fabric illustrated by the present time fluviatile *Phormidium* oncolites (Monty and Mas, 1981) appears less conservative. Radial sparite crystals enclose patches of micrite-coated filaments. This local observation suggests that radial sparite fabric is an early diagenetic phenomenon, post-dating an earlier micritic coating.

Group A-I : Intermediate size filaments (10-40 µm)

They are subdivided according to the growth habit of the filaments in the laminae.

Type 1. They comprise algal filaments enclosed in a sparitic coating (Fig.4a,b). The filament made of dark micrite with a constant diameter (10 µm) is encased in a saw-tooth sparite coating (25-30 µm of total diameter). These unbranched filaments are commonly erect but occasionally they grow in a prostrate position at lamina base. Bundles of filaments reaching several mm of diameter are juxtaposed and form irregular laminae (Limagne-Gannat). Bundles are separated by voids remaining empty or filled by small clasts.
These filaments may be compared to *Scytonema* filaments with alternating prostrate and erect growth habits (Golubic, 1976b, p.137 ; Monty, 1976). Filaments of similar size referred to *Phormidium* are illustrated in Monty and Mas (1981). They are surrounded by elongate crystals of sparite which differs from those described here.

Type 2. Erect filaments are preverved as clear sparitic tubes outlined by a thin, poorly defined micritic envelope (Fig.4c,d). The filament size ranges from 20 to 45 µm, averaging 35 µm. Long erect, slightly curved, the branching filaments radiate from a basal point. They form turbinate bushes, isolated or juxtaposed forming a discontinuous lamina (Languedoc-Corbières, Talairan). They also occur as isolated filaments within micritic laminae (Languedoc-Montpellier).
Some of the cyanophytes from the order Nostococales exhibit morphologies in which branching is distributed in an intercalary position along the filaments leading to a similar radiate branching habit. However a few of them are calcified.

Type 3. The horizontal filaments are clear sparitic tubes rimed by a thick well defined micritic coating with an average diameter of 20-30 µm (Fig.5). Circular cross sections of the filament and short spindle-like oblique sections are scattered among a micro-sparitic matrix. Never erect they lie pararel to slightly oblique within the sediment bedding. Such filaments are abundant in many stromatolitic-laminated carbonates from brackish transitional environments where they never form well defined buildup.
This type of horizontally deposited filaments were first reported in Europe by Colom (1961, 1967), Oliveros et al. (1960) and Bignot (1972) who compared them to *Melosira*

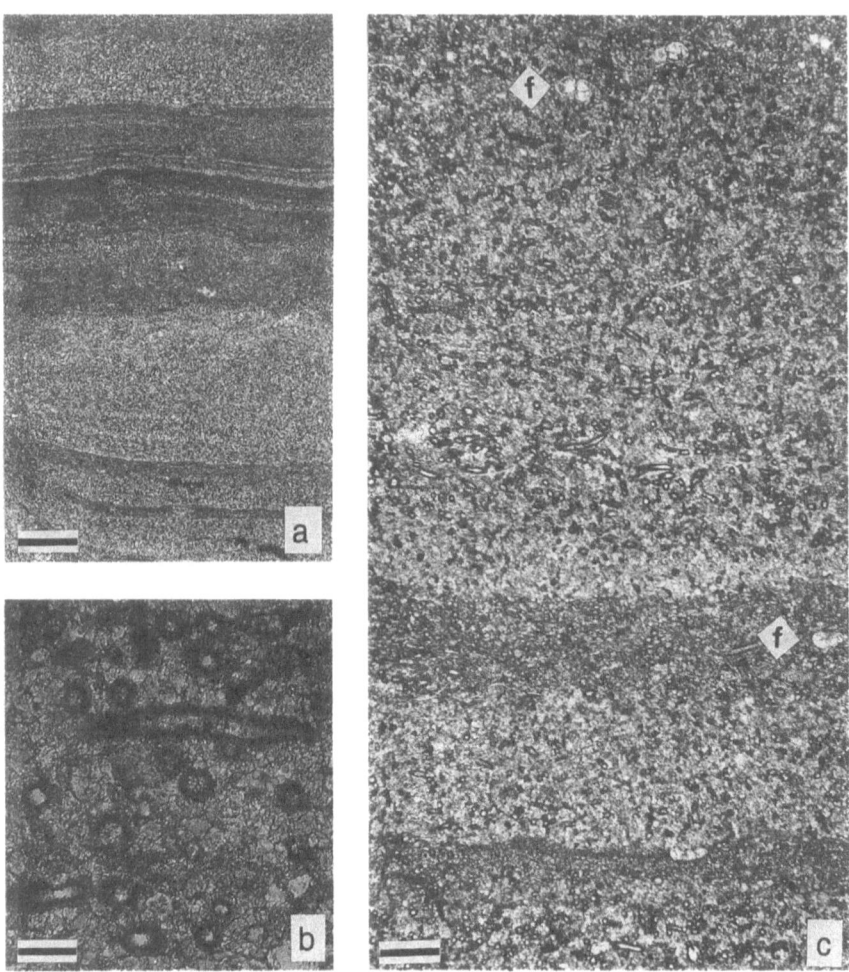

Figure 5 : Well preserved intermediate size filaments : Type 3 ; a) flat stromatolitic laminations comprising lots of filaments randomly distributed, scale bar = 7 mm ; b) close view of the filaments with transverse subcircular sections and oblique sections more or less flat-lying, parallel to the laminae. Filaments are filled with sparry calcite and outlined by a micrite coating ; scale bar= 75 µm ; c) Stromatolitic laminae are materialized by an alternation of layers with dark and light matrix. Rotalimorph foraminifers (f) suggest a lagoonal environment (Spanish-Pyrenées-Berga, Early Ilerdian) ; scale bar = 500 µm.

(Diatom) or the cyanophyte *Microcoleus*. More recently they have been attributed to *Girvanella palustris* in a study of the so-called lacustrine limestones from the late Cretaceous of Southern France (Colin and Vachard, 1977) and Yugoslavia (Bignot, 1981). As the filament habit differs from that of typical interwoven *Girvanella*, this precise denomination seems irrelevant. Moreover field evidences and sequencial distribution are inconsistent with the freshwater environment suggested (Colom et al., 1973). On the contrary they may be good indicators of a brackish lagoonal or salt-lake environment (Plaziat, 1984), as also indicated by the presence of dwarf (brackish) Foraminifers (Fig.5c) or Istrian Melanian gastropods. However we recognize that despite this brackish habitat indication, we do not know any modern analogs of these structures.

Type 4. Branched sinuous filaments are composed of dark micrite lacking differentiation of a central zone or an envelope (Fig.6). Their diameter varies from 18 to 40 µm ; the outer surface of the filament looks uneven. Filaments merge from a bulge on the filament or on its tips. They built thick laminae of interwoven filaments (fluviatile

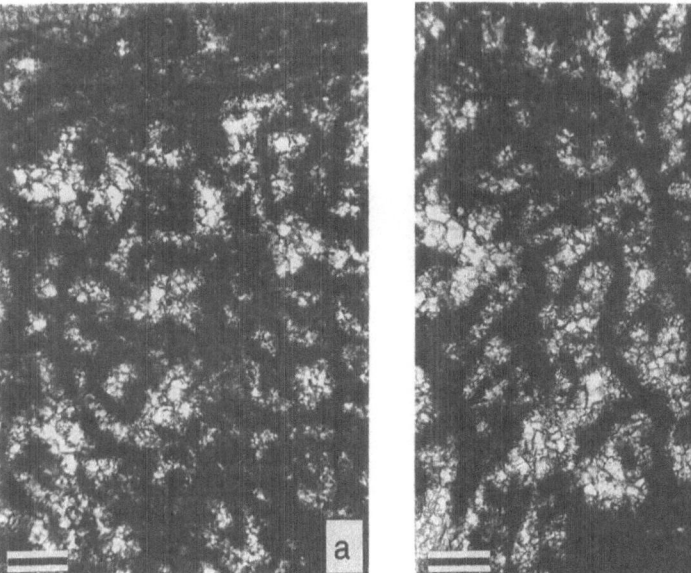

Figure 6 : Well preverved intermediate size filaments : Type 4 ; a) and b) Branching sinuous micritic filaments, more or less erect or prostrate in the upper left of the photograph (a). They are embedded in a sparry calcite cement. Filaments bulges may be seen. They are found in small, 0,5-0,7 m, bioherms isolated in fluviatile channels (Provence-Manosque, Revest-St-Martin, Eocene-Oligocene) ; scale bar= 150 µm.

bioherms, Provence-Manosque, Revest St Martin). They are erect and dense and poorly defined laminae result from occasional prostrate growth. No sediment between the filaments, but clear sparite filling.

Comparisons may be proposed with Stigonematales (Golubic, 1976b, p.139) where small bulges appear along straight portions of the filaments and originate branching. Among Proterozoic stromatolites microstructures the "vermiform" microstructure described by Walter (1972) can be compared to such a micritic filamentous network.

Group A-II : Thin filaments (7 to 25 μm)

They present different modes of preservation and will be subdivided according to the filaments habit.

Type 1. Filaments are erect, unbranched and densely packed parallel and vertical in the lamina. They exhibit a large diversity of preservation types from long empty molds (Fig.7a) filled with clear micro-sparite, crossing several laminae, to dark micritic filaments in sparitic layers. Sometimes filaments are composed of an external dark micritic envelope and a clear central tube (Fig.7b,c). Generally they form continuous repetitive carpet-like laminae. They are especially frequent among fluviatile and lacustrine environments (Languedoc).

Such filaments have been misinterpreted with *Rivularia hematites* (Freytet and Plaziat, 1965 ; Donsimoni and Giot, 1977). Filaments of similar structure have been reported from Northern Spain (Anadon and Zamarreno, 1981) and from Himalaya (Pleistocene,

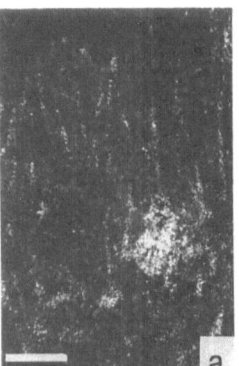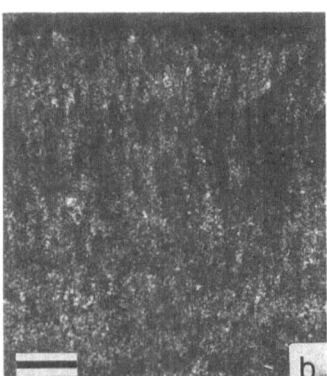

Figure 7 : Small size filaments : Type 1; Unbranched, erect subparallel filaments in dense carpet-like laminae : a) sparitic infill of the filaments (Limagne-Chadrat, lake environnement, thick pseudocolumnar biostromes) ; b) filaments preserved as dark micrite (Limagne-Gannat, deeper lake environnement, central part of a bioherm, 1-3 m in diameter) ; c) filaments preserved as micritic rods or outlined by a micritic coating (Provence-Fox-Amphoux, Campanian, fluviatile small oncolites). Scale bar= 150 μm.

Fort et al., 1982). *Phormidium* is more likely responsible for such a laminated fabric (Monty and Hardie, 1976 ; Monty 1976, p.205 ; Freytet and Plet, 1991)

Type 2. Radiating micritic filaments are grouped in bundles appearing as fans in thin section (Fig.8a,b). The filaments have no well defined boundaries. They display a large range of sizes (18-30 µm). Within the lamina the juxtaposed fans of densely packed filaments retain large clasts. They are very frequent in stromatolitic fluviatile buildups (Languedoc-Corbières, Azillanet, Combebelle).
This pattern of growth is suggestive of the fans of *Phormidium* filaments in recent freshwater settings (Monty, 1976). To some extent they also resemble the tufted filaments described in the Ebro Basin (Anadon and Zamarreno, 1981) or in the Early Cretaceous of Spain (Monty and Mas, 1981). There may be a relation between fan habit and trapping of large clasts.

Type 3. Micritic filaments (7 to 12 µm), poorly preserved are radially arranged within a sparitic hemisphere (Fig.8c,d,e,f). They appear as dark lines within a mosaic of calcite crystals. The hemispherical tussocks are usually discrete but may coalesce in a more continuous lamina (Limagne-Gannat and Languedoc-Corbières). Concentrical discontinuous dark laminae included in the tussock may be interpreted as distinct microbial layers or sediment-rich layer (clay) deposited during periods of cessation of the microbial growth.
These filaments described as "cloudy" by Donsimoni and Giot (1977) are referred to *Rivularia sp.* forming small hemispheric colonies (Golubic, 1976b, p.137 ; Monty, 1976, p.204). The filaments are poorly preserved as rods in a mosaic of elongated crystals. In Burgundy similar hemispherical constructions are found in running streams : they display a *Phormidium-Schizothrix* assemblage locally including *Gongrosira* (Freytet and Plet, 1991, Pl.3). In Cretaceous oncolites and laminar stromatolites, flabellate growth of filaments have been attributed to *Phormidium* (Monty and Mas, 1981).

Group A-III : Very small filaments (< 7 µm)

We record here microstructures where the filaments are poorly preserved as clear or dark rods. The different types are based on the laminar fabric, which are obviously due to microbial growth. Recrystallization may be responsible for the poor preservation of the filaments but microbial control on crystallization may also contribute to a fossilization which obliterates the filamentous structure.

Type 1. Small micritic filaments or rods radially arranged are scattered in a microsparitic lamina (Fig.9a,b). The lamina upper boundary is usually delineated by a thin micritic film (see Group B). In some examples, they comprise small columns (1-2 mm high) with concentric subconical dark films. Poorly preserved molds of erect filaments pass through the laminae (Fig.9c). This is a composite microstructure with

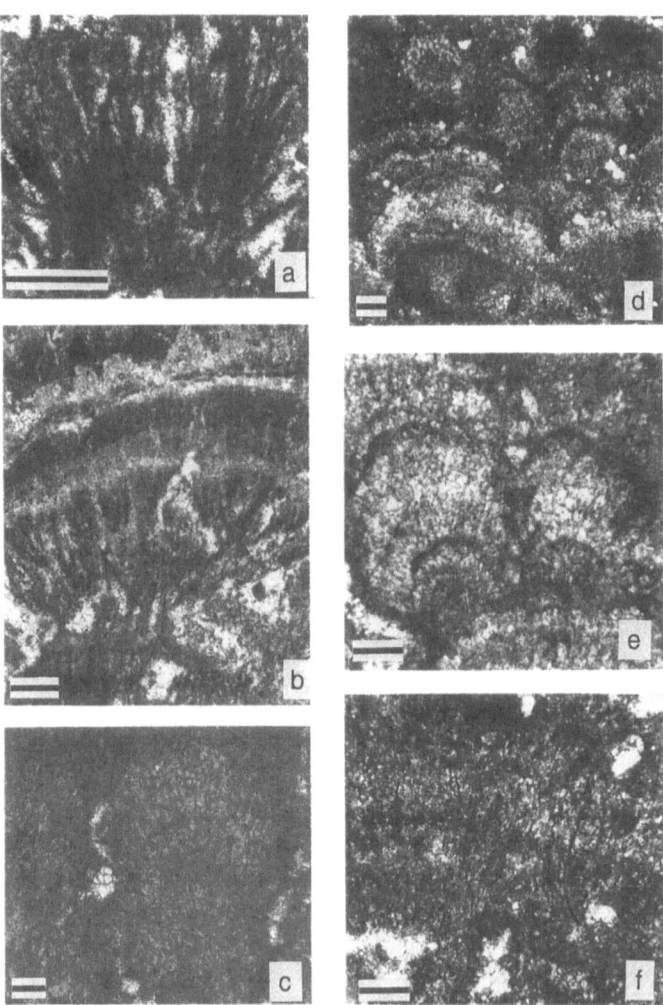

Figure 8 : Small size filaments : Type 2 and Type 3 ; a) Type 2 : flabellate growth of micritic filaments (Languedoc-St Chinian, Tudéry, oncolites in fluviatile channels, Early Eocene) ; b) Type 2 : the flabellate growth may be alternating with coalescent growth in carpet-like discontinuous laminae (Languedoc-Corbières, Combebelle, oncolites in fluviatile channels, Early Eocene) ; c) Type 3 : hemispheric tussocks composed of radially arranged filaments, clearly separated by a sparitic matrix (Languedoc-Corbières, Talairan, bioherms along the lake shore) ; d) and e) Type 3: hemispheric sparitic tussocks, alternating with dark micritic concentric laminae and superimposed randomly (Spanish-Pyrénées, Berga, bioherms in fluviatile channels) ; f) Type 3 : close view of a tussock showing the radiating filaments outlined by micritic coating (Limagne-Gannat, small columnar bioherms in lake environnement). Scale bar= 250 μm

two superimposed features : radial filaments molds and the concentric-conical films. The microcolumns have subcircular transverse section and are juxtaposed in a layer, encased in a matrix composed of microsparite and small clasts.

A modern analog of this type may be found in microbial mats from the Bahamas (Monty, 1965b, 1976) where horizontal films are built by *Schizothrix* filaments in alternation with mats built by radial filaments of *Scytonema sp*. In our example these are not well calcified and they are only poorly preserved as rods. In modern constructions, Freytet (1989b) and Freytet and Plet (1991) report a progressive transition from flat laminae to conical microcolumns due to a decrease in *Phormidium* filaments replaced by the flexuous small *Schizothrix*. This was related to the light restriction underneath big stable buildups or under the shadow of a bridge. This has also been described in some parts of fluvial stromatolites and interpreted as a diminution in light intensity (Freytet and Plaziat, 1965). Conical small columns have been found in a dried river dam (Languedoc-Montpellier, Bertrand-Sarfati unpublished data) and in the Pleistocene stromatolites from East African lakes (Casanova, 1986, Pl. 56) where they seem to be occasional.

Type 2. Small hemispheres with a radial-fibrous pattern and dark concentric laminae are frequent (Fig.9d). They can be isolated on the surface of different laminae (Provence-Manosque, Revest St Martin) or coalescing hemispheres form continuous or discontinuous layers (Montpellier-Languedoc). The hemispheres are composed of 8 to 20 concentric layers formed by alternating dark micritic and clear microsparitic layers. Each couplet averages 10-20 μm in thickness. Dark radial rods cut through several superposed laminae.

Mechanisms of growth may be compared to that of colonies of *Rivularia haematites*. The dark layer is due to a stop in the growth of the filaments emphasized sometimes by the development of another microbiota (Caudwell, 1987). Such hemispherical colonies have been attributed to *Rivularia* in Plio-Pleistocene travertines (Casanova, 1981a, b) and occasionally in lacustrine stromatolites from East Africa (Casanova, 1986).

Type 3. They are sub-hemispherical calcitic bodies with undulating extinction, encasing long rods (filament ghosts, Fig.10a,b). The linear filaments are 4 to 8 μm thick and radiate within the crystal. The discrete hemispheres are isolated, encrusting small twigs, or coalesce to form discontinuous laminae in small fluviatile bioherms (Provence-Manosque, Revest St Martin). In many pseudocolumnar coating of bioherms from the Limagne, very persistent laminae (lateral extent and thickness) are organized in palissadic sparite alternating with dark micritic films or laminae. Rare filaments can be recognized among the sparitic crystals (Fig.10d).

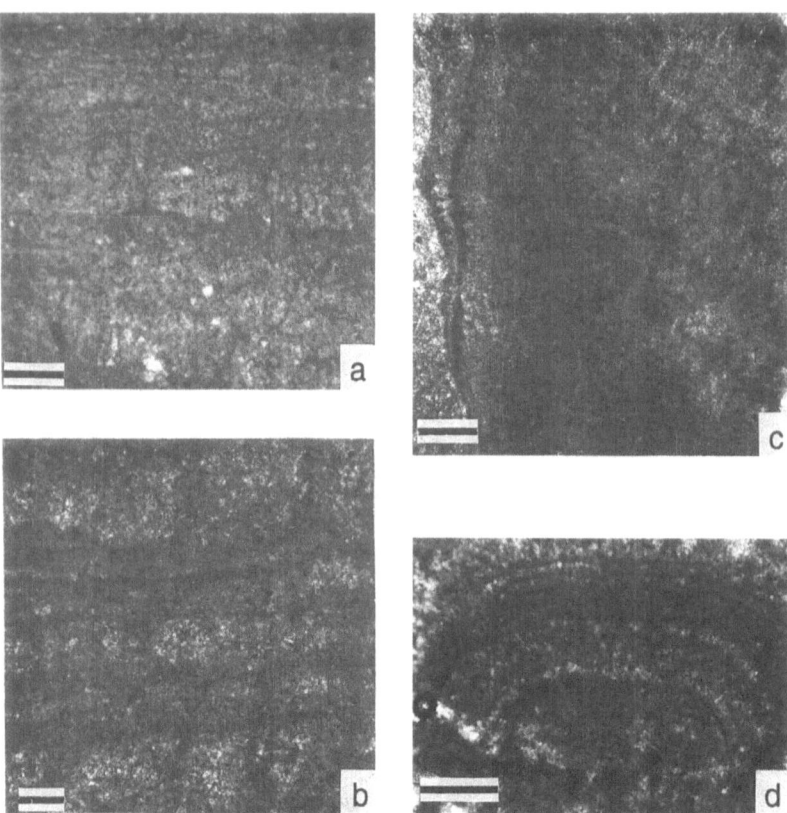

Figure 9 : Poorly preserved filaments, very small ; a) Type 1 : the laminae with small filaments are alternating with thin micritic films of the group B, Type 1 (Languedoc-St Chinian, Tudéry, early Eocene) ; b) Type 1 : microsparitic laminae with thin filaments, alternating with dark micritic films (Languedoc, Montpellier, Middle Eocene, large oncolites in fluviatile channels) ; c) microcolumns showing the concentric conical micritic films traversed by radial filaments (Languedoc-Corbières, Montlaur, lower Eocene, fluviatile oncolites) ; d) Type 2, small isolated hemispheres ; radial-filamentous pattern and concentric micritic laminae (Provence-Manosque, Revest St Martin, Eocene, small bioherms in fluviatile channels). Scale bar= 250 µm.

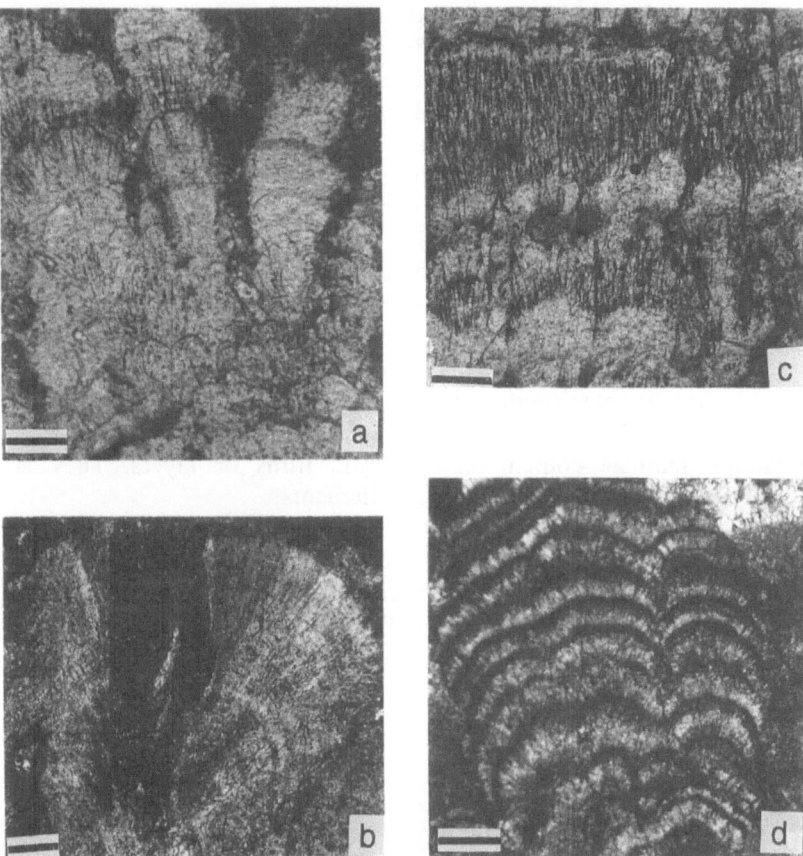

Figure 10 : Poorly preserved or dubious filaments, Type 3 ; a) fan shaped, sometimes microcolumnar large crystals of calcite encasing residual filaments reduced to inclusion-rich dark lines ; b) hemispherical monocrystals (Languedoc-Manosque, Revest, small bioherms in fluviatile channels) ; c) similar fabric with undubitable remnants of filaments in a recent travertine (The Alps, le Planay, St Jean-de-Belleville) ; d) like in the preceding example, the hemispheric pattern grade to coalescing continuous laminae with palissadic sparite encasing filaments or rods. These sparitic laminae alternate with another type of microbial structure, described in the Group B, Type 1. (Limagne-Gannat, stromatolitic envelope of large bioherms in lacustrine environnements). Scale bar= 400 µm except d= 1,5 mm.

Analogs of the hemispherical colonies have been described by Casanova (1986, Pl.20) in stromatolites from east african lakes and compared to what Chafetz and Folk (1984) called "bacterially constructed shrubs". Radial-fibrous calcite domes compared to

Rivularia and containing thin radiating rods are described from Recent stromatolites in freshwater ponds on Aldabra reef (Braithwaite et al., 1989). Laminar fabric analog exists in a Recent alpine spring travertine near St Jean de Belleville, (Fig.10c Bertrand-Sarfati, unpublihed data) with calcitic crusts underlaying the uppermost living layer. In the crystalline crust, very similar linear rods are the remnants of filaments. Living or recent *Oocardium* constructions present a transition from hemispheres to columns occasionally branching with a few micritic dark laminae intercalated between long radial-fibrous sparite. Smaller colonies are easily identified but older ones, even unmodified by diagenesis become very similar to the Limagne microstructures (Freytet and Plet, 1991 ; Freytet and Verrechia, 1989). Colonies attributed to *Oocardium* and illustrated by Wallner (1935, Pl1a,b) have also a similar crystalline fabric.

Group B : Micritic layers without visible filaments

The stromatolite laminae contain dark micritic films or layers. They are treated separately because they never include visible filaments.

Type 1. Very thin (3-5 µm) and very dark micritic layers comprising micro-crystals and organic matter are called films (Bertrand-Sarfati, 1976). These films lie flat or slightly undulating   capping other laminae (Fig.9a,b,c). They occur occasionally or in

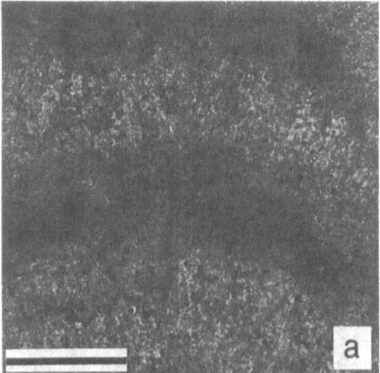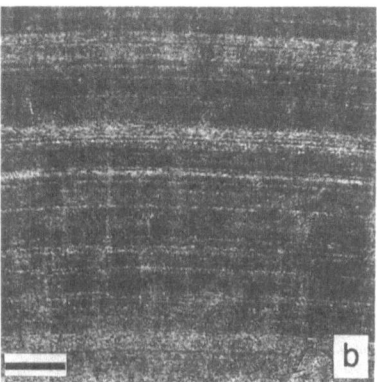

Figure 11 : Poorly preserved or dubious filaments : a) Type 1, Group B : micritic laminae without filaments interpreted as due to another type of microbial coenose (Languedoc-St Chinian, Tudéry, fluviatile oncolites) alternating with sparitic laminae, with poorly preserved filaments ; b) Type 1, Group C : fibrous and laminated sparitic fabric, fine laminae clear and dark grouped in clearer or darker macrolaminae (15-20 laminae). Color is due to abundance of dark inclusions. A radial fabric (diagenetic) is superimposed, (Limagne-Gannat, flat crusts in a lacustrine environnement). Scale bar= 400 µm.

series especially around large oncoids (Languedoc-Montpellier). Micritic laminae thicker than films alternate with the sparitic palissadic laminae, in lacustrine bioherms (Limagne-Gannat, Fig.10d,11a). Such films have been described by Monty (1967, 1976) from Recent stromatolite heads in the Bahamas. Thin intertwined filaments of *Schizothrix* are responsible for the calcite-rich dark films. The micritic laminae may be either set of films compacted or more complex diagenetic features.

Group C : Fibrous and laminated sparite microstructure

Alternating clear and dark couplets constitute the lamination (Fig.11b). Both units of the couplets are very thin averaging 50 µm. The dark layers are inclusion-rich while the clear layer are inclusion-poor. The laminae are very regularly superposed, in undulating crusts. Darker or lighter macrolaminae, are irregularly superimposed to the simple couplets. A fibrous striation crosses radially the laminated fabric and passes through several macrolaminae.
The dark inclusions of the dark and clear laminae have been attributed to bacteria (Donsimoni and Giot, 1977). Casanova (1986, Pl. 32) illustrates very finely laminated structures, he also attributes to bacterial growth.

Group D : Non microbial biogenic microstructure

Several structures obviously non microbial, are involved in the stromatolite constructions in the Late Cretaceous to Recent, nonmarine settings.

Type 1 : arched microstructure. Some stromatolites transverse sections comprise layers with scattered semicircular fenestrae (half-moon voids, Fig.12a-d). Microbial laminae are alternating with series of voids or juxtaposed to voids. The microbial coating is sometimes reduced to a thin arched lamina or develop in small columns and thick laminated layer. This structure has been called "celluleuse" (Freytet and Plaziat, 1965) and misunderstood with a microbial structure. In tangential sections these crescentic voids are contorted hemispherical tubes, obviously built by dwelling organisms colonizing the stromatolite surface (Fig.13).
Langeron (1902) first referred to Chironomid (*Diptera*) larvae when studying Modern travertines of the French Jura springs. Chironomid travertines from Southern Germany streams are also described by Wallner (1935 c,d,e) in non laminated stromatolites (*Vaucheria geminata*, *Plectonema phormidioides*) with a clear seasonal development. They have also been recorded occasionally in Recent stromatolites from european streams (Symoens, 1952 ; Bertrand, 1954 ; Monty, 1976) or Burgundy (Freytet and Plet, 1991) and in the Paris Basin (Freytet and Verrechia, 1989) associatiated with microbial laminae.

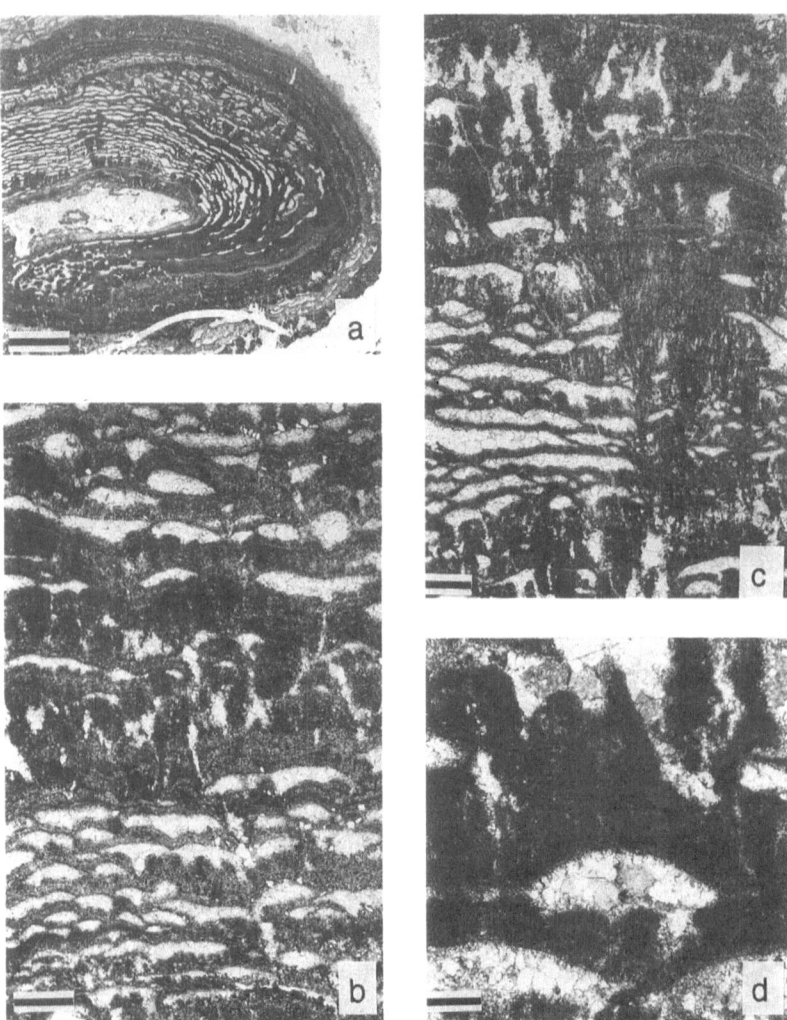

Figure 12 : Non microbial , arched structures ; a) arched ("cellular") structure forms a laminated coating on an Unio shell; polarity of the arched structures is induced by the more active accrection on the upper side of the shell; (erosional truncation, top of the photograph); scale= 1,2 cm ; b) associated growth of Chironomid larvae tubes and stromatolite, regularly alternating except in the centre where stromatolite laminae are thicker; scale= 2 mm ; c) stromatolite and Chironomid larvae alternation is more complex, regular on the left, the stromatolite is more continuous, on the right, preventing colonization by the larvae, on top, stromatolites are dominant; scale= 2 mm ; d) detail of a larvae arch and a flabellate filamentous stromatolitic laminae, (Languedoc-Corbières, St André, oncolites in fluvial channels) ; scale=500 µm.

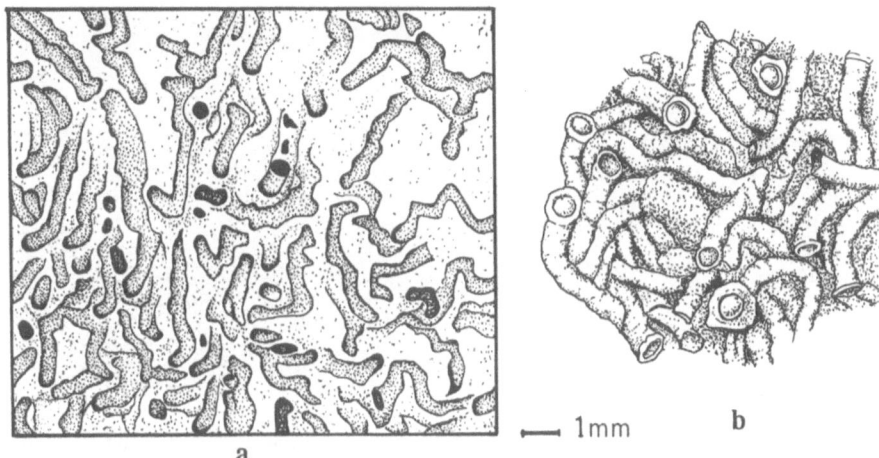

Figure 13 : Non microbial structure, Type 1; a) tangential section of the Chironomid larval tubes (Languedoc, Montoulier, Pliocene travertine); b) Chironomid larvae tubes (*Lithotanytarsus emarginatus*) on the surface of a spring travertine in the Pyrénées (after Bertrand, 1954, p.237).

Type : scattered spheres. Spheres (100-250 µm) with or without well defined microsparitic or sparitic envelopes are embedded within stromatolite lacustrine buildups of Limagne (Fig.14a,b). They are usually filled by sparry calcite. These spheres occur in clusters or scattered in some specific laminae.

They were attributed to *Chlorellopsis* (Dangeard, 1931) according to similarities with the spherulitic structure of the Green River stromatolites (Bradley, 1928). However the american spheres are much smaller and form the main component of the laminae. In the Limagne stromatolite their closer association with encrusted Caddies-fly larvae, rather suggests an interpretation as eggs of insects (Bertrand-Sarfati et al., 1966 and Donsimoni et al., 1975). Some Recent stromatolites from the River Seine display also similar spheres (Freytet, 1989b). Another interpretation is proposed, considering some of them as fungal spores (Lindqvist, this book).

Type 3 : bushes of large filaments. Very long (120-170 µm in diameter) erect filaments constitute very porous club-shaped or columnar constructions of centimetric to decimetric vertical extention associated or building larger metric edifices in lacustrine environment (Fig.15c ; Limagne-Gannat, Montaigu-le-Blin). The axial empty tube is coated by one or two layers of homogeneous micrite (Fig.15d, e). Filaments loosely interwoven with frequent dichotomous branching, present a fairly constant diameter (Fig.15a, b). Their visible length may exceed 2 cm in constructions of more than 5 cm thickness. Lamination is expressed by thin veneers of micrite or sediment (Fig.15c, e) related to a clear change in the filament growth habit : more prostrate except some isolated erect filaments crossing the micritic coating. Sometimes the filaments may be entirely filled by radially arranged micrite.

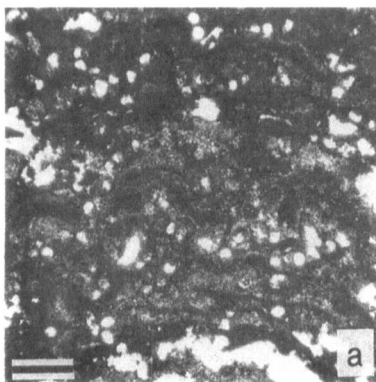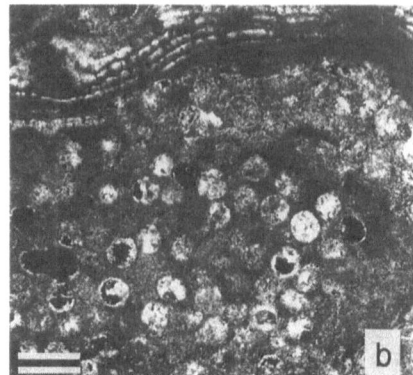

Figure 14 : Non microbial laminae : Type 2 : spheres ; a) spheres are scattered randomly among very irregular undulated stromatolitic laminae comprising micritic layers alternating with poorly preserved sparitic laminae (no filaments are visible). The spheres have a clear sparitic infill and no micritic outline. They are interpreted as eggs of insects (Limagne-Gannat, small columnar bioherms in lacustrine environnement) ; scale bar= 1,5 mm ; b) detail of a group of spheres with a sparitic envelope closely packed in a micritic matrix and supporting a few stromatolitic laminae (in a stromatolitic small isolated column, using aggregated groups of spheres as a substrate, Limagne-Montaigu-le-Blin); scale bar= 300 μm.

In previous works these filaments have been erroneously interpreted as calcified mosses (Bertrand-Sarfati et al., 1966) or as *Rivularia* filaments (Donsimoni and Giot, 1977). In respect to the size and habit of these filaments they are more probably remains of a green or brown Algae (Golubic, 1976a, p. 121). In large bioherms from an Upper Miocene lake of southern Germany , such large branching tubes coated by micrite (180 μm in diameter) were attributed to the green Algae *Cladophorites* by analogy to the Modern *Cladophora* (Riding, 1979). *Vaucheria sp.* constitutes tufts of thinly coated long filaments lying parallel to the current, presenting the same loose organization (Fig.15b). However they clearly differ by the prostrate attitude of the filaments controled by the current flow on the ledge of the travertine dams (Bertrand-Sarfati, unpublished data, Le Lez-Montpellier; Freytet and Plet, 1991, Pl.4).

COMPARISON WITH PROTEROZOIC MICROSTRUCTURES

Obvious analogies can be found between some Proterozoic stromatolitic microstructures and some of the types described previously. We have selected four examples of typical Proterozoic stromatolitic microstructures which can offer analogs in the Cretaceous-Tertiary record. The main originality of Proterozoic microstructure is the fact that one morphological type comprises only one type of microstructure even when biostromes reach 10 meters high.

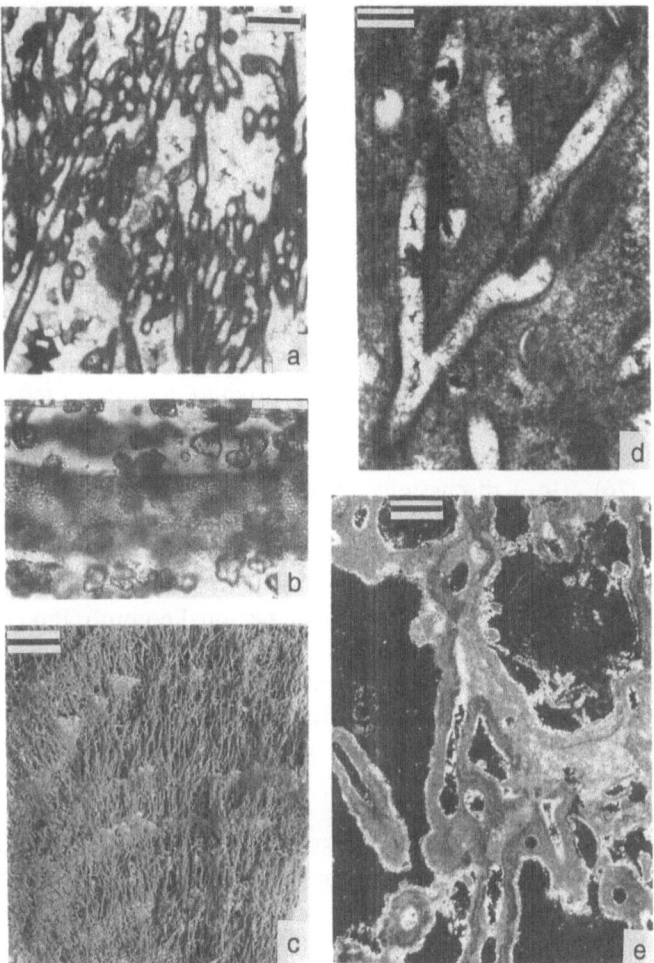

Figure 15 : Non microbial laminae, large filaments ; a) erect, ramified filaments coated by dense black micrite ; the spaces between filaments and filament centers are filled by sparry calcite cement (Limagne-Gannat, large hemispherical bioherms in lake environment) ; scale= 280 µm; b) Living filament of *Vaucheria* with scattered calcite crystals (Burgundy); scale= 20 µm; c), d) and e) large filaments in Limagne-Montaigu-le-Blin, bioherms built by columns : c) part of a column. A rough lamination is materialized by a stop of growth of some of the filaments (not all) ; scale= 1,5 mm; d) Ramified filaments with a micrite coating, remain empty or filled with spar cement ; interspaces are filled with calcite ; scale= 80 µm ; e) The filaments are coated by two or three micrite layers, and a veneer of small sparite crystals lines the void. Mud is deposited at the cavity bottom giving rise to the faint lamination visible in c) ; scale= 200 µm.

Microstructures of Proterozoic stromatolites have been tentatively used as diagnostic features to discriminate stromatolites at the form level (Semikhatov, 1978). Few attempts to classify microstructures have been made taking into account the distinction between diagenetic features and biogenic structures (Bertrand-Sarfati, 1976), however, the use of a simply descriptive classification has been proposed by Komar (1976, 1989).

Tussocky microstructures

Tussocky microstructures compose "regular lamination defined by the juxtaposition of separated hemispherical tussocks of different size (from 30 µm up to 1mm) sometimes with a concentrical growth pattern" (Bertrand-Sarfati, 1976, p. 253). Tussocks may be outlined by a dark film, and sometimes display concentric dark lines that can be interpreted as growth discontinuities. This microstructure has been firstly recognized in different sections of the Proterozoic cover of the West african craton (Bertrand-Sarfati, 1972). Later on they have been found in Australia, in the Early Proterozoic Nabberu Basin (Preiss, 1976). In a discussion about Australian stromatolites, Grey (1984) differenciates several types of tussocks in regard to their shapes and mode of superposition. In some Early Proterozoic stromatolites from Canada, laminae displaying hemispherical colonies are recorded by Semikhatov (1978). Another example is also recognized in China (Cao Rui-ji, personal comm.). Proterozoic tussocks display different degree of recrystallisation from well defined cushion-like tussocks (Fig.16a) to sparitic tussocks with remnants of linear rods (Fig.16c). In an isotopic and mineralogical study of some carbonates from the Atar area (West African Craton, Mauritania) a shift in the isotopic ratio, associated with the stromatolites with tussocky microstructure allow Fairchild et al. (1990) to interpret tussocks as firstly aragonitic prior to recrystallization in calcite.

A comparison with modern cyanophyceae stromatolite builders (Bertrand-Sarfati and Pentecost, 1992) leads to the conclusion that among eleven diagnostic characters of *Rivularia*, seven are common with these tussocks. Interesting comparisons can also be made with tertiary stromatolite fabrics, where filaments are still visible in some of the tussocks (Fig.16f) and grade to sparitic laminae when less visible (Fig.16e). Couplet-like coalescing tussocks found in the Proterozoic can also be compared with couplets from the Tertiary stromatolites (Fig.16b,16e).

Microstructure in alternating micrite-microsparite laminae

This microstructure is one of the most frequently found in Proterozoic stromatolites ; it was defined as "micritic mats" (Bertrand-Sarfati, 1976). Laminae may be composite with a gradational passage from a microsparitic base to a micritic top ; they may end also with a dark micritic film or layer, clearly outlined (Fig.17a). A possible analog is presented by one of the Tertiary stromatolite microstructure. Here, the lower part of the mat is microsparitic and passes to an upper micritic fringe outlined by a dark micritic film (Fig.17e). The faint filaments still visible in the clearer part refer to a type described in this paper and interpreted in terms of microbial growth (Fig.9a). Another

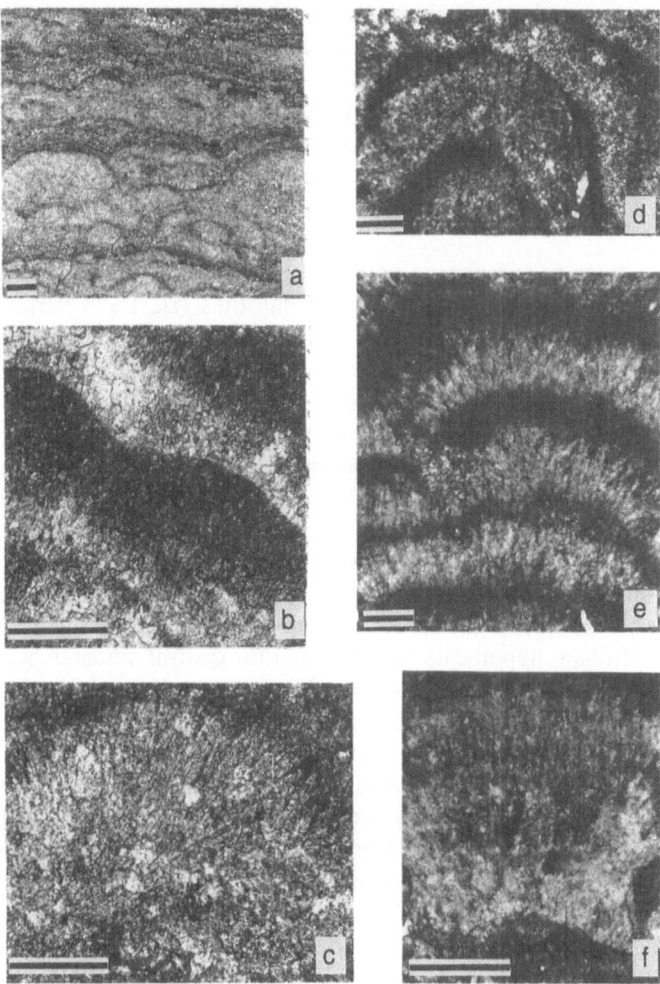

Figure 16 : Tussocky microstructure, in Proterozoic (a,b,c) and Tertiary (d,e,f) stromatolites ; a) cushion-like tussocks outlined by a dark film, *Serizia radians,* upper Proterozoic, Atar-Mauritania; b) coalescing cushions giving rise to doublet-like undulate laminae, *Tungussia cumata,* upper Proterozoic, Algeria ; c) detail of a tussock showing the radially arranged rods within the sparite, *Serizia radians,* upper Proterozoic, Atar-Mauritania; d) tussocks in isolated colonies, outlined by a sparitic laminae, filaments are still visible, Spanish Pyrénées, Berga, early Ilerdian ; e) coalescing cushions grading to doublets, filaments are clearly visible in these Limagne-Gannat Oligo-Miocene stromatolites ; f) isolated hemispheric tussock with obvious filaments (Limagne-Gannat, Oligo-Miocene). Scale bar = 400 μm.

possible comparison can be made between couplet-like, clear and dark layers from the Proterozoic (Fig.17c) and couplet made of a composite mat : a sparitic base with trapped detrital grains grading to a microsparitic upper part in which scattered filaments are still visible (Fig.17f). In these Proterozoic microstructure it is impossible to distinguish through diagenesis and compaction what was the initial microbial habit.

Microstructures in films

A specific microstructure comprising very thin dark micritic films regularly superposed has been recognized in some Proterozoic stromatolites (Bertrand-Sarfati, 1972, 1976). They alternate with clear sparitic layers of different thickness. They built different types of stromatolites according to the depth of the water. The most striking are *Conophyton* in which the films are building high cylindrical-conical columns. In the vertical peripheral part of the columns, lamina is a thin, very coherent film that extends up to 2 meters in height (Bertrand-Sarfati and Moussine-Pouchkine, 1985). Films are supported by the buildup, probably rapidly lithified. Within this vertical film, the filaments may be erect toward the light, therefore composing parallel vertical pattern. The central zone with the conical part of laminae, represents a small cylinder (1cm diameter) where "thickened" laminae, reached the highest point of their growth. There, the mat with filaments, erect toward the light, become folded and/or collapsed within the conical central zone.   Another hypothesis to explain this central conical zone, implies the existence of a trapped buble of gaz (Golubic, 1973 ; or in Arctic lakes, Love et al., 1983 and Wharton, this book). When it escaped, the mat collapsed and folded. Modern analogs with a very much smaller size, but displaying similar phototropic mechanism of growth are well known in Yellowtone park and are related to the phototropic response of a *Phormidium* associated to other bacteria (Walter et al, 1976). The result is the appearance of a conical apex. In ancient *Conophyton* a similar fabric of filamentous cyanobacteria has been illustrated from cherty stromatolites of the Proterozoic of USSR (Schopf and Sovietov, 1976). In both case the thinness of the film is related to the thinness of the filamentous micro-organisms and their growth parallel to the lamina surface, vertical in peripheral parts. Other stromatolites in the Proterozoic encompass also dark micritic films, but are not conical (Fig.17b). These may be compared to regularly superposed films from Limagne-Gannat (Fig.17d), or to the isolated films which are associated with other laminae (see group B).

Microstructure with preserved filament molds

This microstructure was first described in Proterozoic stromatolites of Siberia (Schapovalova, 1974). The example from Siberia comprises repetitive micritic laminae without clearly defined boundaries, striated by filaments molds. In an attempt to classify stromatolite microstructures, Komar (1976) named this microstructure Canalophorida later he precised that they should be comprised in the supertype *Filiformita* (Komar, 1989, Pl.2,4-5-7).

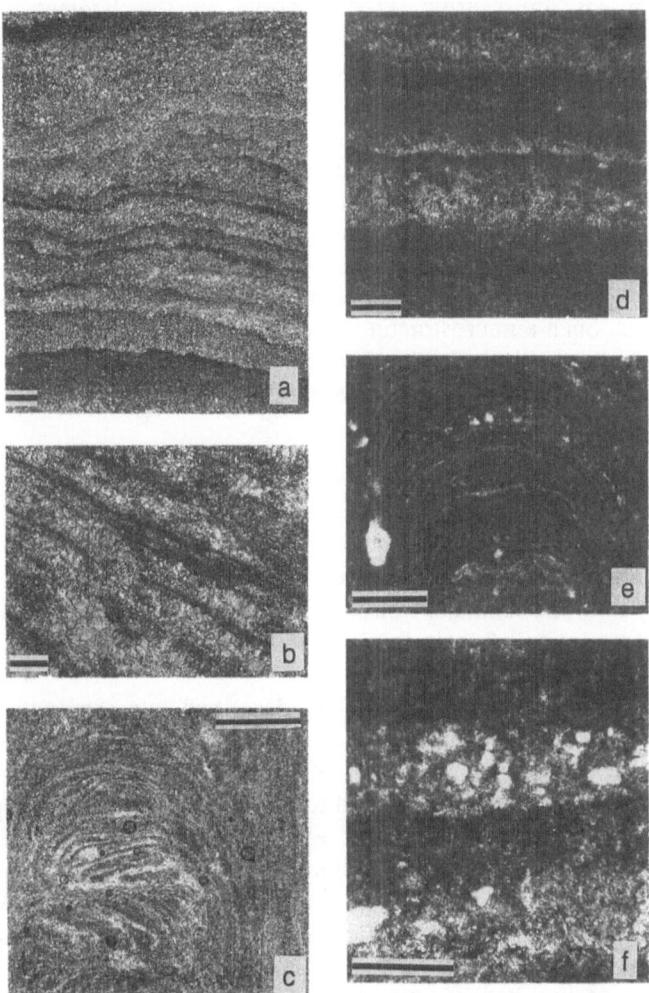

Figure 17 : Micritic mats and micritic films in Upper Proterozoic (a,b,c) and Tertiary (d,e,f) stromatolites ; a) superposed composite mats with a gradation from sparite to microsparite and alternating with a micritic discontinuous laminae, *Inzeria lindina*, Proterozoic, Zaïre ; b) superposed thin films alternating with laminae of sparitic cement, *Baicalia mauritanica*, Proterozoic, Atar-Mauritania ; c) central conical zone of *Conophyton*, folding and collapse of the apical cone as well as cement filling voids, are clearly visible ; Proterozoic, Atar-Mauritania ; d) micritic composite mat grading from sparite to micrite (arrow) and alternating with groups of thin films, Languedoc-Corbières, Montlaur ; e) regularly superposed clear and dark couplets, the dark laminae may be compared to films, Limagne-Gannat, Oligo-Miocene ; f) composite mat, the basal sparitic part of the mat contains trapped detrital grains, while the micritic upper part clearly shows filamentous texture, Languedoc-Corbières, Jonquières. Scale bar= 400 µm.

The filament molds are sparitic, erect, normal to the laminae and branching. Subcircular in section they have a diameter of 35-50 µm. In the micritic matrix, smaller, less visible filaments with a size range of 8-12 µm in diameter, display a sheath and a central tube. The difference between these two populations of filaments may be due to fossilization or reflects two distinct species. This example (personally studied on a sample given by Dr Komar) presents an alternation of clear sparitic and dark micritic thick laminae. Filaments molds can be either restricted to the micritic lamina or cut through several couplets (Fig.18a,c). These filament molds averaging 40 µm, in diametre, are filled by sparite and have no micritic coating. Some of them stop at the lamina boundary. This can be interpreted in term of diagenetic alteration of the filament in the clear lamina. Such a microstructure can be compared to that of intermediate size filaments (Fig.18c,d) or with the largest filaments (Fig.15). Therefore, discussion arises about the value of size criterion in interpretating the filaments origin.

DISCUSSION AND CONCLUSION

We proposed a tentative classification of the diverse nonmarine fossil stromatolite microstructures, based on biogenic features according to the filament sizes and the filament habits within the laminae. For each type a comparison with modern analogs supports the interpretation of their fossil counterparts in terms of microbial growth habit.

This inventory may be applied to previously described microstrutures. For instance, microstructures of lacustrine stromatolites from the Pliocene Ridge Basin, California (Link et al., 1978) correspond to several of our types : sparitic palissadic laminae alternating with micritic laminae correspond to our class A-IV, and our type 3 is similar to the microstructure figured by Link et al. in their Figure 5e. The short, large, unbranched filaments illustrated in their figure 5c, fall in the range of our type 1 of Group A-I. We suggest that the use of a more precise code of definition of the microstructures would certainly facilitate comparisons of stromatolites of diverse ages and environments.

Regarding the relationships between morphologies and environments, it is impossible to distinguish freshwater, brackish and marine environments on the basis of the microstructure of their respective stromatolites. However it is well known that freshwater stromatolite microstructures are especially well preserved and contain more diverse microbial remains, mainly because of their early calcification. For several authors including Golubic (1973) incrustation processes appear to be limited to specific environments rather than to specific species. However specific aptitude to encrustation, recorded in many different groups varies from general to exceptional. The difficulty remains about the evaluation of the place of carbonate precipitation during algal or microbial life or during early diagenesis (Golubic and Fischer, 1975).

Some microstructures in Late Cretaceous-Tertiary freshwater stromatolites from France include features obviously similar to Proterozoic ones. The comparison is worthwile in

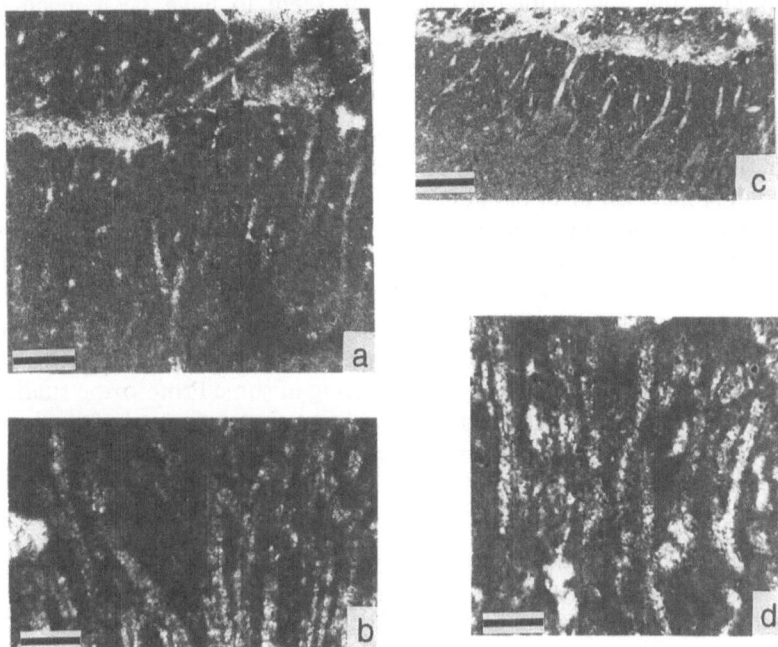

Figure 18 : Preserved filaments molds in Proterozoic (a,c) and Tertiary (b,d) stromatolites ; a) and c) filaments of the type *Canalophorida*, filaments are preserved as sparite infill, they are radial to the laminae surface. They are embedded in micritic thick lamina, c) same example, some filaments are clearly stopping at lamina boundary (borings ?), Souther Ural Avzyan formation Upper Proterozoic, (thin section Komar collection, Moscow,) ; scale bar= 400 µm. b) and d) sample displaying different stages of alteration of the filaments, in b) they are clearly visible in their sparitic coating, in d) they have disappeared and the sparite infill only remains. (Sample figured in Fig.6a,b) Limagne-Gannat ; scale bar= 50 µm.

the case of stromatolites which display laminae with well preserved microbial habit and altered fabric. Therefore the interpretation of Proterozoic microstructures becomes more convincing. The basic dark-and-clear couplet may be interpreted in terms of different biocenoses, diagenesis masking the remnants of the original feature. Both Proterozoic and freshwater stromatolite microstructures are dominated by processes of precipitation-calcification rather than grains-trapping and binding.

A major discrepancy between Proterozoic and Modern (Late Cretaceous to Present time) stromatolites is the microstructural and morphological stability of the ancient stromatolites. In the Proterozoic, the same simple or complex microbial biocenose appears to have produced similar morphotypes on areas of hundreds of km^2, according to chemical and hydrodynamical factors stable in space and time. This supports the use of morphologic systematics in Proterozoic basinal biostratigraphy. In modern stromatolites the environmental conditions seem to have been much more variable resulting in several changes of microstructures within the same bioherm. Moreover the complexity of biogenic components reinforces this variability.

Like modern stromatolites, Proterozoic constructions resulted from microbial/algal coenoses. However ancient morphologies and laminated structures look more stable and simpler than modern ones. Is it a matter of environment stability (predictability), of low biologic competition or of simplicity of assemblages ? Conversely we must point out that Modern continental environments are excessively diverse and moreover that microbiota now develop in association with animals (Chironomid tufas) and vegetals (mosses, algae, phanerogams). Modern continental stromatolites are so deeply influenced by their biogenic substrate morphologies that we insist on the necessity of microstructural and biologic approaches excluding the para-systematic one based only on pure macro-morphology as it was prevailing in some Proterozoic studies.

The present contribution intends also to show the limits of a formal "biologic" description in the stromatolite systematics, especially for freshwater buildups.

AKNOWLEDGEMENTS

We are very grateful to Stan Awramik, Bruce Purser and Claude Monty for their constructive critics and their help concerning the improvement of manuscript regarding the English language.

REFERENCES

Adolphe J.P. and Rofes G. (1973) "Les concrétionnements calcaires de la Levrière", Bull Association Française d' Etudes du Quaternaire 2, 79-87.
Adolphe J.P. (1981) Observations et expérimentations géomicrobiologiques et physico-chimiques des concrétionnements carbonatés continentaux, actuels et fossiles, Thèse Doct Etat, Univ Paris.
Anadon P. and Zamarreno I. (1981) "Paleogene nonmarine algal deposits of the Ebro Basin, Northern Spain", in C.L.V. Monty (ed), Phanerozoic stromatolites, Springer-Verlag, Berlin, pp. 140-180.
Bertrand M. (1954) Les insectes aquatiques d'Europe, Encyclopédie Entomologique Le Chevallier, Paris.
Bertrand-Sarfati J. (1972) Les stromatolites du Précambrien Supérieur du Sahara Nord-Occidental ; inventaire, morphologie et microstructure des laminations, corrélations stratigraphiques, Publication C.N.R.S, Centre de Recherches sur les zones arides, Paris.
Bertrand-Sarfati J. (1976) "An attempt to classify Late Precambrian microstructures", in M.R. Walter (ed), Stromatolites, Elsevier, Amsterdam, pp. 251-259.
Bertrand-Sarfati J., Freytet P. and Plaziat J.C. (1966) "Les calcaires concrétionnés de la limite Oligo-Miocène des environs de Saint Pourçain sur Sioule : (Limagne d'Allier) : rôle des Algues dans leur édification, analogie avec les stromatolites et rapport avec la sédimentation" Bull Société Géologique de France 7, 652-6224.
Bertrand-Sarfati J. and Moussine-Pouchkine A. (1985) "Evolution and environmental conditions of *Conophyton-Jacutophyton* associations in the Atar dolomite (Upper Proterozoic, Mauritania)", Precambrian Research 29, 207-234.
Bertrand-Sarfati J., Casanova J., Duringer P., Massoubre M., Freytet P., and Plaziat J.C. (1990) "Nonmarine stromatolites in the Western Europe Oligo-Miocene Rift System", 13th Intern Sedim Congress, Nottingham, Abstr, pp. 30-31.

Bertrand-Sarfati J. and Pentecost A. (1992) "Tussocky" microstructure, a biologic event in Upper Proterozoic stromatolites : comparisons with modern freshwater stromatolite builders" in M. Schidlovski, S. Golubic, M.M. Kimberley and P.A. Trudinger (eds), Early Proterozoic biological events, Springer-Verlag, Berlin, pp. 468-477.

Bignot G. (1972) Recherches stratigraphiques sur les calcaires du Crétacé et de l'Eocène d'Istrie et des régions voisines. Essai de révision du Liburnien, Trav Lab. de Micropaléontologie, Paris.

Bignot G. (1981) "Illustration and paleoecological significance of Cretaceous and Eocene *Girvanella* limestones from Istria (Yugoslavia, Italy)", in C.L.V. Monty (ed) Phanerozoic Stromatolites I, pp. 134-139.

Bradley W.H. (1928) "Algal reefs and oolithes of the Green River formation", Geological Survey, U.S., Prof pap. 154, 203-223.

Braithwaite C.J.R., Casanova J., Frevert T. and Whitton B.A. (1989) "Recent stromatolites in landlocked pools on Aldabra, Western Indian ocean" Paleogeo. Paleoclim. Paleoecol. 69, 145-165.

Casanova J. (1981a) Etude d'un milieu stromatolitique continental : les travertins plio-pléistocène du Var (France). Lithogenèse des formations carbonatées, Thèse, Lab Géologie Quaternaire, Luminy-Marseille.

Casanova J. (1981b) "Morphologie et biolithogenèse des barrages de travertins", Act Coll Ass Geographes Fr, Memm Ass Fr de Karstologie 3, 45-54.

Casanova J. (1986) Les stromatolites continentaux : paléoécologie, paléohydrologie, paléoclimatologie : application au rift Gregory, Thèse Aix-Marseille.

Casanova J. and Lafont R. (1985) "Les Cyanophycées encroûtantes des eaux courantes du Var", Verh Internat Verein Limnolologie 22, 2805-2810.

Casanova J. and Nury D. (1989) "Biosédimentologie des stromatolites fluvio-lacustres du fossé Oligocène de Marseille". Bull Société Géologique de France. 8, 1173-1184.

Castanier S., Maurin A. and Perthuisot J.P. (1989) "Production bactérienne expérimentale des corpuscules carbonatés sphéroïdaux à structure fibro-radiaire. Réflexions sur la définition des ooides". Bull Société Géologique de France 8, 589-596.

Caudwell C. (1983) "Les Rivulariacées actuelles : interprétation possible de la structure zonée des concrétions stromatolitiques à *Rivularia hematites*", Geobios 16, 169-177.

Caudwell C. (1987) "Etude expérimentale de la formation de micrite et de sparite dans les stromatolites d'eau douce à *Rivularia*", Bull Société Géologique de France 8, 299-306.

Chafetz H.S. and Folk R.L. (1984) "Travertines : depositional morphology and the bacterially constructed constituents", Journ Sedimentary Petrolology 54, 289-316.

Colin J.P. and Vachard D. (1977) "Une "Girvanelle" dulçaquicole du Cénomanien du Sud-ouest de la France ; *Girvanella palustris*", Rev Paleobotanique Palynologie 23, 293-302.

Colom G. (1961) "La paléoécologie des lacs du Ludien-Stampien inférieur de l'ile de Majorque", Rev. de Micropaléontologie 4, 17-29.

Colom G. (1967) "Les lacs du Burdigalien supérieur de l'ile de Majorque (Baléares) et le rôle des Mélosires (Diatomées) dans la formation de leur varves", Bull Société Géologique de France 7, 825-843.

Colom G., Freytet P. and Rangheard Y. (1973) "Sur des sédiments lacustres et fluviatiles stampiens de la Sierra nord de Majorque (Baléares)". Ann. Sci. Univ Besançon, Géol 20, 167-174.

Dahanayake K., Gerdes G. and Krumbein W.E. (1985) "Stromatolites, Oncolites, and Oolites Biogenically Formed in situ". Naturwissensh, Springer-Verlag, Berlin, pp. 513-518.

Dangeard P. (1938) "Le genre *Vaucheria* spécialement dans la région sud-ouest de la France", Le Botaniste 39, 183-254.

Donsimoni M. (1975) Etude des calcaires concrétionnés lacustres de l'Oligocène et de l'Aquitanien du bassin de Limagne (Massif Central,France), Thèse 3° cycle, Paris VI.

Donsimoni M. , Giot D., Lang J. and Lucas G. (1975) "Rôle des organismes et des réactions physico-chimiques dans la genèse des calcaires concrétionnés de Limagne", 9 th Intern Sedim Congress, Nice 2, 51-56.

Donsimoni M. and Giot D. (1977) "Les calcaires concrétionnés lacustres de l'Oligocène supérieur et de l'Aquitanien de Limagne (Massif Central)", Bull B.R.G.M. I-2, 131-169.

Emeis K.C., Richnow H.H. and Kempe S. (1987) "Travertine formation in Plitvice National Park, Yugoslavia : chemical versus biological control", Sedimentology 34, 959-609.

Fairchild I.J., Marshall J.D. and Bertrand-Sarfati J. (1990) "Stratigraphic shifts in Carbon isotopes from Proterozoic stromatolitic carbonates (Mauritania) : influences of primary mineralogy and diagenesis", in A. Knoll and J.H. Ostrom (eds), Preston Cloud volume, Geol. Society America Bull. 290-A, 46-79.

Fort M., Freytet P. and Colchen M. (1982) "Structural and sedimentological evolution of the Thakkola Mustang grabben (Nepal Himalayas)", Z Geomorph N F, Stuttgart 42, 75-98.

Freytet P. (1989a) "Contribution à l' étude des tufs du Bassin de Paris : typologie des édifices tuffacés (stromatolitiques) des chenaux fluviatiles (aspect macroscopique)", in Les tufs et travertins quaternaires des bassins de la Seine et de la Somme et des régions limitrophes, Bull centre de Géomorphol. Caen 38, 9-28.

Freytet P. (1989b) "Contribution à l'étude des tufs calcaires (édifices stromatolitiques) du Bassin de Paris: les organismes constructeurs, aspect microscopique", in Les tufs et travertins quaternaires du bassin de la Seine et de la Somme, et des régions limitrophes, Bull Centre de Géomorphol., Caen 38, 35-53.

Freytet P. (1992) "Exemple de fossilisation de restes végétaux (algues, feuilles) par de la calcite, en milieu fluviatile et lacustre, dans l'actuel et dans l'ancien", Bull. Société Botanique de France 139, 69-74.

Freytet P. and Plaziat J.C. (1965) "Importance des contributions algaires dues à des Cyanophycées dans les formations continentales du Crétacé supérieur et de l'Eocène du Languedoc", Bull. Société Géologique Fr. 7, 679-694.

Freytet P. and Plaziat J.C. (1972) "Les constructions algaires continentales stromatolitiques. Exemples pris dans le Crétacé supérieur et le Tertiaire de France et d'Espagne du Nord", 24th Intern. Geol. Congr., Montreal 7, 524-534.

Freytet P. and Plaziat J.C. (1982) Continental carbonate sedimentation and pedogenesis. Late Cretaceous and Early Tertiary of Southern France, Schweizerbart'sche Verlags Publ, Stuttgart.

Freytet P. and Moissenet E. (1983) "Présence de restes algaires identifiables dans les croûtes calcaires plio-quaternaires du Nord-Est de l'Espagne", C. R. Acad. Sci, Paris 296, 1563-1566.

Freytet P. and Verrecchia E. (1989) "Les carbonates continentaux du pourtour méditerranéen : microfaciès et milieu de formation", Méditerranée 23, 5-28.

Freytet P. and Plet A. (1990) "Contribution à l'étude des tufs du Bassin de Paris : typologie des édifices tuffacés (stromatolitiques) des chenaux fluviatiles (aspect macroscopique)". Géobios.

Gauthier-Lièvre L. (1955) "Le genre *Vaucheria* en Afrique du nord", Bull Soc Histoire Naturelle Afrique du Nord, Alger 46, 309-339.

Golubic S. (1973) "The relationship between blue-green algae and carbonate deposits", in N.G. Carr and Whitton B.A. (eds), The biology of Blue-green algae, Blackwell Oxford Press, pp. 434-472.

Golubic S. (1976a) "Organisms that build Stromatolites", in M.R. Walter (ed), Stromatolites, Elsevier, pp. 113-126.

Golubic S. (1976b) "Taxonomy of extant stromatolite building cyanophytes", in M.R. Walter (ed), Stromatolites, Elsevier, pp. 127-140.

Golubic S. (1991) "Modern stromatolites a review", in R. Riding (ed), Calcareous algae and Stromatolites, Springer-Verlag, pp. 541-561.

Golubic S. and Marcenko E. (1965) "Uber Konvergenzerscheinungen bei Standortsformen der Blaualgen unter extremen Lebensbedingungen, Schweitz", z Hydrologie 27, 207-217.

Golubic S. and Fischer A.G. (1975) "Ecology of calcareous nodules forming in little Connestoga Creek near Lancaster, Pensylvania", Verh. Int. Verein Limnol. 19, 2315-2323.

Golubic S. and Barghoorn E.S. (1977) "Interpretation of microbial fossils with special reference to the Precambrian", in H. Flügel (ed), Fossil Algae, Springer-Verlag, Berlin, pp 1-14.

Golubic S. and Campbell S.E. (1981) "Biogenically formed aragonite concretions in Marine *Rivularia*", in C.L.V. Monty (ed), Phanerozoic Stromatolites I, pp. 209-229.

Grey K. (1984) "Biostratigraphic studies of stromatolites from the Proterozoic Earaheedy Group, Nabberu Basin, Western Australia", Geological Survey, W. Australia Bull 130, pp. 123.

Hugueney M. , Tachet H. and Escuillie F. (1990) "Caddis-fly *pupae* from the Miocene indusial limestone of St-Gérand-le-Puy, France", Palaeontology 32, 49-502.

Kalkovski E. (1908) "Oolith und stromatolith imm norddeutschen Bundsandstein", Z, geol. Ges, Hannover 60, 68-125.

Kann E. (1978) "Systematic un Okologie der Algen österreichister Berbäche", Archiv fur Hydrobiol., suppl. Bd. 53, 405-643.

Komar V.A. (1976) "Classification of stromatolites according to microstructures", in Paleontology of Precambrian and early Cambrian, All Union Symposium, Novosibirsk, pp. 41-43 (in Russian).

Komar, V.A. (1989) "Classification of the microstructure of the Upper Precambrian stromatolites", in K.S. Valdiya (ed), Stromatolite and stromatolitic desposits, Himalayan Geology 13, 229-238.

Krumbein W.E. (1986) "Biotransfer of minerals by microbes and microbial mats", in B.S.G. Leadbetter and R. Riding (eds), Biomineralization of Lower plants and animals, Clarendon Oxford, pp 55-72.

Langeron M. (1902) "Contribution à l'étude de la flore fossile de Sezanne. Nouvelles considérations sur les formations travertineuses anciennes et contemporaines", Bull. Société nat. Autun 15, 33-370.

Leinfelder R.R. and Hartkopf-Fröder C. (1990) "In situ accretion mechanism of concavo-convex lacustrine oncoids ('swallow nests') from the Oligocene of the Mainz basin, Rhineland, FRG", Sedimentology 37, 287-301.

Lindqvist (1993) " Lacustrine stromatolites and oncoids : Manuherikia Group, (Miocene), New Zealand. This book .

Link M.H., Osborne R.H. and Awramik S. (1978) "Lacustrine stromatolites and associated sediments of the Pliocene Ridge Route Formation, Ridge Basin, California", Journ. Sedimentary Petrology 48, 143-156.

Love F.G., Simmons G.M., Parker B.C., Wharton, R.A., and Seaburg K.G. (1983) " Modern conophyton-like microbial mats discovered in Lake Vanda, Antarctica" Geomicrobiol. J 3: 33-48.

Love K.M. and Chafetz H.S. (1988) "Diagenesis of laminated travertine crusts, Arbuckle mountains, Oklahoma", Journ. Sedimentary Petrology 58, 441-445.

Lowenstam H.A. (1986) "Mineralization processes in Monerans and Protists", in B.S.G. Leadbetter and R. Riding (eds), Biomineralization of Lower plants and animals, Clarendon Oxford, pp. 1-17.

Lowenstam H.A.and Wiener S. (1989) On biomineralization. Oxford Univ Press, Oxford.

Monty C.L.V. (1965a) Geological and environmental significance of Cyanophyta, Thesis Princeton Univ.

Monty C.L.V. (1965b) "Recent algal stromatolites in the Windward lagoon, Andros Island, Bahamas", Ann Société Geologique de Belgique 88, 269-276.

Monty C.L.V. (1967) "Distribution and structure of recent stromatolitic algal mats, Eastern Andros Island, Bahamas", Ann Société Géologique Belge 90, 55-100.

Monty C.L.V. (1976) "The origin and development of cryptalgal fabric", in M.R. Walter (ed), Stromatolites, Elsevier, Amsterdam, pp. 193-205.

Monty C.L.V. and Hardie L.A. (1976) "The geological significance of the freshwater blue-green algae calcareous marsh", in M.R. Walter (ed), Stromatolites, Elsevier, Amsterdam, pp. 447-477.

Monty C.L.V. and Mas J.R. (1981) "Lower Cretaceous (Wealdian) blue-green algal deposits of the province of Valencia, Eastern Spain", in C.L.V. Monty (ed), Phanerozoic Stromatolites I, Springer-Verlag, Berlin, pp. 85-120.

Oliveros J.M., Escandel B. and Colom G. (1960) Temas geologicos de Mallorca, Mem. Inst. Geol. y Min, Espana 61.

Pentecost A. (1978) "Blue-green algae and freshwater carbonate deposits", Proc. Royal Soc. London 200, 43-61.

Pentecost A. (1985) "Association of Cyanobacteria with tuffa deposits : identity, enumeration, and nature of the sheath material revelated by histochemistry", Geomicrobiology Journ. 4, 285-289.

Pentecost A. (1987) "Growth and calcificaion of freshwater Cyanobacterium *Rivularia haematites*". Proc. Royal Soc. London 202, 125-135.

Pentecost A. (1991) "Calcification processes in Algae and Bacteria", In R. Riding (ed) Calcareous Algae and stromatolites, Springer-Verlag, Berlin, pp. 2-20.

Pentecost A. and Riding R. (1986) "Calcification in cyanobacteria", in B.S.G. Leadbeater and R. Riding (eds) Biomineralization in Lower plants and animals, Clarendon press, Oxford, pp. 73-90.

Pentecost A. and Terry (1988) "Inability to demonstrate calcite precipitation by bacteria isolated from travertine", Geomicrobiology Journal 6, 185-194.

Pia J. (1934) "Die Kalkbiddung durch Planzen". Beiheft z Bot. Zentralbl, A 52, 1-72.

Plaziat J.C. (1966) "Contribution à l'étude stratigraphique du Lutétien continental des Corbières (au sud de Lagrasse, Aude)". Feuille de Capendu au 50000°, Bull. Carte Géologique France 278, 41, 225-236.

Plaziat J.C. (1974) "Observations paléolimnologiques sur les lacs situés entre le massif de Mouthoumet et la Montagne Noire (détroit de Carcassonne, Aude-Hérault). Remarques sur les ingressions marines à prélude lacustre", C R 96° Congr Société Savantes, Toulouse 1971 2, 71-93.

Plaziat J.C. (1984) Le domaine Pyrénéen de la fin du Crétacé à la fin de l'Eocène, Thèse Paris-Sud, Orsay.

Preiss W.V. (1976) "Proterozoic stromatolites from the Nabberu and Officer basins, Western Australia and their biostratigraphic significance", Geol. Survey of South Australia, Rp. of Investig. 47, pp. 51.

Riding R. (1979) "Origin and diagenesis of lacustrine algal bioherms at the margin of the Ries crater, Upper Miocene, Southern Germany", Sedimentology 26, 645-680.

Riding R. (1983) "Cyanoliths (Cyanoids) : oncolites formed by calcified Cyanophytes". In   T.M. Peryt (ed), Coated Grains, Springer-Werlag, Berlin, pp. 276-283.

Schäffer A. and Stapf K.R.G. (1978) "Permian Saar-Nahe Bassin and Recent Lake Constance (Germany): two environments of lacustrine algal carbonates", in Matter and M. Tucker (eds), Modern and ancient Lake sediments 2, 83-107.

Schapovalova I.G. (1974) Stratigraphy and stromatolites of Riphean sediments of the northern edge of the Iodomo-Maïkogo depression, Siberian fil Academ Sc of URSS, Iakoutsk, Geol Inst, pp. 113 (in Russian).

Schneider J. and Schröder H.G. (1980) "Calcification des Cyanophycées et leur contribution à la génèse des sédiments calcaires". Cristallisation, déformation, dissolution des calcaires, Réunion SGF, AGSO, SFMC, Bordeaux, pp. 421-428.

Schneider J., Schröder H.G. and Lecampion-Alsumard Th. (1983) "Algal micro-reefs coated grains from freshwater environments", in T.M. Peryt (ed), Coated Grains. Springer-Verlag, Berlin, pp. 284-298.

Schopf J.W. and Sovietov Y.K. (1976) "Microfossils in Conophyton from the Soviet Union and their bearing on Precambrian stratigraphy", Science 193, 143-146.

Semikhatov M.A. (1978) "Some Aphebian carbonate stromatolites of the Canadian Shield", in Lower boundary of the Riphean and stromatolites of the Aphebian, Trudy Geol. Akad. Nauk., SSSR 312, 111-147 (in Russian).

Strasser A. (1986) "Ooids in Purbeck limestones (lowermost Cretaceous) of the Swiss and French Jura", Sedimentology 23, 95-104.

Thompson J.B., Ferris F.G. and Smith D.A. (1990) "Geomicrobiology and sedimentology of the mixolimnion and chemocline in Fayetteville Green Lake, New-York, Palaios 5, 52-75.

Tilden J. (1897) "Some new species of Minnesota Algae which live upon calcareous or siliceous matrix", Bot. Gazette, Chicago 23, 95-104.

Wallner J. (1934) Uber die Beteiligung kalkablagernder Pflanzen bei der Bildung südbayerischer Tuffe, Biblioteca botanica, Stuttgart.

Wallner J. (1935a) Zur Kenntnis der Gattung Oocardium, Hedwidgia 75, 130-136 .

Wallner J. (1935b) "Diatomeen als Kalkbildner", Hedwidgia 75, 137-141.

Wallner J. (1935c) "Uber die Bedeutung der sog. Chironomidentuffe für die Messung der jährlichen Kalkproduktion durch Algen", Hedwidgia 74, 176-181.

Wallner J. (1935d) "Uber die Beteiligung kalkablagernder Algen am Aufbau der Chironomidentuffe", Beihenfte z Bot. 54 A, 143-150.

Wallner J. (1935e) "Zur weiteren Kenntnis der sog. Chironomidentuffe", Botanishe Archiv 37, 128-145.

Walter M.R. (1972) Stromatolites and the biostratigraphy of the Australian Precambrian and Cambrian, Spec pap in Palaeontology, 11, Palaeontol. Association London.
Walter M.R., Bauld J.and Brock T.D. (1976) "Microbiology and morphogenesis of columnar stromatolites (*Conophyton, Vacerrilla*) from Hot springs in Yellowstone national Park", in M.R. Walter (ed), Stromatolites, Elsevier, Amsterdam, pp. 273-310.

STROMATOLITES FROM THE EAST AFRICAN RIFT: A SYNOPSIS

J. CASANOVA
BRGM   Department of Geochemistry, PO Box 6009 ; 45060 Orléans, Cédex 2 ;
FRANCE

ABSTRACT

During the last ten million years, abundant stromatolites have formed in the East African Rift. They have colonized a variety of environments ranging from fluvio-lacustrine to hydrothermal settings and corresponding to different hydroclimatic situations. The building microorganisms and the fabrics are exceptionally well preserved due to the lack of diagenetic recrystallization. This made possible a very detailed and precise analysis of relationships between morphology and environmental conditions. The observed variations in the bulk shape as well as in the detailed morphology of the stromatolites can be related to local changes in the environment such as depth, relief and slope of the substrate, geochemistry of the surrounding waters, available light, current direction and turbulence.

INTRODUCTION

Since the main uplift of the Ethiopian and Kenyan domes, during the Lower-Middle Miocene (Baker et al., 1972), and up to the present day, encrusting microbial communities have colonized all continental aquatic environments of the East-African Rift (Fig.1). The oldest firmly dated stromatolites have been described in the northern part of the Rift and are known to be Late Miocene in age (10.5 Ma) in the Ch'orora Formation (Tiercelin et al., 1979). Except a travertine dam presently forming in the Engarescero River (SW Lake Natron, Tanzania), all stromatolite occurrences in the Rift are fossil. Mentioned in numerous regional studies (Taieb, 1974, Gasse, 1975, Street, 1979, Renaut, 1982), stromatolites have especially been used as indicators of former high lake level. Owing to the fact that detailed studies have been completed only recently (Casanova, 1986a, 1986b, 1991; Casanova and Hillaire-Marcel, 1992a; Tiercelin et al., 1987), the great number of occurrences and an improved understanding of the relationships between morphology and geological and hydroclimatic settings make it timely to present a review of East-African Rift stromatolites. Lake Tanganyika stromatolites (Casanova and Thouin, 1990; Casanova and Hillaire-Marcel, 1992b) represent the only known western occurrence, and this paper will, therefore, concentrate on the eastern branch of the Rift.

J. Bertrand-Sarfati and C. Monty (eds.), Phanerozoic Stromatolites II, 193–226.
© 1994 *Kluwer Academic Publishers.*

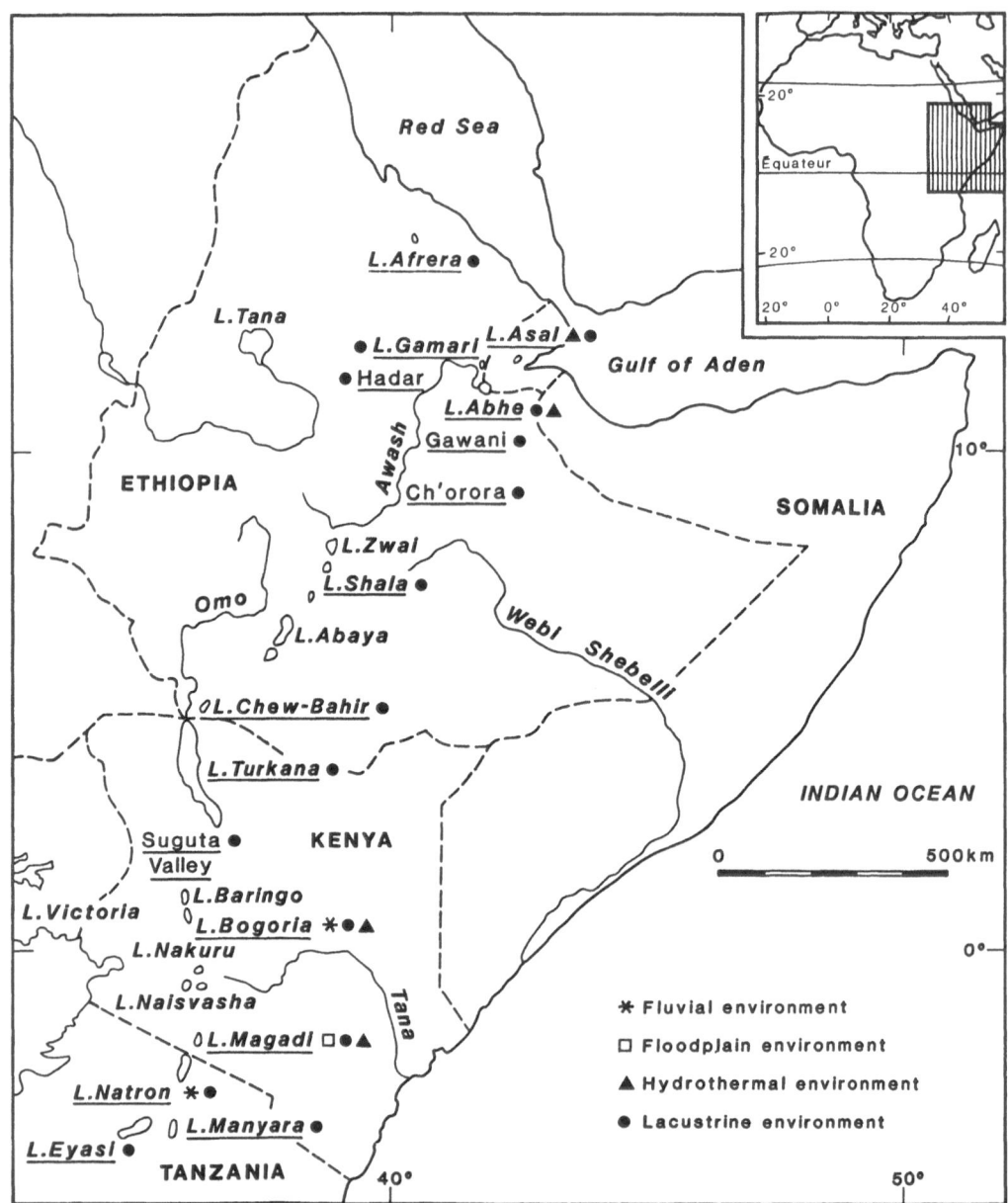

Figure 1. Environmental distribution of East African Rift stromatolites.

GEOLOGICAL SETTING

The NNE-SSW trending East-African Rift system constitutes the salient geological feature of East Africa. It is a complex network of faults, forming grabens 50 to 200 km wide, which cut through East Africa from the Red Sea south to the lower Zambezi Valley. The Rift faulting corresponds to the axial collapse of the upper part of a basement bulge, in such a way that the depression corresponding to the Rift Valley is flanked by elevated plateaux (Grove 1986).

The East-African Rift is divided into two branches. The eastern branch extends from the triple junction, centred on Djibouti, south to the Kenya-Tanzania border: it begins in the north with the Afar Rift, extends to the Ethiopian and Kenyan Rifts to the SSW, and ends at the North-Tanzanian divergence. The eastern branch is characterized by a large number of small lakes, in contrast, the western branch extends from Uganda to Mozambique, and is characterized by a succession of large lakes: Mobutu, Edwards, Kivu, Tanganyika, Rukwa and Malawi.

The lakes of these two branches are fundamentally different. The most striking geological contrast lies in the fact that volcanism has been abundant in the east and relatively weak in the west (Baker et al. 1972). The morphology of the lakes and their capacity to act as sediment traps depend on this asymmetric occurrence of volcanism (Yuretich 1982; Cohen 1989). The amount of vertical displacement of faults (throw ca.2000 m) is equivalent in the two branches. In the west, the existence of long, continuous lineaments and large troughs has induced the formation of large lacustrine basins, whereas the small lacustrine basins in the eastern branch were caused by numerous parallel faults and volcanoes.

The relative paucity of volcanism in the western branch favours the existence of large axial rivers parallel to the lineaments. The basement being crystalline, the drainage pattern is of the overland flow type. In the eastern branch, the volcanism disrupts the drainage and favours smaller transverse rivers. Meteoric waters rapidly infiltrate the volcano-sedimentary substrate and contribute to the groundwater supply. As a consequence of tectonic and volcanic influences, lakes are large and deep (max.1470m in Lake Tanganyika) in the west, but small and shallow (max.100m in Lake Turkana) in the east.

In the western Rift, the weathering of schists and phyllites rich in biotite and ferromagnesian minerals produces freshwater lakes where K^+, Mg^{++} and SO_4^{--} ions are dominant. The sedimentation rate in these lakes is low (30-50 cm.10^{-3} yr in Lake Tanganyika). In the eastern Rift, the weathering of alkaline volcanic rocks rich in plagioclase and amphibole produces saline lakes where Na^+ and HCO_3^- ions are dominant. The sedimentation rate is high (100-300 cm.10^{-3} yr in Lake Turkana).

Sediment composition differs strongly from lake to lake; the topographically lowest lakes contain mainly sediments of detrital origin, whereas, flank or crest lakes contain mainly organic sediments. In general, however, clay minerals of variable nature, a few evaporites, Mg-calcite, aragonite and dolomite form in the western lakes, while smectite, kaolinite, evaporites (zeolites, halite, trona, nahcolite etc) and low-Mg calcite form in the eastern lakes.

Tectonism and volcanism define the physical and chemical limits in which lakes and sedimentation will evolve, but the diversity of sediments in the East-African Rift suggests a major influence of both the altitude and the geographic setting of each lake. This climatic influence and its change over time (on a 100-103yr scale) are recognized as fundamental to explain the variations in Rift basin sedimentation.

Table 1. Location, morphologies and ages of East African Rift stromatolites. (1); Abell et al. 1982;   (2); Casanova 1986a ;   (3); Casanova 1987b;   (4); Casanova unpublished ; (5); Casanova et al. 1988;   (6); Casanova & Hillaire-Marcel 1992a ;   (7); Casanova and Tiercelin 1982;   (8); Demanges and Stieltjes 1975 ;   (9); Dixit 1984; (10); Fontes and Pouchan 1975; (11); Fontes et al. 1980; (12); Gasse 1975 ; (13); Gasse et al. 1974; (14); Gasse and Fontes 1989 ; (15); Grove et al. 1971; (16); Grove et al. 1975 ; (17); Hillaire-Marcel and Casanova 1987; ; (18); Hillaire-Marcel et al. 1982 ; (19); Hillaire-Marcel et al. 1986; (20); Johnson 1974 ; (21); Johnson and Raynolds 1976; 22); Renaut 1982 ; (23); Saliége 1979; (24); Street 1979 ; (25); Taieb 1974; (26); Tiercelin et al 1979 ; (27); Tiercelin et al. 1987; (28); Vincens et al. 1986

Sites	Environment	Morphologies	Ages	References
Lake Afrera 13°15'N 42°55'E	Lacustrine	Encrustations	7.3 Ka	13
Lake Asal 11°40'N 42°25'E	Hydrotherm.	Chimneys		8
	Lacustrine	Block encrustations	8.6 - 3 Ka	14
Assaita 11°30'N 41°30'E	Lacustrine	Oncoids	9.4 - 8.6 Ka	13
Lake Gamari 11°30'N 41°40'E	Lacustrine	Bioherms	6.7 Ka	13
Lake Abbe 11°10'N 41°50'E	Hydrotherm.	Chimneys	6.3 - 1.5 Ka	10 11
	Lacustrine	Block encrustations	2 Ka-Modern	13
		Planar encrustations		12
		Cylindrical encrustations		23
Hadar 11°10'N 40°25'E	Lacustrine	Oncoids	3.5 Ma	18
Gawani 11°30'N 40°25'E	Lacustrine	Pebble encrustations	9 Ka	25
Chorora 8°55'N 40°20'E	Lacustrine	Encrustations	10.5 Ma	26
Lake Shala 7°30'N 38°30'E	Lacustrine	Block encrustations	Late Pleistocene	16 24
Lake Chew-Bahir 5°N 37°E	Lacustrine	Block encrustations	5.7 Ka	15
		Oncoids		16
Lake Turkana 4°N 36°20'E	Lacustrine	Planar encrustations	3 - 2.5 Ma	20
		Oncoids	1.9 - 1.4 Ma	21
		Bioherms	ca. 10 Ka	1
		Cylindrical encrustations		
Suguta Valley 2°N 36°30'E	Lacustrine	Bioherms	>30 Ka	2
		Planar encrustations	31 - 27 Ka	5
		Pebble encrustations		
		Oncoids		
		Cylindrical encrustations		
Lake Bogoria 0°15'N 36°05'E	Fluvial	Oncoids	Early Holocene	2
		Bryophyte encrustations		
	Hydrotherm.	Pool-Rim dams	>5 Ka	22
		Planar encrustations		2
	Lacustrine	Block encrustations	5 - 4.5 Ka	22
		Planar encrustations		28
		Cylindrical encrustations		27
Lake Magadi 2°S 36°15'E	Floodplains	Oncoids	6.1 - 2.6 Ka	7
	Hydrotherm.	Chimneys	13 - 10 Ka	17
Lake Natron 2°30'S 35°50'E	Fluvial	Bryophyte encrustations	Modern	2
Lake Magadi- Lake Natron	Lacustrine	Bioherms	ca. 240 Ka	19
		Pebble encrustations	ca. 135 Ka	2
		Block encrustations	13-10 Ka	17
		Planar encrustations		3
		Oncoids		
		Cylindrical encrustations		
Lake Manyara 3°30'S 35°50'E	Lacustrine	Bioherms	ca. 90 Ka	9
		Oncoids	28-22 Ka	2
		Pebble encrustations		6
Lake Eyasi 3°30'S 35°E	Lacustrine	Bioherms	>30 Ka	4
		Block encrustations	10 Ka	

The distribution of stromatolites in the East-African Rift covers a large range of time and space (Table 1). However, the occurrences can be summarized in four main types (Casanova 1986b, fig.6), according to their environmental settings: fluvial environment, floodplain environment, hydrothermal environment and lacustrine environment (Fig.1).

MORPHOLOGICAL CLASSIFICATION OF RIFT STROMATOLITES

About ten different morphologies have been identified among the stromatolites of the East African Rift (Table 1). Their size and microstructure, as well as their microbial content, vary greatly according to specific local conditions within each environmental setting. Therefore the following classification, based on field observations of the gross-morphology, is strictly descriptive.

Chimneys are vertical buildups related to the thermal fluid circulations channeled by faults. Pool-rim dams are characterized by their circular shape and low relief, and, together with chimneys, are the main constructions found in the hydrothermal environment.

Oncoids (0.1-30cm in diameter) display concentric lamination around a nucleus. They differ from coated grains and pebble encrustations by the thickness of lamination which is greater than the size of the nucleus.

Bioherms are commonly isolated and locally coalescent buildups of variable morphology whose height and length are of the same order (0.1-2m in size). Although most initiated on a planar substratum, some bioherms have grown over a substrate asperity, caused by topography, pebbles or former stromatolites.

Block and pebble encrustations, millimetre to centimetre-thick, result from a uniform coating over steeply sloping substrates. Consequently the shape of the stromatolites strongly reflects that of the substrate. Block encrustations form on the largest (>20cm) detrital elements and on vertical rocky walls. On terrigeneous sediments of deltas, beaches and alluvial fans, only the 2-20cm fraction (gravels, pebbles) is coated; the external shape of these stromatolites are characterized by a high degree of inheritance, and their thickness never exceeds that of the nuclei.

Planar encrustations are centimetre-thick coatings which occur on planar or gently sloping rocky surfaces. They differ from block encrustations only because they formed thick stromatolites (up to 1m in the Suguta Valley), definitely independent of the morphology of the substrate.

Cylindrical encrustations, centimetre to decimetre-thick, grow around detrital or *in situ* vegetal remains (twigs, branches, trunks). The decay of the organic substrate leaves an empty internal mold reproducing the morphology of the plant. Massive bryophyte encrustations found in travertines represent a peculiar case of this facies, each moss leaf being coated by a single encrusting phase.

MINERALOGIC COMPOSITION OF STROMATOLITES

East African Rift stromatolites (Abell et al. 1982, Casanova 1986a) are composed of a dominant carbonate phase, a small percentage of organic matter and a variable detrital fraction; the average values (120 analyses) being respectively 93%, 1.3% and 5.7% dry weight.

The organic fraction is mainly of microbial origin although it also contains sparse fish, insect and vegetal remains. The lowest organic percentages (8 samples average: 0.55%) have been obtained from bacterial lacustrine stromatolites from the Suguta Valley.

The detrital component, transported by rivers and run-off waters and trapped by the Benthic Microbial Communities (BMC), comes from three main petrographic provinces: the crystalline basement, volcanic formations and Quaternary sediments. Qualitative and quantitative variations in the detrital content of lacustrine stromatolites make it possible to reconstruct the paleo-drainage of each basin (Casanova 1986a, 1987b). The detrital fraction, which reaches high percentages at river mouths, is localized in internal cavities, between the cyanobacteria filaments, and around the periphery of colonial microbial forms. Nevertheless, the rhythmicity of the stromatolitic laminae is never controlled by the detrital supply. Hence the lamination strictly reflects the BMC growth.

The carbonate analyzed by X-ray diffraction and atomic absorption is calcite (<0.5% mole $MgCO_3$), although some bacterial colonies from Lake Natron are composed of low-Mg calcite (6% mole $MgCO_3$). In all cases the carbonate has been precipitated *in situ* and has never been related to trapping and binding phenomena; given the close relationship between carbonate and microbial filaments, the carbonate is thought to have a biological origin.

ATTRIBUTES OF STROMATOLITES THROUGH TIME AND SPACE

Stromatolite records in East Africa cover the last 10 Ma. The morphologies, growth patterns and fabrics developed in the stromatolites are very similar during this period and were controlled by ecological factors. Such environmental control is revealed through three levels of observation.

Environmental setting

The grossmorphology of stromatolites is, above all, determined by the type of environment in which they form. For instance, bioherms develop only in lakes; chimneys and pool-rim dams are characteristic of hydrothermal springs, while massive bryophyte encrustations are related to running waters. Some forms like oncoids and planar or cylindrical encrustations are ubiquitous but the stromatolite assemblages and their spatial distribution are specific to each environment. Hydrothermal constructions are always related to fault orientations; travertines develop along drainage patterns, whereas lacustrine stromatolites form at definite altitudes along shorelines which can be reconstituted   even after vertical tectonic displacement (see Hillaire-Marcel and Casanova 1987, fig.11).

Geological and hydroclimatic setting

Within the general framework given by the four main types of environment, each buildup episode has its own characteristics. Hence, within a single lake basin, several positive hydroclimatic episodes generate as many distinct stromatolite generations. Each generation displays specific morphologies, macrofacies and microbial biota, which are surprisingly uniform for the entire basin within a given period and ecological niche. For example, the Natron-Magadi basin has been occupied, in its recent history, by three successive water bodies with similar volume and hydrologic properties (Hillaire-Marcel et al. 1986, Hillaire-Marcel and Casanova 1987) related to the hypsometric, petrographic and tectonic pattern of the drainage basin. The three corresponding generations of stromatolites, however, show specific morphologies (Casanova 1987b), that have been interpreted as a response to specific climatic conditions.

The duration of the lacustrine episode in which stromatolites have formed is also of great importance. If the stability of the paleolake is relatively short (of the order of 102-103 years), stromatolites uniformly cover hard substrates and form a poorly diversified crust. In that case, only the surface ornaments can be used to differentiate successive generations (see, for example, Vincens et al. 1986). On the other hand, long lasting lacustrine episodes (of the order of 103-104 years) lead to stable ecological niches according to depth (Fig.2), which favour large bioherms showing distinct features.

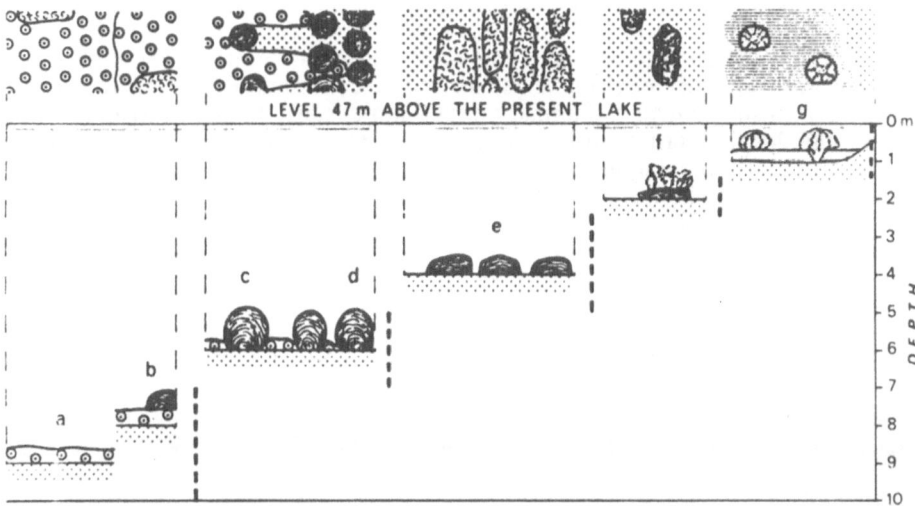

Figure 2 Lake Natron. Ecological zonation of the second-generation stromatolites (ca. 135 Ka) : (a) polyphased oncolitic crust overlapped by bulbous bioherms, at first scattered (b,c) then dominant (d); (e) large flattened bioherms poorly ornamented; small isolated bioherms showing a pronounced ornamentation of "petaloid" (f) then "cerebroid" (g) facies. From Casanova 1987b.

Local ecological parameters

Independently of the regional environmental and hydroclimatic context, each generation of stromatolites exhibits variations in their morphology due to local ecological parameters.

Substrate

The substrate plays the predominant role in the initial morphology of the stromatolite. Its dip is important: gently sloping surfaces are most likely to permit the establishment of large stromatolites. The lithology is of little importance since any hard substrate is colonized by BMC when favourable growth conditions occur. On the other hand, soft substrates are never colonized. This feature has been verified for the entire Rift and seems to characterize fresh- to slightly salt-water stromatolites (Casanova 1986a), contrary to what is observed in hypersaline waters (Walter et al. 1973, Krumbein and Cohen 1974, Moore et al. 1984).

Bathymetry

In lacustrine environments, water depth strongly influences the stromatolite morphology (Fig.2) as well as BMC growth. This is particularly evident in the case of columnnar growth. Almost all lacustrine stromatolites go through a columnnar phase during their growth, whereas in other environments, microbial communities are characterized by planar mats or bushy growth. Columnar growth, not specific to any individual BMC, is related to the local micro-environment since all mats may, within the same hand specimen, change from columnnar to other growth types and back again. Stable isotopes in carbonates often indicate physico-chemical shifts contemporaneous to such morphological variations. As a rule, columnar growth takes place in the top three metres of water; columnns display great diversity in shape and size. In the intermediate water layer (-3 to -7m), growth is often columnar but the columns tend to coalesce. The resulting morphologies appear to be the same everywhere in the basin for that bathymetric zone. In deeper waters (down to -30m), columnar growth is generally absent. In other respects, numerous oncoids and cylindrical encrustations experienced columnar growth upwards, while their lower part exhibits planar laminae. This suggests that columnar growth occur preferentially in environments with the most light, oxygen and nutrients, as a result of biologic competition within microbial mats.

Hydrodynamism

Growth stages of oncoids were studied in several areas of the Rift by means of shape parameters (Casanova 1985). These data suggest that the environmental factor which has the greatest influence on oncoid morphology is hydrodynamism. When oncoids are classified according to the hydrodynamic energy of the environment in which they form (Casanova 1985, table VII), the ratios of upper versus lower cortex thickness increase parallel with decreasing in energy. In agreement with Monty and Mas (1981), subspherical to spherical morphologies do not imply "turbulent" conditions; however, they do require a relatively continuous displacement of the oncoid. In the case of episodic though repetitive displacement, the lamination shows preferential growth direction (see for example Braithwaite et al. 1989, fig. 20). Hydrodynamic influence is of less importance for stromatolites developed on a firm substrate, but the column orientation as well as bioherm thickening are observed to occur in the direction facing the major sublacustrine current.

FLUVIAL CHANNEL AND SPRING ENVIRONMENT

Rivers are typically short and transverse to the eastern Rift Valley; most of them are ephemeral and characterized during the rainy season by torrential flow. Perennial rivers generally contain turbid sodic and slightly calcic waters with a Mg/Ca ratio >2 (Renaut 1982; Gueddari 1984). Benthic Microbial Communities are poorly developed and are generally not calcified. Hence, the travertines presently forming in the Engarescero River (SW Lake Natron, Fig.3a) are exceptional; there, waters are oversaturated with respect to calcite (pH=9.1, [Ca^{++}]=12.8 mg/l), but calcification of cyanobacterial filaments is very weak (Fig.3b).

Morphology

As is usually the case in the fluvial environment (Golubic 1969; Casanova 1981; Chafetz and Folk 1984), the fossil spring mounds and waterfall dams are composed of planar and cylindrical encrustations, coated pebbles, oncoids and massive encrustations of bryophyte bushes. Two types of oncoids are associated with travertines where they locally form metre-sized accumulations near Lake Bogoria (Tiercelin et al. 1987) (Fig.3c). About 70% of the cylindrically shape oncoids, formed from unbroken *Melania* (gastropods) shells; lithoclasts, bioclasts and vegetal debris form equal proportions of the remaining nuclei. Together with reed-bed encrustations and carbonate muds, these oncoids characterize low energy environments. Other oncoids, spherical in shape, are related to the flow directions of the run-off waters. In that case, lithoclasts and bioclasts represent 45% and 34% of the nuclei, while gastropods and vegetal debris constitute only 13% and 8%. These oncoids occur with pebble encrustations embedded in coarse sediments, reflecting a high energy environment.

Microstructure

The microstructure and microbial content of East African travertines is similar to that described elsewhere. This confirms that travertine formation depends mainly on the local micro-environment and is little influenced by climate. Three main microstructures can be distinguished.

Spongious fabric

This fabric arises from successive spongious laminae, each 500-1000μm thick, and made of erect micritic threads 15-20μm in diameter separated by micritic films. Each thread represents the remnant of one single filament within its own sheath. Well preserved samples show that these filaments (Fig.3e), characterized by a 5.7-7.1μm wide trichome and a conical apical cell, belong to *Phormidium incrustatum*. Identical running-water BMC, dominated by *P.incrustatum*, have been described in Europe (Fritsch 1950, Symoens 1957, Geurts 1976, Casanova 1981, Monty and Mas 1981), North America (Tilden 1897, Howe 1931, Golubic and Fischer 1975) and Africa (Symoens 1968).

Radial sparitic fabric

Superposed radial sparitic fabric laminae are individually 600-1500μm thick, made of juxtaposed hemispheric or fan-like colonies and separated by undulating micritic

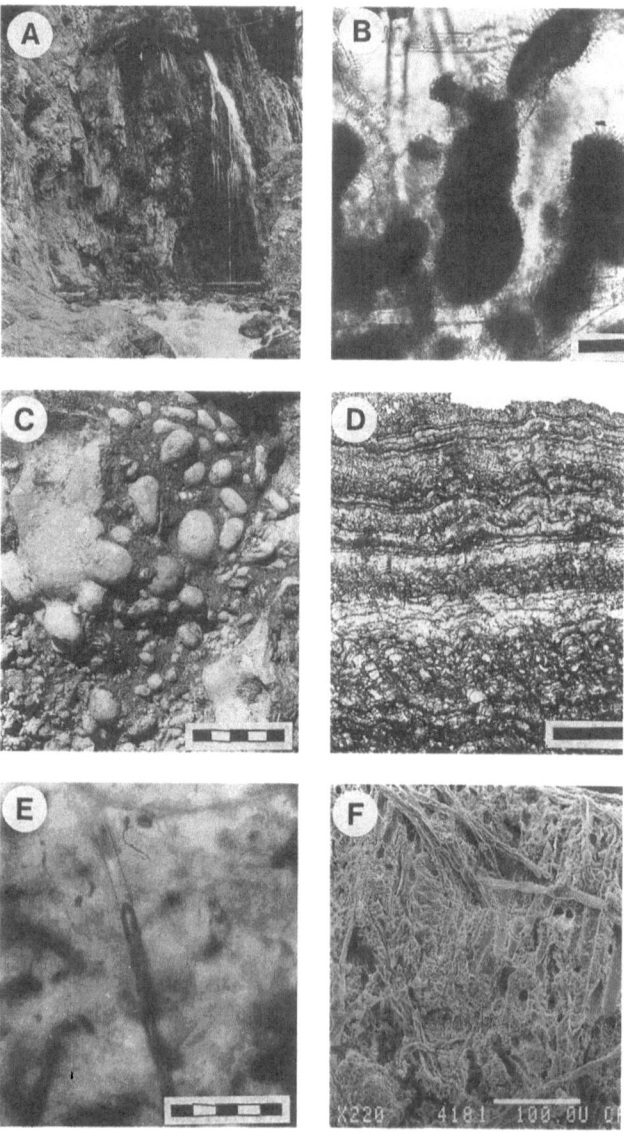

Figure 3 (a-f). Stromatolites from fluvial channel and spring environments. a) Lake Natron, Modern travertines from Engarescero River made of massive bryophyte encrustations. b) Lake Natron, micrite generated by *Homeothrix janthina* epiphytic of *Chantransia* sp. Scale bar = 100 μm. c) Lake Bogoria, Accumulation of oncoids associated with a travertine dam. Scale bar = 5 cm. d) Lake Bogoria, Radial sparitic fabric generated by colonies of *Rivularia haematites*. Planar encrustations in travertine; scale bar = 400 μm. e) Lake Bogoria, *Phormidium incrustatum* filament preserved in travertine. Scale bar = 50 μm. f) Lake Natron, Collapsed filaments and molds within a micritic fabric. SEM. Scale bar = 100 μm.

laminae (Fig.3d). Each sparitic colony exhibits a second-order concentric lamination and contains numerous radiating filaments (1-8µm wide) which are thicker at the base than at the top. The fabric of the colonies, as well as the ultrastructure of crystals, are characteristic of calcified *Rivularia haematites* growths (Wallner 1935, Monty 1976, Schneider 1977, Casanova 1981). The intercalated micritic laminae are built by tiny filaments growing horizontally, as seen in certain living species of *Schizothrix* (Monty and Mas 1981). This facies represents an ecotone in the *R. haematites/Schizothrix* association frequently encountered in modern freshwater stromatolites (Kann 1959, Schafer and Stapf 1978).

Thrombolitic fabric

This microfacies mainly corresponds to massive encrusted bryophyte bushes. Discrete lamination likely corresponds to different growth stages of mosses. Micrite is precipitated around filamentous (Fig.3f) or coccoid cyanobacteria species, epiphytic on bryophyte leaves. In the modern Engarescero River travertine (Fig.2a), the main building organisms which generate this facies are *Homeothrix janthina*, often epiphytic on *Chantransia* (Fig.3b), and *Rivularia dura*. Such a cyanobacterial-moss association has frequently been described in European travertines (Symoens 1968, Geurts 1976).

HYDROTHERMAL ENVIRONMENT

The thinning of the lithosphere and the occurrence of numerous vertical faults lead to very active hydrothermalism within the Rift basins, such as springs, geysers and seepages. Thermal waters (pH=7-9.9 and T=35-100°C)(United Nations 1971, Fontes et al. 1980, Renaut 1982), are characteristically rich in Na^+, CO_3^{--} and HCO_3^-, and moderately rich in dissolved silica (< 200 mg/l) and fluor (< 135mg/l). Springs, some of which issue from lake floors, are linked to the tectonic network. Gels composed of amorphous silica, porcellanite, fluorite and aragonite precipitate around certain effluents. A variety of BMC also occurs: (1) spheroidal mats composed of a decimetre-thick yellowish-white gel, surrounded by a green millimetre-thick layer; (2) rubbery planar mats composed of a basal red layer several centimetres thick, a middle green layer of variable thickness and superficial red film; and (3) hairy green mats in the channels. The main microbial constituants belong to the genera *Oscillatoria* and *Synechoccus*. Maximum temperatures of 50 °C and 74 °C have been measured within the cyanobacteria and the bacteria colonies, respectively. No mineral precipitation has been observed in these modern mats.

Morphology

Fossil deposits in the form of pool-rim dams, chimneys, oncoids and planar encrustations are commonly associated with this hydrothermal environment (Fig.4).

Chimneys are aligned along fractures and occur in either lake floor (Lake Abhe) or shoreline settings (Lake Magadi, Fig.4d and 4e). They grow vertically and can reach several tens of metres in height but only a few metres in width. Some of them still show traces of thermal activity (fumeroles, geysers) even though the constructions themselves are fossil (Fig.4b).

Figure 4 (a-e). Stromatolites from hydrothermal environments. a) Lake Bogoria, typical pool-rim dam from subaerial setting. b) Lake Bogoria, active geyser nearby a central chimney covered by modern silicate gels. c) Lake Bogoria, On hand specimen, stromatolites show coarse lamination. Scale = 5 cm. d) Lake Magadi, Chimney built on the shore of the paleolake around 12-10 Ka. e) Close-up of the vegetal frame used by the BMC as a support. Scale = 10 cm.

A typical pool-rim dam on Lake Bogoria (Fig.4a and 4c) is composed of a pool, 10m in diameter, surrounded by a 60cm high wall concave towards a central, low chimney (70cm high, 120cm in diameter). During its growth period, the pool is filled by waters coming out of the chimney, the height of the wall determining the level of the overflow. This growth mechanism indicates that accretion did not occur in the lake but on the shore.

Microstructure

Two types of microfacies distinguish the hydrothermal stromatolites of Lakes Bogoria and Magadi (Fig.5), corresponding to different hydrologic situations. In the Bogoria basin, deep groundwater was the main source of thermal springs and the ^{18}O content of the precipitated carbonate indicates a paleotemperature of $79\pm7°C$ during formation, comparable to the thermal range of modern springs (Tiercelin et al. 1987). In the Magadi basin, the contribution of surface waters (ca. 75%) was greater than deep sources arrivals, and the paleotemperature during formation (ca. 25°C) was close to that of contemporaneous lacustrine stromatolites (Hillaire-Marcel and Casanova 1987).

Bogoria Type

Hydrothermal stromatolites from Lake Bogoria, regardless of their grossmorphology, exhibit a very uniform microstructure made of repetitive spongious sparitic laminae, 200µm to several cm thick (Fig.4c). This coarse lamination reflects the stability of the environment during growth, as shown by the homogeneity of the ^{18}O and ^{13}C content (Tiercelin et al. 1987). Laminae is composed of criss-crossed or bushy calcitic needles (Fig.5a), each being a single, lanceolate crystal (50-500µm in length). Growth porosity is variable; it is generally high (up to 80%), but it may be locally occluded by secondary cementation or completely obliterated due to crystalline competition for space. Under the scanning electron microscope, acicular calcite crystals appear as microbial colonies (Fig.5b) generated by filamentous bacteria, 0.3µm in diameter, organized in polygonal networks (Fig.5b and c) and associated with other isolated spherical or rod-shaped bacteria (Fig.5d). The fine preservation of bacterial bodies and mucilaginous gels shows that the fabric is primary and that the low-Mg calcite results from direct precipitation rather than from the recrystallization of former aragonite.

Magadi Type

Magadi chimneys are composed of carbonate encrustations built over a vegetal frame (Fig.4e). In thin section, the presence of cylindrical or planar voids testifies to the existence of former vegetal supports in the form of stems and leaves, respectively. Because the calcite crystals grow perpendicularly to the surface of these intertwined supports, there is no preferential growth direction other than outward from the substrate and encrustation appears as laterally connected microstromatolites. The fabric is made of fibroradial sparitic calcite containing numerous tufted or bushy cyanobacterial filaments. A second-order microstructure interrupts this fabric as dark micrite which occurs either on each cyanobacterial bush (Fig.5f) or directly on the surface of the sparitic crystals. These micritic films, which appear sporadically but uniformly in each encrustation, define the microstromatolite lamination. Because the cyanobacterial colonies continue their normal growth between micritic phases, the latter very likely

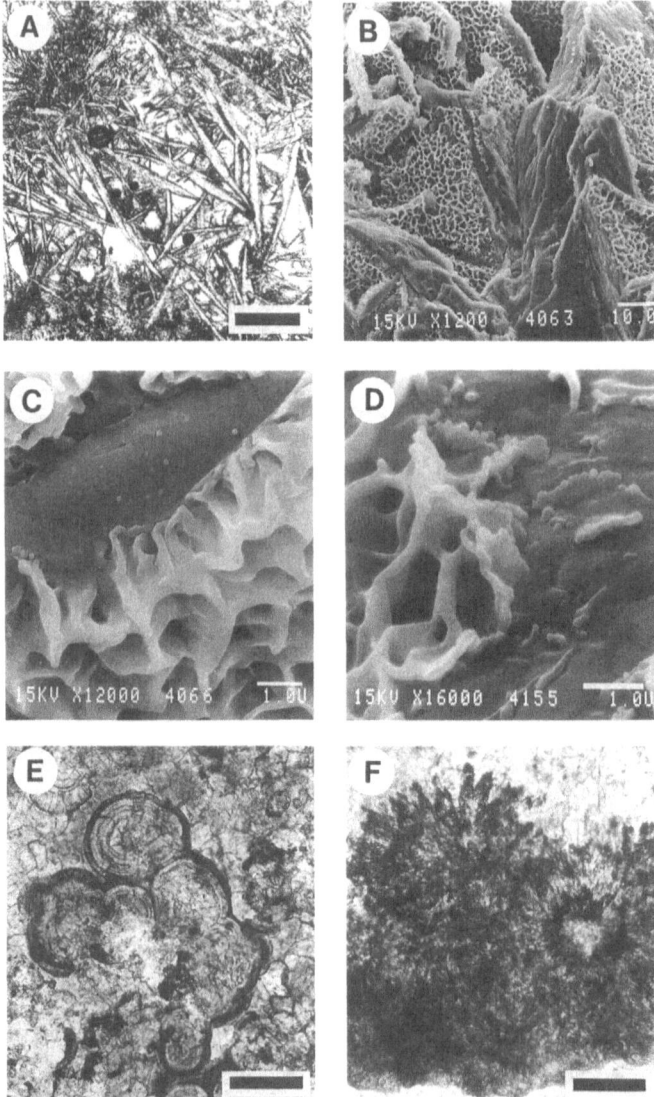

Figure 5 Hydrothermal stromatolite microfacies of Bogoria (a-d) and Magadi (e-f) types. a) Mesh of calcitic needles which form the Bogoria type microstructure. Scale bar = 500 μm. b) Polygonal networks of bacterial colonies generating the acicular crystals. SEM. Scale bar = 10 μm. c) Close-up of the bacterial network embedded in mucilagineous gels. SEM. Scale bar = 1 μm. d) Isolated punctiform and rod-shaped bacteria on a crystal surface. SEM. Scale bar = 1 μm. e) Sub-spherical ooids growing within the growth porosity. Scale bar = 100 μm. f) Cyanobacterial bushes coated by dark (bacterial ?) micrite. Scale bar = 100 μm.

correspond to periodic bacterial blooms, favoured by qualitative variations in the hydrologic budget.

A late microfacies occurs within the growth porosity in the form of spheroidal ooids, 100-150µm in diameter (Fig.5e). They display concentric laminae developed around black organic dots, from which numerous microfilaments radiate. Since ooid growth often starts from the surface of a sparite crystal, their shape can be hemispherical, lobate or interpenetrated. These structures are similar to those obtained by Monty and Van Laer (1984) from bacterial cultures. Accordingly, black dots represent degraded bacteria in diverse stages of cell ondivision and the microfilaments represent the organelles of the bacteria cell.

FLOODPLAINS ENVIRONMENT

Floodplains appear as extended areas covered by an ephemeral water sheet coming from direct rainfall and runoff during exceptionally rainy seasons. At Lake Magadi (Casanova and Tiercelin 1982) this hypohaline water layer, of very low hydrodynamic energy, is about one metre deep in the middle of the basin and a few centimetres on the surrounding flats. Only oncoids have been observed to form in this environment (Fig.6).

Morphology

Oncoids associated with coarse terrigeneous deposits occupy large areas on the floodplains of Lake Magadi (Fig.6a and b). These mainly disc-shaped structures, 1-3cm in diameter, are characterized by concave upper and lower faces as well as by a lateral thickening (Fig.6c). The upward growth, two to five times greater than the downward growth, indicates in situ accretion with very little displacement, whereas the hypertrophic lateral growth represents a response by building organisms to maximum moisture in a thin water layer subjected to intense evaporation.

Microstructure

Oncoids from Magadi, regardless of their morphology and distribution in the basin, possess identical microstructures resulting from two main fabrics, which can occasionally be found in the same lamina.

Cyanobacterial facies

The lamination is composed of light-coloured microsparitic laminae, 100-500µm thick, containing many erect filaments and separated by micritic films, 10-20µm thick. The filaments appear either as large isolated forms (1.2-1.5µm in diameter) (Fig.6d) or as smaller threads grouped together in a common sheath (0.5-1µm in diameter); they may grow through or be abruptly interrupted by the micritic layers. This microfacies is very similar to that of the lacustrine stromatolites from the Magadi-Natron basin and their isotopic composition is of the same order (Casanova 1986a). Hence, this cyanobacterial facies may reasonably be considered the equivalent of the "lacustrine doublet" microfacies (see Section I.II.1).

Figure 6 (a-f). Lake Magadi, Floodplain oncoids. a) View of the southern floodplain cross-cut by hydrothermal channels. b) Oncoids associated with coarse terrigeneous deposits. c) Various oncoid morphologies showing the pronounced thickening towards the margins of the oncoids. Scale bar = 1 cm. d) Erect filaments individually coated. SEM. Scale bar = 1 µm. e) Fan-like sparitic microstromatolites very abundant in the upper part of oncoids. Scale bar = 100 µm. f) Rod bacteria embedded in the sparitic crystals of the microstromatolites. SEM. Scale bar = 1 µm.

Bacterial colonies

Finely laminated, fan-like sparitic crystals (Fig.6e) are intercalated between the filaments of the light-coloured laminae. The growth of these microstromatolites is normal to the base of the lamina and contemporaneous with cyanobacterial growth. Under the scanning electron microscope, each microstromatolite appears as an individual bacterial colony (Fig.6f). These colonies are abundant at the outermost part of the upper cortex of the oncoids and, therefore, seem to be related to the evaporation of the ephemeral water body.

LACUSTRINE ENVIRONMENT

All lakes are siliciclastic in the East African Rift. The lake waters in the eastern branch are, however, richer in Ca^{++} and HCO_3^- due to the leaching of alkaline volcanic rocks, rich in minerals such as plagioclase, augite, aegerine and nepheline. Certain of these lakes, like Langano and Abiyata in the Ethiopian Rift or, Naivasha and Baringo in the Kenyan Rift, are freshwater lakes, but the majority are saline and contain sodic carbonated-bicarbonated brines (salinity < 300 mg/l in Lake Magadi).

The chemical composition of the lakes is closely related to the hydrologic budget and to the precipitation/evaporation ratio of each basin. This climatic variability has also been clearly expressed in the past, and it is possible to distinguish two extreme situations: "humid" and "arid" periods. Humid periods are time intervals characterized by organic matter accumulating in deep freshwater lakes that are stratified due to the weakness of the winds. Because of forest expansion and increased terrestrial organic productivity, the organic matter transported down to the lake by runoff waters is of continental origin.

Figure 7. Lake Manyara. Paleoshoreline corresponding to the second- generation stromatolites (28-23 Ka).

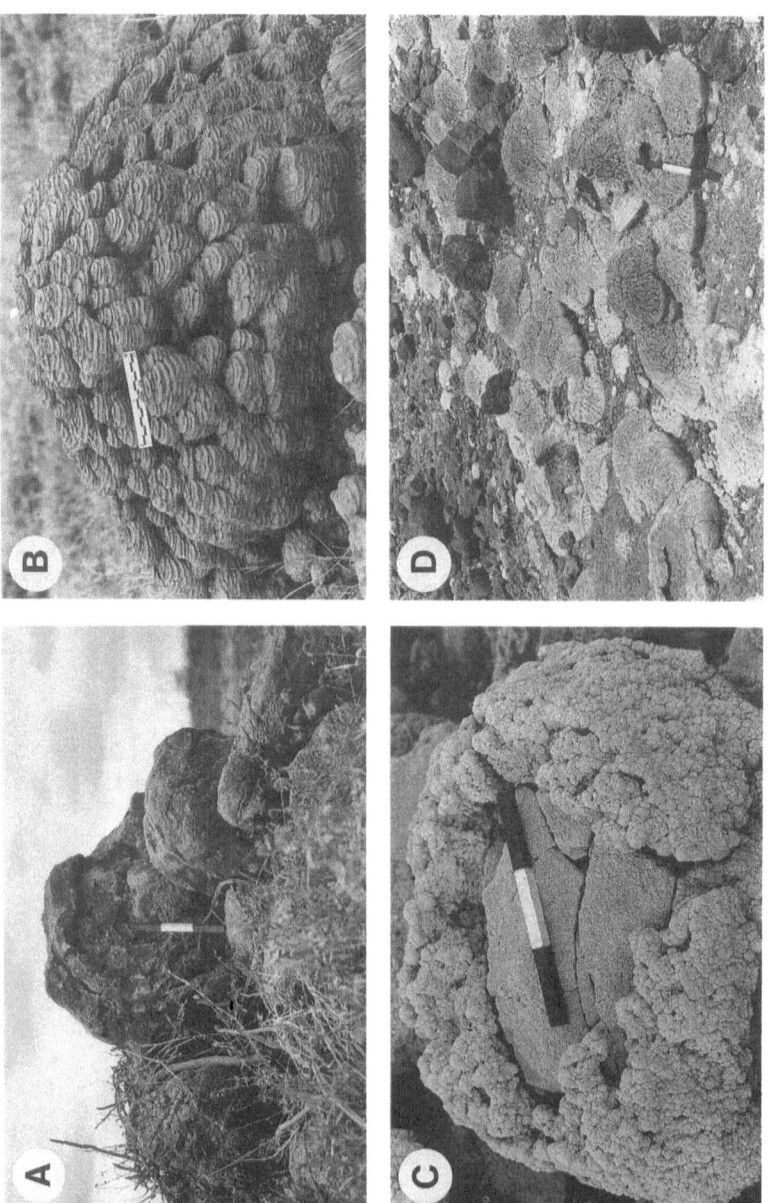

Figure 8 (a-d) Lacustrine stromatolites. a) Lake Manyara, Accumulation of composite bioherms from the first generation (140-90 Ka), showing their diversity in shape and size. Scale bar = 30 cm. b) Lake Manyara, Bioherm from the second generation (28-23 Ka). Note the well defined cylindrical columns, characteristic of the bathymetric zone 0-4 m. Scale bar = 10 cm. c) Lake Bogoria, Block incrusted by the first (5.3 Ka) and the second (4.9 Ka) generation of stromatolites. Scale bar = 30 cm. d) Suguta Valley, "ox tongue" facies bioherms (ca. 28 Ka). Scale bar = 30 cm.

The presence of large amounts of soil reduces fine particle supply to the basins. During the arid periods, evaporites precipitate in shallow hypersaline lakes deprived of stratification due to greater wind activity. Related to both the forest regression and the nutrient concentration in waters, the organic matter that accumulates in these lakes is of microbial origin.

Lacustrine stromatolites in the Rift are most often associated with humid hydroclimatic phases and grow along the shorelines of former lakes (Fig.7). Among a great variety of morphologies, bioherms are particularly characteristic of this type of environment (Fig.8).

Morphology

Bioherm shape and spatial distribution reflects microenvironments; some form elongated bars parallel to the shore, as in Lake Natron basin (Casanova 1987b), or coalescent reefs along a slope break as in the Suguta Valley (Casanova 1986a). The following examples illustrate the diversity of bioherms in the Rift.

The composite bioherms from the first generation (ca. 90 Ka) of Lake Manyara are characterized by sinuous outlines, reaching 2m in length and 50cm in height (Fig.8a). A cross section through the bioherms shows several successive growth phases composed of oncoids, stromatolite fragments and lithoclasts coated with a single envelope; the last phase delimits the overall shape of the bioherm. These bioherms therefore record hydrodynamic cycles in the paleolake, with successive run-off periods interrupted by phases of stability.

Bioherms from the second generation (28-23 Ka) of Lake Manyara have essentially grown on the underlying bioherms, since they occupy the same environmental position. They occur as small circular mounds, <1m in height, regularly spaced and up to 2m in diameter (Fig.8b). These bioherms correspond to a single relatively uniform growth period as is demonstrated by their regular lamination. The build-up starts with well-defined cylindrical vertical columnns (5-15cm in diameter) which tend to expand and coalesce upwards.

In the Suguta Valley "ox tongue" facies bioherms (Fig.8d), 10-40cm in length, display stem prints around their base evoking former aquatic plant tufts. The surface of the bioherms exhibits small cupule-shaped depressions with sub-circular edges. These pits do not seem to be related to detrital fluxes or grazer activity, but rather to episodic necrosis of the microbial mat and increased size of successive growths.

Cemented oncoids from the second-generation stromatolites of Lake Natron (ca. 135 Ka), crop out on the bottom of paleochannels perpendicular to the shore. In the 7 to 10m depth interval, this facies extends over large areas and is characterized by a cyclic alternation of two oncoid populations (Fig.9). The cycle starts with small (<2mm diameter) spherical forms, deposited with a large proportion of detrital sediment. Growth is isotropic as the oncoids are continuously shifted by sub-lacustrine currents. These oncoids increase in size and evolve upward into strongly asymmetric stromatolite forms (2-3mm wide, 2-16mm high), indicating a decrease in hydrodynamic energy which immobilizes oncoids and thus favours upward, phototropic growth.

Figure 9 Lake Natron, Polished section of an oncolitic crust from the second generation, showing the variation in size and growth related to hydrodynamic fluctuations of lacustrine currents. Scale bar = 5 cm.

Microstructure

Doublets of sparitic/micritic laminae

The most frequent microstructure observed in lacustrine stromatolites results from the superposition of doublets composed of a light-coloured sparitic lamina and a dark micritic lamina (Fig.10a and b). Each light-coloured lamina, 20-1500µm thick, contains numerous erect filaments distributed in grasslike mats. The dominant species of the BMC is represented by elongated filaments, 1-1.6µm in diameter, forming interlaced tufts (Fig.11c). Filaments are often found together in a common sheath and resemble present-day *Schizothrix*. In most cases, they are also accompanied by isolated filaments, 1.8-2.5µm and 3-8µm in diameter. These oscillatoriacean forms likely represent different species of the present-day *Phormidium* (Fig.11b). The dark laminae range from 5 to 900µm in thickness and are composed of one or several micritic films rich in organic matter. Under the scanning electron microscope, these films are composed of clusters of bacteria, mucilagineous gels and rare cyanobacterial filaments. The sparitic microstructure is interpreted to have been generated by the periodic growth of a single population dominated by a *Schizothrix/Phormidium* association, similar to that observed in stromatolites from Lake Annecy (Casanova 1987a), light laminae represent, therefore, growth phases favourable to Cyanobacteria, whereas the dark laminae correspond to a slowing down or a cessation of cyanobacterial growth, contemporaneous with bacteria blooms.
Most stromatolites display a microstructural gradation between two end members depending on the thickness of the dark laminae. Where the light laminae are thicker than the dark ones, cyanobacterial filaments pass through the micritic films (Fig.10b),

and the doublets contain high percentages of detrital particles (< 24.5%) and organic matter (< 1.6%), whereas the carbonate is depleted in ^{13}C (3.1<$\delta^{13}C$<3.8 ^{o}/oo PDB) and ^{18}O (2.4<$\delta^{18}O$<3.7 ^{o}/oo PDB). Accordingly, the light laminae formed during extensive runoff periods, which favoured both suspension transport as well as increased content of dissolved nutrients, and consequently microbial growth. Waters were frequently renewed by meteoric precipitation depleted in heavy isotopes. In the other extreme, the dark laminae that are at least as thick as the light ones often rest unconformably on the light laminae (Fig.10a). This suggests that microbial growth in a light lamina was interrupted, then followed by erosion, before formation of the overlying dark lamina. In these cases, the doublets are characterized by small percentages of detrital (ca. 3.5%) and organic matter (ca. 0.6%), while the carbonate is considerably enriched in ^{13}C (4.0<$\delta^{13}C$<5.4 ^{o}/oo PDB) and ^{18}O (3.5<$\delta^{18}O$<5.4 ^{o}/oo PDB). Consequently, each dark lamina is interpreted to reflect a break in microbial productivity in stagnant water. The high ^{18}O content indicates a water residence time long enough to enrich the surrounding waters in heavy isotopes through evaporation (Hillaire-Marcel and Casanova 1987). The high ^{13}C content indicates an increase in photosynthetic activity, which preferentially consumes $^{12}CO_2$ and $H^{12}CO_3$ and leads to an enrichment of heavy carbon in interstitial waters and in the precipitated carbonates. Pentecost (1978) have shown, from living BMC, that an increase in photosynthetic activity is due to an increase in light intensity and temperature.

The basic doublet represents an ecological cycle of the microbial mat, which is interpreted here in terms of seasonal contrast: the light lamina corresponds to the rainy season and the dark lamina to the dry season. Such a seasonal record in BMC has already been observed in living stromatolites from southern France (Casanova and Lafont 1985), and seasonal isotopic variations in microbial carbonates have been reported by Monty and Hardie (1976) from modern stromatolites of the Bahamas and the Everglades.

Radial sparitic laminae

This microfacies is composed of irregular laminae, 600-1500µm thick, made of flabellate filament bushes embedded in clear elongated calcite crystals (Fig.11a). Each bush defines a distinct colony of cylindrical shape when joined, or upside-down cone shape when separated by growth porosity (up to 40%). The filaments, 1.2-1.5µm thick, are well preserved as organic streaks or hollow tubes within the crystals (Fig.11a). They are never coated by micritic sheaths like those described by Monty and Mas (1981) in the radial sparitic fabric generated by *Phormidium* associations. The bushes evoke those developed by *Calothrix/Dichothrix*-colonies in Lake Constance (Schafer and Stapf 1978).

This facies is independant of the sparitic/micritic doublets and seems to correspond to a change in the BMC composition. Second-generation stromatolites (ca.135 Ka) of Lake Natron show this change clearly: a drastic shift in both the fabric and the microbial growth, coincides with a drop (of 1.6 ^{o}/oo vs PDB) in the ^{18}O content of carbonate, suggesting a change in paleorainfall regime toward more humid conditions. The laminae are undulating as a result of successive growth of bushlike microbial colonies. On top of the sequence, each colony is surrounded by a micritic envelope probably produced by boring microbes, each sparitic lamina being then outlined by a 20-40µm thick micritic layer. Finally, the radial sparitic fabric disappears abruptly to be replaced by the doublet microfacies.

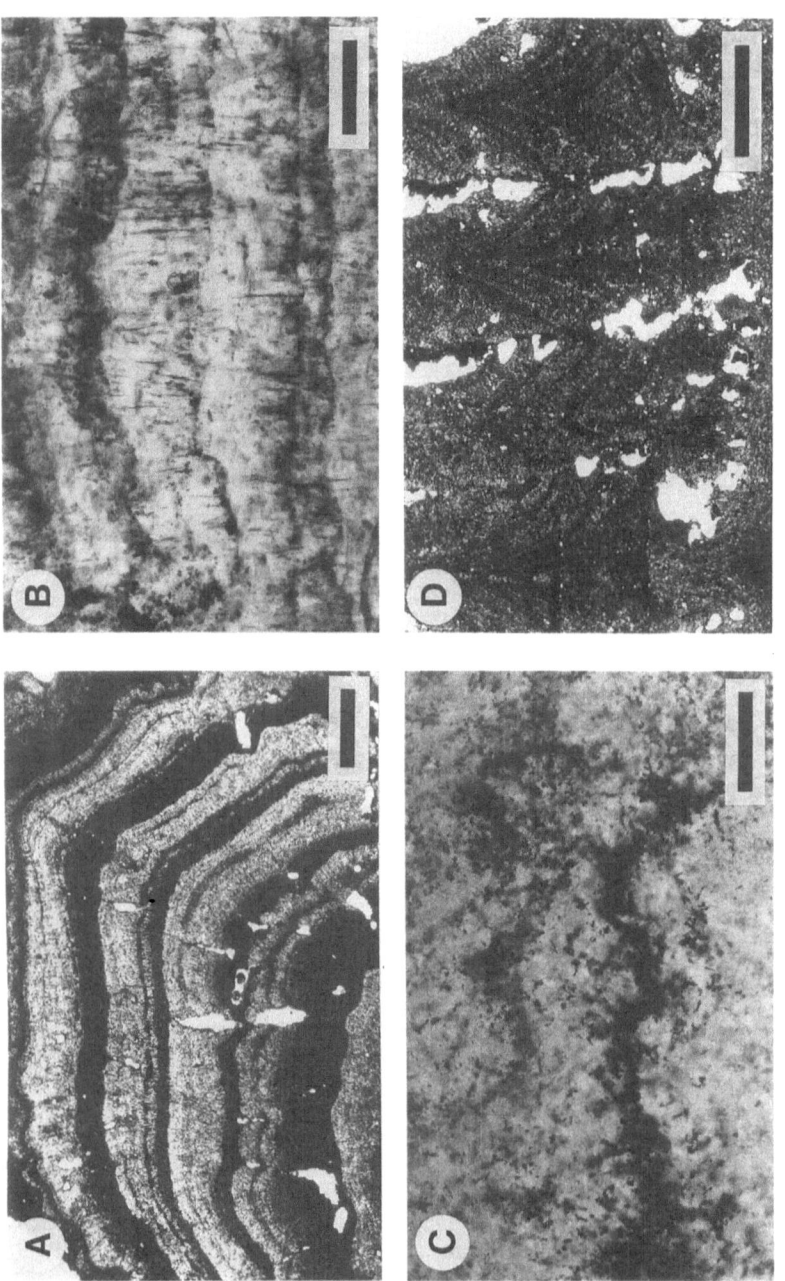

Figure 10 (a-d) Lacustrine stromatolite microfacies. a) Lake Manyara, Doublets composed of a light-coloured sparitic laminae and dark micritic laminae. The thickness of the dark laminae, formed during the dry season, can be of the same order than that of the light-coloured laminae. Scale bar = 1 mm. b) Lake Natron, Thick light-coloured laminae separated by thin dark ones, with most of the filaments passing through the micritic films. Scale bar = 100 μm. c) Suguta Valley, Simple micritic laminae. Clusters of bacteria on top of the micritic layers give rise to the second order lamination. Scale bar = 100 μm. d) Lake Manyara, *Conophyton*-like fabric: "Christmas-tree-like" shape of the columns of the third generation (22 Ka). Scale bar = 3 mm.

Simple micritic laminae

This microfacies characterizes all the stromatolites of the Suguta Valley (Fig.12a and b) regardless of geographic location. In hand specimen, the stromatolites show little or no lamination; when lamination does occur it is coarse (a few mm to a cm thick) and very likely corresponds to major ecological variations. The crust is compact but occasionally shows vesicles, elongated in shape, that are radially distributed from the growth surface in a centrifugal direction. High density of vesicles may confer a bushy structure to the stromatolites (Fig.12b).

In thin section, a "simple repetitive lamination" sensu Monty (1976, fig.1) is observed. The texture is uniformly micritic (Fig.11d). Laminae consist of light-coloured micritic layers 50-200µm thick, the upper parts of which are made of clots of small, dark-coloured, presumably bacterial filaments, <1µm in diameter (Fig.10c). The superposition of laminae sets, separated by a physical discontinuity and sometimes accentuated by secondary microsparite, defines the gross layering.

Locally, the laminae can be organized in columns (Fig.11d) cross-cut by vertical tubular vesicles suggesting a gaseous origin. These vesicles in most cases cross-cut the whole stromatolite, although some may be restricted to a limited series of laminae. Consequently, the vesicles are thought to be contemporaneous or immediately posterior to the lamination. The bases of the vesicles consists of dense swarms of bacteria that are nearly opaque in transmitted light. This suggests that the origin of the gaseous releases could be related either to the metabolism of the stromatolite building bacteria or to the organic decay of the mat.

Lacustrine stromatolites from the Suguta Valley are interpreted to be purely bacterial in origin and in this respect have no known equivalent elsewhere in the East African Rift. Only a few large filaments have been seen in their sections. Filamentous cyanobacteria (3-4µm in diameter) are randomly distributed and embedded in the bacterial mats, while remaining in an erect growth position. Accordingly, the inhibition of cyanobacterial growth is not related to colonization; more likely the physico-chemical environmental conditions were unfavourable to cyanobacteria, even while bacteria bloomed. Because of the lack of other contrasting types of encrustations in the Suguta Valley, it is not possible to ascertain the exact nature of those conditions.

Conophyton-like fabric

This stromatolite facies is composed of narrow columns, conical in vertical section and growing from planar laminae coating the substrate. The lamination consits of light-coloured micritic to microsparitic laminae separated by dark micritic films. The thickness of the laminae increases from the planar base (20-40µm) to the top of the columns (25-350µm), whereas the thickness of the films remains relatively constant (5-15µm). Both are built by intertwined tufts of small filaments (<1µm in diameter) growing vertically within the light laminae and horizontally in the films. In the laminae the tufts are surrounded by micrite passing laterally to larger crystals (10-20µm) containing isolated small filaments, while the films are made up of bundles of non-mineralized filaments. These tiny prostrate filaments can be compared to the habit of some modern species of *Schizothrix* (see Monty 1976, fig.4 a,b,f). This fabric can be explained by the seasonal alternation of mineralized and non-mineralized growth periods. At the base of the columns the films are almost planar; at regular intervals along the base they bend upward to form chevrons, which gradually become more and

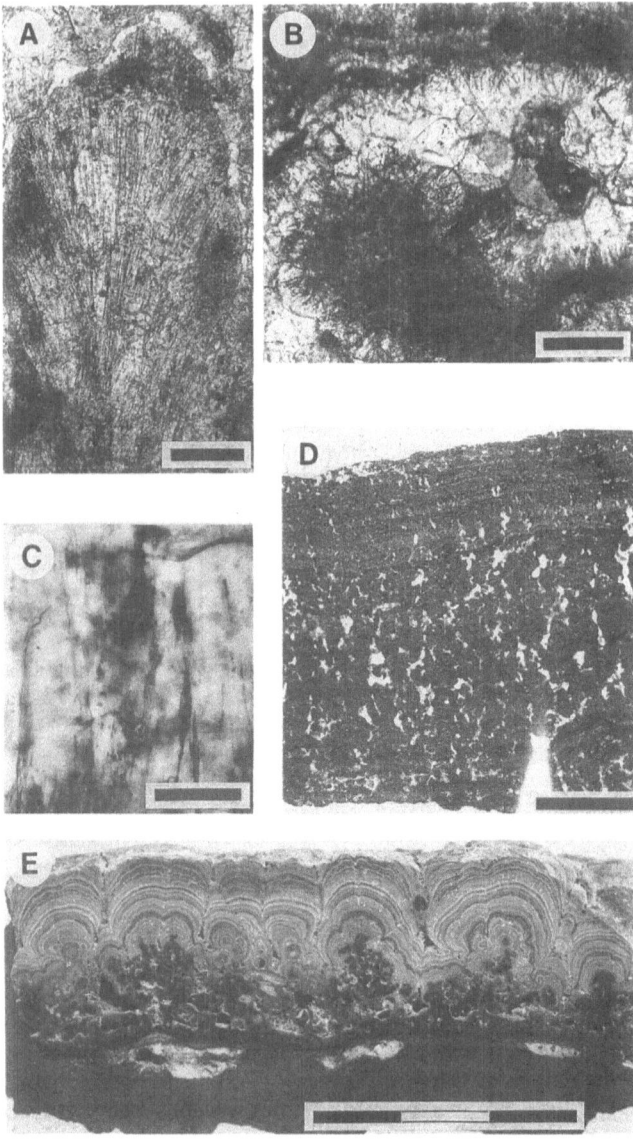

Figure 11 (a-e) Lacustrine stromatolite macro- and microfacies. a) Lake Natron, Radial sparitic fabric. Bushes, probably of *Calothrix/Dichothrix*-colonies embedded in clear elongated calcite crystal. Scale bar = 50 μm. b) Lake Natron, *Phormidium* mat covering the walls of a void, the cementation of which is very early as shown by the fine preservation of cyanobacterial filaments. Scale bar = 100 μm. c) Lake Natron, *Schizothrix* filaments forming interlaced tufts. Scale bar = 50 μm. d) Suguta Valley, Microlamination organized in columns delineated by vertical vesicles. Scale bar = 1 cm. e) Lake Natron, Polished slab of a planar encrustation from the third generation (12-10 Ka) showing a succession of bacterial colonies overlain by laminated columns. Scale bar = 3 cm.

more exaggerated with growth until they produce columns. The tufted filaments seem to have grown preferentially, then exclusively, on these chevrons, probably because of spatial competition. The resulting columns have a Christmas-tree shape (Fig.10d) and are separated by interspace regions, usually devoid of any sediment. The chevron angle is very regular along the columns (50-60°), but it increases upwards, with the apex becoming less angular, to give rise to coalescent rounded domes at the top.

This stromatolite facies characterizes all the encrustations of the third generation (ca 22 Ka) from Lake Manyara (Casanova and Hillaire-Marcel, 1992a); no comparable fabric has been found elsewhere in the Rift. It very likely represents the terminal phase of the second generation stromatolites (28-23 Ka). Stromatolites of the third generation extend only in shallow waters (0 to -2 m), and their high $\delta^{18}O$ and $\delta^{13}C$ values suggest a reduced rate of water renewal and an increase in the nutrient content of paleolake water subjected to strong evaporation.

Layers of shrubs

This microstructure, composed of shrub-shaped microbial colonies, has been observed only in the third generation stromatolites (13-10 Ka) from the Natron-Magadi basin (Casanova 1987c). Colonization of the substrate by these stromatolites is characterized by three successive phases (Fig.11e): (1) black basal layer, 0.5-2mm thick, composed of cross-bedded sparitic colonies; (2) dark digitate zone, 7-20mm thick, made of shrub-like sparitic colonies; and (3) sparitic/micritic doublet facies. The first two phases represent two growth modes of a single population of bacteria colonies, similar to those described in bacterial travertines from Italy and west-central USA (Chafetz and Folk 1984, fig. 23). Under the scanning electron microscope, the sparitic colonies are made of "gothic-arch" calcite crystals, that after gentle etching in dilute HCl, are seen to have entombed well-preserved bacterial cells (see Folk et al. 1985, fig.9). The third phase is generated by cyanobacteria.

The chemical composition of these bacterial carbonates differs from other Lake Natron stromatolites in having a slightly higher Fe and Mg content. Localized at the periphery of the colonies in the form of dark envelopes, this ferromagnesian enrichment is not related to the clear sparitic carbonates, but rather suggests secretions related to bacterial metabolism. Some cylindrical encrustations formed around isolated trees contain only the bacterial phases: such decimetre-sized deposits (9 650±100 yr B.P.) are contemporaneous with neighbouring stromatolites (9 710±100 yr B.P.) and indicate only a lack of the third, cyanobacterial stage of colonization. Hence, it seems that this tripartite succession of building organisms corresponds to a phenomenon of biological competition, with the bacteria acting as the initial encrusting flora.

DISCUSSION

Paleoecology of rift stromatolites

Despite the absence of living examples, the stromatolites that have colonized the rifts of East Africa during the last 10 Ma represent an extraordinary biosedimentologic record

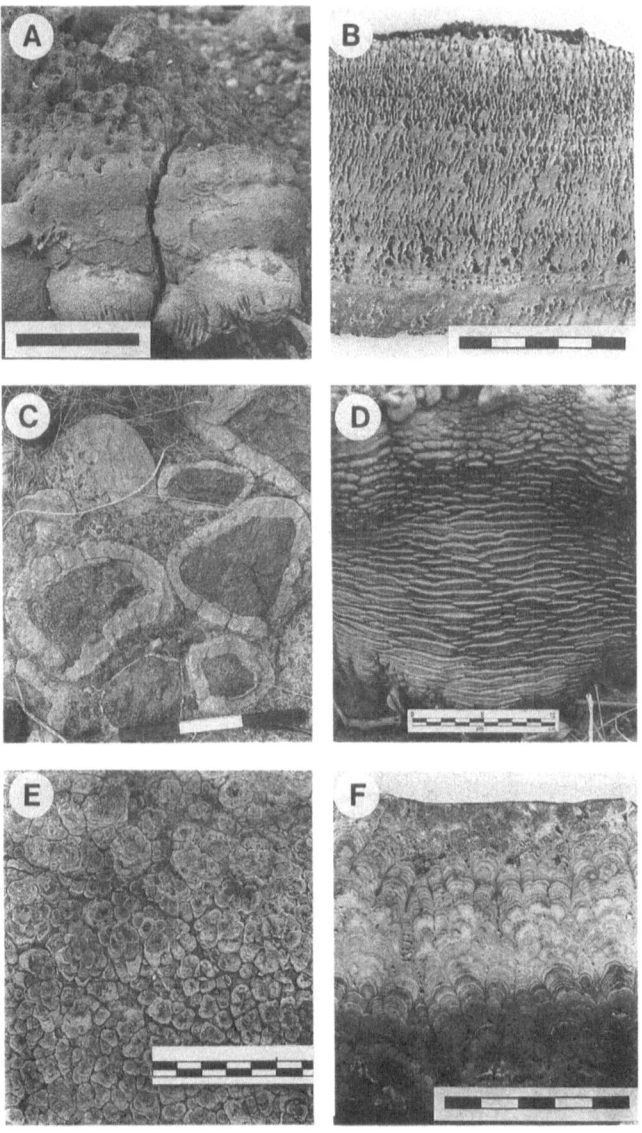

Figure 12 (a-f) Lacustrine stromatolite macrofacies. a) Suguta Valley, Close-up of an "ox tongue" facies bioherm, showing stem prints around its base and cupule-shaped depressions on its surface. Scale bar = 10 cm. b) Suguta Valley, Bushy structure related to dense vesicular porosity. Scale bar = 5 cm. c) Suguta Valley, Conglomerate made of pebble encrustations. Scale bar = 30 cm. d) Lake Manyara. Ridges developed on vertical substrate facing the major sublacustrine current. e) Lake Natron, Surface ornamentation given by columns in the bathymetric zone 1.5-2.5 m. Petal-shaped bioherm of the second generation (ca. 135 Ka). Scale in cm. f) Polished surface of the previous facies. Scale bar = 5 cm.

documenting the relationships between micro-organisms and carbonate minerals in continental settings. Though much data has been gathered so far, only a small part of the stromatolite history of the Rift has been revealed. Numerous sites undoubtedly remain to be discovered or have been only partially studied; at the same time, erosion and tectonism, as well as volcanic and sedimentary overlaps, lead to obliteration of stromatolite outcrops.

Fed by torrential rivers, freshwater modern lakes of the Rift (e.g., Lake Baringo) contain waters that are too turbid to allow stromatolite growth. Fossil lacustrine stromatolites form continuous shorelines that can be used to reconstruct high lake levels contemporaneous with positive hydroclimatic phases. These paleolakes contained calcic carbonate waters and their levels remained stable during the growth of the stromatolites. Continuous ecological zonations show that the potential bathymetric extension of the stromatolites ranges commonly from 0 to -30m (exceptionally down to -50m in Lake Tanganyika; Casanova and Thouin 1990, Casanova and Hillaire-Marcel 1992b). All hard substrates are colonized by microbial mats and mineralization results from in situ carbonate precipitation. Bioherms are the morphologies most currently developed in lacustrine environments. Columnar growth, as a result of biological competition in the BMC, seems to be typical of this environment. In the bathymetric profiles, columns reach their maximum development in the surface water layer (Fig.12d, e and f) where conditions are favourable in terms of light intensity, oxygenation and nutrient supply.

East African Rift stromatolites are characterized by: (1) a large diversity of forms and assemblages from various environments; (2) well-defined gross morphologies recording clear relationships with ecological parameters; and (3) exceptional preservation of micro-organisms and fabrics, due to the absence of diagenetic recrystallization. It appears, therefore, that continental environments, rather than the poorly diversified modern marine setting (see for example Dill et al.1986, Burne and James 1986), provide better analogues of Precambrian stromatolite morphologies. In the East African Rift, a stromatolite biostratigraphy based on morphological evolution has no fundamental biological significance since morphology has been controlled only by the environment, especially hydroclimatic conditions. On the other hand, the features of stromatolites formed in the Rift are very similar to those observed in modern settings elsewhere (Schafer and Stapf 1978; Osborne et al. 1982; Winsborough and Seeler 1984), and in most Cenozoic sites in general (see for example Bertrand-Sarfati et al. 1966; Anadon and Zamarreno 1981; Freytet and Plaziat 1982; Casanova and Nury 1989). The morphological diversity and the distribution of Cenozoic stromatolites is not related to a biological evolution of BMC: the *Phormidium-Schizothrix-Calothrix/Dichothrix* association that builds freshwater stromatolites in the Rift, also characterizes fluvio-lacustrine settings during the Cenozoic (Monty and Mas 1981). A further argument in favour of ecological control over morphology lies in the fact that the same BMC may generate planar laminae or columns in the same hand specimen, because of different micro-environments. The biological stability of Cenozoic BMC raises the problem of the temporal meaning of successions of stromatolite assemblages observed in the Precambrian: do these successions represent a biological evolution of BMC, or global changes in climatic and ecologic conditions?

Partly due to selective erosion, lacustrine stromatolites represent the majority of the known occurrences in the Rift, whereas fluvial structures are rare. All stromatolites that have been analysed are made of low-Mg calcite and do not contain aragonite, dolomite or evaporites common to saline lakes elsewhere. Furthermore, fabrics as well as micro-organism associations are characteristic of freshwater environments (Casanova 1986a). Accordingly, the stromatolites of the Rift, formed either on floodplains or in fluvio-lacustrine environments, document the presence of clear and dilute freshwaters. The absence of stromatolites during arid periods, characterized by alkaline rivers and lakes, can be explained by the Ca deficitency of siliciclastic basins in the Rift. Jones et al. (1977, fig.3) measured a 90% loss of Ca and Mg in the step from stream flow to the various groundwaters feeding the lacustrine brines of the evaporitic Magadi basin. Modern saline lakes of the Rift as a rule contain only traces of Ca and Mg (Eugster 1980; Yuretich 1982; Gueddari 1984), which is reflected in the absence of calcic evaporites (gypsum, aragonite, dolomite, hydromagnesite) but precipitation of sodic carbonate evaporites and sodic zeolites (Eugster 1967; Yuretich 1982; Tiercelin et al. 1982). The microbial mats associated with these saline waters are never mineralized and differ in that respect from the stromatolites of other evaporitic basins (see Buchbinder 1981 and references therein).

East African climato-stratigraphy: the stromatolite record

Fluctuations in the water levels of closed lake basins of East Africa have been recorded since the 18th century, and recently high lacustrine levels have been used as indicators of former climatic changes (e.g. Butzer et al. 1972; Owen et al. 1982; Street-Perrott and Harrison 1984; Kutzbach et al. 1985). Comparison of these paleoclimatic inferences with numerical simulations of global climate tends to support current models of general atmospheric circulation (Street-Perrott 1987).
Study of high lacustrine levels in the Rift is therefore essential in reconstructing the evolution of sedimentary basins in this intertropical area as it relates to paleohydrology and paleoclimatology. The specific contributions of stromatolitic shorelines towards this are numerous:
(1) Stromatolites reveal more information than most other lacustrine deposits. As biogenic carbonates, they preserve both a temporal record (^{14}C, Th/U), as well as the isotopic imprint of the environment in which they form (^{13}C, ^{18}O). The oxygen 18 and carbon 13 content of stromatolitic carbonates gives an indication of the paleohydrology, when compared to the values for modern waters and carbonates. Long residence time for paleolake waters and isotopic equilibrium between the dissolved inorganic carbon and the atmospheric CO_2 are generally inferred (Hillaire-Marcel and Casanova 1987). Parallel variations in ^{13}C and [18]O during stromatolite growth can be linked with variable development of microbial mats in the paleolakes due to succession of heavy precipitation and intense evaporation.
(2) The existence of ecological zones provides a continuous recording of the limnologic evolution from the littoral zone to the deep zone (sensu Dussart 1966). The specific morphology of stromatolites within a basin or for a given hydroclimatic episode makes it possible to use them as local stratigraphic markers.
(3) Stromatolites document hydroclimatic optima in the absence of available contemporaneous lacustrine sediments. For example, it has been shown that stromatolites keep excellent pollen records (Vincens et al. 1986).
(4) Taking into account the need for precise data on seasonal components of the climate in continental environments (Manabe et al. 1975; Van Campo 1983; Kutzbach et al. 1985), stromatolites, by means of rhythmic growth of the microbial mat, record seasonal

contrasts expressed as SR (seasonal ratio), derived from the lamina thickness, and their variation in time.

Although it is logical that the preservation of fossil stromatolites is inversely proportional to time, and hence that the shorelines of the oldest lacustrine levels are most difficult to reconstruct, isotopic and radiochronologic data allow a detailed reconstruction of the major climatic trends in East Africa for the last 4 Ma.

Between 3.9 and 2 Ma, stromatolites from Hadar (Hillaire-Marcel et al. 1982) and Turkana (Abell et al. 1982) provide evidence for humid and cool climatic conditions. Water infill of the basins took place by means of abundant precipitation that was depleted in ^{18}O, coming from the Ethiopian plateaux.

Between 1.9 and 1.4 Ma, stromatolites from Turkana (Abell et al. 1982) indicate a shift in $\delta^{13}C$ and $\delta^{18}O$ to more positive values, characteristic of a warmer and dryer climate. This climatic change is also indicated by changes in fossil vertebrates and molluscs, by the geochemistry of volcanic tuff weathering, as well as by pollen in sediments. The Upper Pleistocene was characterized by a return to cooler and wetter conditions (Abell et al. 1982).

During the Terminal Pleistocene-Holocene, three main hydroclimatic optima occurred ca. 240 Ka, between 135 and 140 Ka, and between 9 and 13 Ka (Casanova 1986a; Hillaire-Marcel et al. 1986; Casanova and Hillaire-Marcel 1992a). Thus, a humid episode seems to characterize East Africa during each glacial-interglacial transition period (transitions 8-7, 6-5 and 2-1 of the ^{18}O curves in oceanic environments). The paucity of first generation stromatolites (ca. 240 Ka) in the Natron-Magadi basin does not permit a detailed analysis of seasonality. However, the values measured (1< RS <10), as well as the ^{18}O content ($0.6 < \delta^{18}O < 2.2$ ^{o}/oo PDB), argue in favour of a cool and humid climate.

This climatic tendency persisted through the 135 Ka humid episode (second generation of the Natron-Magadi basin and part of the first generation of Lake Manyara), during which negative $\delta^{18}O$ values (down to -0.6 ^{o}/oo vs PDB) are recorded at Lake Natron. The end of this episode corresponds to a change toward dryer conditions: the top of the Lake Natron stromatolitic sequence is characterized by a seasonal contrast clearly dominated by a long dry season and positive $\delta^{18}O$ values (up to 4.9 ^{o}/oo PDB). Stromatolites in Lake Suguta (ca 121 Ka) do not allow a seasonality analysis, being of bacterial origin; nevertheless their pollen content indicates the presence of dry conifer forests in the basin, as well as the existence of a marked dry season.

A sequence of high lake levels, spanning the 28-23 Ka period in the Manyara basin (second and third stromatolite generations) corresponds to unstable paleoclimatic conditions (Casanova and Hillaire-Marcel 1992a). Seasonality analysis of these stromatolites shows a cyclical alternation of periods during which the seasonal contrast successively favours the dry or rainy season (0.08< RS <14).

The 13-9 Ka hydroclimatic episode, recorded in stromatolites of the third generation in the Natron-Magadi basin, was typified by regular, fine rainfall throughout a long wet season, and by a short dry season. The seasonal contrast recorded in stromatolite laminae averages 10, and occasionally reaches 16. Paleoprecipitation of ca. 750 mm/yr at lake level has been estimated; this represents ca. 175% over the modern value (427 mm/yr).

A more recent (5.3-4.7 Ka) high lake level is also recorded by stromatolites around Lake Bogoria (Casanova 1987b). Palynological assemblages in stromatolites and in correlative lacustrine sediments show a transition from a humid forest cover to a dry coniferous forest with *Podocarpus, Juniperus* and *Olea*. A seasonality ratio <1 is recorded by the stromatolites. From a hydrological viewpoint, this episode was characterized by a long dry season interrupted by thunderstorms and heavy rains distributed over a short rainy season. This short event have marked the onset of drier conditions over East Africa.

ACKNOWLEDGEMENTS.

I would like to express my appreciation to my colleagues A. Maurin, C. Monty, C. Hillaire-Marcel, J. Sarfati, S. Awramik and B. Pratt for their helpful comments during various stages of the study; I would also like to thank the latter for his assistance in the English version of the manuscript. Field work on East African stromatolites was supported by the Compagnie Française des Pétroles (Paris, France) and the GEOTOP Center (Montréal, Canada).

REFERENCES

Abell, P.I., Awramik., S.M., Osborne, R.H., and Tomellini, S. (1982) "Plio-Pleistocene lacustrine stromatolites from Lake Turkana, Kenya: morphology, stratigraphy and stable isotopes", Sediment. Geol. 32, 1-26.

Anadon, P, and Zamarreno, I. (1981) "Paleogene nonmarine algal deposits of the Ebro Basin, Northeastern Spain". In: C. Monty (ed), Phanerozoic stromatolites: Case histories, Springer Verlag, Berlin Heidelberg New York, pp. 140-154.

Baker, B.H., Mohr P.A., and Williams M.A.J. (1972) "Geology of the Eastern Rift System of Africa", Geol. Soc. Am. Sp. Pap. 136, 1-67.

Bertrand-Sarfati, J., Freytet, P. and Plaziat, J.C. (1966) "Les calcaires concrétionnés de la limite Oligocène- Miocène des environs de Saint-Pourcain-sur Sioule (Limagne d'Allier): rôle des algues dans leur édification; analogie avec les stromatolithes et rapports avec la sédimentation", Bull. Soc. Géol. Fr. 7 VIII, 652-662.

Braithwaite, C.J.R., Casanova. J., Frevert, T., and Whitton, B.A. (1989) "Recent stromatolites in landlocked pools on Aldabra, western Indian Ocean", Palaeogeogr., Palaeoclimatol., Palaeoecol., 69, 145-165.

Buchbinder, B. (1981) "Morphology, microfabric and origin of stromatolites of the Pleistocene precursor of the Dead Sea, Israel". in: C. Monty (ed) Phanerozoic Stromatolites : Case histories. Springer Verlag, Berlin Heidelberg New York, pp. 180-196.

Burne, R.V., and James, N.P. (1986) "Subtidal stromatolites from Shark Bay, western Australia". 12th. I.A.S. Congress, Camberra", abstracts, 366.

Butzer, K.W., Isaac, G.L., Richardson, J.L., and Washbourn-Kamau, C. (1972) "Radiocarbon dating of East African lake levels", Science 175, 1069-1076.

Casanova, J. (1981) "Etude d'un milieu stromatolitique continental: les travertins Plio-Pleistocenes du Var (France)", Thèse 3me Cycle Aix-Marseille II University

Casanova, J. (1985) "Les oncolites du Rift Est-Africain: morphométrie et paléoenvironnements", Actes du 110e Congrès national des Sociétés Savantes 6, 345-357.

Casanova, J. (1986a) "Les stromatolites continentaux: paléoécologie, paléohydrologie, paléoclimatologie. Application au rift Gregory", Thèse Aix-Marseille II University

Casanova, J. (1986b) "East African Rift Stromatolites". In: Frostick L.E. et al (eds) Sedimentation in the African Rifts, Geol. Soc. Spec. Publ. 23, 195-204.

Casanova, J. (1987a) "Microbiology and geochemistry of modern lacustrine stromatolites, lake Annecy, France", 4th Intern Symp on fossil Algae Cardiff.

Casanova, J. (1987b) "Les stromatolites et hauts niveaux lacustres pléistocènes du bassin Natron-Magadi (Tanzanie-Kenya)", Sci. Géol. 40, 135-153.

Casanova, J. (1991) "Biosedimentology of Quaternary stromatolites in intertropical Africa", Jour. of African Earth Sci. 12, 409-415.

Casanova, J. and Hillaire-Marcel, C. (1992a) "Chronology and paleohydrology of Late Quaternary high Lake Levels in the Manyara Basin (Tanzania) from isotopic composition (^{18}O, ^{13}C, ^{14}C, Th/U) of fossil stromatolites". Quater. Res. 38, 205-226.

Casanova, J. and Hillaire-Marcel, C. (1992b) "Late Holocene hydrological history of Lake Tanganyika, East Africa, from isotopic data on fossil stromatolites", Palaeogeogr., Palaeoclimatol., Palaeoecol., 91, 35-48.

Casanova, J. and Lafont, R. (1985) "Les cyanophycées encroûtantes du Var (France)", Verh. Internat. Verein. Limnol. 22, 2805-2810.

Casanova, J. and Nury, D. (1989) "Biosédimentologie des stromatolites fluvio-lacustres du Fossé Oligocène de Marseille", Bull. Soc. Géol. Fr. 5, 1173-1184.

Casanova, J. and Thouin, C. (1990) "Biosédimentologie des stromatolites holocènes du Lac Tanganyika (Burundi)" Bull. Soc. Géol. Fr. 6, 1173-1184.

Casanova, J. and, Tiercelin J.J. (1982) "Construction stromatolitiques en milieu carbonaté sodique: les oncolites des plaines inondables du lac Magadi (Kenya)", C. R. Acad. Sci. Paris 295, 1139-1144.

Casanova, J., Hillaire-Marcel, C., Page, N. Taieb, M. and, Vincens, A. (1988) "Stratigraphie et paléohydrologie des épisodes lacustres du Quaternaire récent du Rift Suguta (Kenya)", C. R. Acad. Sci. Paris 307, 1251-1258.

Chafetz, H.S. and, Folk, R.L. (1984) "Travertines: depositional morphology and the bacterially constructed constituents", Journ. Sedim. Petrol. 54, 289-316.

Cohen, A. (1989) "Facies relationships and sedimentation in large rift lakes and implications for hydrocarbon exploration: examples from lakes Turkana and Tanganyika", Palaeogeogr., Palaeoclimatol., Palaeoecol. 70, 65-80.

Demange, J. and Stieltjes, L. (1975) "Géologie de la région sud-ouest du T.F.A.I. (région du lac Abhé-lac Assal)", Bull. Bur. Rech. Géol. Min. 2, 83-120.

Dill, R.F., Shinn, E.A., Jones, A.T., Kelly, K. and Steinen, R.P. (1986) "Giant subtidal stromatolites forming in normal salinity waters", Nature 324, 55-58.

Dixit, P.C. (1984) "PLeistocene lacustrine ridged oncolites from the lake Manyara area; Tanzania, East Africa", Sedim. Petrol. 39, 53-62.

Dussart, B. (1966) "Limnologie. L'étude des eaux continentales". Gauthier Villars Paris, (Géobiologie-Ecologie-Aménagement).

Eugster, H.P. (1967) "Hydrous sodium silicates from Lake Magadi, Kenya: Precursor of bedded chert", Science 157, 1177-1180.

Eugster, H.P. (1980) "Lake Magadi, Kenya and its precursor". in: A. Nissenbaum (ed) Hyper-saline brines and evaporitic environments, Developments in Sedimentology 28, Elsevier Amsterdam, 195-230.

Eugster, H.P. and Hardie, L.A. (1978) "Saline lakes". in: A. Lerman (ed) Chemistry, Geology and Physics of lakes, Springer Verlag New York, 237-293.

Folk, R.L., Chafetz,H.S. and Tiezzi, P.A. (1985) "Bizarre forms of depositional and diagenetic calcite in hot-spring travertines", Central Italy. Soc. Econom. Paleont. Mineral., 349-369.

Fontes, J.C. and Pouchan, P. (1975) "Les cheminées du lac Abhé (T.F.A.I.): stations hydroclimatiques de l'Holocène", C. R. Acad. Sci. Paris 280, 383-386.

Fontes, J.C., Pouchain, P., Saliege, J.F. and Zuppi, G.M., (1980) "Environmental isotope study of groundwater system in the Republic of Djibouti". in: Arid-Zone Hydrology Investigations with Isotopes Techniques", Vienna Intern. Atom. Energy Agency, 237-262.

Freytet, P. and Plaziat, J.C. (1982) "Continental carbonate sedimentation and pedogenesis. Late Cretaceous and Early Tertiary of Southern France", Contribution to Sedimentology Schweizerbart'sche Verlags, Edit. Stuttgart.

Fritsch, F.E. (1950) "Phormidium incrustatum, an important member of the lime-encrusting communities of flowing water", Biol. Jaarb. 70, 27-39.

Gasse, F. (1975) "L'évolution des lacs de l'Afar central (Ethiopie et T.F.A.I.) du Plio-Pléistocène à l'Actuel", Thèse Paris VI University, 390p.

Gasse, F., Fontes, J.C. (1989) "Palaeoenvironments and Palaeohydrology of a Tropical closed Lake (Lake Asal, Djibouti) since 10,000 yr B.P.", Palaeogeogr. Palaeoclimatol. Palaeoecol. 69, 67-102.

Gasse, F., Fontes, J.C. and Rognon, P. (1974) "Variations hydroclimatiques et extension des lacs holocènes du désert Danakil", Palaeogeogr., Palaeoclimatol., Palaeoecol. 15, 109-148.

Geurts, M.A. (1976) "Genèse et stratigraphie des travertins de fond de vallée en Belgique", Acta Geogr. Lovaniensia 16: 9-70.

Golubic, S (1969) "Cyclic and non cyclic mechanisms in the formation of travertine". Verh. intern. Verein. Limnol. 17, 956-961.

Golubic, S. and Fisher, A.G. (1975) "Ecology of calcareous nodules forming in Little Conestoga Creek near Lancaster Pennsylvania", Verh. intern. Verein. Limnol. 19, 2315-2323.

Grove, A.T. (1986) "Geomorphology of the African Rift System". in: L.E. Frostick et al (eds) Sedimentation in the African Rifts", Geol. Soc. Spec. Publ. 25, 9-16.

Grove, A.T. and Goudie, A.S. (1971) "Secrets of Lake Stephanie's past", Geogr. Mag. 43, 542-547.

Grove, A.T., Street, F.A. and Goudie, A.S. (1975) "Former lake levels and climatic change in the Rift Valley of Southern Ethiopia", Geogr. Mag. 141, 177-202.

Gueddari,. M (1984) "Géochimie et thermodynamisme des évaporites continentales. Etude du lac Natron en Tanzanie et du Chott El Jerid en Tunisie", Thèse Strasbourg 143., Sci. Géol. Mém. 76.

Hillaire-Marcel, C. and Casanova, J. (1987) "Isotopic hydrology and paleohydrology of the Magadi (Kenya)-Natron (Tanzania) basin during the Late Quaternary", Palaeogeogr. Palaeoclimatol. Palaeoecol. 58, 155-181.

Hillaire-Marcel, C. Taieb, M., Tiercelin, J.J. and Page, N. (1982) "A 1.2 Myr record of isotopic changes in a late Pliocene Rift lake Ethiopia", Nature 296, 640-642.

Hillaire-Marcel, C., Carro, O. and Casanova, J. (1986) "[14]C and Th/U dating of Pleistocene and Holocene stromatolites from east african paleolakes", Quatern Res 25, 312-329.

Johnson, G.D. (1974) "Cenozoic lacustrine stromatolites from hominid-bearing sediments East of lake Rudolf Kenya", Nature 247, 520-523.

Johnson, G.D. and Raynolds, R.G.H. (1976) "Late Cenozoic environments of the Koobi Fora Formation: the upper member along the western Koobi Fora Ridge". in: Coppens, Y., Howell, F.C., Isaac, G.L. Leakey REF (eds) Earliest man and environments in the Lake Rudolf basin, Univ Chicago Press, Chicago, pp. 115-123.

Jones, B.F., Eugster, H.P. and Retting, S.L. (1977) "Hydrochemistry of the lake Magadi basin Kenya", Geochem. Cosmochem. Acta 41, 53-72.

Kann, E. (1959) "Die eulitorale Algenzone im Traumsee (Oberoîsterreich)", Arch. Hydrobiol. 55, 129-192.

Krumbein WE, Cohen Y (1974) "Biogene, klastische und evaporitische sedimentation in einem mesothermen monomiktischen ufernahen See (Golf von Aqaba)". Geol. Rundschau 63: 1035-1065.

Kutzbach, J.E. and Street-Perrot, F.A. (1985) "Milankovitch forcing of fluctuations in the level of tropical lakes from 18 to 0 kyr B.P.", Nature 317, 2130-2134.

Manabe, S. and Holloway, J.L. (1975) "The seasonal variation of the hydrologic cycle as simulated by a global model of atmosphere", J. Geophys. Res. 80, 1617-1649.

Monty, C.L.V. (1976) "The origin and development of cryptalgal fabrics". in: Walter M.R. (ed) Stromatolites, Development in Sedimentology 20 Elsevier Amsterdam, pp. 193-249.

Monty, C.L.V. and Hardie, L.A. (1976) "The geological signifiance of the freshwater blue-green algal calcareous marsh". in: Walter MR (ed) Stromatolites, Development in Sedimentology 20, Elsevier Amsterdam, pp. 447-477.

Monty, C.L.V. and Mas, J.R. (1981) "Lower Cretaceous (Wealdian) blue-green algal deposits of the province of Valencia, Eastern Spain". in: C. Monty (ed) Phanerozoic Stromatolites: Case histories, Springer Verlag, Berlin Heidelberg New York, pp. 85-120.

Monty, C.L.V. and Van Laer P. (1984) "Experimental radial calcite ooids of microbial origin and fossil counterparts", 5e Congr. Europ. Sédiment. Marseille 1984, 296-297.

Moore, L. Knott, B. and Stanley, N. (1984) "The stromatolites of Lake Clifton, Western Australia", Search 14, 309-314.

Osborne, R.H., Licari, G.R. and Link, M.H. (1982) "Modern lacustrine stromatolites, Walker Lake, Nevada", Sedim. Geol. 32, 39-61.

Owen, R.B., Barthelme, J.W., Renaut, R.W. and Vincens, A. (1982) "Paleolimnology and archaeology of Holocene deposits north-east of Lake Turkana Kenya", Nature 298: 523-529.

Pentecost, A. (1978) "Blue-green algae and freshwater carbonate deposits", Proc. R. Soc. London B 290, 43-61.

Renaut, R.W. (1982) "Late quaternary geology of the lake Bogoria fault-through Kenya Rift Valley", Thesis London University.

Saliège, J.P. (1979) "Détermination expérimentale du fractionnement isotopique $^{13}C/^{12}C$ au cours des processus naturels", Diplôme Ingénieur C.N.A.M. Paris.

Schafer, A. and Stapf, K.R.G. (1978) "Permian Saar-Nahe Basin and recent lake Constance (Germany): two environments of lacustrine algal carbonates". in: Matter, A. and Tucker, M.E. (eds) Modern and Ancient lake sediments, Blackwell Sci. Publ. Int Ass. Sed. Sp. Publ. 2, pp. 83-107.

Schneider, J. (1977) "Carbonate construction and decomposition by epilithic and endolithic microorganisms in salt- and freshwater". in: Flugel E (ed) Fossil Algae, Springer, pp. 248-260.

Street-Perrott, F.A. (1987) "Monsoon climates and water-budget variations", XII INQUA Intern Congr Ottawa, Abstract 271.

Street-Perrott, F.A. and Harrison, S.P. (1984) "Temporal variations in lake levels since 30,000 yr B.P.- an index of the hydrological cycle". in Ewing M (ed) Climate processes and climate sensitivity, Geophys. monogr. 29, pp. 118-129.

Street, F.A. (1979) "Late Quaternary lakes in the Ziway-Shala basin, Southern Ethiopia", Thesis Cambridge 457p.

Symoens, J.J. (1957) "Les eaux douces de l'Ardenne et des régions voisines: les milieux et leur végétation algale", Bull. Soc. R. Bot. Belgique 89, 111-314.

Symoens, J.J. (1968) "Classification zonale des eaux tropicales et types d'eaux du bassin du Bangweolo-Luapula-Moero", La minéralisation des eaux naturelles Bruxelles II.1, 53-94.

Taieb, M. (1974) "Evolution quaternaire du bassin de l'Awash (Rift éthiopien et Afar)", Thèse Paris VI University.

Tiercelin, J.J., Michaux, J. and Bandet, Y. (1979) "Le Miocène supérieur du sud de la Dépression de l'Afar, Ethiopie: sédiments, faune, âges isotopiques", Bull. Soc. géol. France 3, 255-258.

Tiercelin, J.J., Perinet, G., Le Fournier, J., Bieda, S. and Robert, C. (1982) "Lacs du Rift est-africain, exemples de transition eaux douces-eaux salées: le lac Bogoria, Rift Gregory, Kenya", Mém. Soc. géol. France 144, 217-230.

Tiercelin, J.J., Vincens, A., Barton, C.E., Carbonel, P., Casanova, J., Delibrias, G., Gasse, F., Grosdidier, E., Herbin, J.P., Huc, A.Y., Jardiné, S., Le Fournier, J., Mélières, F., Owen, R.,B., Pagé, P., Palacios, C., Paquet, H., Péniguel, G., Peypouquet, J.P., Raynaud, J.F., Renaut, R.W., Renéville, P., Richert, J.P., Riff, R., Robert, P., Seyve, C., Vandenbroucke, M. and Vidal, G. (1987) "Le demi-graben de Baringo-Bogoria, Rift Gregory, Kenya: 30 000 ans d'histoire hydrologique et sédimentaire", Bull. Centres Rech. Explor. -Prod. Elf-Aquitaine 11, 249-540.

Tilden, J.E., (1897) "Some new species of Minnesota Algae which live in a calcareous matrix", Bot. Gaz. 23, 95-104.

United Nations (1971) "Geology, geochemistry and hydrology of hot springs of the East African Rift system in Ethiopia", Unpubl. rep. E.T.H 26, 1-170.

Van Campo, E. (1983) "Paléoclimatologie des bordures de la mer d'Arabie depuis 150,000 ans. Analyse pollinique et stratigraphie isotopique", Thèse Montpellier University.

Vincens, A., Casanova, J. and Tiercelin, J.J. (1986) "Paleolimnology of lake Bogoria (Kenya) during the 4500 B.P. high lacustrine phase". in: L.E. Frostick et al (eds) Sedimentation in the African Rift, Geol. Soc. Sp. Publ. 23, pp. 315-322.

Wallner, J., (1935) "Zur Kenntnis der Kalkbildung in der Gatung Rivularia", V. Beith. Z. Bot. Zbl. 54/A, 128-134.

Walter, M.A., Golubic, S., Preiss, W.V. (1973) "Recent stromatolites from hydromagnesite and aragonite depositing lakes near the Coorong Lagoon, South Australia", J. sed. Petr. 43, 1021-1030.

Winsborough, B.M. and Seeler, J.S. (1984) "The relationship of diatom epiflora to the growth of limnic stromatolites and microbial mats", 8th Diatom Symposium, 395-407.

Yuretich, R.F. (1982) "Possibles influences upon lakes development in the East African Rift Valley", J. Geol. 90, 329-337.

LACUSTRINE STROMATOLITES AND ONCOIDS: MANUHERIKIA GROUP (MIOCENE), NEW ZEALAND

J. K. LINDQVIST
Institute of Geological & Nuclear Sciences ; Private Bag 1930, Dunedin; New Zealand

ABSTRACT

Stromatolites, oncoids, ooid sand, and palustrine carbonates comprise a minute part of largely siliciclastic Miocene lake deposits in the Central Otago region of southern South Island. Oncoids range from millimetres to over 35 cm in diameter and 20 cm high. They have formed around grains, plant stems (now decayed), molluscs, and quartz clasts. Large oncoids developed by encapsulation of small growths and fragments. Fixed stromatolite pavements and bulbous heads up to 20 cm high locally established on ooid sand and terrigenous mud substrates and locally encrusted remnant silcrete boulders and schist basement exposed along a fault-escarpment shoreface.

The oncoids and stromatolites are well laminated, with prominent millimetre and centimetre-scale columnar or pseudocolumnar structure. Laminae are typically 10-500 µm thick. Well preserved microfossils comprise calcified *Scytonema*-like cyanobacterial tufts and abundant 100-150 µm diameter spheroidal endogonaceous fungal spores. Similar spheroidal forms have been identified by previous authors as peloids, coccoid algae, and insect eggs.

Lacustrine mudstone, shale and muddy sandstone associated interbedded with the microbial carbonates contain freshwater bivalves (*Hyridella*), hydrobiid gastropods, smooth-shelled ostracods (*Paralimnocythere* and *Gomphocythere*) and local concentrations of fish and bird bones.

The stromatolites and oncoids are interpreted to have grown in the shallows of an ephemeral, calcium-enriched, and tectonically isolated sub-basin of the Manuherikia paleolake system where the terrigenous sediment supply was low.

INTRODUCTION

Lacustrine carbonate rocks comprising stromatolites, oncoids and ooid sand, together with palustrine marl, occur within Miocene Manuherikia Group exposed in the upper Manuherikia Valley, Central Otago, South Island, New Zealand (Fig.1). Oncoids and stromatolites, the focus of this paper, are very similar in external morphology to

J. Bertrand-Sarfati and C. Monty (eds.), Phanerozoic Stromatolites II, 227–254.
© 1994 *Kluwer Academic Publishers.*

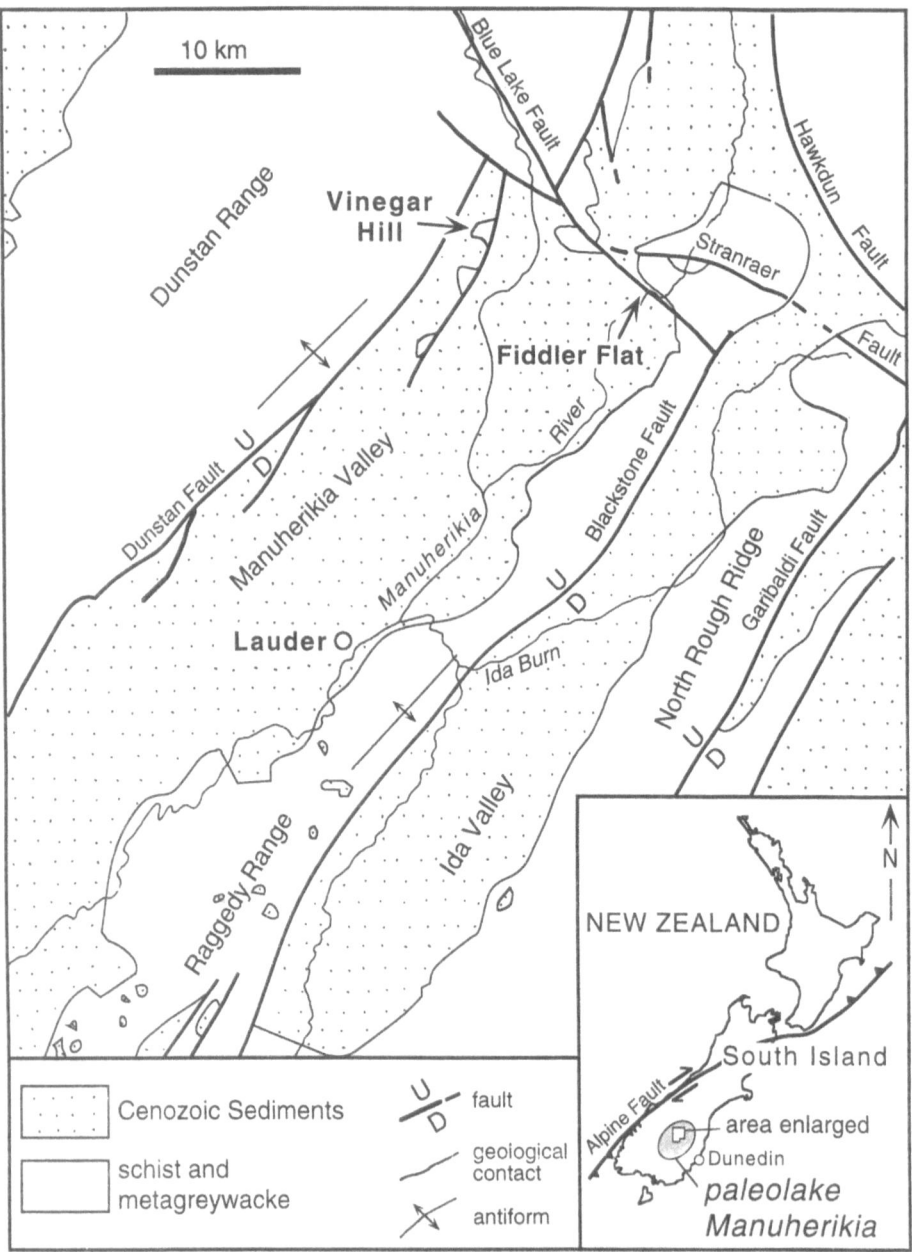

Figure 1: Location of Vinegar Hill, Lauder, and Fiddler Flat in Manuherikia Valley, Central Otago. Geology modified after Mutch (1963) and New Zealand Geological Survey (1977).

nonmarine microbial carbonate growths described by many authors (e.g. Bertrand-Sarfati et al., 1966; Link et al., 1978; Schafer and Stapf, 1978; Freeman 1982, Freytet and Plaziat, 1982; Osborne et al. 1982; Cohen and Thouin, 1987; Burne and Moore, 1987; Moore 1987; Braithwaite et al. 1989; Grey et al., 1990; Smith and Mason, 1991; and Casanova and Hillaire-Marcell, 1992). Many such examples are associated with hard-water lakes and carbonate sediments, but a few are reported from largely non-carbonate sequences (e.g. Masson and Rust, 1983). The New Zealand examples occur in a predominantly terrigenous sequence of illite/kaolinite mudstone and coarse quartzose clastic sediments.

An unusual feature of the New Zealand oncoids and stromatolitic rocks is the good preservation of microfossils. Together with calcified remains of filamentous cyanobacteria typical of microbial stromatolitic carbonate sediments, peloid-like fossils identified as fungal spores of possible endogonaceous affinity are very abundant. The purpose of this paper is to describe the stromatolites and oncoids from 3 localities (Vinegar Hill, Lauder, and Fiddler Flat) paying special attention to oncoids and microfossils from the Vinegar Hill area.

GEOLOGICAL SETTING

The late Cretaceous - mid Cenozoic was a period of marine transgression and submergence of much of the present New Zealand landmass. Late Oligocene to early Miocene sediments deposited along the southeastern margin of the South Island typically consist of marine deltaic coal measures, shallow-marine mixed carbonate/clastic sandstone, and cool-water limestone.

By early Miocene time the area of land had increased and a large inland lake (paleolake Manuherikia) and tributary river system covered at least 5,600 km^2 in the Central Otago region (Fig.1). Large deltaic bodies characterised by interchannel lignite-bearing successions (up to 100 m thick) and channel fill deposits of quartzose sandstone and sandy conglomerate built up at the major points of sediment entry into the lake system (Douglas, 1986).

Manuherikia Group overlies deeply kaolinised early Mesozoic chlorite schist and metasedimentary rocks. The Manuherikia rocks were subject to post-Miocene deformation and now preserved in several northeast-trending, asymmetric synformally folded basins that have gently dipping eastern limbs and steeply dipping western limbs that reflect deep basement faulting (Beanland et al., 1986; Turnbull et al., 1993) (Fig.1). Remnant sediments, including silcrete, occur locally on the dip slopes of uplifted, antiformally deformed schist blocks.

Except for infrequent bivalve and ostracod shell lenses, rare palustrine carbonates, stromatolites, oncoids, and thin ooid sands in the Upper Manuherikia Valley and nearby areas, lacustrine sediments that overlie the deltaic successions are generally non- or weakly-calcareous. Evidence of desiccation occurs locally in floodplain and lake-margin sediments, but no evidence of severe aridity such as evaporite minerals or pseudomorphs has been reported.

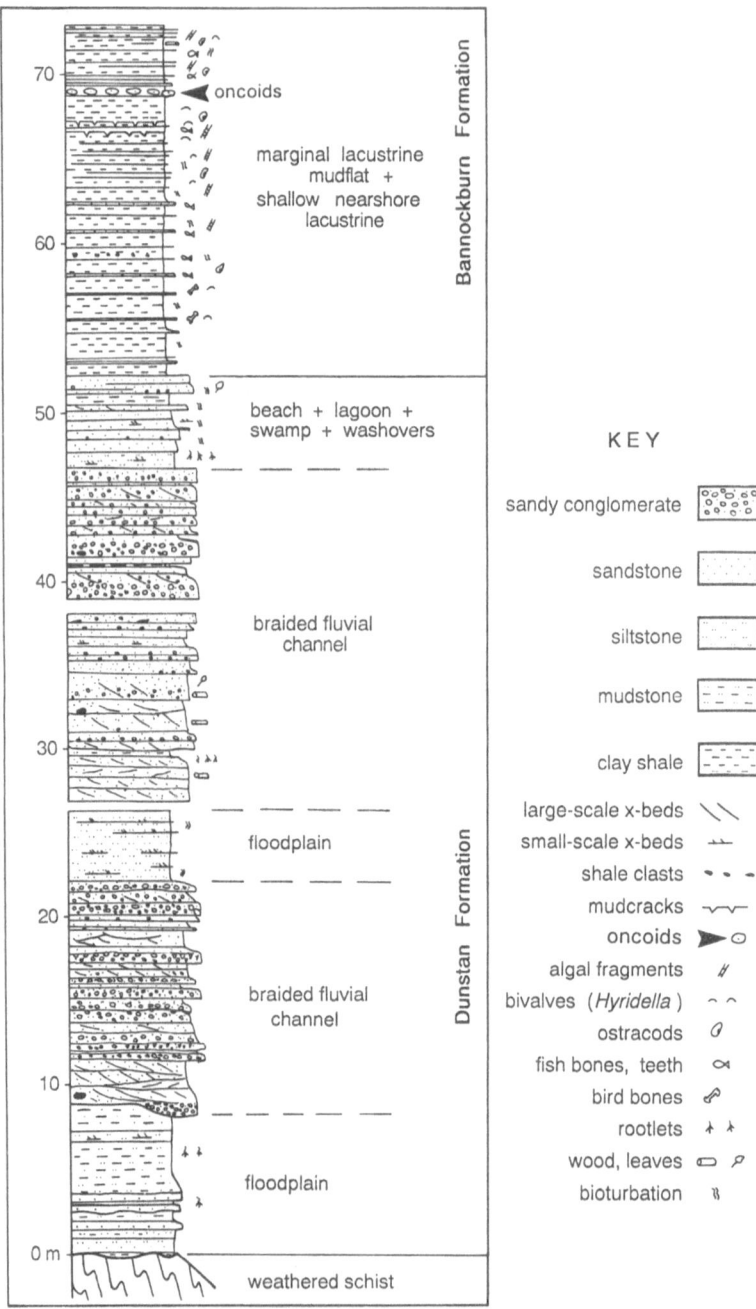

Figure 2: Composite stratigraphic column showing position of the oncoids in Bannockburn Formation at Vinegar Hill. Modified after Douglas (1986) and unpublished data.

Figure 3: a) View of the central part of the Vinegar Hill oncoid bed (arrowed). Hammer, lower left, gives scale. Clayshale and mudstone dominating this part of the sequence contains a low diversity freshwater fauna of unionid bivalves (*Hyridella*), rare gastropods, ostracods, and bird and fish remains; b) Close view of the oncoids arrowed in photo a, overlying a thin bed of microbial carbonate and quartzose sand (lower contact arrowed).

VINEGAR HILL SECTION

Vinegar Hill is on the west side of Manuherikia Valley. A locally thin Manuherikia Group sequence caps a narrow schist splinter between en-echelon segments of Dunstan Fault (Fig.1). Oncoids occur near the top of the section exposed by gold mining operations (Fig.2). The gold-bearing fluvial lower unit (Dunstan Formation) overlying kaolinitised schist basement consists of silty clay overlain by quartzose sandstone and sandy conglomerate. Overlying fine-grained quartzose sands and greenish blue to bluish grey mudstone and shale (Bannockburn Formation) were deposited in littoral and nearshore lake environments. Clay minerals composing the mudrocks are illite and kaolinite ± minor smectite.

Figure 4 : Large lobate-spheroidal oncoids in the east margin of the Vinegar Hill exposure have enlarged by coalescence and encapsulation of smaller forms. The basal sand layer contains oncoid debris (lower contact arrowed, lower left).

Compared with exposures of the Manuherikia Group lacustrine rocks elsewhere in Central Otago, vertebrate and invertebrate fossils are common. Birds bones, of probable anatid (duck) affinity have been collected (Douglas et al., 1981; Fordyce, 1991) and spines, vertebra, ear bones, and jaws of bony fish are abundant, often concentrated within thin sand (? wave reworked) laminae. Two genera of smooth-shelled freshwater ostracods, *Paralimnocythere* and *Gomphocythere* are locally abundant (identified by P. De Decker, Australian National University, pers. comm.). Both the ostracods and freshwater unionid bivalves (*Hyridella*) form thin shelly laminae or completely decalcified impressions in shale. *Hyridella* also occurs in life position. Pollen samples from the lacustrine beds contain abundant freshwater algae *Botryococcus* and *Pediastrum* (Mildenhall and Pocknall, 1989). Vertical 1-2 mm diameter *Trichichnus*-like trace fossils are abundant in lacustrine shales. Marine or brackish taxa are absent.

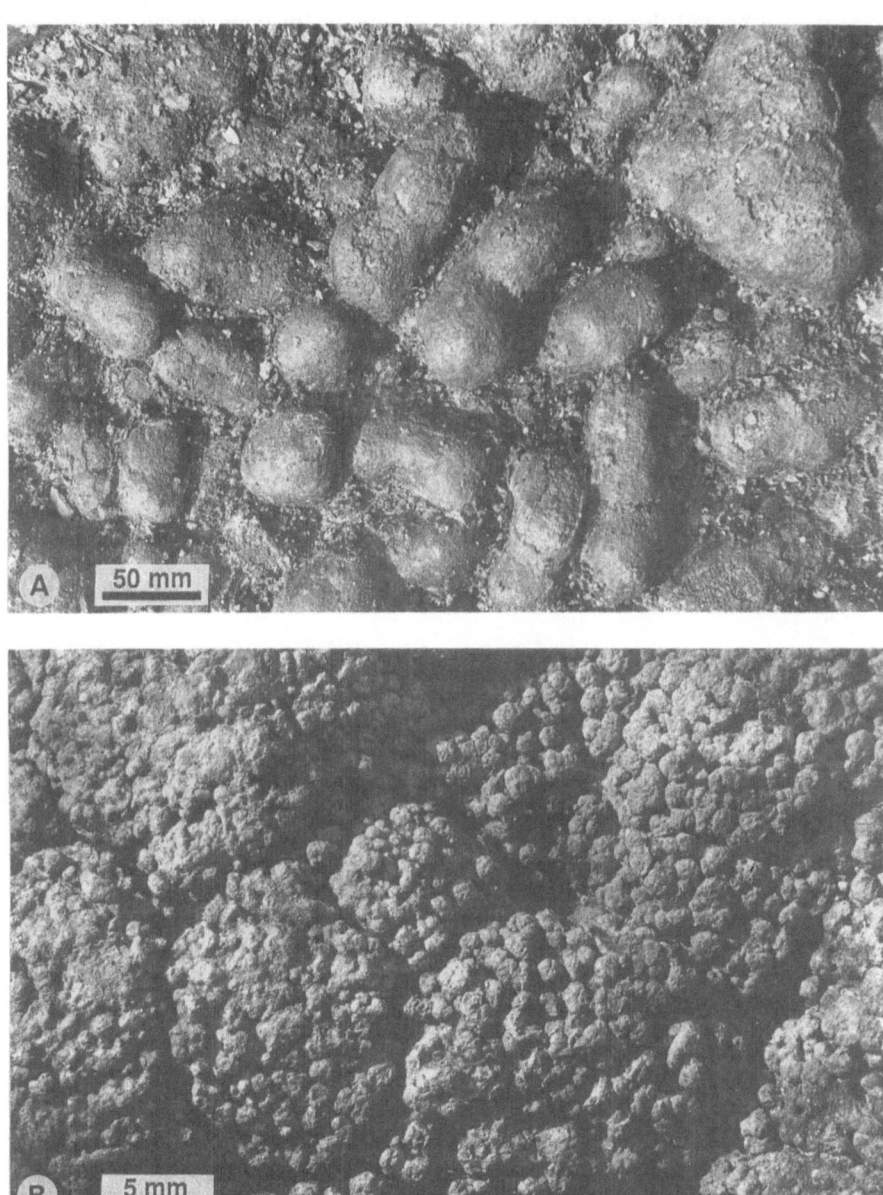

Figure 5: a) Plan view of close-packed oncoids excavated at the western end of the Vinegar Hill bed. Relatively small cylindrical, dumb-bell and spheroids are most common at this location; b) Vinegar Hill oncoid surfaces commonly show two orders of pseudocolumns. Small 0.25-2.5 mm diameter pustules are superimposed on low-relief 5-20 mm diameter hemispheres.

Desiccation cracks and thin mudstone breccia occur 4 m beneath the oncoids but are absent in immediately adjacent beds. From pollen data (Mildenhall, 1989; Mildenhall and Pocknall, 1989) the Vinegar Hill sequence is of early-middle Miocene age.

Oncoid bed

Oncoids form a single densely packed 6-20 cm bed that extends across the full 50 m width of the exposure (Fig.2-7). They overlie a 2-6 cm thick bed of granular microbial carbonate and quartz sand that sharply overlies terrigenous mudstone. The basal layer includes abundant 2-6 mm diameter tubular microbial carbonate debris considered to have originated as encrusted plant stems (Fig.3). Sand-sized ooids are locally common in coarse laminae within adjacent mudstone beds but no larger oncoidal forms occur elsewhere in the section.

The cream to pale brown oncoids attain a maximum size of 40 cm diameter and 20 cm high. The largest oncoids, which developed by coalescence and encapsulation of smaller forms, and have lobate botryoidal shapes, occur in the eastern end of the exposure, spaced 0.3 to 1 m apart (Fig.4). Dumb-bell and elongate cylindrical oncoids

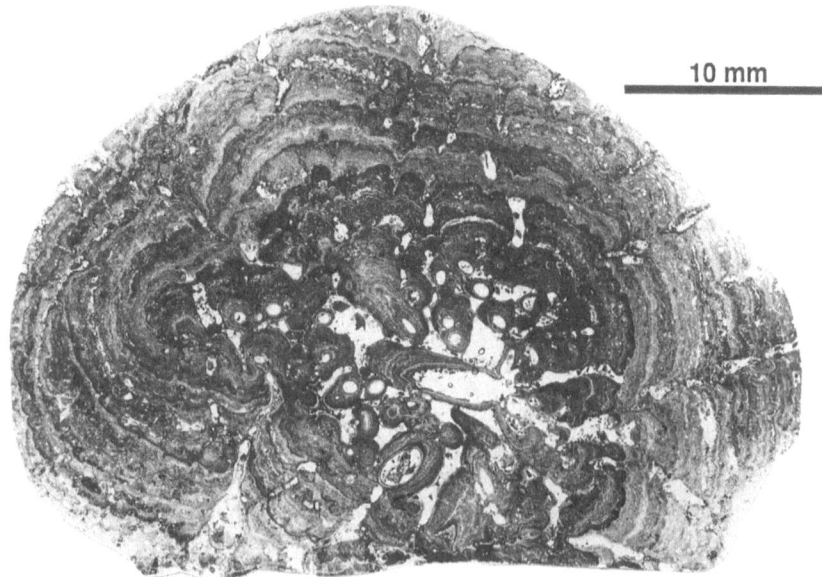

10 mm

Figure 6: Vinegar Hill oncoid in thin section. Laminae are composed of pale brown - pale grey and colourless micrite. Small-scale pustular structures commonly reflect the growth form of "*Scytonema*" tufts preserved in the darker laminae. Oncoid has a core of tubular microbial carbonate fragments. Note that groups of laminae rarely encircle these oncoids without significant changes in thickness.

5-12 cm in length and 4-6 cm diameter are more abundant across the western 30 m width of exposure (Fig.5).

Oncoids surfaces (Fig.5, 6) characteristically show a three-ordered pseudocolumnar structure, with millimetre diameter pustules and 8-25 mm diameter low-relief rounded

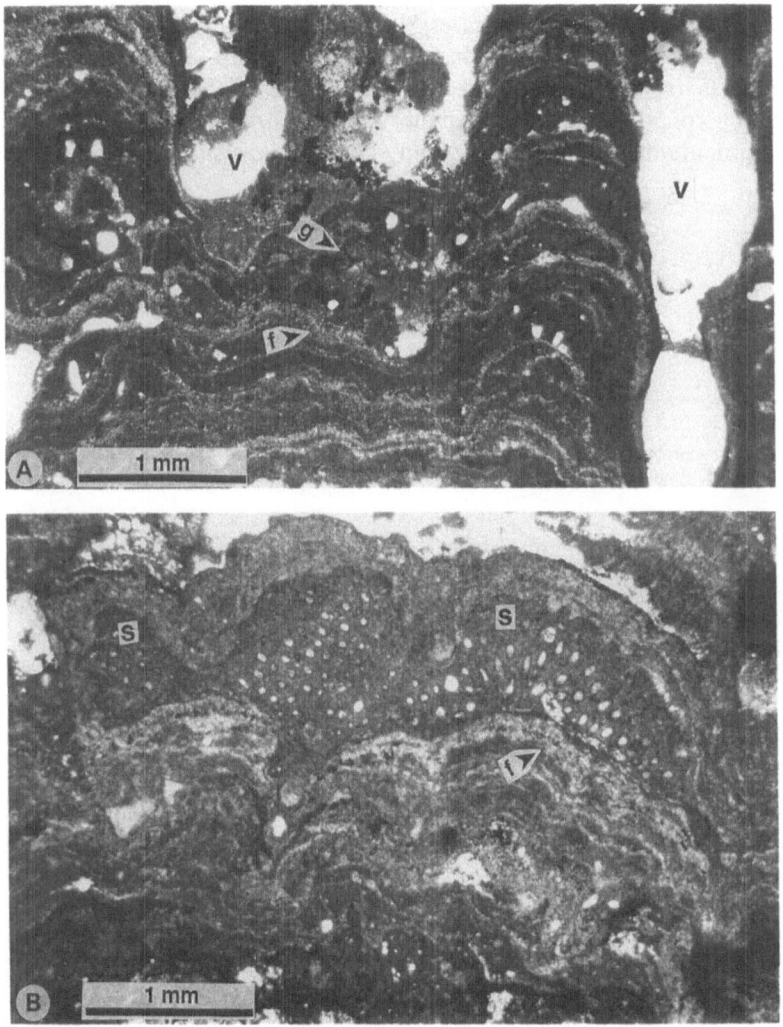

Figure 7: Lamination in Vinegar Hill oncoids. Note bridged radial voids between pseudo-columns (v), a cluster of fungal spore moulds (g), relatively thick pustular lamina formed by a *"Scytonema"* tuft (s), and radial filamentous structure of thin pale laminae (f).

polygons superimposed on the principal encapsulating surface (cf. 3rd, 2nd and 1st orders of curvature of Walter (1976, p. 689).

In thin-section, laminations consists of alternating 10-500 μm thick cloudy pale brown or cloudy grey micrite (relatively dark) and pale grey micrite or microspar (Fig.6 and 7). In places, brown laminae form prominent small-scale discontinuous pustules, 0.25-2.5 mm wide and 0.25-1.5 mm high. The pustules enclose clusters of 40-55 μm internal diameter cylindrical microfossils identified as moulds of cyanobacterial filament sheaths (see below). The thin, colourless laminae commonly display indistinct radial structure (normal to the growth surface) and include fan-shaped relics of a second cyanobacterial taxon that has 4-7 μm diameter filaments. Also common in the darker laminae are spheroidal 100-150 μm diameter fungal spore microfossils (see below). Laminae commonly display cyclicity; cloudy grey laminae sharply overlie and

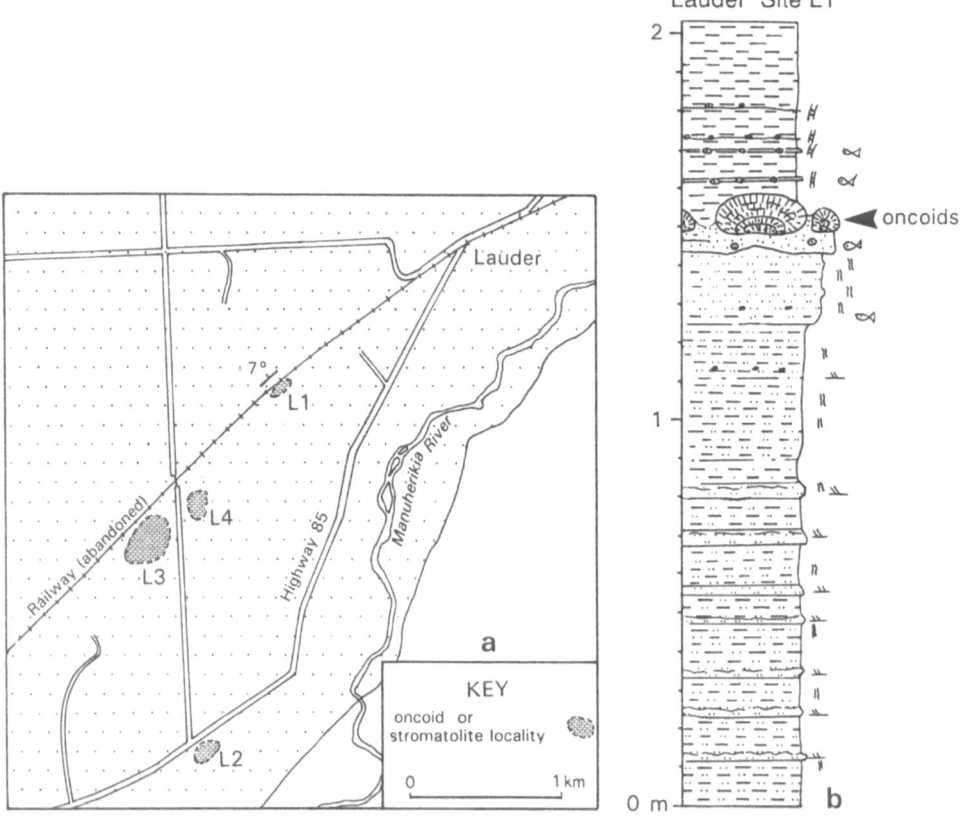

Figure 8: a) Map of Lauder area, Manuherikia Valley, showing stromatolite and oncoid sites discussed in text. Modified after Saunders (1971). See Figure. 1 for general location, and explanation of symbols. b) Stratigraphic column showing oncoids in Bannockburn Formation lacustrine sediments at Lauder Site L1. Lithological symbols are explained in Fig. 2.

grade (outward) into thin (10-30 μm) colourless coarser grained laminae. Individual couplets occur in groups of 5-20 or more of similar thickness, separated by thicker (0.5-1.5 mm) grey or pale brown pustular layers composed largely of "*Scytonema*" tufts (Fig.7).

Narrow radial cavities between 2nd order pseudocolumns (Fig.6 and 7a) frequently contain quartz silt and very-fine sand grains. Quartz debris is otherwise lacking within laminae. The central cores of elongate oncoids, composed of single or clustered hollow cylinders 2-8 mm diameter, resemble tubular carbonate fragments in the basal sand layer, suggesting they initially encapsulated plant stems, possibly reeds that grew along the lake margin. Rare irregular cone-shaped central voids suggest that some oncoids may have encrusted gastropods.

Because of irregularities in growth, individual laminae or groups of laminae rarely encircle any oncoid without moderate-severe changes in thickness (Fig.6).

LAUDER AREA

Oncoids and domal stromatolites are preserved on the east (up-dip) side of Manuherikia Valley, south of Lauder township (Fig.8a,b). They occur within Bannockburn Formation lacustrine sediments a few metres above schist basement and as local residual accumulations within the modern soil. Observed in-situ at two localities only, the stromatolitic growths differ significantly in structure over short distances but appear to have been part of a single development.

Oncoids, site L1

Oncoids exposed in a railway cutting at Site L1 (Fig.8a,b) are sparsely distributed along a single horizon, similar to the eastern portion of the Vinegar Hill exposure.

Underlying the oncoid bed is a coarsening-up (?shoaling) sequence of pale olive coloured micaceous mudstone and well-sorted quartzose silt. It contains rare microbial carbonate sand, vertebrae and spines of small fish, and commonly displays disrupted and mottled bedding suggestive of recurrent subaerial exposure and desiccation. Well-sorted very-fine quartz sand forming the immediate base of the oncoid bed contains 5-10% ooids, and coarse oncoidal carbonate fragments. The 0.5-0.8 mm diameter ooids are composed of 20-50 μm thick concentric radial-fibrous calcite laminae that enclose detrital quartz nuclei. Laminated clay-shale overlying the oncoid bed includes thin laminae of transported ooids identical with those in the bed below.

The few large oncoids remaining intact are 10-32 cm diameter, up to 18 cm high, and circular in plan view. Low-relief hemispheric "columnar" structure is well developed. Close-spaced hemispheres are 5-15 mm in diameter and up to 10 mm in height, branching at acute angles. Quartz sand is commonly trapped within narrow (<1-4 mm wide) spaces between columns. The latest growth stages of in-situ oncoids are strongly asymmetrical towards the top, suggesting they were immovable during their later growth phases.

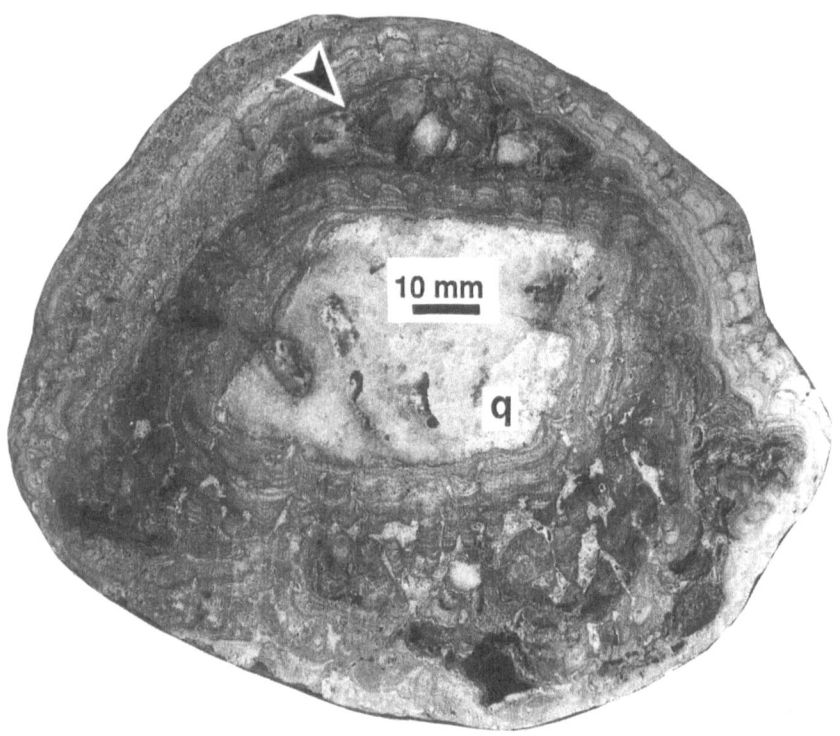

Figure 9 : Sections of an oncoid from Lauder site L2 which has a core is of quartz-cemented sandy conglomerate (q) and incorporates a lens of quartz gravel (arrowed).

Oncoids, site L2

Elsewhere in the Lauder area oncoids and stromatolites occur as sparse erosion-resistant accumulations within the modern soil. Spheroidal oncoids at the most easterly Site L2 commonly have nuclei of angular schistose quartz or silcrete pebbles and cobbles, and probably formed directly on a thin gravel lag overlying schist basement. The silcrete clasts consist of quartz-cemented quartz sandstone or pebble conglomerate. Oncoid laminae incorporate patches of angular quartz sand and granules (Fig.9). As seen on oncoid surfaces, centimetre-scale low-relief pseudocolumnar structures have strongly developed cerebroid or polygonal (commonly pentagonal) outlines (cf. Abell et al., 1982). In Sawn section matching macroscopic groups of laminae 10-15 mm thick, and individual laminae within the groups, are recognisable among oncoids by their characteristic colour banding and texture. In thin-section, Site L2 oncoids resemble those at Vinegar Hill but contain more (1-2 %) adventitious quartz, both within

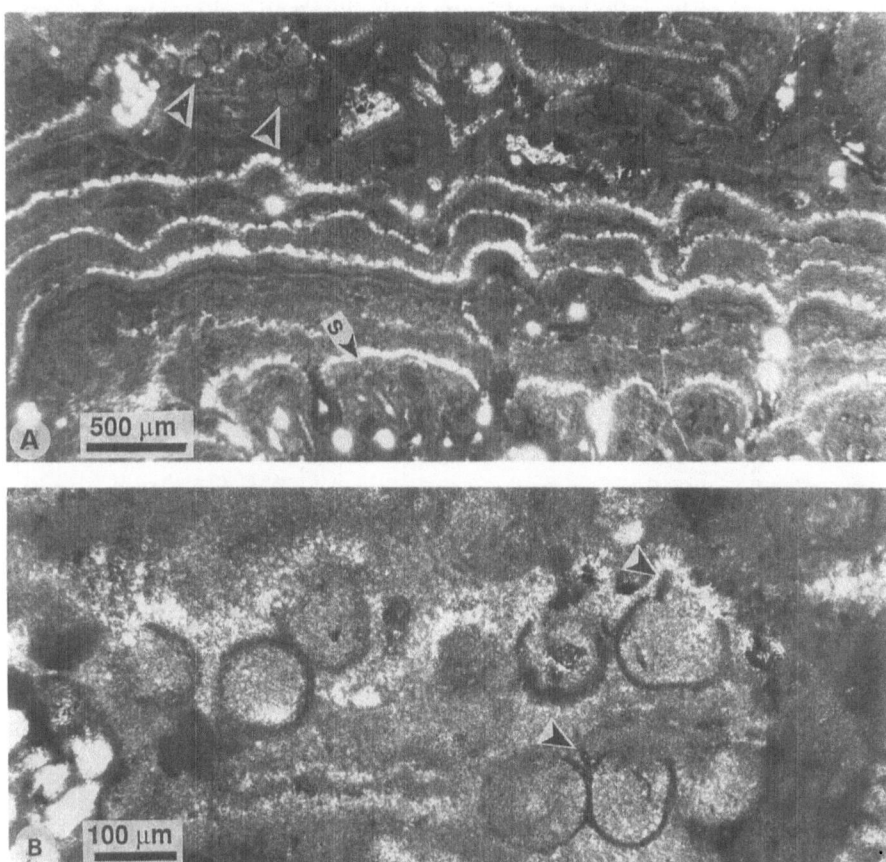

Figure 10 :   Lauder Site L2 oncoids in thin section; a) Cyclic grey/brown and colourless micrite laminae show regular (?periodic) growth increments within groups of laminae. Note a group of fungal spores (arrowed, upper left), and a *"Scytonema"* tuft, lower left (s); b) Group of fungal spore moulds, as arrowed in photo b. Note possible traces of hyphae (arrow, right).

concentric laminae and trapped in narrow (<1 mm wide) cavities between columns. As in the Vinegar Hill oncoids, similar cycles of grey or brown and colourless micrite are present. Cyanobacterial filament moulds and spheroidal fungal spore moulds are abundantly preserved (Fig.10).

Bulbous stromatolites, site L3

Small, 4-15 cm diameter bulbous stromatolites at Site L3 have stout 1-5 cm diameter pedestal-like bases (Fig.11). They are semi-circular in plan view, and consist of single growths or clumps of laterally merged individuals. Internally, prominent 3-8 mm thick concentric layers are separated by relatively smooth discontinuities. Prominent radial voids reflect the development of close-spaced 1-4 mm diameter pustules.

Preserved along the base of each pedestal is a thin (2-3 mm) cemented concentration of ooids that have quartz sand nuclei, comparable with those at Sites L2 and L4. The Site L3 stromatolites probably developed by expansion of microbial carbonate crusts that locally bound the ooid sand substrate.

Stromatolite pavement, site L4

At Lauder Site L4, 300 m northeast of L3, scattered stromatolite blocks up to 80 cm across and 10-15 cm thick are the remnants of a former laterally extensive pavement.

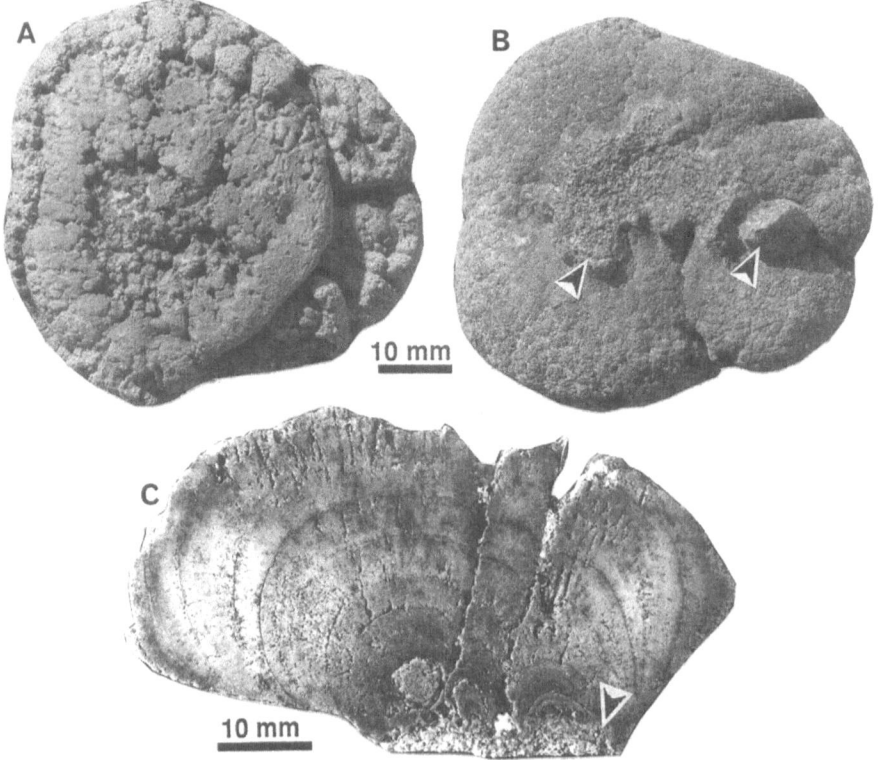

Figure 11: Stromatolite heads from Lauder Site L3; a, b) Plan and underside views of a specimen composed of four individuals, alternations of relatively smooth and pustular surfaces. Ooid sand is attached to the pedestal bases (arrow); c) Vertical section of a specimen showing development of concentric laminae above the basal zone of encrusted ooid sand (arrow).

Figure 12: Part of the remnant stromatolite pavement at Lauder Site L4 composed of low-relief linked columns and domes. Pocket knife is 9.3 cm.

The blocks (Fig.12) are composed of linked hemispheres and thickened dome-like masses up to 12 cm high and 15 cm diameter, spaced 20-30 cm apart. Well developed 8-30 mm diameter partly-linked columns have smooth or low-relief millimetre-scale pustular surfaces. Thin-sections show concentrations of small gastropods (0.3-0.5 mm diameter) and quartz- centred ooids trapped in intercolumnar cavities that are mainly less than 2 mm wide.

STROMATOLITES NEAR FIDDLER FLAT

Near Fiddler Flat (Fig.1 and 13) two contrasting developments of stromatolitic carbonate are separated by a branch of Blue Lake Fault which is inferred to have been active during early Manuherikia Group deposition (Lindqvist and Craw, 1992).

Encrusted silcrete boulders and schist basement, sites FF1 & FF2

On the upthrown block, north of Blue Lake Fault, stromatolite carbonate 2-100 mm thick encrusts a silcrete boulder and pebble lag overlying a gently sloping surface cut into schist bedrock (Fig.13 and 14). The silcrete clasts, composed of quartz-cemented sandstone and sandy conglomerate, are 10-60+ cm in diameter and spaced up to 50 cm apart (Fig.14). The encrusting carbonate is petrographically similar to the Vinegar Hill and Lauder Site L2 oncoids. Thin-sections show comparable discontinuities between cm-thick groups of laminae. Small, 5-30 mm diameter oncoids, some with angular vein-quartz or schist pebble centres, partially fill gaps between the larger silcrete boulders. Areas of schist lacking boulder cover are directly encrusted by stromatolitic carbonate. Oncoids, silcrete boulders, and schist are all encrusted by the latest stromatolite growth phase and draped by a 2-10 cm thick bed of green clay and 20+ cm of quartzose silt.

Stromatolite lens, site FF3

Southward of Blue Lake Fault, a thicker, tilted Manuherikia Group sequence on the downthrown block comprises 150 m of fluvial quartzose conglomerate, sandstone, mudstone and lignite; 10 m of pedoturbated palustrine mudstone and marl; and 70 m of lacustrine mudstone and clayshale (Fig.13).

The lacustrine succession includes a poorly exposed encrusted oncoid lens up to 8 cm thick overlying a 2-3 cm thick base of calcareous silty sand. The basal sand contains abundant 1-2 mm diameter tubular carbonate fragments comparable to those found at Vinegar Hill, together with fragments of molluscs and bones of small fish. The top of the oncoid lens is encrusted by a 5- 10 mm thick layer of laminated microbial carbonate, and draped by green mudstone. The overlying mudstone and shale succession contains thin (2-8 cm) siltstone beds and local concentrations of unionid bivalves, small hydrobiid gastropods, ostracods, and vertebrae and teeth of small fish.

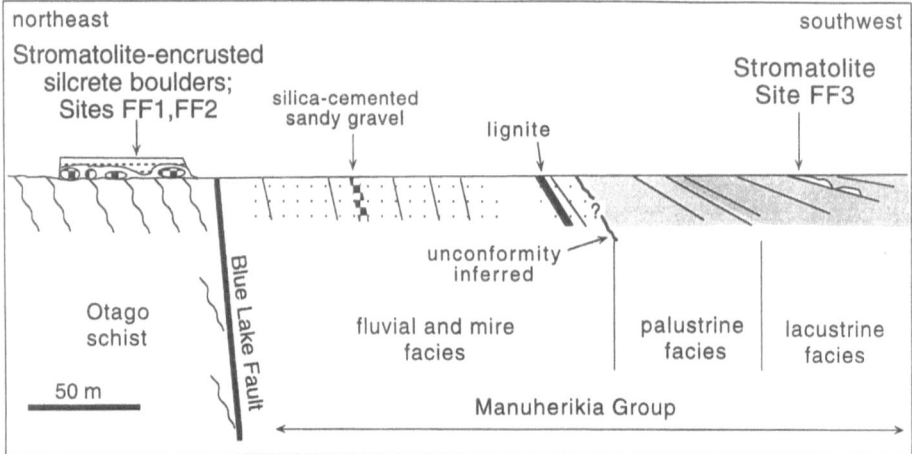

Figure 13: Diagrammatic northeast-southwest section across Blue Lake Fault near Fiddler Flat, showing contrasting developments of stromatolites on silcrete boulders and schist on the northern escarpment, and lacustrine shale on the downthrown side (not strictly to scale). Present-day juxtaposition of the two settings indicates that the sense of movement on Blue Lake Fault has reversed since the middle Miocene.

Bird bones, of probable anatid affinity, are also locally present.

Remarks

Stratigraphic and structural relationships indicate that part of a schist escarpment that developed north of Blue Lake Fault was essentially stripped of Tertiary sediment cover during the middle Miocene. The lag of encrusted silcrete boulders thus marks a significant unconformity (Lindqvist and Craw, 1992). Stromatolitic carbonate encrusting the boulders and schist basement at Sites FF1 and FF2 is thought to correlate with the stromatolite lens within the fine-grained shallow basinal succession south of Blue Lake Fault at site FF3. Stromatolite development appears to have punctuated a phase of lake expansion similarly recognised at Vinegar Hill and Lauder.

MICROFOSSILS

Cyanobacteria

Dense clusters of hollow or micrite-filled micrite tubes representing calcified filament sheaths of cyanobacteria occur within the pale brown or grey laminae of oncoids and stromatolites from all three areas. The tubes are calcified gelatinous sheaths that originally encased the cyanobacterial trichomes (cf. Monty, 1976, Wright and Wright, 1985). One prominent taxon, possibly *Scytonema*, has a flabellate form similar to modern examples documented by Black (1933), Monty (1976), and Monty and Hardie

Figure 14: An unconformity cut into weakly kaolinitised schist basement at Fiddler Flat Site FF2, north of Blue Lake Fault, is overlain by silcrete boulders (arrowed). Both the silcrete boulders and schist surface are encrusted with microbial carbonate.

(1976). The comparatively robust tubes are 40-55 μm internal diameter and up to 1 mm
high. Wall thickness is 27-30 μm (Fig.7b and 15). Adjoining walls are usually mutually
attached, sometimes slightly separated.

In thin-sections of Vinegar Hill and Lauder Site 2 oncoids, conspicuous well preserved
radiating tufts are very abundant within dark brown laminae. The *"Sytonema"* tufts
appear responsible for the ubiquitous millimetre-scale pustular surfaces. An additional
filament type seen during SEM examination of Vinegar Hill oncoids is 15-25 μm in

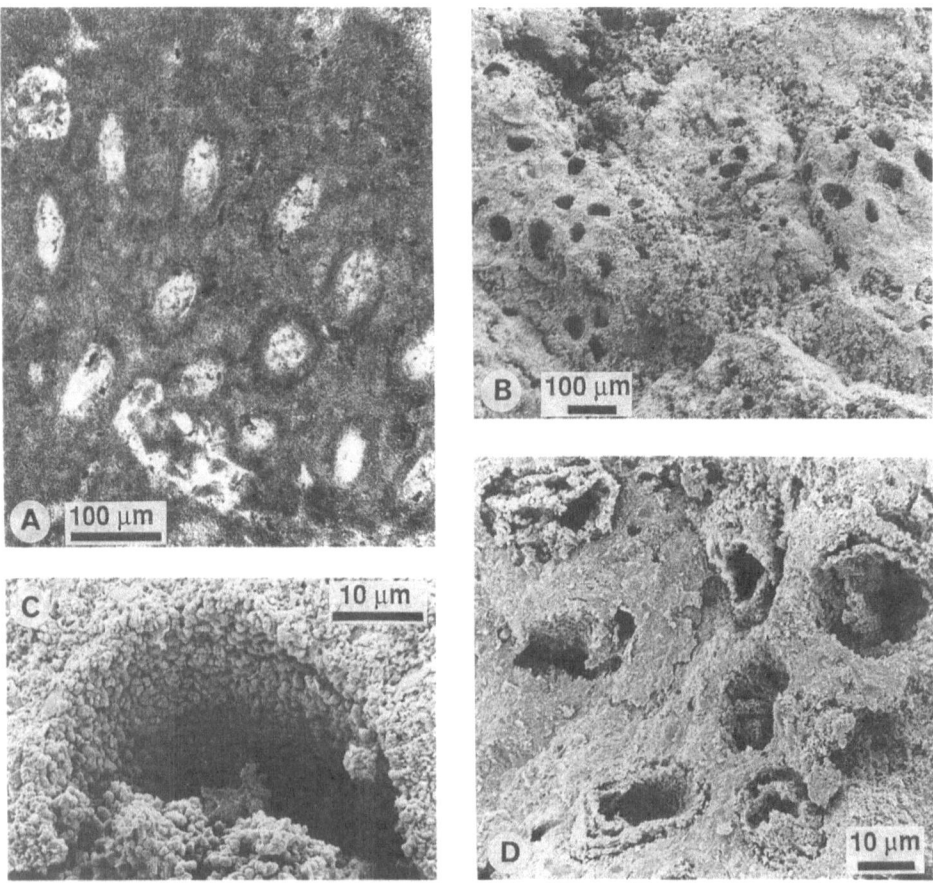

Figure 15: Filamentous cyanobacteria moulds (*"Sytonema"*), in Vinegar Hill oncoids; a)
Transverse section showing a tuft of calcified filament sheaths composed of pale brown micrite
surrounded by coarser- grained micrite and infilled with microspar. Plane-polarised light; b)
SEM view of groups of hollow tubes. The surrounding matrix contains a smaller (4-6 μm internal diameter) filamentous taxon; c) SEM view of a single filament mould composed of micron-
size calcite euhedra; d) SEM view of 40 μm outside diameter *"Sytonema"* moulds showing
traces of 1-3 μm thick inner tubes that possibly represent part of the trichome wall structure.

nternal diameter, and has a secondary internal tube with 1-3 µm thick walls (Fig.15d). The inner-most tube possibly represents a calcified trichome outer membrane. Traces of significantly smaller calcified filaments 4-6 µm internal diameter and with 2-5 µm thick walls have been seen (under SEM) within micritic matrix associated with "*Scytonema*" tufts and (in thin sections) within otherwise homogenous, pale grey or colourless laminae that show palisade structure. They probably represent a third cyanobacterial taxon but have yet to be studied in detail.

Fungal spores

Fungal spores, provisionally placed in the family Endogonaceae, are abundant in oncoids from Vinegar Hill and Lauder Site L2.

Description: Spores are spheroidal or pip-shaped, 100-150 µm in length. In thin section (Fig.7a and 10a, b) the spores appear as diffuse to sharply demarcated ovate to spherical bodies, partly or wholly infilled with microcrystalline calcite that is usually slightly coarser than the surrounding micrite. Where the spores are sharply defined, 1-3 µm wide voids commonly separate the interior and exterior spore moulds. The voids are also evident under SEM examination (Fig.16e) and probably represent the original spore wall. Locally the void space between interior and exterior moulds is filled with coarser micro-spar cast. Traces of the original spore wall are locally preserved as diagenetically altered brown to opaque ?limonite films and elsewhere may be replaced by dark, cloudy micrite. Irresolvable carbonaceous patches 4-7 µm in diameter preserved within the micrite matrix of the interior moulds may be the remains of the original spore contents. Subtending hyphae, clearly visible under the SEM, are not easily resolvable in thin section (Fig. 10b).

The spores occur in groups of 10-15 or more within concentric laminae. On broken surfaces the spore moulds are observable with a 10X hand-lens as glossy-surfaced spheroids. Interior moulds are readily separated from the matrix by gentle grinding and hand-picking the washed residue.

As viewed under the scanning electron microscope, external moulds of spores have a single subtending hypha at least 25 µm long and 5-6 µm diameter. Hyphal walls are composed of two or more layers of annularly oriented fibres (Fig.16d). Hyphae have parallel side walls, apparently continuous with the spore walls and become wider at trumpet-shaped points of attachment to spores, where 8± prominent diverging ribs extend up to 10 µm out along the exterior spore wall (Fig.16). Smaller scale anastomosing ridges 0.5-1.8 µm apart, oriented mostly at right angles to spore axis (as defined by the subtending hyphae) have crest lengths up to 12 µm but usually 3-5 µm. Surfaces of the ridges are covered by 0.1-0.3 µm diameter stubby rod-shaped single calcite crystals that represent either bacterial microfossils or moulds of pits in the external spore wall (Fig.16g).

Remarks: From considerations of gross morphology and the presence of a single subtending funnel-shaped hypha, the fossil spores (chlamydospores) are provisionally placed in the endogonacean genus *Glomus*. According to I. R. Hall (Invermay Research Centre, Dunedin, pers. comm.) the spores most closely resemble the modern species *Glomus fasciculatum* and *G. macrocarpum* in gross morphology and tendency to occur as aggregates (sporocarps).

Regularly arranged fibres in the spore wall (as seen in the Vinegar Hill fossil hyphae; Fig. 16d) are a widespread feature of the Endogonaceae (Stubblefield et al., 1985). Many characters presently used to classify extant endogonaceous species such as spore colour, nature of the spore wall, and number of chlamydospores in a sporocarp are not

Figure 16: S.E.M. views of fungal spore moulds ("*Glomus*") on fracture surfaces of Vinegar Hill oncoids; a) External spore mould with attached subtending hypha; b, c) Detail of longitudinal ridges at the hyphal attachment area of spore a; d) Detail of calcified hyphal walls of spore a, showing traces of annular fibrous structure. e) Group of spore moulds. Note deformation of the internal mould in centre of view, and void space (arrowed) between the interior and exterior moulds; f) Exterior spore mould, as in upper right of photo e, showing structure at the hyphal attachment area; g) Detail of photo f, showing anastomosing ridge structure of the spore outer wall surface, and sub-micron scale pitting.

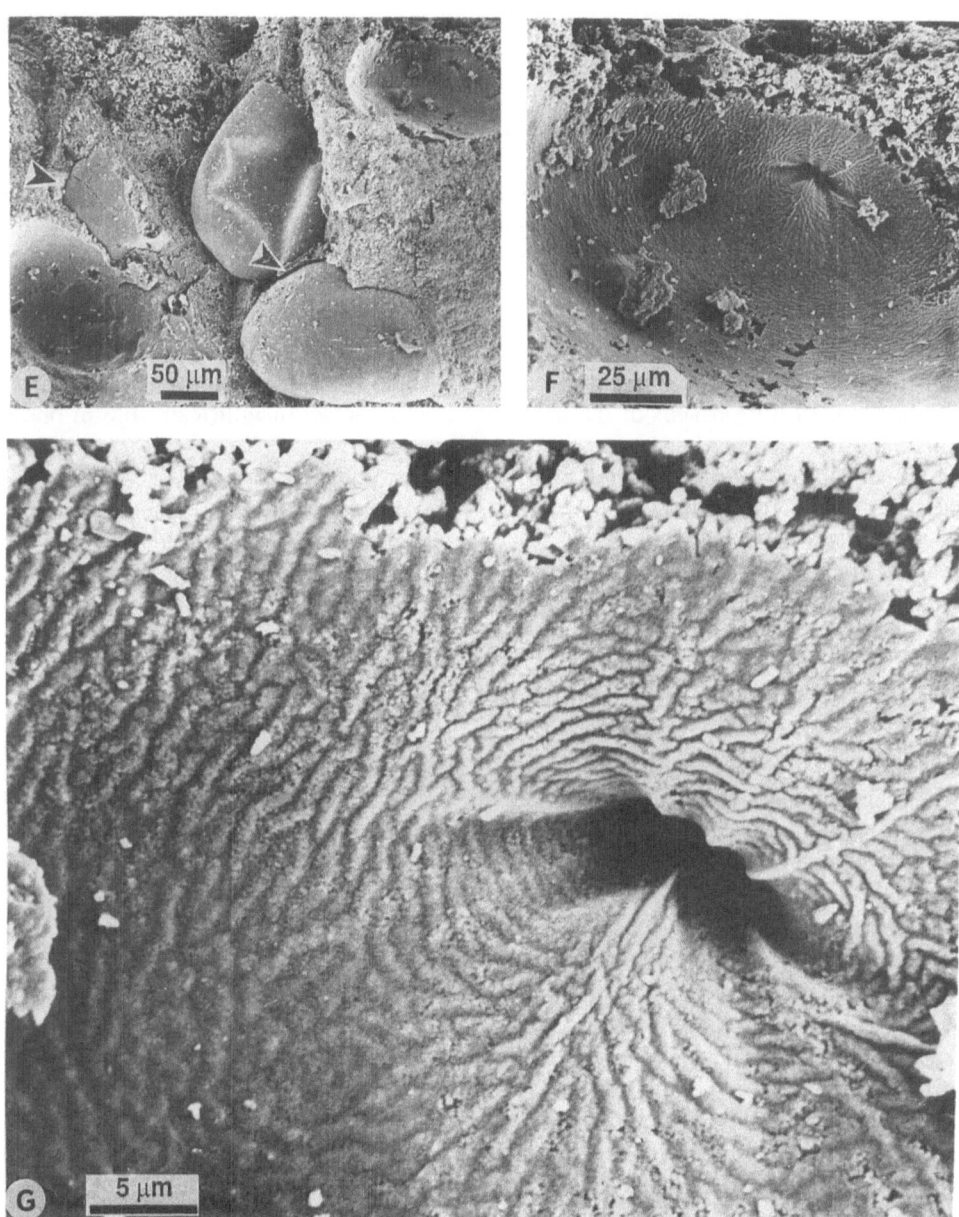

applicable to the fossil material described here. It is noteworthy, however, that this group of fungi, in which nine living genera are recognised, are currently identified by their spore morphology (Hall, 1984).

DISCUSSION

Microbial mat ecology

Microfossil evidence indicates that precipitation of micrite in the Manuherikia Group oncoids and stromatolites was associated with photosynthetic activities or degradation of a microbiota that included filamentous cyanobacteria. Although the processes of calcification around elements of a microbial community have generally been related to the removal of CO_2 from saturated water during photosynthesis, with consequent elevation of pH, non-photosynthetic processes such as chemo-organotrophic bacterial degradation of organic matter may also have contributed to the precipitation of $CaCO_3$ (cf. Burne and Moore, 1987; Paull et al., 1992). At times endogonaceous fungi heavily infested the microbial community. Other groups, such as hydrobiid gastropods, probably grazed the surfaces of the living mats, but had an insignificant role in oncoid or stromatolite construction.

Endogonaceae are well known for their role in forming mycorrhizal associations with the root systems of higher plants. Their symbiotic function in transferring mineral nutrients to the host plant in return for organic compounds has been the subject of intensive agricultural research (e.g. Abbott and Robson, 84; Powell and Bagyaraj, 1984). They are known over a broad ecological range, from aquatic to desert environments. Fossil Endogonaceae-like fungal spores are occasionally found, commonly associated with tree roots, and date back to the Devonian (Kidson and Lang, 1921; Butler, 1939; Wagner and Taylor, 1981; Stubblefield et al., 1985; and Sherwood-Pike, 1988).

While there is evidence that fungi formed a significant part of the microbial community, the physiological relationship between fungi and cyanobacteria is unknown. As in the lichen association, it is possible that the relationship was symbiotic, the cyanobacteria providing nutrients, and the fungus providing a beneficial service such as protection from desiccation or high light intensity. Possible interpretations are (1) "*Glomus*" had a symbiotic or parasitic relationship with the living microflora, or (2) functioned as a saprophyte on dead cyanobacterial tissue within the incipiently cemented substrate. It seems unlikely that "*Glomus*" functioned as a symbiont. Mineral nutrients required by the cyanobacterial community probably would have been freely available in the surrounding lake waters and therefore there would have been little need to recycle nutrients from the oncoid interior.

The possibility that the spores originated as infections in higher plants is rejected because of their abundance and intrinsic nature within the stromatolite host. The absence of rootlets in immediately adjacent beds suggests that no higher plants were growing in close association.

Given the well established role of fungi in the decomposition of organic matter in both marine and nonmarine aquatic environments (E. B. G. Jones 1988) it seems surprising that structures of fungal origin are rarely mentioned in the stromatolite literature. Algae and fungi associated with plant rhizomorphs have been noted in caliche deposits (Klappa, 1977, 1979; B.J. Jones, 1988; Jones and Ng, 1988).

Most reports of fungi in carbonate depositional environments deal with their destructive role as substrate borers. Jones and Pemberton (1987) suggested that fungi play a far more important role in carbonate diagenesis in the vadose environment than previously suspected and fungal spores (much smaller than those reported in this study) have since been reported in sinkhole oncoids from the Cayman Islands, British West Indies (Jones, 1991). Extending their proposition to sub-aquatic settings it seems likely that evidence of fungal activity in the formation of freshwater and possibly also marine microbial carbonates may have been largely overlooked.

Microfossils from freshwater oncoids superficially similar to the New Zealand fossil spores and variously identified as 'peloids' (Freytet and Plaziat, 1982) and coccoid algae or insect eggs (Donsimoni et. al., 1975) may also have a fungal origin (see Bertrand-Sarfati et al., this volume). The term 'peloid' is commonly applied to nondescript cloudy spheroidal micrite grains in both modern and ancient marine and nonmarine deposits (e.g. Monty and Hardie, 1976, Coniglio and James, 1985, Chafetz, 1986). Although a bacterial origin has been proposed for some marine peloids (Chafetz, 1986) I suspect that fungi may be involved in the formation of other examples, such as clotted peloid fabrics ("structure grumeleuse" of Bathurst (1975)) reported from stromatolites and thrombolites.

Carbonate diagenesis generally plays havoc with nonskeletal microorganisms and thus preservation of delicate cell structure is unusual in carbonate rocks (Hofmann, 1973). Several factors probably effected the excellent preservation of the spore moulds described above: (1) the very fine grain size of the precipitated carbonate, (2) decay-resistance of the spore membrane, and (3) absence of severe carbonate dissolution and recrystallisation.

SEM and light microscope examinations show that the grain size of calcite crystals (commonly 0.2-1 µm) forming both the cyanobacterial microfossils and the surfaces of the fungal moulds is significantly finer than the enclosing oncoid matrix. Freshwater authigenic carbonates are typically fine grained. This, coupled with factors such as the decay resistance of fungi spore wall polymer (chitin rather than cellulose based; Swain and Cooper-Driver, 1981; Klappa, 1979); the low permeability of enclosing shales; and the locally shallow depth of burial (less than about 500 m) probably contributed to the unusually good preservation of delicate cellular structures in carbonate. In addition, the abundance of bivalve and ostracod shell material in adjacent beds would have buffered the dissolution effects of meteoric water.

Stromatolite and oncoid development

Growth irregularities, particularly evident in oncoids from Lauder Site 2, probably resulted from overturning, rolling and abrasion during vigorous current action. At Lauder Site L2 current action was sufficiently strong to maintain a substrate of granule gravel which sometimes adhered to the surface of the growing oncoid (Fig.10). Current action is also indicated by local close- packing of oncoids at Vinegar Hill (Fig.5a). These features indicate that oncoids developed in a shallow environment, perhaps extending from the shoreline to a metre or more deep.

Oncoid development is therefore likely to have been influenced by small fluctuations in lake level, such that spurts in microbial mat growth and calcification during times of immersion would have alternated with pauses in growth during emergence and desiccation. Correlation of corresponding centimetre-thick groups of laminae, or layers, between similar sized oncoids from Vinegar Hill and Lauder suggests they represent a single development, but with several major growth phases. It seems likely that the 10-500 µm thick laminations represent seasonal fluctuations in temperature and light intensity, whereas longer-period climate-induced changes in lake level and nutrient supply may be recorded in cm-scale layers separated by micro- unconformities.

Analogous multiple growth phases have been described in oncoids from Namibia (Smith and Mason, 1991) and in stromatolites from Lakes Natron and Magadi, East Africa, where growth stages are separated by unconformities marking hiatuses that span thousands of years (Hillaire-Marcel and Casanova, 1987).

Prominent growth discontinuities present in the Vinegar Hill and Lauder Site L2 oncoids, and also the silcrete- and schist-encrusting stromatites and Fiddler Flat Sites FF1 and FF2, contrast with the more-uniformly laminated stromatolite heads and pavement at Lauder Sites L3 and L4. At Site L3, the delicate pedestal-based bulbous stromatolitic heads apparently remained in an upright position throughout their growth history, presumably because they were isolated from strong current action. Relatively uniform laminations in stromatolites from these areas indicate steady growth conditions at greater depths, perhaps as deep as 50 m, as suggested for the development of stromatolites that have encrusted hard substrates in Lake Tanganyika (Cohen and Thouin, 1987; Casanova and Thouin, 1990; Casanova and Hillaire-Marcel, 1992).

It is unknown whether changes in lake hydrology (such as climate-controlled increases in Ca concentration and salinity; (cf. Casanova and Thouin, 1990) that induced microbial carbonate development extended throughout paleo-Lake Manuherikia or were restricted both spatially and temporally to the study area. The absence of reports of stromatolitic carbonates elsewhere in Central Otago region and a demonstrated relationship between local basement uplift (Lindqvist and Craw, 1992) and later stromatolite development suggests that deposition may have been confined to a tectonically isolated, carbonate-saturated sub-basin.

CONCLUSIONS

Manuherikia Group stromatolitic sediments occur within a lacustrine succession of predominantly terrigenous sediments. The stromatolites and oncoids represent a short-lived but locally well preserved phase of microbially induced carbonate precipitation and minor detrital sediment entrapment in the littoral-shallow nearshore lacustrine environment. Correlation of laminations between oncoids in local areas show they probably formed simultaneously at each site, but with discrete growth phases.

Morphological differences between apparently correlative oncoids and stromatolites were probably controlled by substrate (soft mud, ooid sand, pebble beach or rock) and depth (shoreline down to tens of metres), and seasonal and longer term (?climatic) fluctuations in depth and hydrology.

Well-preserved calcified filament moulds show that the oncoids and stromatolites formed as cyanobacteria-dominated lithoherms. Spores of probable endogonaceous affinity indicate that fungi also formed a significant part of the microbial community, however their ecologic role has yet to be established.

Manuherikia Group microbial carbonates provide good examples of the potential of stromatolites in providing conditions suitable for the preservation of a tiny sample of past microbiota (cf. Hofmann, 1973, p. 351), as well as providing important new information on the Miocene tectonic development of the Central Otago region (Lindqvist and Craw, 1992). Although associated faunal elements such as smooth-shelled ostracods and hyridellid molluscs indicate that paleolake Manuherikia was fresh, not saline, for much of its history, the stromatolites and oncoids are interpreted to have formed under alkaline conditions in an ephemeral, tectonically isolated sub-basin of the Manuherikia paleolake system where the flux of terrigenous detritus was low.

ACKNOWLEDGEMENTS

I thank D. Craw and B.J. Douglas and for discussions in the field; I. R. Hall for information pertaining to the Endogonaceae; S. M. Lindqvist, D. T. Pocknall, I. M. Turnbull, P. J. Forsyth, C. L. Monty and J. Bertrand-Sarfati for suggesting improvements to the manuscript; W. Watters for X-ray diffraction results; and S. Bishop, B. Burt, P. J. Forsyth, and N. Orr for technical assistance. Use of photomicroscope equipment in the Department of Geology, University of Otago, is also gratefully acknowledged.

REFERENCES

Abbott, L. K. Robson, A. D. (1984) "The effect of mycorrhizae on plant growth", In C. L. Powell and D. J. Bagyaraj (eds.), VA Mycorrhiza, CRC Press, Florida, pp. 113-130.

Abell, P. I., Awaramik, S. M., Osborne, R. H., and Tomellini, S. (1982) "Plio-Pleistocene lacustrine stromatolites from Late Turkana, Kenya: Morphology, stratigraphy and stable isotopes", Sedimentary Geology 32, 1-26.

Bathurst, Robin G. C. (1975) Carbonate cements and their diagenesis, Developments in sedimentology 12, Elsevier, Amsterdam.

Beanland, S., Berryman, K. R., Hull, A. G., and Wood, P. R. (1986) "Late Quaternary deformation at the Dunstan Fault, Central Otago, New Zealand", Royal Society of New Zealand Bulletin 24, 293-306.

Bertrand-Sarfati, J., Freytet, P. and Plaziat, J. C. (1966) "Les calcaires concrétionnés de la limite Oligocène-Miocène des environs de Saint-Poutçain-sur-Sioule (Limagne d'Allier): rôle des Algues dans leur édification; analogie avec les stromatolites et rapports avec la sédimentation", Bulletin de la Société Géologique de France 7/VIII, 652-662.

Black, M. (1933) "The algal sediments of Andros Island, Bahamas", Philosophical transactions of the Royal Society of London Series B 222, 165-192.

Braithwaite, C. J. R., Casanova, J., Frevert, T. and Whitton, B. A. (1989) "Recent stromatolites in landlocked pools on Aldabra, western Indian Ocean", Palaeogeography, Palaeoclimatology, Palaeoccology 69, 145-169.

Burne, R. V., and Moore, L. S. (1987) "Microbialites: organosedimentary deposits of benthic microbial communities", Palaios 2, 241-254.

Butler, E. J. (1939) "The occurrences and systematic position of the vesicular-arbuscular of mycorrhizal fungi", Transactions of the British Mycological Society 22, 274-301.

Casanova, J. and Hillaire-Marcel, C. (1992) "Late Holocene hydrological history of Lake Tanganyika, East Africa, from isotopic data on fossil stromatolites", Palaeogeography, Palaeoclimatology, Palaeoecology 91, 35-48.

Casanova, J., and Thouin, C. (1990) "Biosédimentologie des stromatolites holocènes du lac Tanganyika (Burundi). Implications hydrologiques", Bulletin de la Société Géologique de France 8/VI, 647-656.

Chafetz, H. S. (1986) "Marine peloids: a product of bacterially induced precipitation of calcite", Journal of Sedimentary Petrology 56, 812-817.

Cohen, A. S. and Thouin, C. (1987) "Nearshore carbonate deposits in Lake Tanganyika", Geology 15, 414-418.

Coniglio, M. and James, N. P. (1985) "Calcified algae as sediment contributors to Early Paleozoic limestones: evidence from deep-water sediments of the Cow Head Group, western New Foundland", Journal of Sedimentary Petrology 55, 746-754.

Donsimoni, M., Giot, D., Lang, J. and Lucas, G. (1975) "Rôle des organismes et des réactions physicochimiques dans la genèse des calcaires concrétionnés de Limagne, 9th International Sedimentological Congress, Nice 2, 51-56.

Douglas, B. J. (1986) Lignite resources of Central Otago, New Zealand Energy Research and Development Committee, Report P104.

Douglas, B. J., Lindqvist J. K., Fordyce, R. E. and Campbell, J. D. (1981) "Early Miocene terrestrial vertebrates from Central Otago", Geological Society of New Zealand Newsletter 53, 17.

Fordyce, R. E. (1991) ÒA new look at the fossil vertebrate record of New ZealandÓ, in Vickers-Rich, P., Monaghan, J. M., Baird, R. F. and Rich, T. H. (eds.), Vertebrate Palaeontology of Australasia, Pioneer Design Studio & Monash University, Melbourne, pp. 1191-1316.

Freeman, T. (1982) "Oncolites from lacustrine sediments in the Cretaceous of north-eastern Spain", Sedimentology 29, 433-436.

Freytet, P. and Plaziat, J. C. (1982) Continental carbonate sedimentation and pedogenesis - late Cretaceous and early Tertiary of Southern France, Contributions to Sedimentology 12.

Grey, K., Moore, L. S., Burne, R. V., Pierson, B. K., and Bauld, J. (1990) "Lake Thetis, Western Australia: an example of saline lake sedimentation dominated by benthic microbial processes", Australian Journal of Marine and Freshwater Research 41, 275-300.

Hall, I. R. (1984) "Taxonomy of the VA Mycorrhizal fungi", in C. L. Powell and D. J. Bagyaraj (eds.), VA Mycorrhiza. CRC Press, Florida, pp. 57-94.

Hillaire-Marcell, C., and Casanova, J. (1987) "Isotopic hydrology and paleohydrology of the Magadi (Kenya)-Natron (Tanzania) Basin during the Late Quaternary", Palaeogeography, Palaeoclimatology, Palaeoecology 58, 155-181.

Hofmann, H. J. (1973) "Stromatolites: characteristics and utility", Earth Science Reviews 9, 339-373.

Jones, B. J. (1988) "The influence of plants and microorganisms on diagenesis in caliche: example from the Pleistocene Ironstone Formation on Cayman Brac, British West Indies", Bulletin of Canadian Petroleum Geology 36, 191-201.

Jones, B. J. (1991) "Genesis of terrestrial oncoids, Cayman Islands, British West Indies", Canadian Journal of Earth Sciences 28, 382-397.

Jones, B. J. and Ng, C. (1988) "The structure and diagenesis of rhizoliths from Cayman Brac, British West Indies", Journal of Sedimentary Petrology 58, 457-467.

Jones, E. B. G (1988) "Do fungi live in the sea?", The Mycologist 2, 150-157.

Jones, P. B. and Pemberton, S. G. (1987) "The role of fungi in the diagenetic alteration of spar calcite", Canadian Journal of Earth Sciences 24, 903-914.

Kidston, R. and Lang, W. H. (1921) "On Old Red Sandstone plants showing structure, from the Rhynie Chert Bed, Aberdeenshire: Part V", Transactions of the Royal Society of Edinburgh 52, 855-902.

Klappa, C. F. (1977) "Origin of subaerial Holocene calcareous crusts: role of algae, fungi and sparmicritisation", Sedimentology 24, 413-435.

Klappa, C. F. (1979) "Calcified filaments in Quaternary calcretes: organo-mineral interactions in the subaerial vadose environment", Journal of Sedimentary Petrology 49, 955-968.

Lindqvist, J. K and Craw, D. (1992) "Microbial reefs, crusts, and Miocene uplift, Manuherikia Group, Central Otago", Geological Society of New Zealand miscellaneous publication 63A, 93.

Link, M. H., Osborne, L H. and Awramik, S. M. (1978) "Lacustrine stromatolites and associated sediments of the Pliocene Ridge Basin, California", Journal of Sedimentary Petrology 48, 169-187.

Masson, A. G. and Rust, B. R. (1983) "Lacustrine stromatolites and algal laminates in a Pennsylvanian coal-bearing succession near Sydney, Nova Scotia, Canada", Canadian Journal of Earth Science 20, 1111-1118.

Mildenhall, D. C. (1989) "Summary of the age and paleoecology of the Miocene Manuherikia Group, Central Otago, New Zealand", Journal of the Royal Society of New Zealand 19, 19-29.

Mildenhall, D. C. and Pocknall, D. T. (1989) "Miocene-Pleistocene spores and pollen from Central Otago, South Island, New Zealand", New Zealand Geological Survey Paleontological Bulletin 59, New Zealand Geological Survey, Lower Hutt, 128 p.

Monty, C. L. V. (1976) "The origin and development of cryptalgal fabrics", in M.R. Walter (ed.), Stromatolites, Elsevier, Amsterdam, pp.193-249.

Monty. C. L. V. and Hardie, L. A. (1976) "The geological significance of the freshwater blue-green algal calcareous marshÓ" in M. R. Walter (ed.), Stromatolites, Elsevier, Amsterdam, pp. 447-477.

Moore, L. S. (1987) "Water chemistry of the coastal saline lakes of the Clifton-Preston Lakeland System, south Western Australia, and its influence on stromatolite formation", Australian Journal of Marine and Freshwater Research 38, 647-660.

Mutch, A. R. (1963) Geological Map of New Zealand 1: 250,000, Sheet 23, Oamaru. Department of Scientific and Industrial Research, Wellington.

New Zealand Geological Survey (1977) Geological Map of New Zealand 1:1,000,000. South Island. Department of Scientific and industrial Research, Wellington.

Osborne, R. H., Licari, G. R., and Link, M. H. (1982) "Modern lacustrine stromatolites, Walker Lake, Nevada", Sedimentary Geology, 32, 39-61.

Paull, C. K., Neumann, A. C., Bebout, B., Zabielski, V., and Showers, W. (1992) "Growth rate and stable isotopic character of modern stromatolites from San Salvador, Bahamas", Palaeogeography, Palaeoclimatology, Palaeoecology 95, 335-344.

Powell, C. L. and Bagyaraj, D. J. (1984) "VA Mycorrhizae: why all the interest", in C. L. Powell and D. J. Bagyaraj (eds.),VA Mycorrhiza, CRC Press, Florida, pp. 1-3.

Saunders, J. L. (1971) "Study of spongiostromata and porostromata from Lauder, Central Otago" Unpublished report, Department of Geology, University of Otago.

Schäfer, A. and Stapf, K. R. (1978) "Permian Saar-Nahe Basin and Recent Lake Constance (Germany): two environments of lacustrine algal carbonates", in A. Matter and M. E. Tucker (eds.) Modern and Ancient Lake Sediments, Special publication of the International association of Sedimentologists 2, pp. 83-107.

Sherwood-Pike, M. A. (1988) "Freshwater fungi: fossil record and paleoecological potential", Palaeogeography, Palaeoclimatology, Palaeoecology 62, 271-285.

Smith, A. M. and Mason, T. R. (1991) "Pleistocene, multiple growth, lacustrine oncoids from the Poacher's Point Formation, Etosha Pan, northern Namibia", Sedimentology 38, 591-599.

Stubblefield, S. P; Taylor, T. N. and Miller, C. E. (1985) "Studies of Paleozoic fungi IV: wall ultrastructure of fossil endogonaceous chlamydospores", Mycologia 77, 83-96.

Swain, T. and Cooper-Driver G. (1981) "Biochemical evolution in early plants", in K. J. Niklas (ed.) Paleobotany, paleontology, and evolution, Praeger Publishers, New York, pp. 103-134.

Turnbull, I. M. (1993) "Pre-Miocene and Post Miocene deformation in the Bannockburn basin, Central Otago, New Zealand", New Zealand Journal of Geology and Geophysics 36, 107-115.
Wagner, C. A. and Taylor, T. N. (1981) "Evidence for Endomycorrihizae in Pennsylvanian age plant fossils", Science 212, 562-563.
Walter, M. R. (1976) "Appendix 1 - Glossary of Selected Terms", in M.R. Walter (ed.), Stromatolites, Elsevier, Amsterdam, pp. 687-692.
Wright, V. P. and Wright, J. M. (1985) "A stromatolite built by Phormidium-like alga from the Lower Carboniferous of South Wales", in D. F. Toomy and M. H. Nitecki (eds.), Paleoalgology: contemporary research and application, Springer-Verlag, Berlin, pp. 40-54.

STROMATOLIC PHOSPHORITES IN THE EOCENE OF THE NEGEV (SOUTHERN ISRAEL)

D. SOUDRY[1] and G. PANCZER[2]
[1]Geological Survey of Israel, 30 Malkhe Yisrael St., Jerusalem 95501 (Israel).
[2]Institut de Geologie, 1 Rue Blessig, 67084 Strasbourg Cedex (France).

ABSTRACT

Stromatolitic phosphorites of late Early to early Middle Eocene age have recently been found in several sites along the upper northeastern flanks of the Ramon anticline, Negev (southern Israel). As a rule, the stromatolites are associated with condensed Senonian and Paleocene rock-suites with telescoped multi-event unconformities. The stromatolite occurs in two growth forms: 1) as stratiform pseudo-columnar "ministromatolites" 1 cm in height, and 2) as polymorphic columnar "microstromatolites" hundreds of microns high, intermittently intercalated within the ministromatolite edifices and growing on micro-diastems within them. Laminae of both the mini and the microstromatolites are wavy, with a high degree of inheritance, with virtually no clastics embedded. The stromatolites where unweathered, are entirely composed of francolite. SEM study shows that the two growth forms are in large part made up of clustered capsule-like phosphate cells and some multi-enveloped globose structures, respectively tentatively interpreted as the possible remnants of early phosphatized decomposer, and primary producer microbial populations. The attributes of the stromatolitic phosphorites and their areal distribution indicate that they accreted in well-sheltered settings at the margins of the Ramon high. The stromatolitic phosphorites grade distally into phosphate nodules set within glauconite-rich late Early Eocene biomicrites. Globose structures and rod-shaped bodies of bacterial affinities are the support of the phosphate fraction and the "cement" of the nodules. Different growth conditions are considered responsible for the development of these different phosphate constructions in the two settings. These nodular and stromatolitic phosphorites in the Eocene of Negev are assumed to be a faint "echo" of the big phosphogenic event which marked the African Atlantic coastal zone during this time.

INTRODUCTION

The most common form of phosphorites in Israel and the most widely known is the granular form. In most cases, this kind of rock is made of sand-sized, pellet-like

J. Bertrand-Sarfati and C. Monty (eds.), Phanerozoic Stromatolites II, 255–276.

phosphate bodies held together by a calcareous, a siliceous, or a phosphatic matrix. In fact, this make-up is not exclusive to the Israeli phosphorites. It also characterizes the bulk of the phosphate rocks of the Phanerozoic (e.g., Notholt et al., 1989; Jarvis, 1992), including those of the Senonian-Eocene phosphate Tethyan belt to which the Negev phosphorites belong. In Israel (Fig. 1), these phosphate rocks occur mainly in the Late Campanian interval of the Negev in association with cherts, porcelanites and carbonates, in a rock-suite known as the Mishash Formation. Nodular varieties, though less frequent, are also known from younger successions -- in the Maastrichtian (Soudry

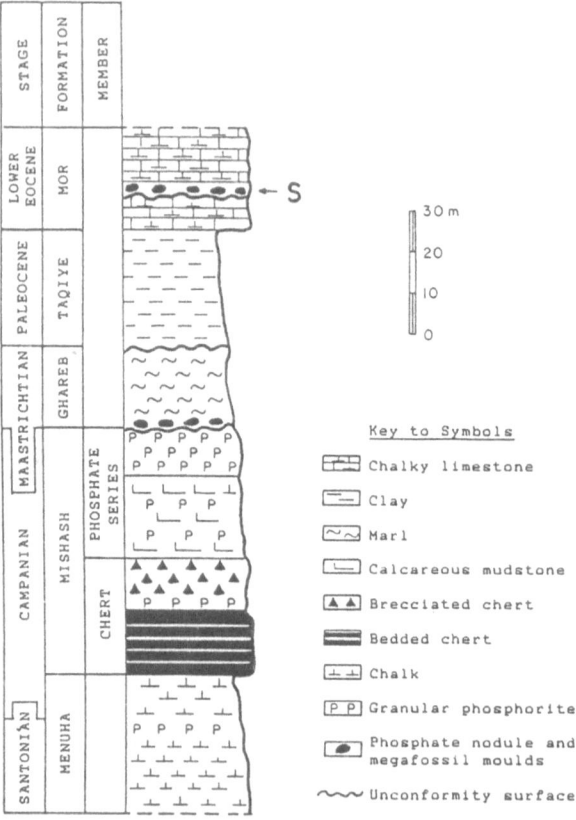

Figure 1. Schematic composite section of the Santonian-Eocene rock-suites in basinal sections of the northern and central Negev (S points to approximate position of the stromatolitic phosphorites in proximal sections around the Ramon high).

and Lewy, 1988), and locally, also in Eocene strata of the Negev (Shiloni et al., 1977). Just recently, a new type of phosphate rock, apparently of Eocene age, was discovered in the central Negev. This type consists of an assemblage of phosphate-permineralized stratiform pseudo-columnar and columnar stromatolites of hundreds of microns to slightly over 1 cm in height.

While stromatolitic phosphorites are widespread in Proterozoic rocks (apparently, this is the main lithic form of phosphate accumulation of that time -- e.g., Bushinskii, 1966; Banerjee, 1971; Chauhan, 1979; Sant, 1979; Cook, 1992), they are by comparison a rare phenomenon in the Phanerozoic. Most examples of phosphate stromatolites of the Phanerozoic are from the Aptian-Albian interval of the Helvetian Domain in southern Europe, where they often accompany starved sediment-suites and depositional discontinuities (e.g., Royant et al., 1970; Delamette, 1981; Krajewski, 1981; Follmi, 1989). An older (and apparently unique) occurrence was reported from the Cambrian of the Georgina Basin, Australia (Southgate, 1980), and Hofmann (1975) described phosphate oncolitic structures in the Ordovician of North Wales, U.K. (see also

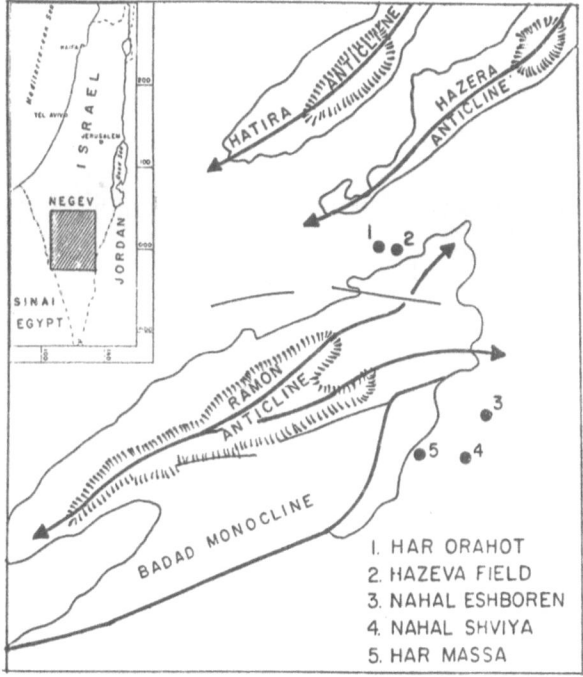

Figure 2. Location map of the Ramon anticline area and outcrop sites of the Eocene stromatolitic phosphorites.

Niedermeyer and Langbein, 1989). No example, to the best of our knowledge, has yet
been documented from the Cainozoic strata.
The stromatolitic phosphate occurrences have been found in several places around the
northeastern flanks of the Ramon anticline in the central Negev, southern Israel (Fig.2).
The present study describes the Eocene stromatolitic phosphorites of the Negev and
their lateral variations, and aims to evaluate the kind of microorganisms generating

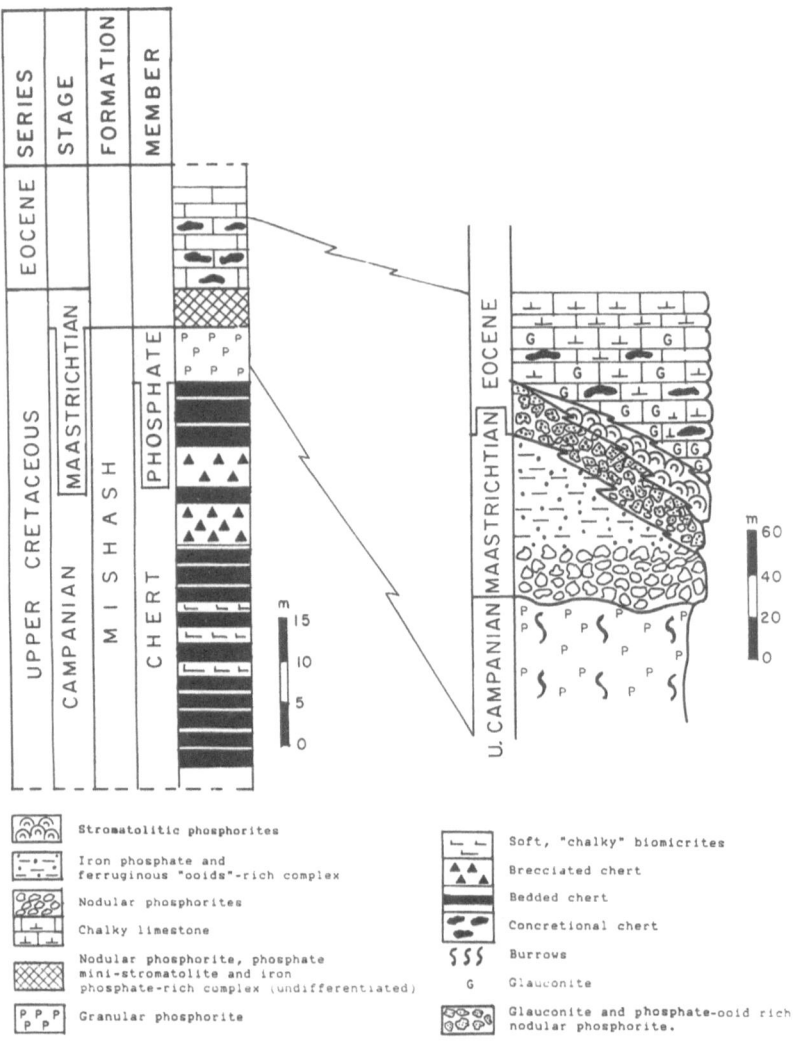

Figure 3. Columnar section of the Late Campanian-Eocene strata in the Har Orahot area and
stratigraphic position of the stromatolitic phosphate complex.

these structures.

SETTING AND FIELD OCCURRENCE

This structure is one of the asymmetrically folded NE-trending series of structures which transect the Negev. Folding commenced in the Late Turonian-Early Senonian (Bentor and Vroman, 1954; Freund, 1965) and continued, in pulses, until Middle Eocene times. Sedimentation patterns during this time-span were deeply affected by the growing anticline structures. Non-deposition phases, truncation phenomena, and overall thinning of rock-successions are more and more prominent when approaching the axial zones of the Negev anticlines.

As a rule, the areas exhibiting the stromatolites have strongly condensed sediment-suites. In these places, the entire succession from the Late Campanian to the Eocene is generally no more than a few tens of meters thick (compared with more than 200 m for laterally equivalent, distal sections of more normal sedimentation), and typically shows repeated telescoped, multi-event unconformities. In all cases, Eocene rocks directly overlie the stromatolite complex, whereas the sediment-column below varies in composition from place to place, according to the degree of stratal condensation. At Har Orahot, where the Campanian-Eocene section is best exposed (Fig.3), the stromatolitic phosphorite occurs as a laterally subcontinuous, 10 cm-thick tabular bed (Fig.4a) lying above a nodular phosphorite of presumably Eocene age. Contacts between the stromatolitic and the nodular phosphorite are unconformable. At the contact between the stromatolite and the nodular phosphorites, clusters of phosphate nodules cemented by phosphate are at places concentrically surrounded by stromatolitic laminae, producing oncolite-like bodies (Fig.4). This Eocene stromatolitic and nodular phosphorite complex overlies an iron phosphorite bed (Panczer, 1990) and a second layer of pebbly phosphorite (Soudry and Lewy, 1988) which together represent, at Har Orahot, the entire Maastrichtian. The contacts are also highly unconformable. This Maastrichtian iron phosphorite and phosphate nodule complex in turn overlies, again unconformably, the economic granular phosphorites of the top of the Late Campanian Mishash Formation.

AGE

The exact age of the stromatolitic phosphorites is uncertain. This is due to the intense stratal condensation of the whole Senonian-Eocene complex in areas of stromatolite growth. A latest Early to early Middle Eocene age was nevertheless obtained (microfossil dating) for the carbonate rocks immediately overlying the stromatolite (Nahal Shviya, Nahal Eshboren). On the other hand, an early Middle Paleocene age was yielded by the youngest rocks found immediately below it (Nahal Eshboren). Thus, theage of the stromatolites should be somewhere between the early Middle Paleocene and the early Middle Eocene. Two points, however, enable narrowingdown this age-interval. First, phosphate nodules of latest Early to early Middle Eocene

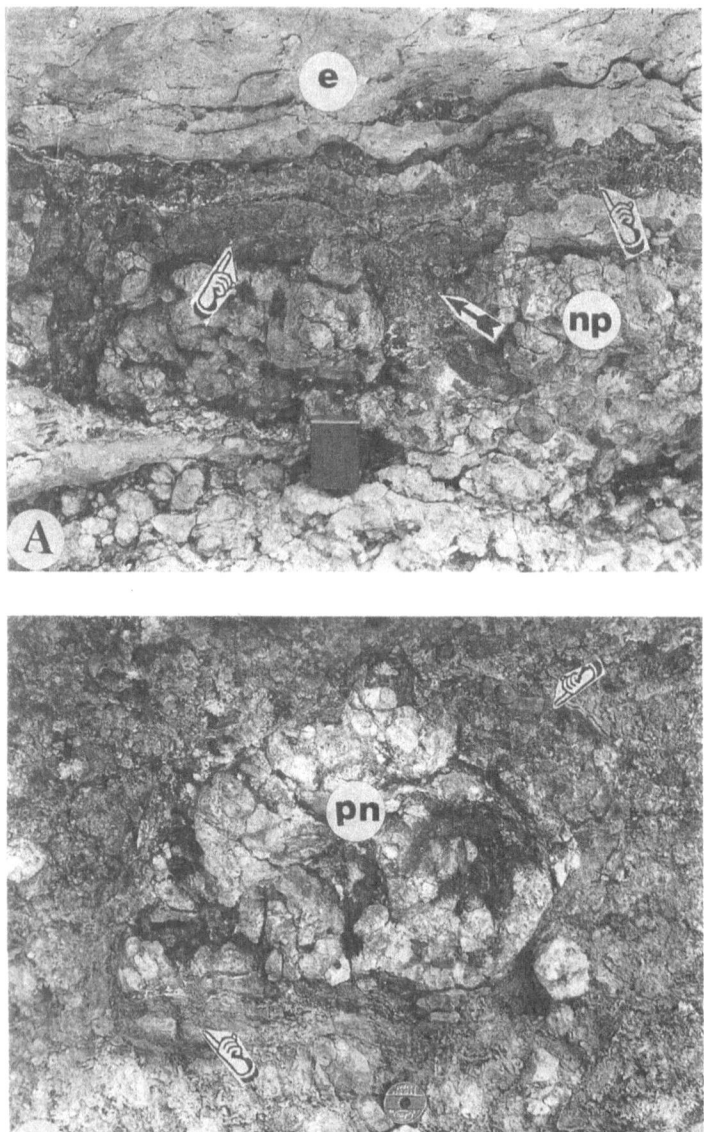

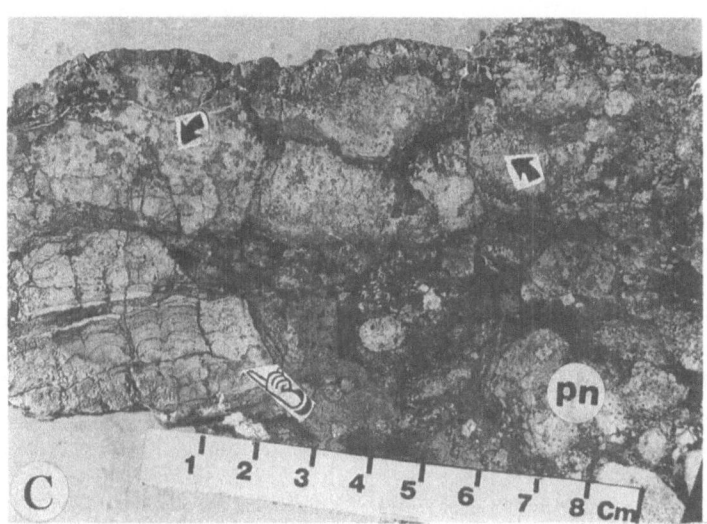

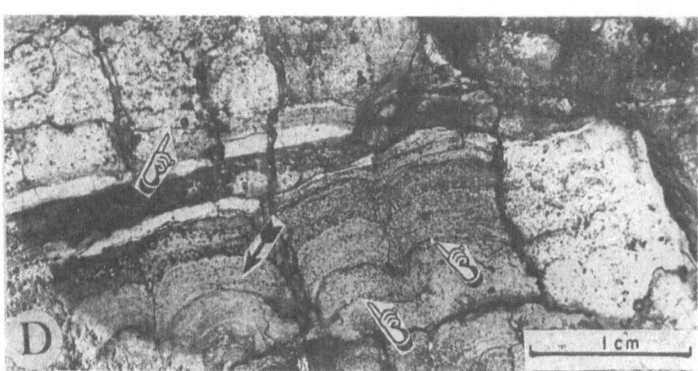

Figure 4 a--d. Macrofabrics of the stromatolite a) The stromatolitic bed (finger) at Har Orahot is largely in growth position. Below it, the Eocene nodular phosphorite (np) and above it, Eocene rocks (e). Contacts between stromatolites and overlying and underlying layers are unconformable. Arrow points to local stromatolite between nodules. The matchbox= 5 cm. b) Oncolite-like body produced by peripheral coating of clusters of phosphate-cemented Eocene phosphorite nodules (pn) by the stromatolite (finger). Token= about 1.5 cm. c) The stromatolitic layer broken and tilted at Nahal Shvyia (polished hand-specimen). Non weathered (finger) and weathered (arrows) stromatolite pieces mixed together with Eocene phosphorite nodules (pn) and iron-phosphate mineral assemblages (dark areas), the latter probably due to intra-Eocene subaerial weathering. d) Close-up view of the pseudo-columnar phosphoritic ministromatolites (polished hand-specimen). Superpostion of contiguous and laterally-linked successive pseudo-columns with a diffuse internal lamination (arrow) separated by discontinuity surfaces (lowermost fingers). The discontinuity surfaces are emphasized by secondary iron phosphate minerals starting to develop along and from these contacts. At the upper left corner (uppermost finger) the stromatolitic lamination has been largely effaced by advanced weathering.

age also occur in laterally equivalent distal sections of the Eocene (see Fig.1 and below). Secondly, no significant phosphate occurrence of earlier age is known either in the Eocene (cf. Martinotti in Shiloni et al., 1977; Benjamini, 1984), or in the Paleocene of the Negev. These facts taken together suggest that the distal phosphate nodules and the proximally-situated stromatolites might be roughly coeval, and hence, a latest Early to early Middle Eocene age is tentatively attributed to the stromatolite body.

BIOFABRICS

Two growth forms of distinct sizes are viewed in the Negev stromatolitic phosphorites: (1) stratiform pseudo-columnar stromatolites (Preiss, 1976), ranging from a few millimeters to more than 1 cm in height (Fig.4d, 5) and here referred to as "ministromatolites" (Edhorn and Anderson, 1977), and (2) columnar microstromatolites (Hofmann, 1969a), hundreds of microns high (Fig.6), intercalated in intervals within the first type. Although the terms "microstromatolites" and "ministromatolites" have previously been synonymously used in the geological literature to describe microscopic

Figure 5. Microfabrics of the stratiform pseudo-columnar ministromatolites. Laterally linked pseudo-columns built by a stacking of crinkled and slightly convex phosphate laminae. Black bands are goethite of mostly sub-recent weathering (after strengite). No clastics occur between and within the phosphate laminae. Clastics, where found, just outline (long arrow) some discontinuity surfaces, or fill erosional pockets (short arrow) cut between adjacent ministromatolites. Note the columnar microstromatolite layer (finger) growing at the top of the central ministromatolite edifice (photomicrograph, plain light).

stromatolitic structures (e.g., Edhorn and Anderson, 1977; Hofmann and Grotzinger, 1985), we separate the two terms here, and reserve "microstromatolite" for accretions requiring microscopic means of observations.

Stratiform pseudo-columnar ministromatolites

The ministromatolites constitute the main growth-form visible in the Negev stromatolitic phosphorites. As such, they are generally seen growing directly attached to phosphate nodules. These constructions consist of a superposition of contiguous and laterally linked successive pseudo-columns each several millimeters high (Fig.4d), with a very fine and diffuse internal lamination (Fig.5). Minor erosional traces are often observed along the discontinuity surfaces separating the consecutive pseudo-columns of the ministromatolites. Such erosional features tend to be more pronounced in the "lows" between laterally adjacent pseudo-columns where they appear as micro-scours (Fig.5). The ministromatolite laminae are wavy to crinkled, show a high degree of inheritance (Hofmann, 1969b), and where unaltered (see below) and entirely made up of francolite. The almost total absence of any skeletal or detrital material between and within the ministromatolite laminae is significant. Clastics, where present, are only found in very small amounts, and only in association with the erosional marks along the discontinuity surfaces between the consecutive pseudo-columns of the ministromatolites (Fig.5).
This detrital input usually consists of glauconite grains, bone fragments and iron-phosphate and goethite ooids of Maastrichtian derivation. Intra-Eocene epigenetic alteration commonly affects the ministromatolites. This results in the replacement of the francolite fraction of the ministromatolite by an assemblage of iron-phosphate minerals (mainly strengite), together with some barite and aluminium-phosphate. Goethite and gypsum of mainly sub-recent weathering (see also Panczer, 1990) are locally associated. These iron phosphates first appear along the discontinuity surfaces separating the consecutive pseudo-columns of the ministromatolites (Fig.4d). Following further weathering, they spread through the whole rock, wiping out large parts of the ministromatolitic fabric.

Columnar microstromatolites.

The microstromatolites (Fig.6) are typically associated with discontinuity features within the ministromatolite body. In most cases, they start growing from the series of micro-diastems separating the consecutive ministromatolite pseudo-columns (Fig.6a, b). In addition, they are seen leaning back against the walls of erosional troughs cut between laterally adjacent ministromatolite pseudo-columns, or, more rarely, anchoredon detrital grains lodged in these troughs (Fig.6a, b). At contacts between pseudo-columns they may form subcontinuous levels of partly truncated successive generations, each made by closely spaced erect microcolumns, which fill up erosional reliefs. The starting point for microcolumn growth is most often a tiny protuberance in the irregularities of the ministromatolite-discontinuity surface (Fig.6a, b) or, more

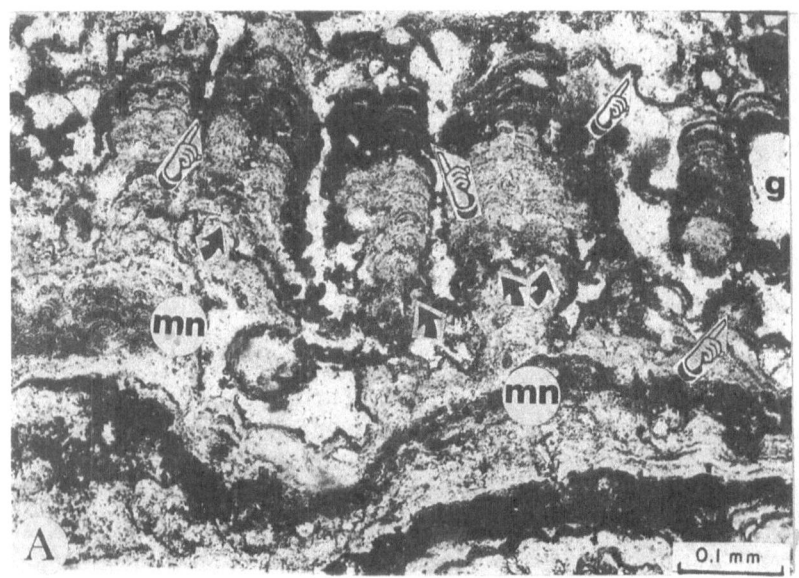

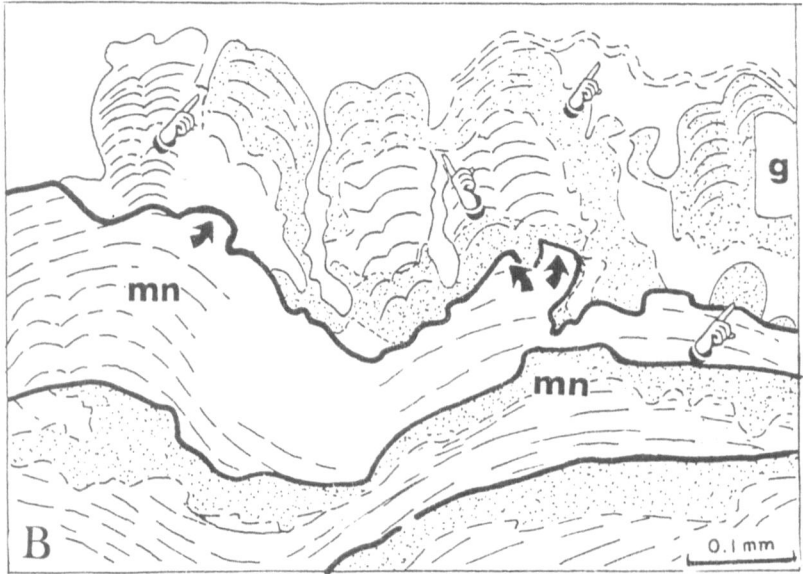

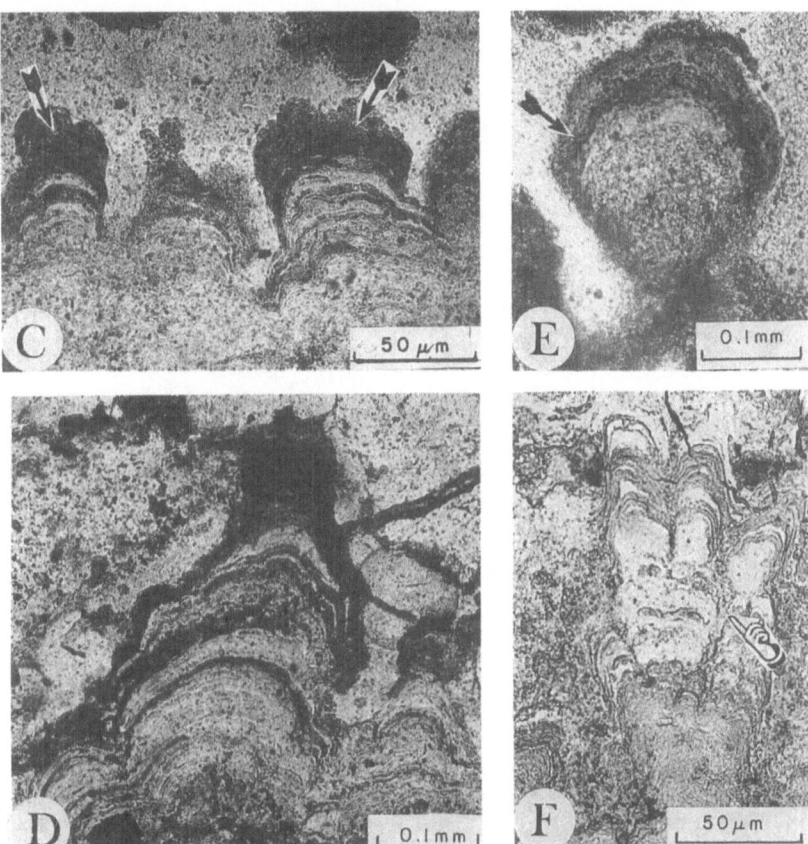

Figure 6 a--f Microfabrics and growth patterns of the columnar microstromatolites. a) and b) Cylindrical forms ; (b) is a schematic drawing of a). Note in a) and b) the tiny protuberances (arrows) from which the microcolumns start growing. These protuberances are tiny erosional irregularities at the summit of a slightly abraded, earlier ministromatolite edifice (mn). Where real columns fail to develop, a low relief form ("node") is generated (lowermost finger in a) and b)). Some of these "nodes" may be artificial and in part may result from tangential cuts of microcolumns. Note local bridging (uppermost fingers in a) and b) between adjacent microcolumns. (g) in a) and b) is a glauconite grain on which a microcolumn is anchored. Black areas in a) (dotted in b) are goethite of mostly sub-recent weathering. (photomicrographs, plain light). c) Roughly nodular forms (arrows), d) Terete forms, e) Turbinate forms (arrow), f) Branching forms (fingered).

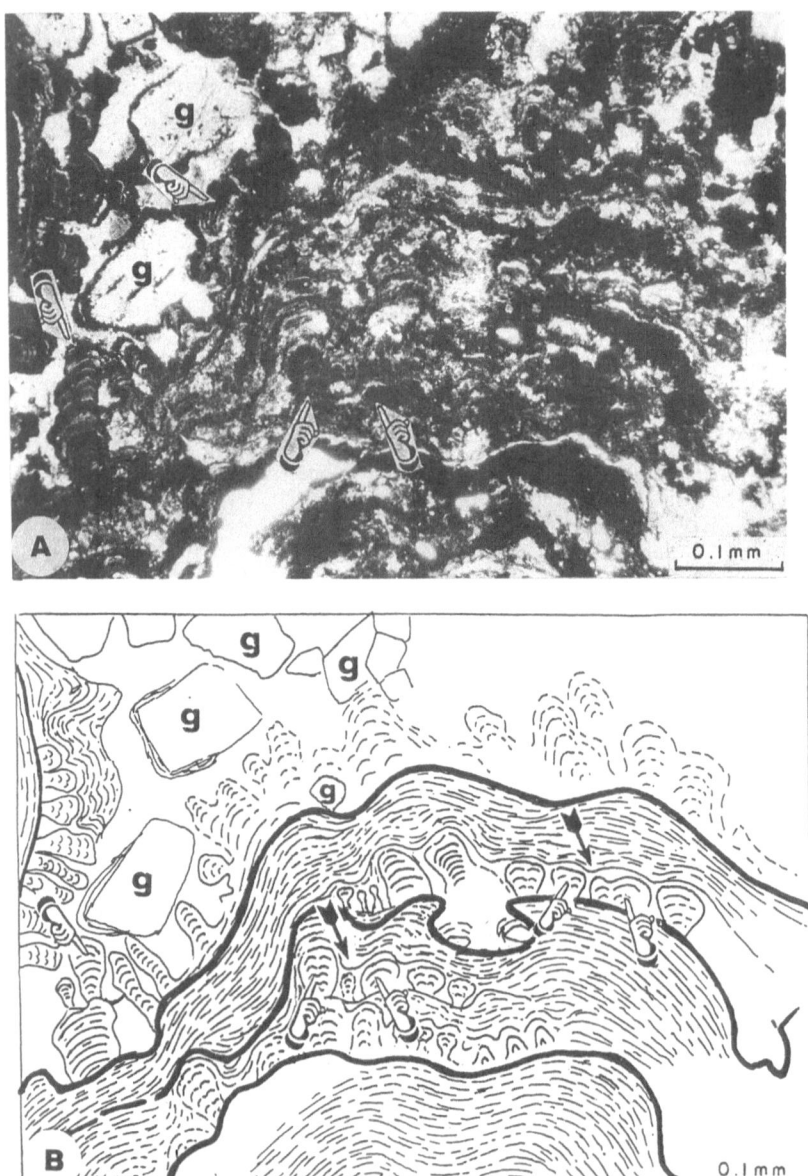

Figure 7 a, b. Ministromatolite edifice evolving from microstromatolite growth forms (a) --photomicrograph, plain light; b) -- schematic drawing of a). Where meeting at mutual contact, microcolumns (lowermost fingers a, b), which start growing from successive ministromatolite-discontinuity surfaces are bridged by a continuous lamina (arrowed in b) to produce superposed ministromatolite edifices. Note the new microstromatolite layer (uppermost finger -- a, b) growing at the summit of the topmost generated ministromatolite structure. (g) are glauconite grains.

rarely, a detrital particle on this surface. In contrast to the ministromatolites, which are quite monotonous in form, the microstromatolites exhibit a variety of growth patterns. In fact they show most of the basic figures encountered in ordinary stromatolites. Cylindrical (Fig.6a, b), roughly nodular (Fig.6c), terete (Fig.6d), turbinate (Fig.6e), and branching (Fig.6f) and non-branching forms are observed, in places in close contiguity (e.g., Fig.6c). These different geometries are perhaps linked to the configuration of the protuberances (e.g., size, shape, spacing) from which the microcolumn growth started. Apart from their dissimilarity in growth patterns, the microstromatolites are virtually identical to the ministromatolites with respect to their internal fabric and their mineralogy. They show the same type of wavy to crinkled fine and diffuse lamination, a total lack of any detrital sediment between and within laminae, and, similarly, are entirely phosphate-mineralized.

Although no clear transition is generally seen between the microstromatolite and the ministromatolite growth forms, there is evidence indicating that, in some cases, the two may be time-and space-related. In places, faint turbinate microcolumns are observed growing from several points along the series of discontinuity surfaces separating the successive pseudo-columns of the ministromatolites (Fig.7a, b). As these inverted columns meet at a mutual contact, they are bridged by a continuous lamina, hence initiating the development of a ministromatolite. Monty and Mas (1981) and many others have shown this type of transition between different growth forms of stromatolites. Indeed, perhaps most of the Negev ministromatolites evolved in this way, and due to a close spacing of protuberances along discontinuity surfaces, real columns failed to develop. Under such conditions, rather contiguous and (barely distinct) low relief growth forms ("nodes") (Hofmann, 1969b) would generate. These in turn would be rapidly bridged, thereby inducing a quick development of the gross morphology of a ministromatolite structure. However, because of the general diffuse nature of the stromatolitic lamination, and the masking effect of the pervasive post-depositional alteration phenomena, it is hard to decide if this is indeed everywhere the case.

Nannofabrics.

SEM examination of untreated fragments containing both types of stromatolites show them to consist of a stacking of contorted apatitic layers (Fig.8a). At first glance, nothing special is disclosed by these stromatolitic layers. They look essentially structureless, and the phosphate matter which forms them (no carbonates are detected by EDS techniques) is usually in the form of randomly-packed nannograins (<<1 micron in size). However, careful examination reveals some organization as well as some structural differentiation in these layers in places. Where phosphate overgrowths are minimally developed, these layers appear to be composed of spongy phosphate areas sandwiched between subcontinuous phosphate films (Fig.8b) -- the latter apparently responsible for the delicate microscopic lamination viewed in the stromatolite. The spongy areas locally disclose partly obliterated micron-sized capsule-like phosphate cells. These occur as solitary forms (Fig.8c) and as spheroidal

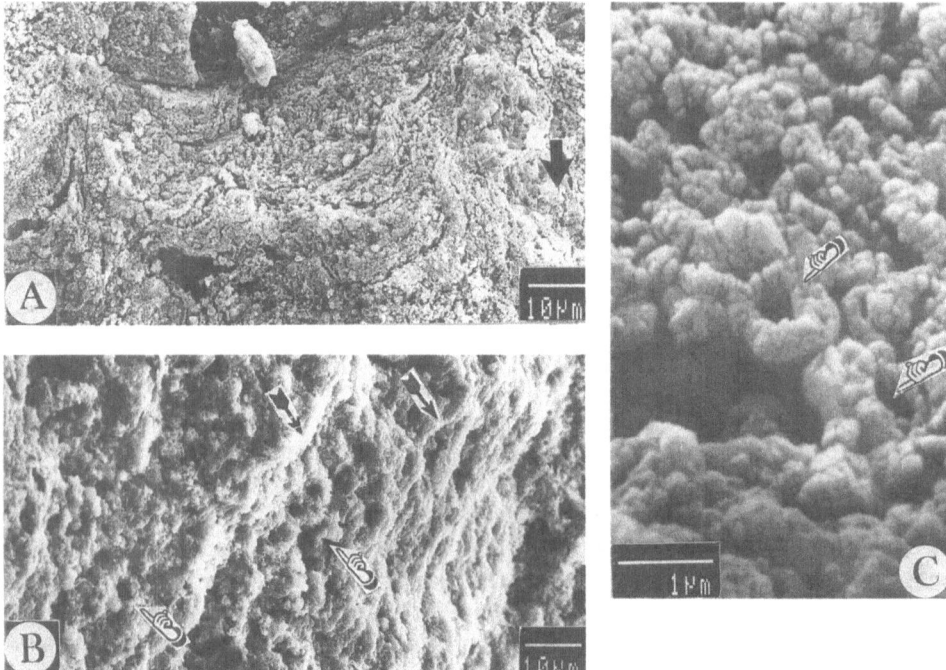

Figure 8 a--e.

cell-colonies of about ten microns in diameter (Fig.8d). In addition, patchy concentrations of encapsulated, globose phosphate structures, 3--7 microns in size, Fig.8e) are locally observed in the spongy areas together with the capsule-like cells. Flocculent phosphate overgrowths often coat the internal and external surfaces of the vesicle-like bodies and the globose structures, leading to their fading and their deformation. Complete occlusion of these structures with coatings leads to their partial disappearance, and ultimately only structureless phosphate is seen.

Distal variations of the stromatolitic phosphorites.

Sporadic concretions of phosphorites occur in Eocene basinal sections, laterally to the stromatolitic phosphorites (Fig.10). These occur as disoriented phosphate nodules, 1--4 cm in size (Fig.9a), and also as a lag deposit of phosphatic megafossil moulds (Shiloni et al., 1977). They are found at the upper part of a succession of soft and indurated glauconite-rich chalks known as the Paran Formation (Benjamini, 1984). Microfossil dating indicates that these phosphate nodules are late Early to early Middle Eocene in age. As noted above, these concretional phosphate occurrences and the stromatolitic phosphorites are considered as coeval and lateral equivalents. The abundant glauconite associated with this concretional phosphorite and the disoriented nature of the latter, the recurrent surficial hardening of associated chalks and their intense bioturbation, all suggest that the concretional phosphorite formed during periods of slow deposition. A

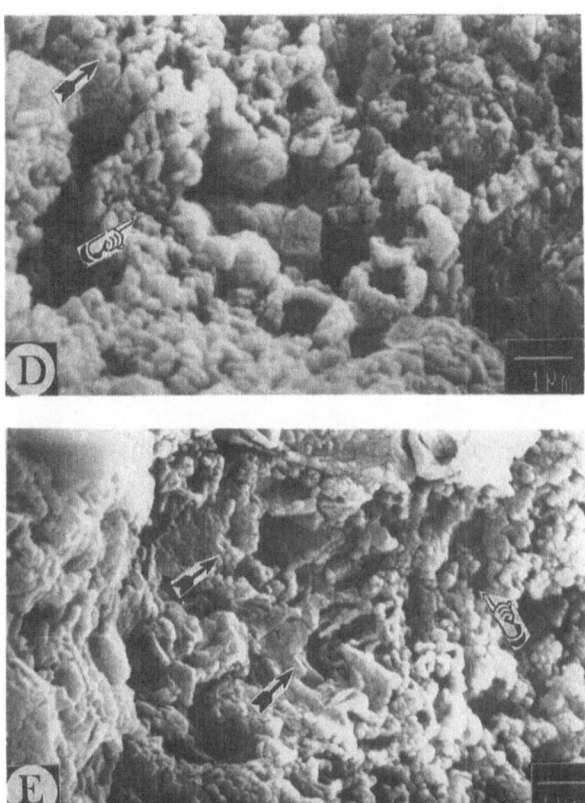

Figure 8 a--e. Nannofabrics of the stromatolitic phosphorites (SEM micrographs). a) Overall view of the stromatolitic lamination in both the mini and the microstromatolites showing a stacking of contorted apatitic laminae. The phosphate fraction constituting these laminae looks essentially structureless. Arrow indicates growth direction. b). Close view of the stromatolitic lamination in both the mini and the microstromatolites showing spongy phosphate areas (fingers) sandwiched between subcontinuous phosphate films (arrows). c) and d) Solitary and colonial capsule-like phosphate cells (fingers) building the spongy areas of the mini and microstromatolitic phosphorites. These cellular structures are visible only where phosphate overgrowths are minimal or lacking. Heavy overgrowths lead to the occlusion of these bodies and convert them into aggregates of structureless phosphate; e) Clustered and multi-enveloped (arrows) globose phosphate structures locally associated with the capsule-like cells in the spongy areas of the Negev stromatolitic phosphorites. Finger points to a ghost structure.

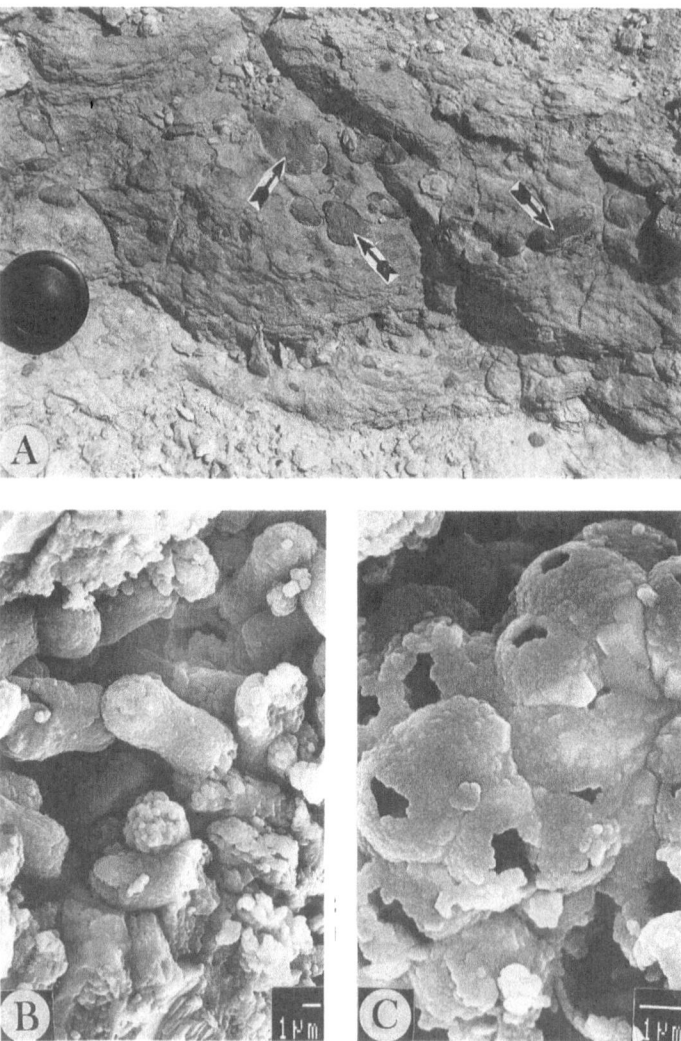

Figure 9 a--c. Macro a) and nannostructures (b, c) of the nodular phosphorites in the Early-Middle Eocene basinal sections of the central Negev. These nodules are assumed to be the distal equivalents of the Ramon stromatolitic phosphorites. a). Field view showing scattered phosphate nodules (arrows) within glauconite-rich chalks (lens cover for scale). b), c). Rod-shaped b) and globose phosphate bodies c) constituting the phosphate fraction of the Eocene phosphate nodules. These structures wrap and fix together glauconite grains and calcareous particles to form the phosphate nodules. Note the typical honeycombed arrangement of the clustered globose structures (c). (SEM micrographs).

previous study of this pebbly phosphorite (Soudry and Lewy, 1988) has shown that its phosphate fraction consists of pustulose sheets composed of 1--10 micron-sized globular bodies of bacterial and possibly cyanobacterial affinities. These wrap and fix together glauconite grains, calcitic planktic foraminifera, and molluscan debris. Spherulites, cerebroid-shaped structures, rod- (Fig.9b) and cocci-like bodies, and globose forms with a honeycombed arrangement (Fig.9c) are the main components of this microbial assemblage.

PALEOENVIRONMENTAL CONSIDERATIONS

To generate the kind of stromatolites that have no detrital or skeletal material trapped at all, manifestly requires very quiet hydrodynamic conditions. Such conditions are concievable in two quite different settings: one a very protected (e.g., lagoonal) shallow environment; the other, a deep-water pelagic setting (below the CCD) where carbonate accumulation is virtually nil (e.g., Scholle et al., 1983). These structures could have well accreted in deep settings. Iron-manganese encrusted microstromatolites with rather similar growth patterns have been described in bathyal (Monty, 1973) and in abyssal environments (Janin and Bignot, 1983). Furthermore, phosphate crusts capping sedimentary discontinuities and built by phosphate-permineralized colonial unicells havebeen recently reported from Senonian hemipelagic settings (Soudry and Lewy, 1990). However, the strict confinement of the stromatolitic phosphorites to the upper margins of the Ramon anticline, the pervasive subaerial weathering which affected them, and their presence as relicts in iron-rich soils (e.g., N. Eshboren), all suggest that these growth forms rather accreted in shallow and well-sheltered settings cut off from detrital inputs. Subaerially weathered and phoscretized stratiform and columnar phosphate stromatolites are also known at the K/T boundary of Greece (Pomoni-Papaioannou and Solakius, 1991). The series of discontinuity surfaces within the Negev ministromatolites may express periodic growth interruptions perhaps related to subtle variations of the environmental parameters. Such breaks might have been triggered, or alternatively, been followed, by slightly more agitated episodes, as indicated by the erosional traces along these surfaces and the transported clastics lining them. Renewal of growth after these periods of rest apparently would take the form of microcolumns rising from "privileged" sites along these discontinuity surfaces. These pioneer microcolumns would later be bridged and the development of a new generation of ministromatolite edifices would thus be initiated.

DISCUSSION

In a recent study of cherty microstromatolites (Transvaal Supergroup, South Africa), Lanier (1986) used a steady-state stromatolitic growth model (Brock, 1971) to evaluate the growth rates of these structures. In this model, primary-producer microorganisms are periodically succeeded by consumer bacterial populations, with a rough balance maintained between primary production and bacterial breakdown. In the Transvaal

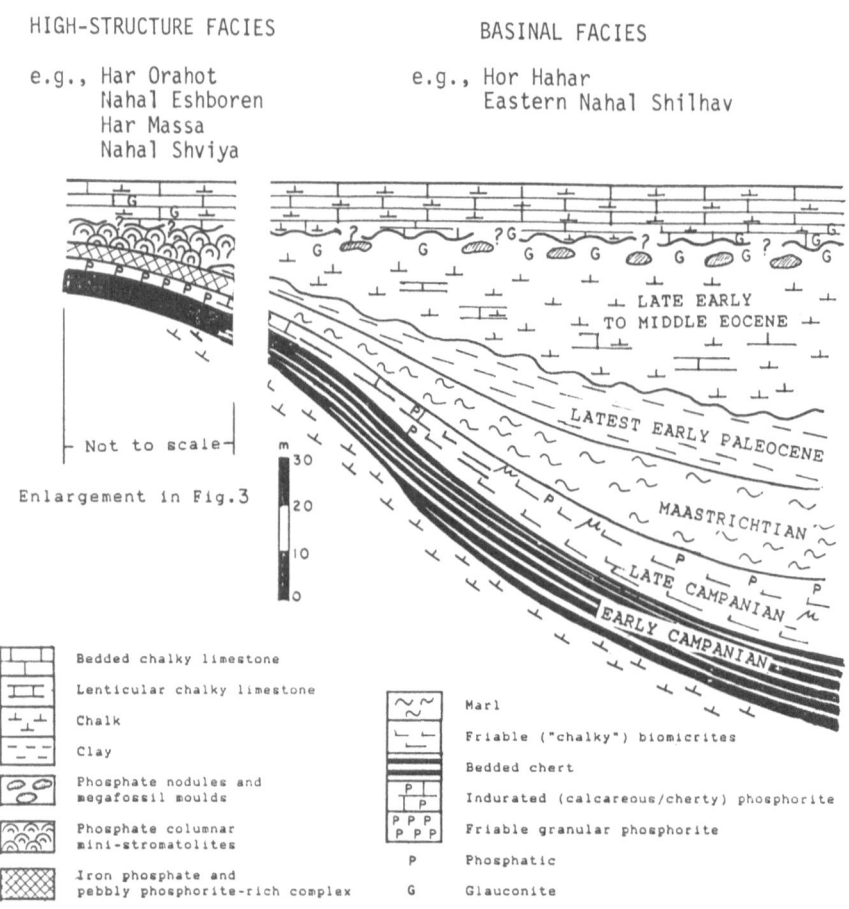

Figure 10. Schematic diagram perpendicular to the Ramon anticline showing the spatial distribution of the stromatolitic and nodular phosphorites in the proximal-distal Senonian-Eocene section.

microstromatolites, patchy concentrations of 1.5--5 micron-sized coccoid microorganisms are considered to be the main primary producers. The Negev stromatolitic phosphorites apparently resemble the Transvaal structures. The clustered multi-enveloped globose bodies building the phosphate stromatolites by intervals (cf. Fig.8e), might recall remnants of colonies of coccoid unicells. Their size, their shape and their overall configuration fit quite well the basic characteristics of many of these forms (e.g., Golubic and Hofmann, 1976; Schopf, 1977). Likewise, the capsule-like phosphate bodies (Fig.8c, d) which make up the bulk of the phosphate fraction of the stromatolites, may be viewed as the fossilized remnants of former bacterial populations which actively degraded the coccoid primary producers. The generally high degree of inheritance of the stromatolitic laminae apparently also implies that coccoid-primary production and bacterial consumption were more or less balanced through time, hence suggesting a rather steady-state mode of growth.

However, a stromatolite accretion according to a steady-state model would impose some constraints in terms of microbial preservation and phosphorus budget. Most probably, phosphatization took place in the near seawater-sediment interface (just below the living mat), on vacant unicell and bacteria envelopes, and within phosphorus-rich pore waters. Many studies have shown that organic soft parts are favoured substrates for phosphate crystallization (Soudry and Champetier, 1983; Muller, 1985; Southgate, 1986), and that the latter is in essence an interface-linked early diagenetic phenomenon (Froelich et al., 1988; Bein et al., 1990). Modern phosphorite formation has also been found to occur in close association with mat-forming microbial communities on the' sea floor (Reimers et al., 1990; Garrison et Kastner, 1990). As concerns the stromatolitic phosphorites of the Negev, two optional sources for pore-water phosphorus enrichment are envisioned. One is the primary producer itself, during post-mortem bacterial decomposition of the cell material. The second is perhaps a synsedimentary redox extraction of iron hydroxide-sorbed phosphate repeatedly pumped from the overlying water column, as was suggested for modern phosphorite formation (Froelich et al., op. cit.). In the first case, a steady-state stromatolitic growth would imply (given the C/P Redfield ratio - 106:1) that although primary production and bacterial breakdown were more or less balanced, only a minimal fraction of the producer/decomposer original biomass was preserved in the sediment. These "privileged" microbial remnants which escaped total lysis may have served as sites of precipitation and sedimentary traps for the amounts of phosphate released by that part of the biomass which was totally degraded. Support for this view may be found in the results from Solar Lake (Red Sea), which show that less than 1% of the original cyanobacterial biomass was converted into carbonate rock (Krumbein et al., 1977). Whatever the mechanism of pore water-phosphorus enrichment adopted, an accelerated supply of dissolved phosphate to the water column is in all cases required. The Early-Middle Eocene interval is in fact a period of major phosphogenesis in North Africa, and the sparse Eocene phosphate sediments occurring in Israel and in other Middle East countries, such as Syria (Omara, 1965) or Egypt (Abul-Nasr and Thunell, 1987), may be faint "echos" of this big event. No clear explanation has been found, however, as to why this Eocene Negev phosphate sedimentation is expressed as columnar stromatolites in proximal positions, and only as patchy microbial bodies "cementing" carbonate sediment in the distal ones. Possible reasons for this could be the relatively higher sedimentation rates in distal positions.

Under such conditions, microbial activity there might have adapted to a "passive" chasmolithic dwelling-mode, rather than to an active (accretional) epilithic living-mode as recorded in proximal situations.

Lastly, worth noting is the striking resemblance of the microstromatolitic phosphorites to the problematic iron-bearing *frutexites* structures reported from various ages in the geological record. These forms have been variously considered as filamentous cyanobacteria (Walter and Awramik, 1979) or bacteria, or purely physicochemical in origin (Hofmann and Grotzinger, 1985). The present microstromatolitic phosphorites are believed to have originated through active growth of unicell-like populations. Except for a few intervals in which they have been preserved (phosphatized) as such, these unicellular-like microbiota were largely converted into microstratified bacterial-like populations. Hence, it is not perhaps completely unlikely that stromatolite constructions now appearing as purely bacterial, began as microbial successions of primary producers which are now undetected because they are totally re-structured.

ACKNOWLEDGMENTS

Thanks are due to S. Moshkovitz, G. Martinotti and R. Siman Tov, of the Geological Survey of Israel for micropaleontological dating, to M. Dvorachek for SEM services, and to Y. Levy and A. Peer for technical assistance. Bevie Katz polished the English of several versions. The constructive comments given by C. Monty and J. Sarfati are gratefully acknowledged. This work was done within the framework of project No. 20448 of the Geological Survey of Israel, and is a contribution to IGCP 325, Palaeogeography of Authigenic Minerals and Phosphorites.

REFERENCES

Abul-Nasr, R.A. and Thunell, R.C. (1987) "Eocene eustatic sea-level changes, evidence from western Sinai, Egypt", Palaeogeogr. Palaeoclimatol. Palaeoecol. 58, 1--9.

Banerjee, D.M. (1971) "Precambrian stromatolitic phosphorites of Udaipur, Rajasthan, India", Geol. Soc. Amer. Bull. 82, 2319--2330.

Bein, A., Almogi-Labin, A. and Sass, E. (1990) "Sulfur sinks and organic carbon relationships in Cretaceous organic-rich carbonates: implications for evaluation of oxygen-poor depositional environments", Am. J. Sci. 290, 882--911.

Benjamini, C. (1984) "Stratigraphy of the Eocene of the 'Arava Valley (eastern and southern Negev, southern Israel)", Isr. J. Earth Sci. 33, 167--177.

Bentor, Y.K. and Vroman, A. (1954) "A structural map of Israel (1:250,000) with remarks on its dynamic interpretation", Bull. Res. Counc. Isr. 4, 125--135.

Brock, T.D. (1971) "Microbial growth rates in nature", Bacteriological Reviews 35, 39--58.

Bushinskii, G.I. (1966) Old Phosphorites of Asia and Their Origin, Acad. Sci. USSR, Isr. Prog. Sci. Transl. (1969).

Chauhan, D.S. (1979) "Phosphate-bearing stromatolites of the Precambrian Aravalli phosphorite deposits of the Udaipur region, their environmental significance and genesis of phosphorite", Precambrian Res. 8, 95--126.

Cook, P.J. (1992) "Phosphogenesis around the Proterozoic-Phanerozoic transition", J. Geol. Soc. London 149, 615--620.

Delamette, M. (1981) "Sur la découverte de stromatolithes circalittoraux dans la partie moyenne du Crétacé nordsubalpin (Alpes occidentales francaises)", C.R. Acad. Sci. Paris 292, 761--764.

Edhorn, A.-St. and Anderson, M.M. (1977) "Algal remains in the Lower Cambrian Bonavista Formation, Conception Bay, southeastern Newfoundland", in E. Flugel (ed.), Fossil Algae, Springer, Berlin, pp. 113--123.

Follmi, K.B. (1989) Evolution of the Mid-Cretaceous Triad, Platform Carbonates, Phosphatic Sediments, and Pelagic Carbonates along the Northern Tethys Margin, Lecture Notes in Earth Sciences 23, Springer, Berlin .

Freund, R. (1965) "A model of the structural development of Israel and adjacent areas since Upper Cretaceous times", Geol. Mag. 102, 189--205.

Froelich, P.N., Arthur, M.A., Burnett, W.C., Deakin, M., Hensley, V., Jahnke, R., Kaul, L., Kim, K.-H., Roe, K., Soutar, A. and Vathakanon, C. (1988) "Early diagenesis of organic matter in Peru continental margin sediments: phosphorite precipitation", Marine Geology 80, 309--343.

Garrison, R.E. and Kastner, M. (1990) "Phosphatic sediments and rocks recovered from the Peru margin during ODP Leg 112", in E. Suess and R. von Huene (eds.), Proc. Ocean Drill. Prog. 112, pp. 111--134.

Golubic, S. and Hofmann, H.J. (1976) "Comparison of Holocene and Mid-Precambrian Entophysalidaceae (Cyanophyta) in stromatolitic algal mats: cell division and degradation", J.Paleontology 50, 1074--1082.

Hofmann, H.J. (1969a) "Stromatolites from the Proterozoic Animikie and Sibley groups", Can. Geol. Surv. Paper 68-69, 1--58.

Hofmann, H.J. (1969b) "Attributes of stromatolites", Can. Geol. Surv. Paper 69-39, 1--58.

Hofmann, H.J. (1975) "*Bolopora* not a bryozoan, but an Ordovician phosphatic, oncolitic accretion", Geol. Mag. 112, 523--526.

Hofmann, H.J. and Grotzinger, J.P. (1985) "Shelf-facies microbiotas from the Odjick and Rocknest formations (Epworth Group; 1.89 Ga), northwestern Canada", Can. J. Earth Sci. 22, 1781--1792.

Janin, M.C. and Bignot, G. (1983) "Microfossiles thallophytiques des concrétions polymétalliques laminées", Rev. Micropaleontol. 25, 251--264.

Jarvis, I. (1992) "Sedimentology, geochemistry and origin of phosphatic chalks: the Upper Cretaceous deposits of NW Europe", Sedimentology 39, 55--97.

Krajewski, K.P. (1981) "Phosphate microstromatolites in the High-Tatric Albian limestones in the Polish Tatra Mountains", Bull. Acad. Pol. Sci. Ser. Sci. Terre 29, 175--183.

Krumbein, W.E., Cohen, Y. and Shilo, M. (1977) "Solar Lake (Sinai). 4. Stromatolitic cyanobacterial mats", Limnology and Oceanography 22, 635--656.

Lanier, W.P. (1986) "Approximate growth rates of Early Proterozoic microstromatolites as deduced by biomass productivity", Palaios 1, 525--542.

Monty, C.L.V. (1973) "Les nodules de manganèse sont des stromatolithes océaniques", C.R. Acad. Sci., Paris 276, 3285--3288.

Monty, C.L.V. and Mas, J.R. (1981) "Lower Cretaceous (Wealdian) blue-green algal deposits of the province of Valencia, Eastern Spain", in C.L.V. Monty (ed.), Phanerozoic Stromatolites, Springer, Berlin, pp. 85--120.

Muller, K.J. (1985) "Exceptional preservation in calcareous nodules", Phil. Trans. Royal Soc. London 311, 67--73.

Niedermeyer, R.O. and Langbein, R.L. (1989) "Probable microbial origin of Ordovician (Arenig) phosphatic pebble coats ('*Bolopora*') from North Wales, U.K.", Geol. Mag. 126, 691--698.

Notholt, A.J.G., Sheldon, R.P. and Davidson, D.F. (1989) Phosphate Deposits of the World, Cambridge University Press, Cambridge.

Omara, S. (1965) "Phosphatic deposits in Syria and Safaga district, Egypt", Econ. Geol. 60, 214--227.

Panczer, G. (1990) "Minéralogie et géochimie de l'altération des phosphorites nodulaires du Neguev, a la discordance Mishash-Ghareb (Israel)", Unpubl. Doct. Thesis Univ. Louis Pasteur, Strasbourg.

Pomoni-Papaioannou, F. and Solakius, N. (1991) "Phosphatic hardgrounds and stromatolites from the limestone/shale boundary section at Prossilion (Maastrichtian-Paleocene) in the Parnassus-Ghiona Zone, Central Grece", Palaeogeogr. Palaeoclimatol. Palaeoecol. 86, 243--254.

Preiss, W.V. (1976) "Basic field and laboratory methods for the study of stromatolites", in M.R. Walter (ed.), Stromatolites, Elsevier, Amsterdam, pp. 5--13.

Reimers, C.E., Kastner, M. and Garrison, R.E. (1990) "The role of bacterial mats in phosphate mineralization with particular reference to the Monterey Formation", in W.C. Burnett and S. Riggs (eds.), Neogene to Modern Phosphorites, Cambridge University Press, Cambridge, pp. 300--311.

Royant, G., Rioult, M. and Lanteaume, M. (1970) "Horizon stromatolithique à la base du Crétacé supérieur dans le Brianconnais ligure", Bull. Soc. Geol. France 12, 372--374.

Sant, V.N. (1979) "Precambrian phosphorite of Rajasthan, India: A Case History", in P.J. Cook and J.H. Shergold (eds.), Proterozoic-Cambrian Phosphorites, 1st. Int. Field Workshop IGCP Project 156, Canberra, Australia, pp. 39--40.

Scholle, P.A., Arthur, M.A. and Ekdale, A.A. (1983) "Pelagic Environment", in P.A. Scholle, D.G. Bebout and C.H. Moore (eds.), Carbonate Depositional Environments, AAPG Memoir 33, 620--691.

Schopf, J.W. (1977) "Biostratigraphic usefulness of stromatolitic Precambrian microbiotas", Precambrian Res. 5, 143--174.

Shiloni, Y., Segev, A., Martinotti G.M. and Raab, M. (1977) "An early Eocene glauconitic bed in Hor Hahar, northern Negev, Israel", Isr. J. Earth Sci. 26, 102--107.

Soudry, D. and Champetier, Y. (1983) "Microbial processes in the Negev phosphorites (southern Israel)", Sedimentology 30, 411--423.

Soudry, D. and Lewy, Z. (1988) "Microbially influenced formation of phosphate nodules and megafossil moulds (Negev, southern Israel)", Palaeogeogr. Palaeoclimatol. Palaeoecol. 64, 15--34.

Soudry, D. and Lewy, Z. (1990) "Omission-surface incipient phosphate crusts on early diagenetic calcareous concretions and their possible origin, Upper Campanian, southern Israel", Sediment. Geol. 66, 151--163.

Southgate, P.N. (1980) "Cambrian stromatolitic phosphorites from the Georgina Basin, Australia", Nature 285, 395--397.

Southgate, P.N. (1986) "Cambrian phoscrete profiles, coated grains, and microbial processes in phosphogenesis: Georgina Basin, Australia", J. Sed. Petrology 56, 429--441.

Walter, M.R. and Awramik, S.A. (1979) "*Frutexites* from stromatolites of the Gunflint Iron-Formation of Canada, and its biological affinities", Precambrian Res. 9, 23--33.

PART III

Mesozoic Deep Marine Stromatolites and Bacterial Marine Phosphorites

AMINO ACIDS IN THE PELAGIC STROMATOLITES OF THE ROSSO AMMONITICO VERONESE FORMATION (MIDDLE-UPPER JURASSIC, SOUTHERN ALPS, ITALY)

L. BALLARINI*, F. MASSARI*, S. NARDI** and L. SCUDELER BACCELLE***
* Department of Geology, Palaeontology and Geophysics, University of Padova, Via Giotto 1, 35100 Padova, Italy.
** Department of Agrarian Biotechnology, University of Padova, Via Gradenigo 6, 35100 Padova, Italy.
*** Department of Mineralogy and Petrology, University of Padova, Corso Garibaldi 37, 35100 Padova, Italy.

ABSTRACT

The Rosso Ammonitico Veronese (RAV) Formation is a condensed pelagic limestone including, at different horizons, laminated facies which have previously been interpreted as stromatolitic. Some samples of these limestones from the Verona area (Western Veneto, Italy) were studied in thin and polished sections and organic compounds were extracted by the analytical method of Chichereau and Trichet (1976). The amino acids of the laminated facies are consistent with the composition of the cell wall of modern and fossil cyanobacteria and thus confirm its stromatolitic nature.
Associated collomorphic banding concretions of Fe-Mn oxides have also been interpreted as stromatolites of fungal origin.

INTRODUCTION

Stromatolitic structures in the pelagic Rosso Ammonitico Veronese facies have been reported by Sturani (1964, 1969, 1971), Jenkyns and Torrens (1969), Wendt (1970), Jenkyns (1971), Bernouilli and Jenkyns (1974), Massari (1979, 1981, 1983), Winterer and Bosellini (1981) and Clari et al. (1984). In addition to planar and hemispheroidal stromatolites, Massari (1981, 1983) and Clari et al. (1984) have also described oncoids made up of a number of more or less overlapping and differently oriented hemispheroids developing around a nucleus, in most cases represented by an early lithified calcareous nodule. Exhaustive reviews of the nature and ecological significance of stromatolites (Monty, 1976, 1977; Walter, 1976) revealed their wide spectrum of genetic and environmental conditions and stated that stromatolites are "algally-controlled structures built by coordinated societies of primitive organisms

J. Bertrand-Sarfati and C. Monty (eds.), Phanerozoic Stromatolites II, 279–294.
© 1994 *Kluwer Academic Publishers.*

(cyanobacteria and bacteria)" (Monty, 1977). In the marine environment, stromatolitic structures are reported from supratidal to deep water aphotic settings (Monty, 1977). In addition, the collomorphic banding concretions of Fe-Mn oxydes occurring as encrusting structures over large areas of the sea bottom, mostly on current swept plateaus, are reported by Monty (1973) as true bacterial stromatolites. It was later noted (Krumbein, 1986) that this type of stromatolite forms due to the activity of fungi in microbialic ecosystems. Since Cenozoic marine cyanobacterial assemblages do not generally calcify their sheaths or mucilage, but simply trap and agglutinate detrital particles, their role and control in the building of stromatolitic and oncolitic structures cannot easily be detected except by relying on the typical laminated fabric and composition of the included fossil amino acids, which may be extracted (Trichet, 1968; Chichereau and Trichet, 1976).

As stated by Degens and Mopper (1979), the various organic compounds present in sediments during the early stages of diagenesis not only interact with each other but also with metal ions and mineral surfaces. Consequently, reaction schemes and resultant organic products become very complex. The randomization of the original structural order of living matter during early diagenesis is then followed by a slow approach to thermodynamic stability in later diagenetic stages. However, in spite of the profound decay of the biomolecules, the composition of the amino acids extracted from the rock may supply significant information on the original nature of the organic matter in the sediments. In particular, Frank et al. (1962) and Chichereau and Trichet (1976) have defined the amino acid composition typical of cyanobacterial mats in both recent and fossil cryptalgal structures.

Following the suggestions of these authors, we examined the amino acids extracted from the Rosso Ammonitico Veronese (RAV) stromatolites, in order to verify the influence of cyanobacterial mats in their formation. Petrographic analysis was performed for additional information on the texture and geometric characteristics of the stromatolitic structures.

Interesting considerations on depositional and diagenetic environments emerged from a comparison with the data of Chichereau and Trichet (1976) on the lacustrine Gannat Limestone (Oligo-Miocene), interpreted from the genetic viewpoint by Bertrand-Sarfati et al. (1966).

MATERIALS AND METHODS

The so-called Rosso Ammonitico Veronese, a stratigraphically condensed pelagic facies consisting of red nodular limestone, is widespread in the Middle-Upper Jurassic of the Venetian Alps and is most represented in the palaeogeographic belt known as the Trento Plateau (Aubouin, 1964). The latter was a pelagic horst block of the southern continental margin of the Jurassic Tethys which evolved from a strongly subsiding Liassic carbonate platform. The Rosso Ammonitico pelagic limestone forms condensed and discontinuous sequences in the Trento Plateau area, particularly rich in microbial structures and locally bearing microcoquinoid layers (Fig.1) consisting of sparcemented

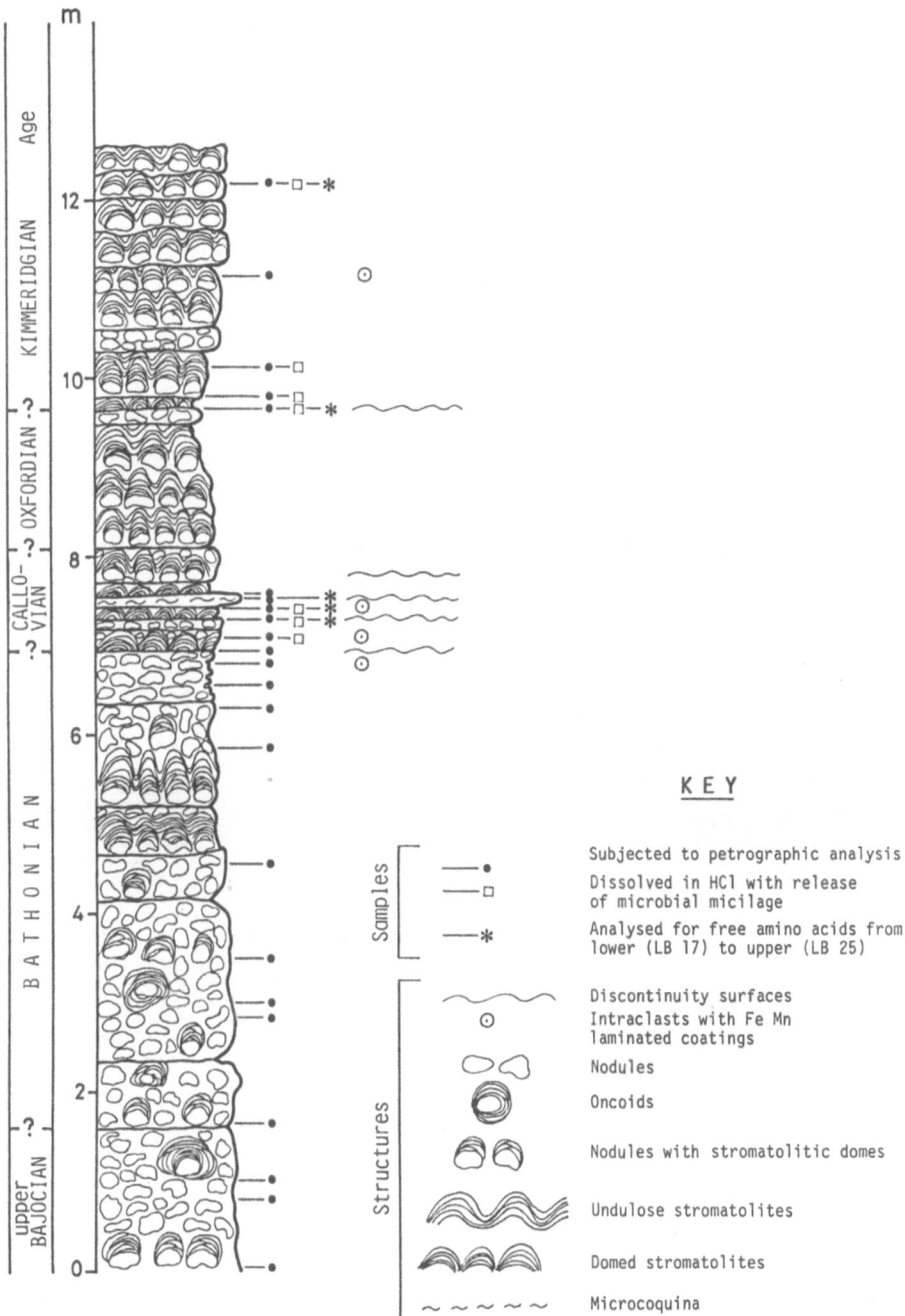

Figure 1 : Examined section of Rosso Ammonitico Veronese near S.Ambrogio di Valpolicella (Verona, Italy).

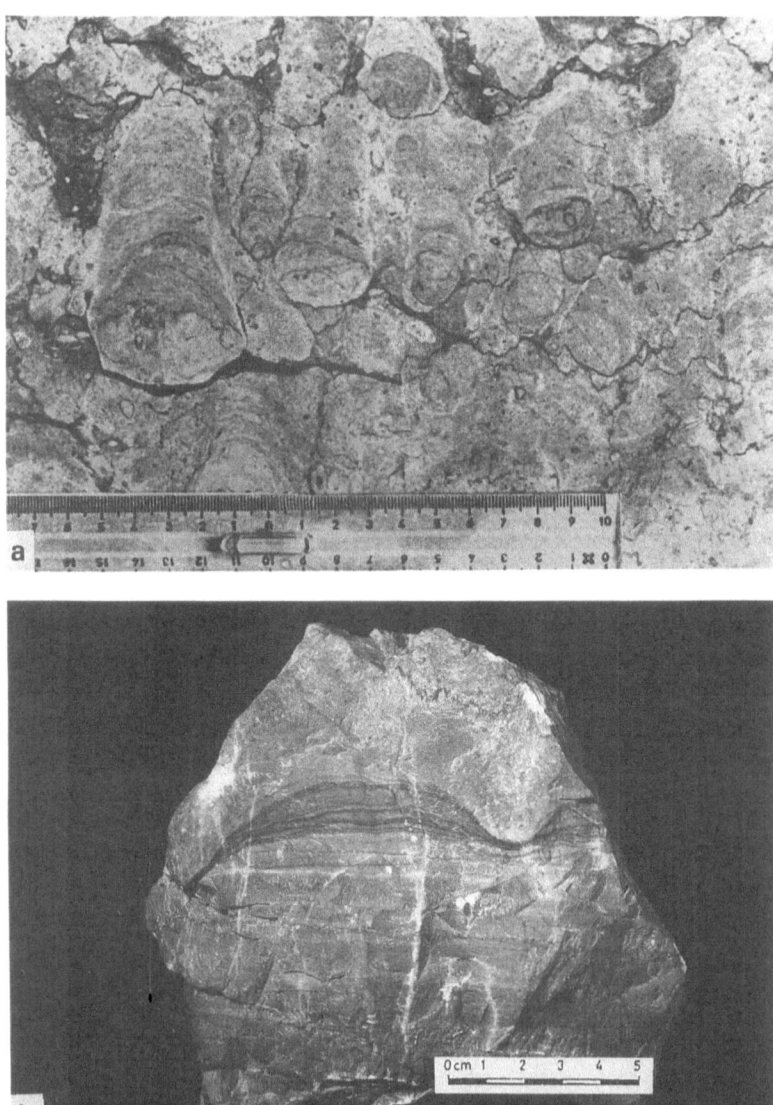

Figure 2 : a) Polished slab in St. Antony's Basilica,Padova, an example of oncolitic-stromatolitic fabric in vertical section. Rosso Ammonitico Inferiore from M. Lessini quarriers north of Verona.
b) Stromatolite in vertical section. Rosso Ammonitico superiore from Erbezzo (Verona).

thin-shelled bivalves ; these have been interpreted as storm deposits (Massari, 1981). The analysed samples were collected from an abandoned quarry of Rosso Ammonitico Veronese (RAV) in the south-western Lessini mountains near S. Ambrogio di Valpolicella (province of Verona). Fig. 1 shows the stratigraphic
sequence and position of samples.
Samples were studied in thin and polished sections. In addition, opaque minerals were identified by microprobe analysis, and organic compounds were extracted using the experimental protocol of Chichereau and Trichet (1976), slightly modified for the LKB autoanalyser. All data recorded were means of three replicates; standard deviations were always within 5%.

STRUCTURAL AND TEXTURAL FEATURES

The most common structures occurring in the Venetian Rosso Ammonitico are represented by stromatolites and oncoids (Figs. 2a,b). In the Oxfordian and Kimmeridgian sediments, these tend to be organized in decimetric cycles in which stromatolitic layers are made up of clustered, domed or wavy structures which grow from a "pavement" composed of various particles such as oncoids, botryoids, simple nodules, intraformational lithoclasts and irregularly shaped aggregates (Massari, 1981, 1983). The cycle may correspond to the soft-ground/hard-ground rhythm of Hollmann (1964), and may record oscillatory changes in the biological and physico-chemical factors controlling sedimentation and bottom stability.
The transition from soft to hard ground is accompanied by a drastic reduction in burrowing activity as a result of the binding and stabilization of bottom sediments by the stromatolites.
In the late stages of the cycle, this activity is mostly reduced to firm-stage tubular burrows (*Thalassinoides*) confined to the interspaces of stromatolitic domes; the burrowing animals were eventually prevented from further activity by hardening of the cryptalgal fabric. Evidence of condensation is particularly strong in the Callovian interval, which is characterized by the presence of closely spaced discontinuity surfaces. Mixing of ammonite faunas, overturned ammonites and abundance of oncoids are typical features of this interval throughout the Lessini area; this points to the episodic activity of strong (storm-induced?) bottom currents which removed soft sediment and induced mechanical reworking.
In thin section, stromatolitic coatings developing on exhumed nodules, ammonite shells or hard grounds consist of an alternation of thick granular laminae and much thinner micritic ones. The former are made up of small pellets, *Globochaete alpina* Lomb., debris of thin-shelled bivalves, sometimes very small gastropods, foraminifera, minute echinoderm debris, and other fine unrecognizable detritus. These particles are usually finer-grained than those occurring in the spaces between the stromatolitic domes; they

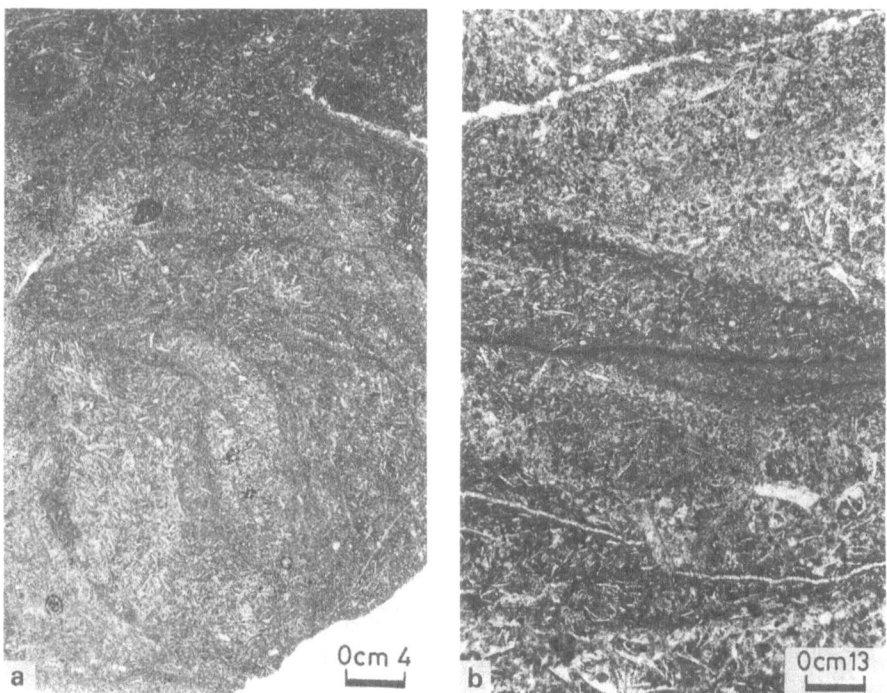

Figure 3 : a) Oncoid with differently oriented and interfingering sets of hemispheroidal laminae. Rosso Ammonitico Inferiore (Bathonian) from Roccolo quarry near S. Ambrogio di Valpolicella (Verona). b) Stromatolitic structure showing alteration of thik granular and thinner micritic laminae. Rosso Amonitico Inferiore (Bathonian?) from Roccolo quarry near S. Ambrogio di Valpolicella (Verona).

may show reverse grading within individual laminae. In addition, these particles are embedded in a matrix made up of coarsely crystalline micrite or even microspar (neomorphism?). The micritic laminae are composed of dense micrite containing rare and very small debris (Figs. 3a,b).

The alternation of texturally different laminae may depend on a balance between the growth of the cyanobacterial mats and the periodic or erratic influx of detrital particles (Monty, 1976). It should be noted that the elongated particles (e.g., thick, unbroken or fragmented bivalve shells) tend to be arranged tangentially to the surface of the micritic laminae, whereas they commonly show edgewise orientation within the granular laminae. This feature, together with the common reverse grade bedding in the granular laminae, suggests an active trapping mechanism interfering with settling particles to form the granular laminae; the presence of erect filaments in the mats allowed size sorting by infiltration of grains between their meshes (Monty, 1976). During long (?) pauses between such periods of active growth, the mats were only able to agglutinate tiny rare particles by a binding rather than a trapping mechanism, perhaps favouring the precipitation of "grumeleuse" micrite.

Fenestral fabrics are completely lacking. Pauses in the growth sequence are somtimes marked by low-angle unconformities truncating the previously formed laminae. Locally deformed or truncated laminae suggest that burrowers reworked the microbial mats and that organisms such as brachiopods, serpulids, small ahermatypic corals, very small endolithic borings of penetrative thallophyta and large borings, possibly related to polychaeta, occur along some bioeroded and iron-stained surfaces. These biostructures indicate early cementation of the stromatolitic fabric (Golubic, 1973; Krumbein, 1979). The growth of the layered coating was probably very slow and interrupted by long pauses, as suggested by the iron-stained and bioeroded micro-unconformities of the fabric (Massari, 1981, 1983).

Most condensed layers, especially those directly overlying unconformities, are very rich in Fe-Mn oncoids as well as in iron-stained bioclasts and intraclasts, which may be considered as lag deposits, implying repeated stages of exhumation, reworking and biomineralization. Mineralizations may appear as outer linings or irregular

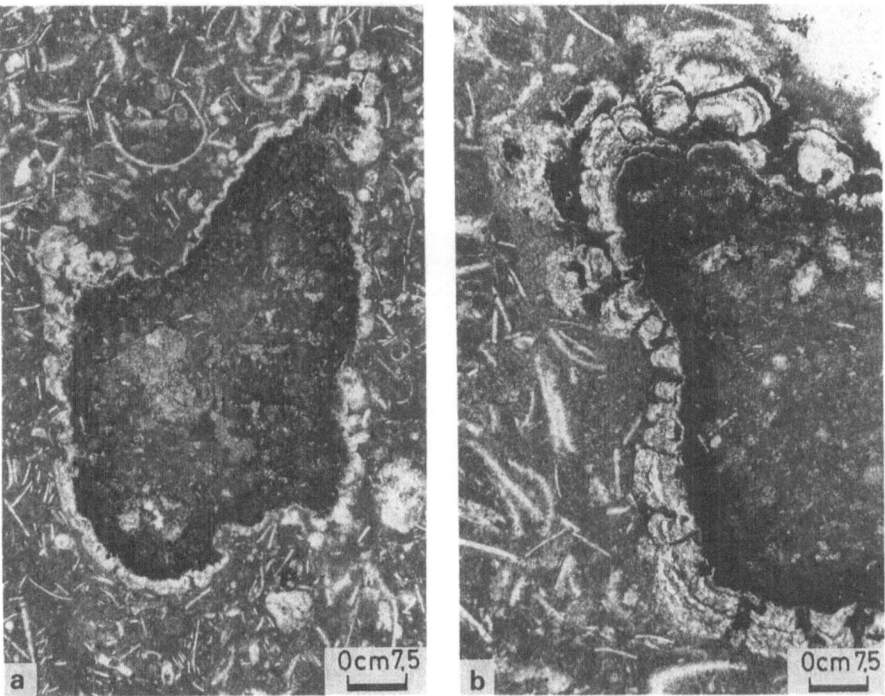

Figure 4: a) Iron-stained and bored intraclast; b) Iron-stained and bored intraclast coated by collomorphic banding concretions (microstromatolites) of carbonate and Fe-Mn oxides (Callovian), from Roccolo quarry near S. Ambrogio di Valpolicella (Verona).

collomorphic banding concretions coating intraclasts or bioclasts: they may also show impregnation of the microporosity, especially in the case of bioclasts such as echinoderm fragments (Figs. 4a,b). These structures correspond to those interpreted by Monty (1973) as microstromatolites and later recognized by Krumbein (1986) as the results of fungal activity in microbialic ecosystems.

Microprobe analysis indicated the presence of Fe-Mn minerals (probably oxides) and phosphate.

Tiny pyrite crystals (10-400 µm) appear irregularly scattered and occasionally concentrated into elongated clouds parallel to the laminae. Pyrite was identified on polished sections; in most cases, it is replaced by goethite and/or hematite which appear as pseudomorphs on the sulphide. The minute dissemination of goethite and hematite are responsible for the characteristic red colour of the rock.

RESULTS AND DISCUSSION

The discussion mainly focusses on the stromatolitic lithotypes, as we are essentially concerned with their genesis.

The pool of basic amino acids, in both free and combined data, from samples collected in the RAV is almost complete (Tables 1-2, 3-4 and Figs. 5,6). The amounts of amino acids were found to be higher in the more recent layers (sample LB25).

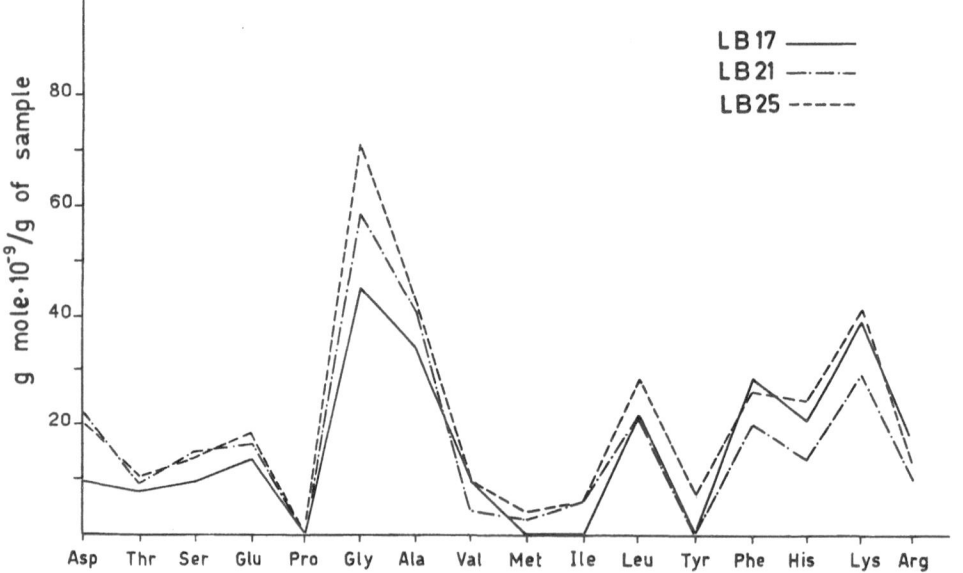

Figure 5 : Free amino acid distribution in stromatolitic samples of Rosso ammonitico Veronese.

A comparison with data from Chichereau and Trichet (1976) on the Oligo-Miocene Gannat Limestone (Central Massif, France) shows that our samples have higher contents in total amino acids. This feature depends only on their higher contents in

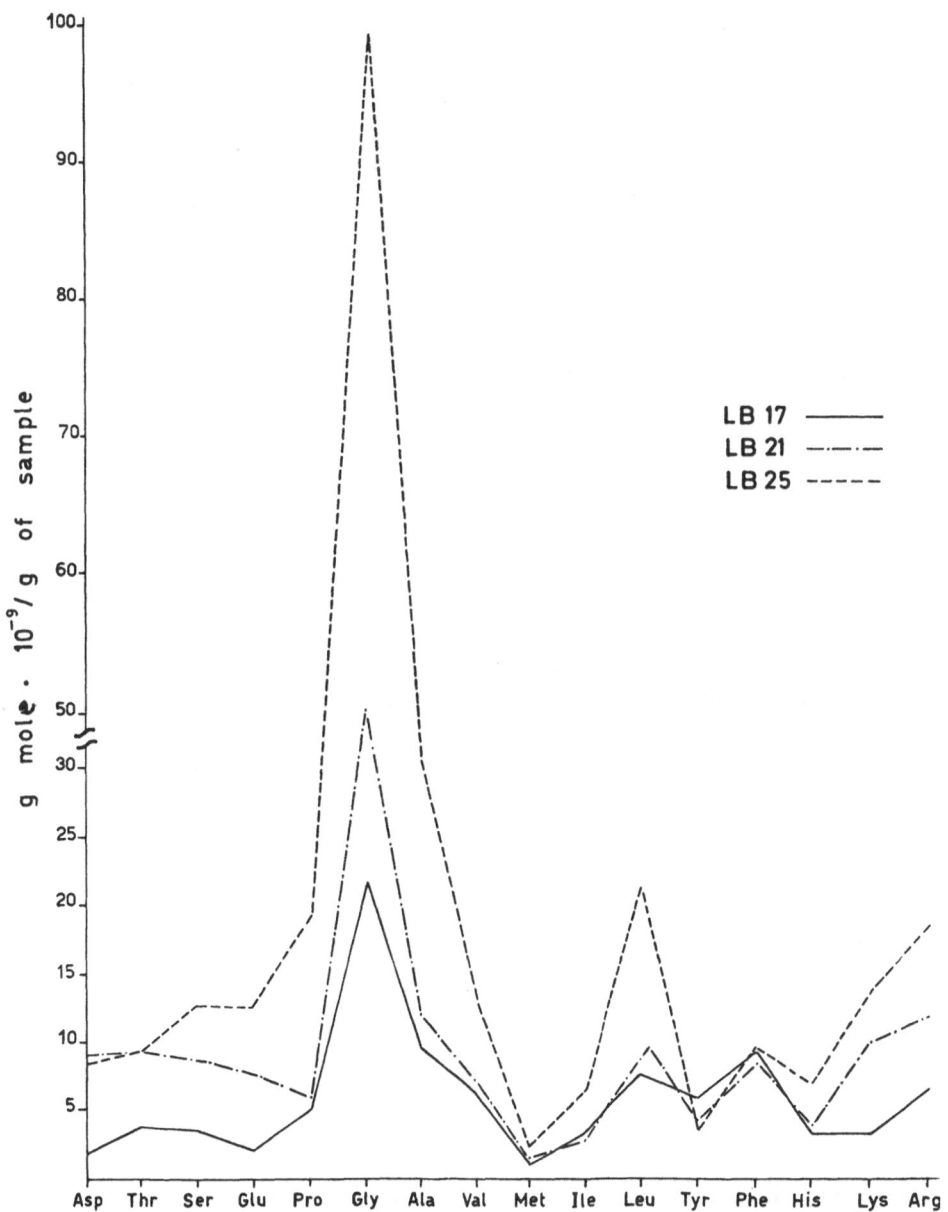

Figure 6 : Combined amino-acid distribution in stromatolites of Rosso Ammonitico Veronese.

Sample no.	Free aminoacids (µg/g)	Combined aminoacids (µg/g)	Total aminoacids (µg/g)
LB 17 (s)	33.50	11.43	44.93
LB 18 (l)	29.22	13.17	42.39
LB 19 (l)	34.37	29.60	63.97
LB 21 (s)	33.46	18.83	52.29
LB 25 (s)	42.01	31.68	73.69
MEAN	34.51	20.94	55.45

Table 1. Concentrations of free and combined amino acids in analysed samples. (s) = samples from stromatolitic layers; (l) samples from microcoquinoid layers.

combined amino acids, since the abundance of free amino acids is similar.

Bearing in mind the great difference in age between the two sedimentary formations (the RAV is 195 Ma old, the Gannat Limestone 22.5 Ma), the difference in the total amount of amino acids is probably related to a different sedimentary and diagenetic history. This may also be confirmed by the larger quantities of combined amino acids in the RAV, the diagenetic pattern of which was probably characterized by conditions favouring delayed transformation of combined amino acids.

SAMPLE	LB 17	LB 18	LB 19	LB 21	LB 25
Cysteic acid	3.58	2.76	3.32	3.04	4.96
Glucosamine	traces	traces	traces	--	traces
Aspartic acid (Asp)	9.74	13.88	17.47	22.28	20.38
Threonine (Thr)	7.82	5.40	9.30	9.21	10.69
Serine (Ser)	9.81	7.20	12.09	14.85	14.30
Muramic acid	--	traces	traces	traces	traces
Glutamic acid (Glu)	13.84	11.16	15.53	15.39	18.77
Proline (Pro)	traces	traces	traces	traces	traces
Glycine (Gly)	45.52	37.01	59.42	58.94	71.36
Alanine (Ala)	34.49	29.69	42.17	41.10	43.48
Valine (Val)	10.01	5.03	8.52	04.74	10.22
Methionine (Met)	--	3.05	--	2.95	4.04
Diaminopimelic acid	traces	traces	traces	traces	traces
Isoleucine (Ile)	traces	4.98	05.05	5.51	06.02
Leucine (Leu)	24.10	21.68	21.87	22.92	28.66
Tyrosine (Tyr)	--	29.88	4.38	--	7.18
Phenylalanine (Phe)	28.72	--	19.93	22.14	26.62
Histidine (His)	20.70	16.13	17.35	16.74	24.71
Lysine (Lys)	39.39	29.85	32.81	29.76	42.98
Arginine (Arg)	17.11	9.61	13.12	9.52	13.33
TOTAL	264.83	227.31	282.33	279.09	347.70

Table 2. Abundance of free amino acids (nmole/g of sample) LB17, LB21, LB25: stromatolitic layers; LB18, LB19: microcoquinoid layers.

As reported by Chichereau and Trichet (1976), basic amino acids are less stable than acid ones, so that the presence of the former in the RAV suggests environmental conditions unable to cause heavy hydrolysis. In particular, lysine is concentrated in free (37.37 µg/g) and combined (9.14 µg/g) amino acids. Instead, arginine (13.32 µg/g in free and 12.22 µg/g in combined amino acids) exhibits higher stability. Methionine occurs in very small quantities in both free (3.49 µg/g) and combined (1.56 µg/g) amino acids. The occurrence of lysine, histidine and arginine may indicate organic material derived from cephalopods (Weiner and Lowenstam, 1978).

Chichereau and Trichet (1976) did not detect free lysine, histidine or arginine, while histidine (4.58 µg/g) was found in only one of their samples in the combined amino acid fraction. Low contents of lysine were detected in all our samples. It is not easy to explain this result.

SAMPLE	LB17	LB18	LB19	LB21	LB25
Aspartic acid (Asp)	1.89	2.91	8.13	9.01	8.26
Threonine (Thr)	3.89	4.70	9.37	9.24	9.28
Serine (Ser)	3.55	6.37	11.93	8.78	12.77
Glutamic acid (Glu)	2.01	3.33	9.67	7.60	12.71
Proline (Pro)	4.94	14.22	16.21	5.96	19.04
Glycine (Gly)	21.85	25.71	90.89	52.51	99.33
Alanine (Ala)	9.49	11.83	21.92	11.92	30.19
Valine (Val)	6.17	5.64	12.12	7.24	13.38
Methionine (Met)	0.92	0.75	2.16	1.43	2.35
Isoleucine (Ile)	3.26	2.82	4.18	2.76	6.34
Leucine (Leu)	7.52	5.78	12.52	8.78	21.22
Tyrosine (Tyr)	5.70	3.74	6.80	4.26	3.41
Phenylalanine (Phe)	9.08	6.89	14.70	8.42	9.44
Histidine (His)	3.16	3.38	6.53	3.77	6.81
Lysine (Lys)	3.91	8.89	14.71	9.90	13.64
Arginine (Arg)	6.51	5.28	19.33	11.80	18.38
TOTAL	93.85	112.24	261.17	163.38	286.55

Table 3. Abundance of combined amino acids (nmole/g of sample). LB17, LB21, LB25: stromatolitic layers;LB18, LB19: microcoquinoid layers.

As far as acid amino acids are concerned, a gradual increase is observed from the oldest to the youngest RAV samples, in line with the expected diagenetic trend. The mean content of both aspartic and glutamic acids in the combined amino acids is not dissimilar from that reported by Chichereau and Trichet (6.90 µg/g). Instead, a difference is found in the derivates of these amino acids: the Gannat samples are more or less lacking in arginine and are very poor in proline, while the contents of the RAV samples reach 12 µg/g and 10 µg/g respectively.

Concerning free amino acids, a great difference in the glutamic acid concentration is noted (Gannat: 62.7 µg/g; RAV: 16 µg/g). The higher value reported for the Gannat samples may be explained by their younger age, implying that diagenetic transformation was less effective. This is in agreement with the fact that the RAV range shows quite a lot of arginine (a biosynthetic derivate of the glutamate family): in the RAV, transformation took place over a longer period of diagenesis. Glutamic acid, alanine and aspartic acid are present in the cell walls of cyanobacteria, in agreement with their contribution to the formation of the RAV.

The abundance of glycine and serine (Alpi et al., 1987) is also consistent with the occurrence of cyanobacteria in the RAV, and the notable amount of glycine (55 µg/g in free and 57 µg/g in combined amino acids, versus Gannat values of 47 and 14 µg/g respectively) suggests large-scale transformation from serine to glycine through the glycolate pathway. However, this transformation is less evident in RAV sample LB18.Glycine and serine values are lower in the Gannat Limestone. In this case, this seems to be due to the fact that it is younger and formed in more oxidizing conditions

SAMPLE	LB17	LB18	LB19	LB21	LB25
Cysteic acid	3.58	2.76	3.32	3.04	4.96
Glucosamine	traces	traces	traces	--	traces
Aspartic acid (Asp)	11.63	16.79	25.60	31.29	28.64
Threonine (Thr)	11.71	10.10	18.67	18.45	19.97
Serine (Ser)	13.36	13.57	24.02	23.63	27.07
Muramic acid	--	traces	traces	traces	traces
Glutamic acid (Glu)	15.85	14.49	25.20	22.99	31.48
Proline (Pro)	4.94	14.22	16.21	5.96	19.04
Glycine (Gly)	67.37	62.72	150.31	111.45	170.69
Alanine (Ala)	43.98	41.52	64.09	53.02	73.67
Valine (Val)	16.18	10.67	20.64	11.98	23.60
Methionine (Met)	0.92	3.80	2.16	4.38	6.39
Diaminopimelic acid	traces	traces	traces	traces	traces
Isoleucine (Ile)	3.26	7.80	9.23	8.27	12.36
Leucine (Leu)	31.62	27.46	34.39	31.70	49.88
Tyrosine (Tyr)	5.70	33.62	11.18	4.26	10.59
Phenylalanine (Phe)	37.80	6.89	34.63	30.56	36.06
Histidine (His)	23.86	19.51	23.88	20.51	31.52
Lysine (Lys)	43.30	38.74	47.52	39.66	56.62
Arginine (Arg)	23.62	14.89	32.45	21.32	31.61
TOTAL	358.68	339.55	543.50	442.47	634.25

Table 4. Abundance or total (free and combined) amino acids (nmole/g of sample).LB17, LB21, LB25: stromatolitic layers; LB18, LB19: microcoquinoid layers.

than the RAV (Gannat values: 2 µg/g serine in free and 1 µg/g in combined amino acids; RAV values: 13 and 17 µg/g).

Cyanobacterial involvement in the biomineralization of the RAV fits the occurrence of the aromatic amino acids tyrosine and phenylalanine, detected in the combined amino acids of all samples (tyrosine is lacking in the free amino acids of RAV samples LB17 and LB21). The Gannat samples yielded only very subordinate levels of tyrosine and phenylalanine. RAV sample LB18 lacks phenylalanine.

As tyrosine and phenylalanine are precursors of humic substances, they may in turn play an important role in calcium carbonate dissolution (Dell'Agnola, 1978; Stevenson, 1982). Furthermore, this suggests that they favour the growth of organisms producing organic acids (citric acid and polypeptides) which in turn are capable of dissolving carbonatic phases (Mitterer, 1972).

SOME CONCLUDING REMARKS

The seafloor on which the Callovian sequence of S. Ambrogio di Valpolicella was deposited must have been occupied both by a bacterial-mat, laminated, microbial ecosystem characterized by a cyanobacterial association responsible for the granular, laminated, centimetric fabric (as shown by the amino acid spectrum) and by bacterial and fungal colonies capable of producing the micrometric collomorphic banding concretions of Fe-Mn oxides round intraclasts and bioclasts more or less intensely pitted by boring organisms. That the Callovian was characterized by extensive microbial activity in response to environmental changes is shown by large-scale biomineral deposits, including specialized metal deposits (Krumbein, 1986).

The differences observed between the RAV and the Gannat Limestone probably reflect different environmental features: (a) the high content of combined amino acids in the RAV suggests the marked input of organic matter, probably linked to environmental conditons favourable to its preservation. This is in line with the persisting subtidal conditions of deposition of the RAV;

(b) the occurrence in the RAV of significant amounts of basic amino acids, which are known to be only fairly stable in sediments, supports the above hypothesis;

(c) on one hand, the occurrence of aromatic amino acids, precursors of humic substances, suggests that they may cause early precipitation of calcium carbonate (Suess and Futterer, 1972), leading *pro parte* to early cementation; on the other hand, they favour the proliferation of microorganisms whose metabolites (organic acids, polypeptides) may cause carbonate dissolution. It cannot be excluded that dissolution was a concomitant cause of the strong condensation of the RAV (Scudeler Baccelle and Nardi, 1991);

(d) the presence of disseminated iron sulphide (later mostly replaced by goethite and/or hematite) indicates that, at least temporarily, anoxic conditions occurred below the sediment/water interface, allowing bacterial reduction of sulphate. The disseminated state of the goethite and/or hematite pseudomorphs after pyrite may be attributed to both oxidizing currents and autotrophic bacterial oxidation of the sulphide: this would imply that iron was not transported far from the initial sulphide but was extensively redeposited close to the pyrite as a secondary oxidized mineral accumulation (Ehrlich, 1964).Nealson (1983, *fide* Krumbein, 1986) also reports possible bacterial and fungal genesis by goethite and hematite. Biological intervention in the formation of these minerals may also explain the horizontal chromatic varieties in the RAV as due to irregular organogenic activity, randomly distributed both on the depositional interface and in the first few decimetres of burial.

The above remarks imply alternating anaerobic and aerobic conditions below the sediment/water interface; in some way, this would be related to alternation of the soft-ground/hard-ground conditions characteristic of the RAV. In the soft-ground stage, the persistent exchange with oxygenated seawater exposed the organic compounds to oxic destruction; only local micro-environments related to bioturbated areas may have favoured local cementation. In the subsequent stage, the bottom was stabilized by cyanobacterial mats, and $CaCO_3$ precipitation (most probably by bacterial activity) near the sediment/water interface may have taken place. According to Berner (1971, 1984), this was best favoured by low sedimentation rates, active sulphate reduction and a high content of reactive organic matter in the sediment. Bacterial and fungal activity may also have been involved in the formation of Fe-Mn concretions. It should be recalled in this regard that marine Fe-Mn nodules and coatings contain a significant number of bacteria (Ehrlich, 1964) and that Monty (1973) described deep-sea Fe-Mn concretions as bacterial stromatolites.

The early cementation of the RAV structures is supported by: (a) lack of compactional features; (b) sporadic occurrence of encrusting organisms such as brachiopods, serpulid tubes and small ahermatypic corals; (c) lack of penetration of stromatolitic structures by burrowers existing in the overlying layers (Massari, 1983).

In conclusion, the stromatolites of the RAV are of both cyanobacterial and fungal origin, and grew in a submarine environment.

The organic matter released by these organisms, as well as that eventually derived from cephalopods, contributed to the formation of the characteristic features of the RAV: (a) primary binding process of particles; (b) early diagenetic mineralization (Défarge and Trichet, 1984); and (c) dissolution (Mitterer, 1972). The latter process, besides generating secondary cavities, contributed to produce repeated small-scale gaps in the RAV sedimentary sequence.

ACKNOWLEDGEMENTS

The authors would like to thank Prof. F.P. Sassi, Director of the "Centro di Studio per la Geodinamica Alpina", CNR, Prof. B. Zanettin, Director of the former "Centro di Studi per i Problemi dell'Orogeno delle Alpi Orientali " CNR; Prof. P. Omenetto, full professor of Mineral Ores of the University of Padova; and Prof. P. Frizzo. Special thanks are also due to Prof. G. Dell'Agnola of the Faculty of Agrarian Sciences of the University of Padova.Cordial thanks go to Prof. G. Calderoni, of the Department of Earth Sciences of the University of Rome. The authors' thanks also go to G. Mezzacasa, G. Facco, F. Todesco and Dr. C. Brogiato for graphs, analyses and thin-section photographs, and to P. Da Roit for help in microprobe analyses. This work was carried out with the support and financial contribution of MPI (60%) - L. Scudeler Baccelle.

REFERENCES

Alpi,A., Pupillo,P. and Rigano, C. (1987)."Fisiologia delle Piante",pp. 415. Soc. Editrice Scientifica, Naples.

Aubouin, J. (1964). "Reflexions sur les facies 'Ammonitico Rosso' ",Bull.Soc.Géol.France 6, 475-501.

Berner, R.A. (1971). "Bacterial processes affecting the precipitation of calcium carbonate in sediments". In: O.O.P. Bricker (ed.), The John Hopkins University - Studies in Geology. Carbonate cements 19, pp . 247-251.

Berner, R.A.: (1984). "Sedimentary pyrite formation: an update", Geochim: et Cosmochim: Acta 48 , 605-615.

Bernouilli D. and Jenkyns, H.C. (1974). "Alpine, Mediterranean and Central Atlantic Mesozoic facies in relation to the early evolution of the Tethys", in R.H. Dott, Jr. and R.H. Shaver (eds.) , Modern and Ancient Geosynclinal Sedimentation. Soc. Ec. Pal . Miner., Spec. Publ,19,pp.128-160.

Bertrand-Sarfati,J., Freitet,P. and Plaziat,J.C. (1966)."Les calcaires concretionnés de la limite Oligocène-Miocène des environs de Saint Pourçain-sur-Sioule (Limagne d'Allier): Rôle des algues dans leur édification; analogie avec les stromatolites et rapports avec la sédimentation", Bull. Soc. Géol. Fr., 7ième sér., VIII, 652-662.

Chichereau, L. and Trichet,J. (1976). "Amino-acid composition of algal (Gannat, Massif Central, France)", Chemical Geology 18, 39-48.

Clari, P.A., Marini,P., Pastorini, M. and Pavia, G. (1984). "Il Rosso Ammonitico Inferiore (Baiociano-Calloviano) nei Monti Lessini Settentrionali (Verona)", Riv. Ital. Paleont.Strat. 90/1, 15-86.

Défarge,C. and Trichet,J. (1984). "Microsctructure de sédiments d'origine cyanobacteriénne au sein desquels précipitent des carbonates de calcium. Application à la compréhension des mécanismes de la biominéralisation",C.R.Acad.Sc. Paris, 299, 711-717.

Degens, E.T. and Mopper, K. (1979) "Early diagenesis of sugars and amino-acids in sediments" , in G.Larsen and G.V. Chilingar (eds.), Developments in Sedimetology,25A, pp.143-205.

Dell'Agnola, G. (1978). " Chimica Agraria", CEDAM, Padova.

Ehrlich, H.L. (1964)."Microbial transformations of minerals", in H. Heukelekian and N.C. Dondero 8eds.), Principles and Applications in Aquatic Microbyology,3,pp. 43-60.

Frank, H., Le Fort, M. and Martin, H.hH. (1962). " Chemical analysis of mucopolymer component in cell walls of the blue-green alga "Phormidium uncinatum", Biochem. and Biophys: Res. Commun, 7/4, 322-325.

Golubic ,S. (1973) " The relationship between blue-geen algae and carbonate deposits", in N.G. Carr and B.A. Whitton (eds.), The biology of Blue-Green Algae, Blackwell Scientific Publications, Oxford, pp. 434-472.

Hollmann, R. (1964). "Subsolution-Fragmente", N.Jb. Geol: Palaont. Abh., 119/1,22-82.

Jenkyns, H.C. (1971). " The genesis of condensed sequences in the Tethyan Jurassic ", Lethaia, 4, 327-352.

Jenkyns, H.C. and Torres, H.S. (1969). " Palaeogeographic evolution of Jurassic seamounts in Western Sicily". Preprint. Coll. Med. Jurassic Stratigr., Budapest.

Krumbein, W.E. (1979) " Calcification by bacteria and algae", in A. Trudinger and D.J. Swaine (eds.), Biochemical Cycling of Mineral-Forming Elements. Studies in Environmental Sciences 3, Elsevier Pub. Comp., Amsterdam, Oxford, New York, pp. 47-68.

Krumbein, W.E. (1986). " Biotransfer of minerals by microbes and microbial mats", in B.S.C. Leadbeater and R. Riding (eds.), Biomineralization in lower plants and animals , Clarendon Press, Oxford, pp. 55-72.

Massari, F.(1979)." Oncoliti e stromatoliti pelagiche nel Rosso ammonitico Veneto", Mem. Scienze Geol. Padov, 32, 1-21.

Massari, F. (1981). " Crypalgal fabrics in the Rosso Ammonitico sequences of the Venetian Alps ", in A. Farinacci and S. Elmi (eds.), Rosso Ammonitico Symp. Proc., Tecnoscienza, Roma, pp. 435-469.

Massari, F. (1983) ." Oncoids and stromatolites in the Rosso Ammonitico sequences (Middle-Upper Jurassic) of the Venetian Alps, Italy, in T.M. Peryt (ed.) , Coated Grains, Springer Verlag, Berlin, pp. 358-366

Mitterer, R.M. (1972). 2 Biogeochemistry of aragonite mud and oolites", Geochim. et Cosmoscochim. Acta, 36, 1407-1422.

Monty, C.L.V. (1973)." Les nodules de manganèse sont des stromatolithes océaniques", C.R.Acad.Sci: Paris, ser. D, 276/25, 3285-3288.

Monty, C.L.V. (1976). " The origin and development of cryptalgal fabrics", in M.R. Walter (ed.), Developments in Sedimentology, 20, Stromatolites, Elsevier , Amsterdam, Oxford, New York, pp.193-249.

Monty,C.L.V. (1977). "Evolving concepts on the nature and the ecological significance of stromatolites: a review", in E. Flugel (ed.), Fossil Algae, pp.15-35.

Nealson, K.H. (1983). " The microbial iron cycle", in W.E. Krumbein (ed.), Microbial Geochemistry, Blackwell Scientific, Oxford, pp.159-190.

Scudeler Baccelle, L. and Nardi, S.(1991). " Interaction between calcium carbonate and organic matter: an example from the Rosso Ammonitico Veronese (Veneto, North Italy)", Chemical Geology,93,303-311.

Stevenson, F.J. (1982). "Humus chemistry: genesis, composition, reactions", John Wiley Sons, New York.

Sturani, C. (1964). "La successione delle faune ad ammoniti nelle formazioni mediogiurassiche delle Prealpi venete occidentali", Mem. Ist. Geol. Min. Univ. Padova, 24,1-63.

Sturani, C. (1969). " Intercalazioni di vulcaniti mediogiurassiche nel 'Rosso Ammonitico' dei Lessini veronesi". Boll. Soc. Geol. It.,88,589-601.

Sturani, C. (1971). "Ammonites and stratigraphy of the 'Posidonia alpina' beds of the Venetian Alps", Mem. Ist. Geol. Min. Univ., Padova, 28, 1-90.

Suess, E. and Futterer, D. (1972). " Aragonitic ooids: experimental precipitation from sea water in the presence of humic acid", Sedimentology,19,129-139.

Trichet, J. (1968). " Etude de la composition de la fraction organique des oolites: comparaison avec celle des membranes des bactèries et des cyanophycèes", C.R. Acad. Sc. Paris, 267, 129-139.

Walter , M.R. (1976). " Stromatolites. Developments in Sedimentology, 20, Elsevier Pub. Comp. Amsterdam, Oxford, New York.

Weiner ,S. and Lowenstam, H.A. (1978). " Well-preserved fossil mollusk shells: characterization of mild diagenetic processes", in P.E. Hare, T.C. Hoerihg and K. King Jr. (eds.), Biogeochemistry of Aminoacids, pp. 95-114.

Wendt, J. (1970). " Stratigraphishe Kondensation in Triadischen und Jiurassischen Cephalopodenkalken der Tethys ", N.Jb. Geol. Pal.Mh., 7, 433-448.

Winterer, E.L. and Bosellini,A. (1981). " Subsidence and sedimentation on the Jurassic passive continental margin, Southern Alps, Italy", Amer.Ass.Petrol. Geol. Bull.,65,394-421.

DEEP-MARINE MICROBIAL STRUCTURES IN THE UPPER JURASSIC OF WESTERN TETHYS

G. DROMART °, C. GAILLARD °° and L. F. JANSA °°°
° Ecole Normale Supérieure de Lyon, 46 allée d'Italie ; 69364, Lyon Cedex O7, France
°° Université Claude-Bernard Lyon-I, Centre des Sciences de la Terre; URA. C.N.R.S.
n° 11, Paléontologie Stratigraphique et Paléoécologie ; 43, bd du 11 novembre 1918,
69622 Villeurbanne Cedex , France
°°°Atlantic Geoscience Centre, Geological Survey of Canada, P.O. Box 1006 ;
Dartmouth, Nova Scotia, B2Y 4A2 , Canada

ABSTRACT

A number of Late Jurassic deep-water carbonates enclosing stromatolites-thrombolites are documented from both Central Atlantic and Mediterranean Tethys margins. Microbial structures have been identified at the base of the Upper Jurassic sequence in southestern France and off Nova Scotia, eastern Canada. Outer shelf and slope depositional environments for these microbial structures are documented from field and subsurface sedimentological - paleoecological studies.
Microbial structures are planar, columnar and columnar reticulate in shape, with mostly well developed internal dome-shaped lamination and occasional clotted, peloidal microfabric. Microbial crusts rest on diverse but always hard substrates and derive from rhythmical sediment binding. Some of them floor cavities. Absence of burrows in the stromatolites, presence of encrusting foraminifers, occurrence of debris of microbial crusts, formation of growth-framework cavities and development of dilational fractures all suggest early submarine lithification of the crusts.
In outer zones of paleoshelves, intensive and various microbial accretions form bioherms (stromatolite-thrombolite bioherms, sponge-stromatolite bioherms). In slope zones, low, planar stromatolites grew upon variably reworked substrates, and occur as scattered microbial forms and biostromes.
The deep-water stromatolites-thrombolites are preferentially located in areas with low sedimentation rate. Microbial communities thrived in environments which extended below storm wave-base to depths of several hundred meters, on unlit or poorly lit marine floors. The extent and form of microbial structures are believed to have been controlled by the rate of sediment accumulation and morphology-size-mobility of the substrates.

J. Bertrand-Sarfati and C. Monty (eds.), Phanerozoic Stromatolites II, 295–318.
© 1994 *Kluwer Academic Publishers.*

INTRODUCTION

Variably laminated carbonate encrustations and structures have been noted within the Upper Jurassic sediments of the Tethys continental margins. They occur at the paleo-outer shelf, shelf edge and slope depositional settings and are organically constructed. They have been described as crusts, stromatolites and thrombolites. Throughout this paper, the non-descriptive term microbial structure is used for these structures. Microbial structures occur in two main lithological forms : 1) laterally extensive, bedded carbonate deposits that share similarities in texture with Ammonitico Rosso lithofacies. This type has been documented in South-East France (Gaillard, 1983 ; Dromart, 1986 ; Dromart and Elmi, 1986 ; Dromart, 1989), South Poland (Szulczewski, 1968 ; Brochwicz-Lewinski et al., 1985), South Hungary (Radwanski and Szulczewski, 1966), North Italy (Massari, 1981 ; Ogg, 1981 ; Clari et al., 1984), West Sicily (Jenkyns, 1971), West Algeria (Dromart, unpublished data), South Spain (Comas et al., 1981), offshore Morocco (Jansa et al., 1984 ; Dromart, 1986), and Portugal (Elmi et al., 1988). 2) lens-, dome-, mound- shaped carbonate buildups (bioherms) with variable macrofaunal contents. The latter were recognized in the French Jura (Gaillard, 1983, 1984 a, b), Germany (Gwinner, 1976 ; Flügel and Steiger, 1981 ; Keupp et al., 1990 ; Meyer and Schmidt-Kaler, 1990), Poland (Laptas, 1974 ; Matyja, 1977 ; Matyzkiewicz, 1989), Romania (Barbulescu, 1972), offshore western Morocco (Steiger and Jansa, 1984 ; Lang and Steiger, 1984 ; Lang, 1989), Portugal (Ellis, 1984 ; Ramalho, 1988), Spain (Martin-Algarra, 1990 ; Pisera, 1991), offshore eastern Canada (Jansa et al., 1982, 1989 ; Ellis et al., 1985 ; Dromart, 1986) and U.S. Gulf Coast (Baria et al., 1982 ; Crevello and Harris, 1984). Some buildups are occasionally incorporated within Ammonitico Rosso lithofacies (e.g. in South France, Dromart 1986 and 1989, and in South Germany, Mathur, 1975). It appears that the majority of these microbial structures were preferentially located during the Jurassic near to the paleoshelf edges, and slopes adjacent to deep basins (Dromart and Elmi, 1986).

Deep-water microbial structures are poorly known in modern environments. Laminated crusts, associated with sponges, have been described from slopes of the Medditerranean Sea but were interpreted as formed by physico-chemical processes (Allouc, 1986, 1990). Most authors agree on a biological origin of such structures but without any clear evidence. Structures most closely resembling Jurassic stromatolites have been encountered in Jamaica at the toe of reefs at depths of a few hundred meters (Land and Goreau, 1970 ; Land and Moore, 1980). They have been more recently documented in the foreslope of the Red Sea (Brachert and Dullo, 1991 ; Brachert 1992).

Studies of the Upper Jurassic carbonate sequence in southeastern France (Jura and Ardèche areas) and off eastern Canada (Nova Scotia shelf) allow us to suggest processes of formation of the deep-marine microbial structures. This paper points to the variation of the microbial structures in relation to the depositional settings and tentatively estimates the influence of environmental parameters.

GEOLOGICAL AND PALEOGEOGRAPHICAL SETTINGS OF MICROBIAL OCCURRENCES IN SOUTHEASTERN FRANCE AND EASTERN CANADA

Western margin of the French Subalpine basin

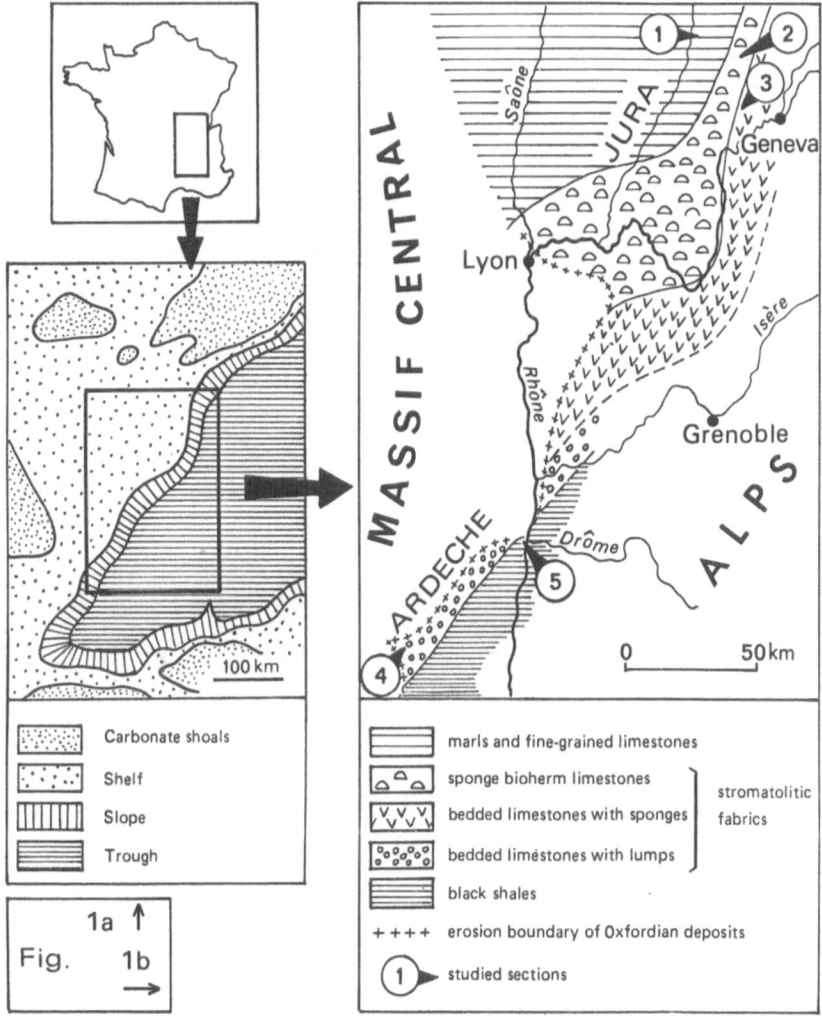

Figure 1 : a) Paleogeography of southeastern France during the Middle Oxfordian (after Enay *et al.*, 1980). b) Depositional zones and facies during the Middle Oxfordian in the South-Jura and Ardèche areas (after Debrand-Passard *et al.*, 1984).

The southeastern part of France was a segment of the continental margin that bordered the Neo-Tethys throughout Jurassic times. Sedimentation was subjected to block-faulting throughout Early Jurassic till the end of Bathonian (Rifting phase). Subsequent widespread drowning of the margin coincided with the presumed initiation of the sea-floor spreading in the nearby eastern Ligurian proto-ocean (Lemoine et al., 1986). This major change originated persistent fairly well-defined depositional zones (Fig. 1)

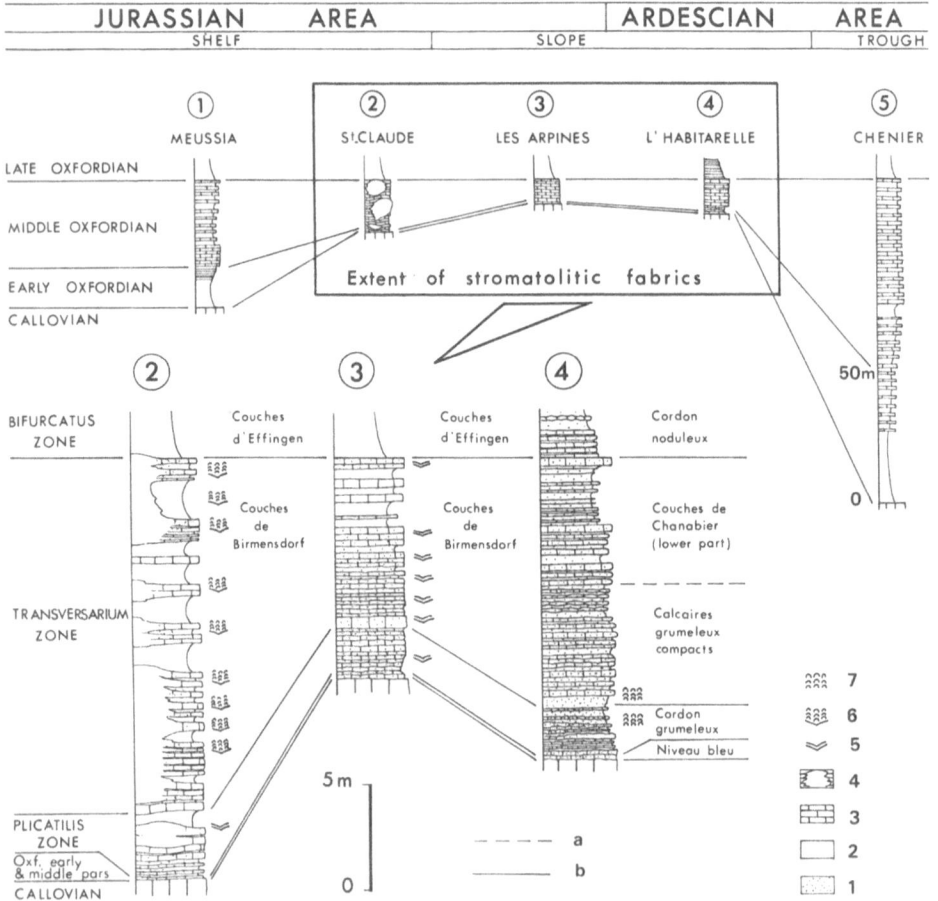

Figure 2 : Stratigraphic columns of Middle Oxfordian of selected outcrops across the Jura and Ardèche (see locations in Figure 1-b).

(Baudrimont and Dubois, 1977 ; Enay et al., 1980 ; Debrand-Passard et al., 1984). During Middle Oxfordian time, a large part of southeastern France was a deep trough in which dark-colored shales accumulated ("Terres Noires" Fm, Subalpine Basin). To the West, the basin had a gentle slope up to outer neritic environments. This wide ramp was marked by a thin sediment accumulation (Fig. 1a). Trough Late Oxfordian, the supply of fine-grained material, presumably sourced from adjacent carbonate platforms, caused the ramp to prograde basinwards.

Field studies in the Jura mountains and Ardèche area have revealed presence of diverse carbonate crusts in the Middle Oxfordian strata. Detailed stratigraphic investigations (Enay, 1966 ; Elmi, 1967) and recent environmental reconstructions (Gaillard, 1983 ; Dromart, 1989) were undertaken in the above-mentioned areas. They led to the recognition of two main depositional zones in the Middle Oxfordian of South-Jura (Fig. 1b). In the western part, a thick sequence of rhythmically bedded marls and biomicrites was deposited in a middle-outer shelf setting (Fig. 2-1). In contrast, a thin sediment accumulation occurred in the eastern part, towards the Subalpine basin, in the outer-shelf/slope zone (Fig. 2-2/3). Environmental conditions were favorable here for colonization by siliceous sponges and development of microbial crusts (Gaillard, 1983, 1984 a-b). At the paleoshelf edge, microbial accretions produced bioherms replete with sponge skeletons (Fig. 2-2). On the upper slope, the sponges were commonly reworked prior to burial with stromatolites being less developed. This resulted in a "lumpy" well-bedded deposits referred to as "couches grumeleuses" (Fig. 2-3).

In the Ardèche area, the Middle Oxfordian sequence is thin and composed of fine-grained limestones that enclose stromatolites along with subspherical, cm-sized, micritic nodules referred to as "lumps" (Dromart, 1986 ; Fig. 2-4). This condensed sequence changes abruptly eastwards into much thicker and more continuous lithologies composed of thin-bedded dusky shales and laminated, argillaceous limestones (Fig. 2-5). The change between lithofacies coincides with certain fault-lines (N-10° E) (Bourseau and Elmi 1980 ; Dromart and Elmi, 1985). The microbial structures developed here upon a gently sloping terrace of the slope, tens of kilometers wide, bounded basinwards by an active syndepositional normal fault.

A schematic cross-section, from the Jura outer shelf down to the Subalpine trough, to include the Ardèche area, is presented in Fig. 12.

Nova Scotia shelf

Jurassic rocks are not subaerially exposed along the eastern North American margin. Geological data are therefore derived from subsurface investigations related to an offshore exploration for hydrocarbons.

The presence of Mesozoic reefs rimming the continental margin is well established, particularly off the east coast of Nova Scotia, eastern Canada (Jansa, 1981). Several oil exploratory wells (Shell Demascota G-32, PetroCanada-Shell Penobscot L-30 and Chevron-Pex-Shell Acadia K-62) located on the Scotian Shelf (Fig. 3a) encountered deep-water microbial boundstones (thrombolite-stromatolite bioherms) within Jurassic

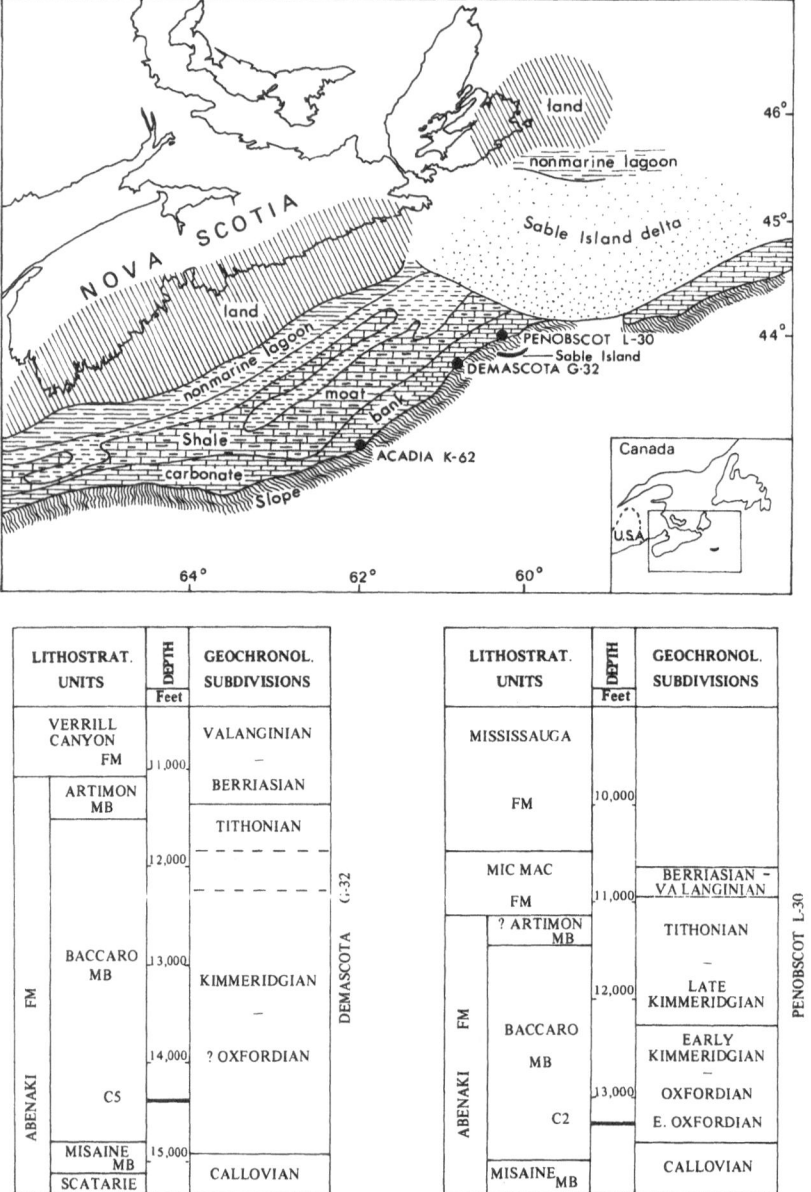

Figure 3 : a) Late Jurassic paleogeography offshore Nova Scotia, eastern Canada (after Eliuk, 1978). Also shown is location of wells near the shelf margin, in which deep-water microbial structures were found. b) Stratigraphic columns and position of cores in wells off Nova Scotia.

limestone sequences (Lower Baccaro Member, Abenaki Formation, Fig. 3b). The microbial boundstones have been described and interpreted as deep-water occurrences by Jansa et al. (1982), Pratt (1982), Ellis (1984), Ellis et al. (1985), Dromart (1986), Jansa et al. (1989) revising Eliuk's (1978, 1981) interpretation that they represent shallow lagoon deposits.

Seismic profiles across the Mesozoic continental edge show that the Upper Jurassic carbonate margin accreted vertically from an earlier, gently-basinwards dipping ramp to a steep geometry (Jansa, 1981). This steep and abrupt style of margin is discontinuous and alternates with more gently-dipping ramps areas. The carbonate deposition was restricted to the outer shelf. Synchronous deposits of the inner shelf were continental clastics (sands, variegated shales and minor coal beds) and the marginal clastics of the Mic-Mac Fm (shale plus intercalated sandstone and limestone beds). Calcareous shales of the Verril Canyon Fm accumulated in front of the carbonate banks of the shelf margin and in part include prodelta deposits of the Mic-Mac Delta System. Very generalized lithofacies distribution for the Late Jurassic off Nova Scotia is shown in Fig. 3a, modified after Eliuk (1978)

The well Demascota G-32 was located near the paleoshelf edge (see seismic lines in Eliuk, 1978, Fig. 12 or Jansa, 1981, Fig. 4). In this area, the carbonate margin developed an abrupt and steep geometry. The well passed through rocks successively older and more basinal in character, such that the Lower Baccaro Member appears from seismic data to be downslope of the earliest shelf barrier. Supporting evidence comes from the conventional core which shows the presence of coarse-grained material transported down-slope from a chaetetid-coral reef (Ellis, 1984 ; Dromart, 1986). Location of the Acadia well is similar to the Demascota G-32 (Fig. 2A in Jansa et al., 1989).

Penobscot L-30 was drilled in the Sable sub-basin, on the flank of a salt diapir (Fig. 6 in Ellis et al., 1985). In this area, seismic data suggest that a ramp-style margin developed (Fig. 2B in Jansa et al. 1989). Through the Lower Baccaro Member, the well penetrated an alternating sequence of intertonguing microbial boundstones and shales (Fig. 13 in Ellis et al., 1985 or Fig. 3 in Jansa et al., 1989). The argillaceous horizons contain quartz silts, plant debris and peloids along with planktonic fossil remains plus abundant and diversified foraminiferal assemblages reflecting deep neritic depositional environments (Ascoli, 1979). Stratigraphycally higher up, the sequence grades into shallower neritic limestones enclosing coral bioherms.

MACROSCOPIC/MICROSCOPIC FABRICS OF MICROBIAL STRUCTURES

Three main types of microbial crusts are distinguished on the basis of their shape and internal texture.

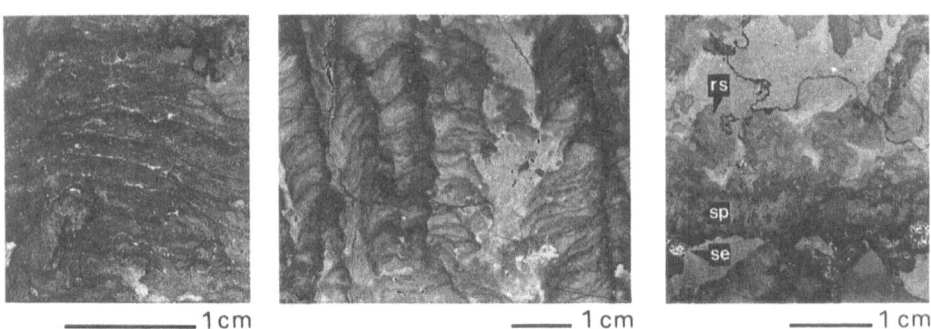

Figure 4 : Polished slabs of Middle Oxfordian stromatolites showing variations in their morphology. South-Jura, France. a) Planar stromatolitic structures ; b) Columnar stromatolites ; c) Columnar reticulate structures (rs : columnar reticulate stromatolitic structure ; sp : sponge ; se : serpulid).

Shape and macroscopic structures

Planar structures

They have low relief not exceeding several centimeters in height with an overall appearance of a flat to slightly undulose, dense crust with clearly visible internal lamination (Fig. 4a).

Columnar structures

The stromatolitic columns range from 5 to 25 mm in width and from 20 to 150 mm in

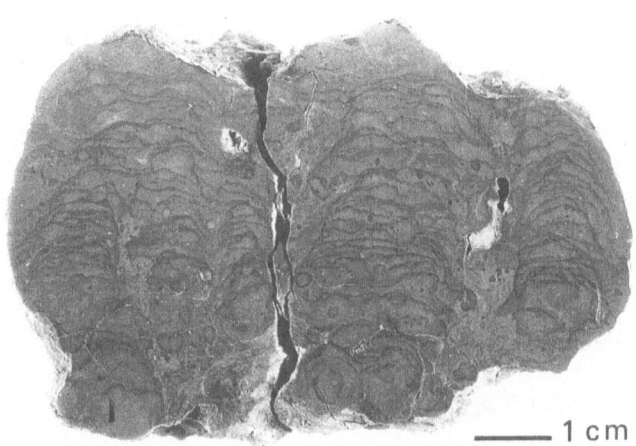

Figure 5 : Polished slab of columnar stromatolites initiated on a stabilized lumpy surface. "Cordon grumeleux", Mid. Oxfordian. Naves outcrop, Ardèche area.

height. The spacing between juxtaposed columns is from 1 to 10 mm. Internal lamination accentuates the dome-shaped structure (Fig. 4b, 5, 11b).

Columnar reticulate structures

They are digitate and branching in shape, with height of structures commonly in excess of 10 cm. These structures resemble a "bush" growth habit (Fig. 4c, 11a). The internal fabric is clotted, with clots of silt size, with an occasional poorly defined lamination. All intermediate forms, from laminated (stromatolites) to non-laminated ones, resembling thrombolites as described by Aitken (1967) and Kennard and James (1986), can be present.

These structures correspond respectively to the stromatolites described as "massifs" (= planar), "columnaires" (= columnar), and "nuageux" (= columnar reticulate) by Gaillard (1983) from the Oxfordian sponge facies of the Jura area.

Microscopic structures

The stromatolites are made up of alternating laminae of micrite with occasional occurrence of peloids and skeletal debris (Fig.6). The lamination is variably distinct,

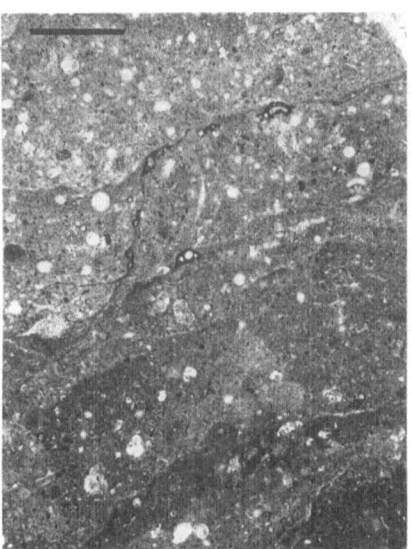

Figure 6 : Stromatolite texture, within stromatolites are protoglobigerinids, radiolarians. Sparse nubeculinellids rest upon the stromatolitic laminae. "Cordon grumeleux", thin sectionMid. Oxfordian. Naves outcrop, Ardèche. Scale bar= 1 mm.

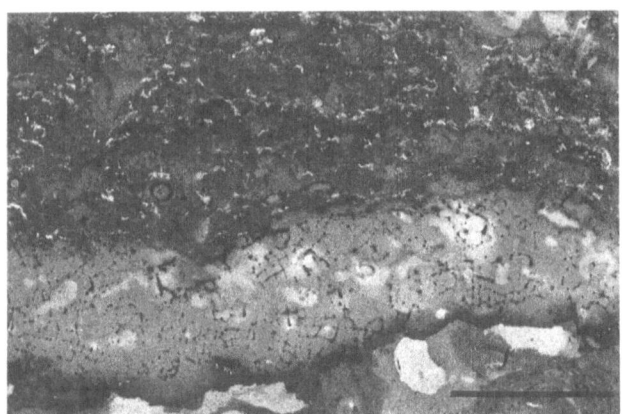

Figure 7 : Polished slab of stromatolite enclosing nubeculinellids, from a sponge bioherm of the Jura area. Stromatolite is developed on the upper surface of a sponge. Encrusting by nubeculinellids is evidence for early lithification of the stromatolite laminae. Scale bar = 1 cm.

locally poorly visible. The laminae are mm-sized in thickness. Each of them displays, from base to top, a gradational darkening that terminates sharply with a grey-black micritic film at the top (Fig. 4a/b, 5, 6). The laminae are laterally variable in thickness. Some of them are coalescent. Convex surfaces are mostly undulatory on a smaller scale. Serpulids and sessile foraminifers (nubeculinellids; Fig.6, 7) are on and in the crust.

In thin section, the individual stromatolitic laminae are micritic, but may have clotted, peloidal (Gaillard, 1983) microfabric. This microfabric consists of indistinct, sometimes spherical clots of micrite, 20 to 60 micrometers in size, embedded in a microsparitic calcite cement. Through each lamina, clots are separated by sparry calcite near its base, but higher up they become more closely spaced and merge near the top (Fig. 8, 11c). The clotted microfabric is extensively developed within thrombolites (Fig. 11e).

The development of a clotted microstructure (Monty, 1976) and the general geometry of the crusts suggest that they were organically formed, even though no preserved algal-bacterial filaments or cells were found within laminae through S.E.M. investigations.

DEVELOPMENT OF THE MICROBIAL STRUCTURES

Substrates

From field evidence the development of microbial crusts required presence of hard substrates. In the Jura, Oxfordian siliceous sponges (hexactinellids and lithistids), whose skeletons were composed of a dense network of spicules, provided suitable

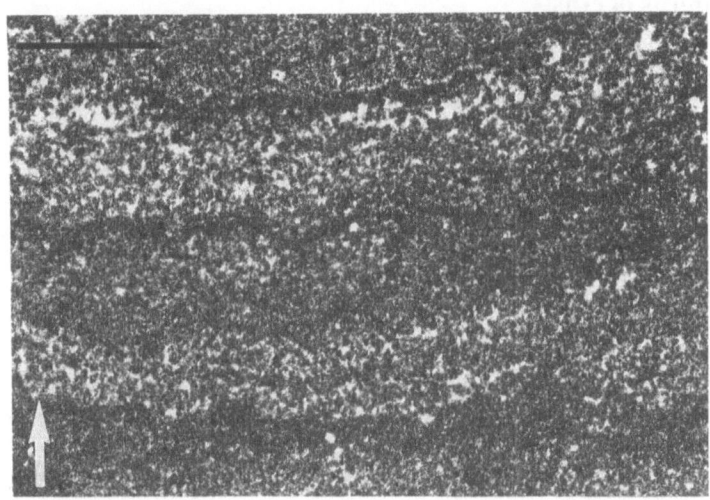

Figure 8 : Photomicrograph of a thin section showing clotted, peloidal microfabric of stromatolite from Jura area. Scale bar = 1 mm.

foundations (Fig. 7, 9a). The large size and dish-shape (15 cm in diameter on average, up to 100 cm) of the sponges favored development of the crusts. In the Ardèche, lime-mud intraclasts (referred to as "lumps") similarly served as substrate. However, their small size along with their round-shape made them less suitable for the development of crusts likeky because the lumps were so more prone to be reworked on depositional floors. Other diverse hard substrates were provided by annelid tubes (Fig. 11d), ammonite (Fig. 9b), bivalve, brachiopod shells and belemnite rostra. In shaly deposits of the "Schistes de Ramalhao" (Kimmeridgian in age) of Portugal, thrombolitic crusts were found to be restricted to limestone blocks of debris-flows (Ellis, 1984).

Figure 9 : Relations between stromatolitic crusts and substrates. a) stromatolite on siliceous sponges. Arrows show epizoans encrusting the downward-facing surface of the sponge (serpulids, bryozoans, thecideans...), Jura area ; b) stromatolite developed on the surface of ammonite shell, Ardèche area ; c) oncolites formed by encrusting of lumps, Ardèche area.

Growth patterns of crusts

The presence of encrusting organisms within the lamination proves that the growth of crusts was rhythmic with phases of microbial accretion and encrustation by protozoans and tiny invertebrates.

Crust development was always upwards. Therefore, mobile substrates (usually of a smaller size) were irregularly wrapped by the crusts, which produced features similar to oncolites (Fig. 9c) (Dromart, 1986). These spherical forms resemble those described by Massari (1983).

Lithification

Several observations demonstrate that the crusts were lithified early :

-- the crusts enclose encrusting organisms such as foraminifers (*Nubeculinella*,...) and serpulids which required a firm substrate (Fig. 6, 7) ;

-- within columnar reticulate forms (e.g. in thrombolite-stromatolite bioherms), a younger thrombolite frequently rests upon older stromatolitic knobs (Fig. 11b) ;

-- in thrombolite-stromatolite bioherms, a number of growth-framework cavities developed between coalescent microbial crusts. Subsequent geopetal filling of the cavities by internal sediment resulted in formation of Stromatactis-like cavities sheltered by microbial forms (Fig. 11b-e) ;

-- near the sponge bioherms, (Fig. 10a) the adjacent marls and biomicrites enclose fragments of stromatolitic crusts and sponges, derived from the bioherms, and referred to as tuberoids (Fritz, 1958 ; Gaillard, 1983). These are rounded in shape and their size is not in excess of 5 mm (Fig. 10b). The tuberoids are frequently rimmed by nubecullids which testifies to the early calcification of the fragments (Fig. 10c) ;

-- minor dilational fractures, restricted to the microbial crusts are common within both carbonate mud-mounds in Ardèche and thrombolite-stromatolite bioherms off Nova

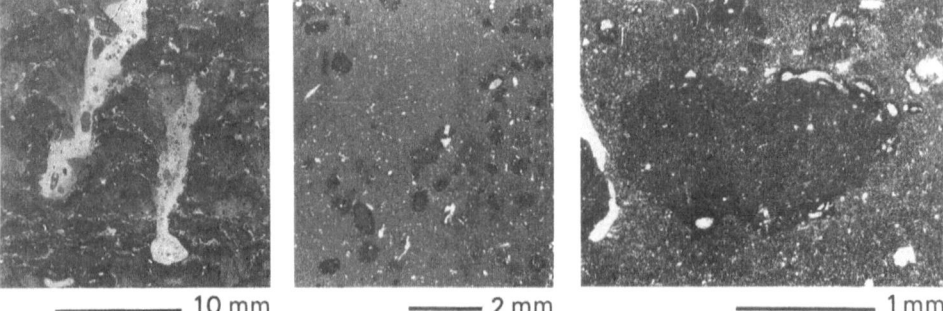

10 mm   2 mm   1 mm

Figure 10 : Photomicrographs of tuberoids that are interpreted as fragments of microbial crusts in the proximity to the Oxfordian sponge bioherms, Jura area. a. partly disrupted columnar stromatolite and tuberoids ; b) tuberoidic facies ; c) tuberoid with surface encrusted by nubeculinellids.

Scotia. Supporting evidence for an early formation of the fractures is that the fractures are either geopetally filled with mud, possible endostromatolites, and microsparitic calcite (Fig. 21 in Dromart, 1986) or lined with fibrous calcite cement (Fig. 4, 9 in Jansa et al., 1989) ;

-- rare borings into the stromatolites and thrombolites suggest that the crusts were firm early but not always calcified before burial.

The rapid lithification of the microbial crusts is poorly understood but we speculate that the microbial activity resulted in chemical precipitation of micrite. Because of this activity, microorganisms could construct a self-supported frame, which resulted in formation of almost exclusively microbial-accreted carbonate buildup (e.g. thrombolite-stromatolite bioherms).

VARIABILITY OF MICROBIALITES IN SHELF AND SLOPE DEPOSITIONAL ENVIRONMENTS

Outer Shelf buildups

Thrombolite-stromatolite bioherms

The bioherms of the Scotian margin developed in an outer shelf setting (Fig. 3). Those of Demascota G-32 are located close to the shelf edge where they are interbedded with fore-reef coral-stromatoporoid deposits. The microbial bioherms of the Penobscot L-30 periodically accreted as low relief buildups on a ramp, in the deep neritic environment, where fine-grained carbonates alternated with calcareous shales.

The mechanical logs together with cores and cuttings indicate that the vertical extent of the buildups in the latter area is from 3 up to 20 meters and possibly more. Even though the geometry is unknown as they are too small to be resolved on seismic sections, it is probable that both these microbial buildups have a shape similar to low-relief mound.

The bioherms are composed of anastomosing, digitate, cm-sized thrombolites (columnar reticulate structures, Fig. 11a) which locally grade into stromatolitic laminae with the appearance of hemispheroidal lamination, especially at the top of individual knobs (Fig. 11b). Single columns and planar crusts are subordinate. The volume of the boundstone made up of dark grey-colored microbial framework is 40 % on average and up to 80 %. Microbial colonies commonly enclose agglutinated ?annelid tubes (*Thartharella*, Fig. 11d), *Tubiphytes*-like tubes (Fig.11e), nubeculinellids, foraminifers and ostracode tests and rare sponge fragments. The microbial framework is enveloped by light-grey skeletal peloidal wackestone.

Sponge-stromatolite bioherms

A different type of bioherm constructed by sponges and stromatolites occurs at a similar paleogeographical setting at the Jura paleoshelf edge (Fig. 12-A). The bioherms are well

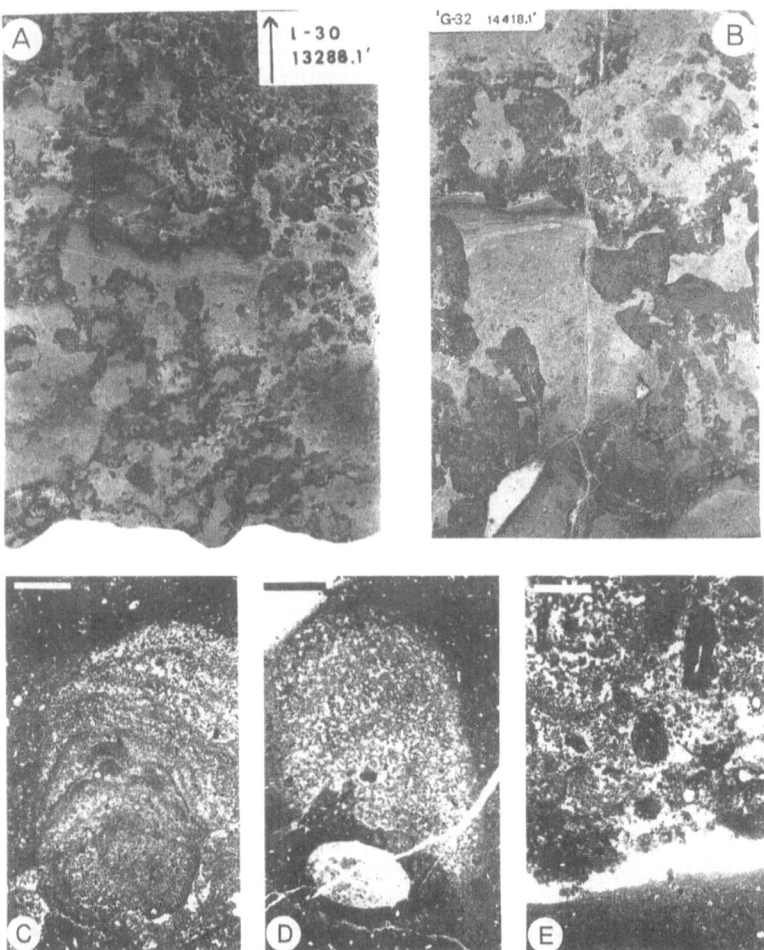

Figure 11 : a) Polished core slab of columnar reticulate thrombolite framestone from the Penobscot L-30 well, Core 2. Thrombolite (dark-grey) with geopetally filled voids enveloped by wackestone and sheltering thin linings of fibrous calcite cement. Arrow is 1 cm high. b) Polished core slab of stromatolitic knobs and anastomosing thrombolites surrounded by a light-grey lime mudstone and intraclastic wackestone, with laminated geopetal internal sediment filling growth-framework cavity in the upper left. Demascota G-32 well, Core 5. Core is 6,5 cm in width. c) Photomicrograph of a thin section of a stromatolite showing laminae of micrite, spongiform clots and peloids with spar-filled interparticle and minor fenestral pores (clotted microfabric). Demascota G-32 well, core 5 (14,417 ft). Scale=1 mm. d) Photomicrograph of a thin section of thrombolite knob initiated upon ? annelid tube (*Thartharella*). Penobscot L-30 well, core 2 (13,293.7 ft). Scale is 1 mm long. e) photomicrograph of a thin section of a thrombolite patch that shows clotted microfabric, enclosed *Tubiphytes* and cavity filled with thin lining of calcite cement underlain by silty lime mud. Penobscot L-30 well (13,286,4 ft). Scale is 1 mm long.

developed in the so-called "Couches de Birmensdorf", Middle Oxfordian in age (Fig. 2-2, 13). The bioherms are lens-, dome- and mushroom- shaped. Their vertical and lateral extent is from 1 m to 10 m. The bulk of the buildups consists of siliceous sponges encrusted and bound by stromatolites (bindstone type, Embry and Klovan, 1967) withina matrix of skeletal micrite and peloids. Volumetrically, sponges are minor component of the bioherms and in average form 22 % of them. Skeletons served as substrates for formation of the microbial crusts which compose 50 % of the rock and appear to have been the most active framebuilder (Fig. 14). These microbial crusts extensively capped the upper sides of sponges and grew up to 20 cm in height. Isolated crusts commonly coalesced upwards, which resulted in formation of rigid structure.
Crusts are of variable shape, but columnar reticulate forms predominate. Nubeculinellid foraminifer communities enclosed by the crusts also contributed to the bioherm building (Fig. 7). Lower sides of sponges are devoid of microbial crusts which are replaced by a rich epifauna composed of annelid worms (*Serpula*, *Terebella*), bryozoans (*Plagioeca*, *Ceriocava*, *Stomatopora*, *Radicipora*...), thecidean brachiopods (*Rioultina*), small sponges (*Neuropora*) and rare bivalves (Fig. 9a).

Slope deposits

Upper slope : sponge-rich deposits

They are developed in the Jura area (Fig. 12-B) and also recognized in the Ardèche area. The stromatolitic crusts are isolated, and poorly developed, being not more than a few millimeters in height. The crusts are of the planar-type only. Associated nubeculinellid foraminifers are scarce. Encrusters on the lower sides of sponges are much less abundant than in the shelf environment, and are mostly restricted to serpulids.

Lower slope : cephalopod-rich deposits

A large variety of microbially constructed carbonates occur in the lower slope paleogeographic setting of the Ardèche area (Fig. 12-C/D). However, we have to stress that "lower slope" is here understood to represent a zone probably only a few hundred meters deep. In this environment there occur oncolites, cm-sized, composed of a micritic core entirely enclosed by a few laminae, loosely and irregularly arranged. The associated stromatolites are knobby, columnar in shape with well-developed internal dome-shaped laminations (Fig. 5). As a base for columns are ammonite shells, belemnite rostra, oncolites and hard-grounds. Skeletal particles of plankton are enclosed by the stromatolites (Fig. 6). Sessile foraminifers attached on the laminae are much more scattered. Within the lithological unit called "Cordon grumeleux" in the Ardèche area, marls contain oncolites and isolated knobs of stromatolite. In contrast, some limestone horizons consist of 10 cm-high biostromes constructed by regularly spaced columnar stromatolites. Individual stromatolite-bearing limestone beds locally merge, forming small carbonate buildups (Fig. 12-D, this text ; Fig. 4D in Dromart, 1989). The buildups are lenticular in shape, no more than 50 cm in height with a smooth hard top

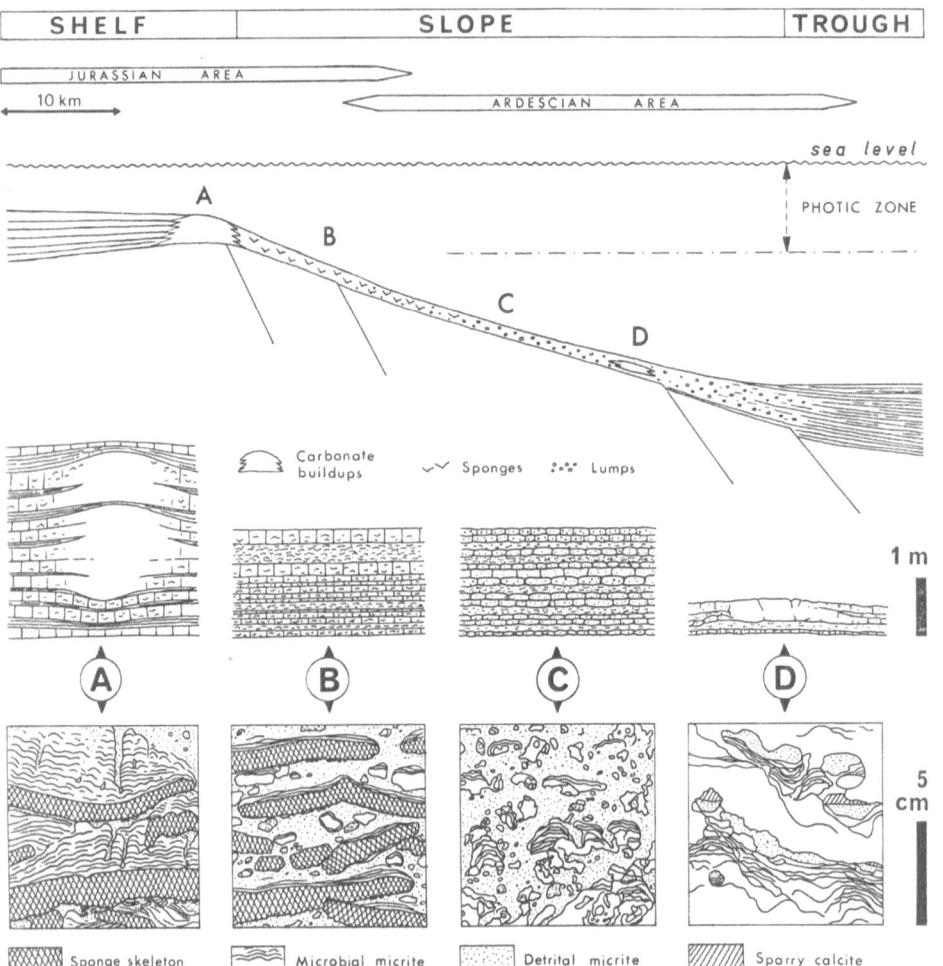

Figure 12 : Idealized composite cross-section from the Jura outer shelf down to the Subalpine trough showing variation of the microbial structures. Middle Oxfordian, southeastern France.

Figure 13 : Outcrop of fine-grained bedded limestones enclosing sponge bioherms. "Couches de Birmensdorf" Fm, Mid-Oxfordian. Chatillon-de-Cornelle outcrop, Jura area. Scale bar=2 m.

Figure 14 : Polished slab showing internal structure of sponge-bioherm with siliceous sponges encrusted by columnar stromatolites. "Couches de Birmensdorf" Fm, Mid-Oxfordian. St Claude-Le Pontet outcrop, Jura area. Scale bar= 5 cm.

surface. Borings into the top of the buildups are scarce. The buildups lack any metazoan framebuilders and are dominantly composed of lime mudstones and wackestones bound by planar stromatolitic crusts.

FACTORS CONTROLLING MICROBIAL GROWTH STRUCTURE

Rate of sediment accumulation

All the studied occurrences of microbial structures show evidence of early lithification and reworking, and are associated with hard-grounds, all characteristics of low sediment accumulation. In contrast, microbial crusts are missing in areas with a high depositional rate such as found in the Jura inner-middle shelf during the Oxfordian, in the Subalpine basin and in the Sable sub-basin off Nova Scotia.

The low sediment input on the slopes permitted colonization of wide areas by the microbial structures because of the presence of a number of firm substrates. Biostromes formed in these areas. The vertical accretion of the microbial structures was limited. The planar and low-columnar stromatolites were dominant here and oncolites are also present.

The presence of bioherms (stromatolite-thrombolite bioherms, sponge-stromatolite bioherms) in outer-shelf zones represents the peak of microbial development similarly associated with a low sedimentation regime. The construction of bioherms here could be the response of microbial communities to a slightly higher sedimentation regime. Vertical accretion of individual microbial structures prevailed within the bioherms, resulting in mainly columnar reticulate forms.

The ideal sequence planar-columnar-columnar reticulate structures reflects the increase of the sedimentation rate as suggested by Gaillard (1983).

Stability of depositional surfaces

Microbial crusts are preferentially developed over stable substrates. Along the French Subalpine basin, such substrate were provided by large-sized sponges which are preserved in growth position. Within bioherms of the Jura shelf, the sponges offered the most suitable conditions of stability. In contrast, on the Jura slope the same sponges were commonly reworked, disrupting the microbial growth.

Within cephalopod-rich deposits of the Ardèche area, the small size of hard substrates and their instability resulted in formation of oncolites and forms with poor vertical development. However, the presence of a few vertical, decimetric, sequences of oncolite - columnar stromatolite - planar stromatolites (Fig. 15) suggests that, from time to time, the sea-bottom was stabilized by the microbial mats. Similar sequences of oncolitic-stromatolitic fabric have been recognized on the Trento Plateau by Massari (1981). These were interpreted as reflecting both decrease of turnover by burrowers and decrease of water turbulence.

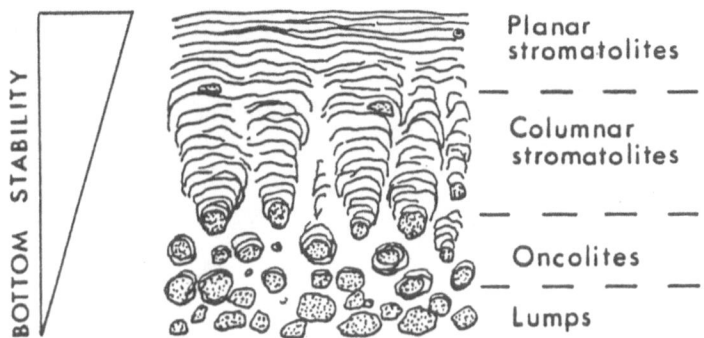

Figure 15 : Ideal sequence (isolated lumps - oncolites - columnar stromatolites - planar stromatolites) reflecting a progressive stabilization of the depositional surface.

Therefore, the stability of the depositional surfaces can be interpreted from the examination of the microbial structures. The stability depended on size of hard foundations and intensity of reworking processes (sea-bottom energy conditions, bioturbation). A diagram (Fig. 16) shows schematic relations between microbial structures, bottom stability, and rate of sediment accumulation.

Light intensity

Sponge-stromatolite biohermal buildups developed on the Jura paleoshelf in depths of around 100 meters, within lower photic zone (estimation extensively discussed in Gaillard, 1983) ("circalittoral" environment ; Peres and Picard, 1959). The upward-facing surfaces of the substrates (sponges) are extensively encrusted by the stromatolites whereas the downward-facing surfaces are colonized by sciaphilic encrusters. Buildups of the Scotian shelf have been interpreted by most authors (Jansa et al., 1982 ; Ellis et al., 1985 ; Dromart, 1986 ; Jansa et al., 1989) to have formed in the depth range 40-100 m or possibly more, based on sedimentologic plus seismic data. The same interval has been suggested by Ellis (1984) for the similar thrombolite-*Thartharella* bioherms found in the Upper Jurassic of Portugal.

In contrast, the Subalpine basin was a deep trough, possibly over 1 000 meters deep (Curnelle and Dubois, 1986). The sea-bottom was below the photic zone across the Ardèche slope with the depth estimated about 400 m, based on distinct lines of evidence such as biotic associations, sedimentologic structures and regional reconstructions (Dromart, 1989). Steiger and Jansa (1984) described Upper Jurassic microbial crusts from the the lower slope of the Mazagan carbonate platform off western Morocco, where they colonized surface of limestone debris-flows in a water depth of 3 000 m.

From these examples it appears that microbial crusts could develop in both photic and non-photic areas.

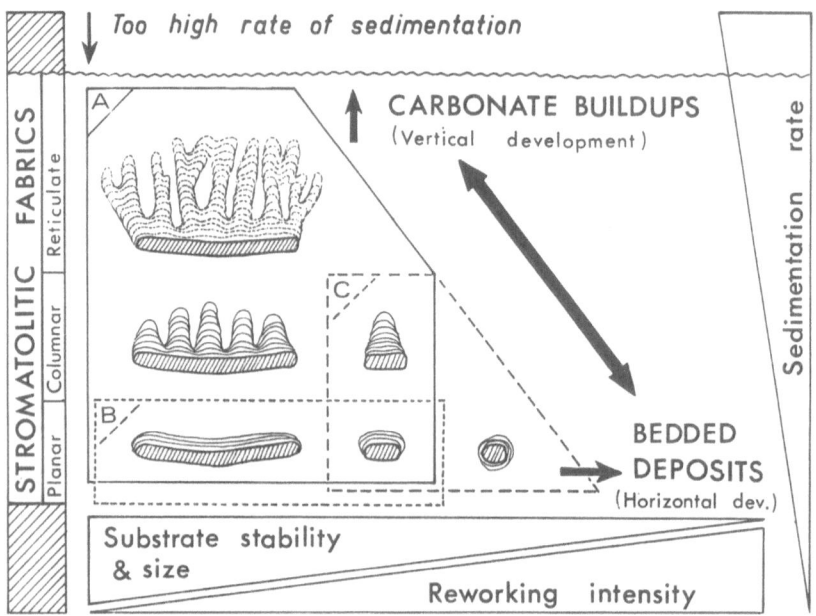

Figure 16 : Scheme of morphological variations of stromatolitic crusts *versus* sedimentation rate, substrate size and stability. A corresponds to outer-shelf buildups, B to upper-slope spongerich deposits and C to lower-slope cephalopod-rich deposits (located in Fig. 3).

This clearly indicates that microbial structures were constructed by heterotrophic (and/or nonphoto-autotrophic) microorganisms in agreement with Monty (1971, 1977) who suggested that microorganisms capable of building stromatolitic structure in the non-photic depositional zones are either bacteria or cyanobacteria.

CONCLUSIONS

Studies of microbial structures across a paleoprofile from the outer shelf zone down to the bathyal zone have disclosed that :

1 - these thrombolite and stromatolite-like structures are not diagenetical fabrics, but organically formed ;

2 - these microbial structures are formed by heterotrophic and/or autotrophic microorganisms (most probably by bacteria) ;

3 - the laminar fabric of the crusts is the result of a rhythmic microbial growth ;

4 - rapid calcification of microbial structures is supported by encrustations by sessile organisms, requiring a firm substrate (nubeculinellids...), incorporated within the growth structures ;

5 - favourable conditions for development of microbial structures are provided by low sedimentation regime ;
6 - initiation of microbial structures requires presence of hard substrates ;
7 - shape and internal texture of microbial structures are strongly influenced by morphology, size, and stability of the substrates and deposition rate.

ACKNOWLEDGMENTS

G. Dromart thanks the World University Service of Canada and the Eastern Petroleum Geology Subdivision (Geological Survey of Canada) for support of his stay at the Atlantic Geoscience Centre in Dartmouth, Nova Scotia. Thanks are extented to the B.R.G.M. (France) for financial assistance for field work in the Ardèche area, incorporated into the French Continental Drilling Project ("Géologie profonde de la France - projet Ardèche"). G. Dromart and L. Jansa thank the C.O.G.L.A. (Canada) for permission to sample cores from oil exploratory wells and the "Centre Jacques Cartier" (Lyon) for a financial contribution which facilitated their cooperative work. C. Gaillard acknowledges financial and technical support from the National Science Research Council of France (C.N.R.S., U.R.A. 11). S. Elmi (Univ. of Lyon) and J. Noble (Univ. of New Brunswick) provided constructive comments on the manuscript. Considerable help was received in photography from N. Podevigne (C.S.T., Univ. of Lyon).

REFERENCES

Aitken, J.D. (1967) "Classification and environmental significance of cryptalgal limestones and dolomites, with illustrations from the Cambrian and Ordovician of southwestern Alberta", Journ. Sedim. Petrol. 37, 1163-1178.

Allouc, J. (1986) "Les encroûtements sous-marins de Méditerranée orientale : une explication génétique", Rev. Inst. Franç. Pétr. 41, 351-359.

Allouc, J. (1990) "Quaternary crusts on slopes of the Mediterranean Sea : A tentative explanation for their genesis", Marine Geol. 94, 205-238.

Ascoli, P. (1979) "Report on the Biostratigraphy (Foraminifera and Ostracoda) and Depositional Environments of the Conventional Core 2 (13285-13315) of the PetroCanada - Shell Penobscot L-30 well, Scotian Shelf", Unpub. Rep., Geol. Surv. Can. EPG S-PAL, Dartmouth, 42-79.

Barbulescu, A. (1972) "Asupra "recifilor" neojurasici din Dobrogea centrala", St. cerc. geol. geof. geor. 17, 93-108.

Baria, L.R., Stoudt, D.L., Harris, P.M. and Crevello, P.D. (1982) "Upper Jurassic reefs of the Smackover Formation, U.S. Gulf Coast", Amer. Assoc. Petrol. Geol. Bull. 66, 1449-1482.

Baudrimont, A.F and Dubois, P. (1977) "Un bassin mésogéen du domaine péri-alpin : Sud-Est de la France", Bull. Centre Prod. Elf-Aquitaine 1, 261-308.

Bourseau, J.P. and Elmi, S. (1980) "Le passage des faciès de bordure ("calcaires grumeleux") aux faciès de Bassin dans l'Oxfordien de la bordure vivaro-cévenole du Massif central français", Bull. Soc. géol. France (7), XXII, 5, 607-611.

Brachert, T. (1992) "Late Jurassic sponge buildups ; environmental interpretation by comparison with microfabrics of modern hardgrounds", Eclogae Geol. Helv. 85, 45-58.

Brachert, T.C. and Dullo, W.C. (1991) "Laminar micrite crusts and associated foreslope processes, Red Sea", Journ. Sedim. Petrol. 61, 354-363.

Brochwicz-Lewinski, W., Gasiewicz, A., Strzelecki, R., Suffczyski, S., Szatkowski, K., Tarkowski, R. and Zbik, M. (1985) "Anomalia geochemiczna na pograniczu jury srodkowej i gornej Poludniowej Polsce", Przeglad Geol. 32, 647-650.

Clari, P.A., Marini, P., Pastorini, G. and Pavia, G. (1984) "Rosso Ammonitico Inferiore (Bajociano - Calloviano) nei Monti Lessini settentrionali (Verona)", Riv. It. Paleont. Strat. 90, 15-86.

Comas, M.C., Oloriz, F. and Tavera, J.M. (1981) "The red nodular limestones (Ammonitico Rosso) and associated facies : a key for settling slopes or swell areas in the subbetic Upper Jurassic submarine topography (Southern Spain)",, in Farinacci A. and Elmi S. (eds) Proc. Rosso Ammonitico Symp. Tecnoscienza, Roma, pp. 113-136.

Crevello, P.D. and Harris, P.M. (1984) "Depositional models for Jurassic reefal buildups, in The Jurassic of the Gulf Rim ; Gulf Coast Section", Soc. Econ. Paleont. Mineral., third Annual Research Conference Proceedings, pp. 57-102.

Curnelle, R. and Dubois, P. (1986) "Evolution mésozoïque des grands bassins sédimentaires français ; bassin de Paris, d'Aquitaine et du Sud-Est", Bull. Soc. géol. France 8, II, 5, 529-546.

Debrand-Passard, S., Courbouleix, S. and Lienhard, M.J . (1984) Synthèse géologique du Sud-Est de la France. Mém. Bur. Rech. Géol. Min. 125.

Dromart, G. (1986) Faciès grumeleux, noduleux et cryptalgaires des marges jurassiques de la Téthys nord-occidentale et de l'Atlantique central : genèse, paléoenvironnements et géodynamique associée", Unpub. thesis, University of Lyon-I, France.

Dromart, G. (1989) "Deposition of Upper Jurassic fine-grained limestones in the Western Subalpine Basin", Palaeogeography Palaeoclimatology Palaeoecology 69, 23-43.

Dromart, G. and Elmi, S. (1985) "Analyse sédimentaire de la séquence grumeleuse inférieure et de ses équivalents latéraux : modalités du passage au bassin des "Terres noires" (Oxfordien)", Doc. Bur. Rech. Géol. Min. 95-11, 91-98.

Dromart, G. and Elmi, S. (1986) "Développement de structures cryptalgaires en domaine pélagique au cours de l'ouverture des bassins jurassiques (Atlantique central, Téthys occidentale)", C. R. Acad. Sci. Paris 303, 2 (4), 311-316.

Eliuk, L.S. (1978) "The Abenaki Formation, Nova Scotia Shelf, Canada. A depositional and diagenetic model for a mesozoic carbonate platform", Bull. Canad. Petrol. Geol. 26-4, 424-514.

Eliuk, L.S. (1981) "Abenaki update : variations along a Mesozoic carbonate shelf, Nova Scotia Shelf, Canada", in Stoakes F.A. (ed.), Annual core and field conference, Canad. Petrol. Geol., pp. 15-19.

Ellis, P.M. (1984) "Upper Jurassic carbonates from the Lusitanian Basin, Portugal and their subsurface counterparts in the Scotian Shelf", Unpub. Thesis, Open University, Milton Keynes, England.

Ellis, P.M., Crevello, P.D. and Eliuk, L.S. (1985) "Upper Jurassic and Lower Cretaceous deep-water buildups, Abenaki Formation, Nova Scotia Shelf", in Crevello P.D. and Harris P.M. (eds), Deep-water Carbonates : Buildups, Turbidites, Debris flows and Chalks - A Core Workshop 6, pp. 212-248.

Elmi, S. (1967) Le Lias supérieur et le Lias moyen de l'Ardèche. Doc. Lab. Géol. Lyon 19.

Elmi, S., Rocha, R. and Mouterde, R. (1988) "Sédimentation pélagique et encroûtements cryptalgaires : les Calcaires grumeleux du Carixien Portugais", Ciências da Terra 9, 69-90.

Embry, A.F. and Klovan, J.E. (1971) "A late Devonian reef tract on northeastern banks Island, N.W.T.". Bull. Canad. Petrol. Geol. 19, 730-780.

Enay, R. (1966) L'Oxfordien dans la moitié sud du Jura français. Nouv. Arch. Mus. Hist. Nat. Lyon VII,.

Enay, R., Mangold, C., Cariou, E., Contini, D., Debrand-Passard, S., Donze, P., Gabilly, J., Lefavrais-Raymond, A., Mouterde, R. and Thierry, J. (1980) Synthèse paléogéographique du Jurassique français. Doc. Lab. Géol. Lyon H.S. 5.

Flügel, E. and Steiger, T. (1981) "An Upper Jurassic sponge-algal buildup from the Northern Frankenalb, West Germany ", in Toomey D.F. (ed.), European fossil reef models. Soc. Econ. Paleont. Mineral. 30, pp. 371-397.

Fritz, G. (1958) "Schammstotzen, tuberolithe und schuttbreccien im weissen Jura der Schwabischen Alb", Arb. Geol. Palaont. Inst. 13, 1-118.

Gaillard, C. (1983) Les biohermes à spongiaires et leur environnement dans l'Oxfordien du Jura méridional. Doc. Lab. Géol. Lyon 90.

Gaillard, C. (1984-a) "Les biohermes à spongiaires du Jura Français, in Géologie et paléoécologie des récifs", Publ. Inst. Géol. Univ. Berne 18/1-18/23.

Gaillard, C. (1984-b) "Bioconstructions jurassiques", in Debrand-Passard S, Courbouleix S. and Lienhard M.J . (eds), Synthèse géologique du Sud-Est de la France. Mém. Bur. Rech. Géol. Min. 125, pp. 276-281.

Gwinner, M.P. (1976) "Origin of the Upper Jurassic limestones of the Swabian Alb (Southwest Germany)", Contr. Sedim . 5, 1-75.

Jansa, L.F. (1981) "Mesozoic carbonate platforms and banks of the eastern North American margin", Marine Geology, 44, 97-117.

Jansa, L.F., Pratt, B.R. and Dromart, G. (1989) "Deep-water thrombolite mounds from the Upper Jurassic of offshore Nova Scotia", Canad. Soc. Petrol. Geol. Mem. 13, 725-735

Jansa, L.F., Steiger T. and Bradshaw, M. (1984) "Mesozoic carbonate deposition on the outer continental margin off Morocco", in Hinz K. and Winterer E.L. (eds), Init. Rep. DSDP 79. U.S. Government Printing Office, Washington D.C., pp. 857-891.

Jansa, L.F., Termier and G., Termier, H. (1982) "Les biohermes à algues, spongiaires et coraux des séries carbonatées de la flexure bordière du "paléoshelf" au large du Canada oriental", Rev. Micropal. 25, 181-219.

Jenkyns, H.C. (1971) "The genesis of condensed sequences in the Tethyan Jurassic", Lethaia 4, 327-352.

Kennard, J. M. and James, N.P. (1986) "Thrombolites and Stromatolites, two distinct types of microbial structures", Palaios 1, 492-503.

Keupp, H., Koch, R. and Leinfelder, R. (1990) "Controlling Processes in the Development of Upper Jurassic Spongiolites in Southern Germany : State of the Art, Problems and Perspectives", Facies, 23, 141-174.

Land, L.S. and Goreau, T.F. (1970) "Submarine lithification of Jamaican reefs", Journ. Sed. Petrol. 40, 457-462.

Land, L.S. and Moore, T.F. (1980) "Lithification, micritization and syndepositional diagenesis of biolithites on the Jamaican Island Slope", Journ. Sedim. Petrol. 50, 357-370.

Lang, B. (1989) "Upper Jurassic Sponge Bioherm Facies of the Northern Frankenalb (Uspring ; Oxfordian) : microfacies, paleoecology, paleontology", Facies 20, 199-274.

Lang, B. and Steiger, T. (1984) "Paleontology and diagenesis of Upper Jurassic siliceous sponges from the Mazagan Escarpment", Oceanologica Acta N.S. 5, 93-100.

Laptas, A. (1974) "The dolomites in the Upper Jurassic limestones in the area of Cracow (Southern Poland)", Ann. Soc. Geol. Pologne 44, 93-100.

Lemoine, M., Arnaud-Vanneau, A., Arnaud, H., Bas, T., Bourbon, M., Dumont, T., Gidon, M., Graciansky, P.C. de (1986) "Etapes et modalités de la subsidence d'une paléomarge passive : Les Alpes occidentales au Mésozoïque", Bull Centre Rech. Prod. Elf-Aquitaine 10, 143-149.

Martin-Algarra, A. (1990) "Oxfordian bioconstructions of stromatolites and sponges in the Sierra de Cazorla (Prebetic zone, Southern Spain)", 13th International Sedimentological Congress. Nottingham, Abst, pp. 329-330.

Massari, F. (1981) "Cryptalgal fabrics in the Rosso Ammonitico sequences of the Venetian Alps", in Farinacci A. and Elmi S. (eds), Proc. Rosso Ammonitico Symp. Tecnoscienza, Roma, pp. 435-470.

Massari, F. (1983) "Oncoids and stromatolites in the Rosso Ammonitico sequences of the Venetian Alps, Italy", in Peryt T.M. (ed.), Coated Grains, Springer, Berlin, pp. 358-366.

Mathur, A.C. (1975) "A deep mud-mound facies in the Alps", Journ. Sedim. Petrol. 45, 787-793.

Matyja, B.A. (1977) "The Oxfordian in the Southwestern margin of the Holy Cross Mountains", Acta Geol. Polon. 27, 41-64.

Matyszkiewicz, J. (1989) "Sedimentation and diagenesis of the Upper Oxfordian cyanobacterial - sponge limestones in Piekary near Krakow", Ann. Soc. Polon. 59, 201-232.

Meyer, R.K.F. and Schmidt-Kaler, H. (1990) "Paleogeography and Development of Sponge Reefs in the Upper Jurassic of Southern Germany. An overview", Facies, 23, 175-184.

Monty, C.L.V. (1971) "An autoecological approach of intertidal and deep water stromatolites", Ann. Soc. Géol. Belgique 94, 265-276.

Monty, C.L.V. (1976) "The origin and development of cryptalgal fabrics", in Walter M.R. (ed.), Developments in Sedimentology, Elsevier, Amsterdam, 20, 193-249

Monty, C.L.V. (1977) "Evolving concepts on the nature and the ecological significance of stromatolites, in Flügel E. (ed.) Fossil Algae : Recent results and developments", Springer, Berlin, pp.15-35.

Monty, C.L.V. (1982) "Cavity or fissure dwelling stromatolites (endostromatolites) from Belgian Devonian mud mounds", Ann. Soc. Géol. Belgique 105, 343-344.

Monty, C.L.V. (1986) "Interactions événements géologiques - stromatolites", Bull. Centre Rech. Prod. Elf-Aquitaine 10, 537-553.

Ogg, J.G. (1981) "Middle and Upper Jurassic sedimentation history of the Trento Plateau (Northern Italy)", in Farinacci A. and Elmi S. (eds), Proc. Rosso Ammonitico Symp. Tecnoscienza, Roma, pp. 479-503.

Peres, J.M. and Picard, J. (1959) "On the vertical distribution of benthic communities", Intern. Oceanogr. Congress Amer. Assoc. Adv. Sci., Washington, pp. 349-351.

Pisera, A. (1991) "Upper Jurassic Sponge Megafacies in Spain : Preliminary Report", in Reitner J. and Keupp H. (eds)., Fossil and Recent Sponges. Springer Verlag. , pp. 485-497.

Pratt, B.R. (1982)" Stromatolitic framework of carbonate mud-mounds", Journ. Sedim. Petrol. 52, 1203-1227.

Radwanski, A. and Szulczewski, M. (1966) "Jurassic stromatolites in the Villany Mountains (Southern Hungary)", Ann. Univ. Sci. Budapest, Sect. Geol. 9, 87-107.

Ramalho, M. (1988) "Sur la découverte de biohermes stromatolitiques à spongiaires siliceux dans le Kimméridgien de l'Algarve (Portugal)", Comun. Serv. Géol. Portugal 74, 41-55.

Steiger, T.S. and Jansa, L.F. (1984) "Jurassic limestone of the seaward edge of the Mazagan Carbonate Platform, Northwest African Continental Margin, Morocco", in Hinz K. and Winterer E.L. (eds), Init. Rep. DSDP, 79. U.S. Government Printing Office, Washington D.C., pp. 449-492.

Szulczewski, M. (1968) "Stromatolity jurajskie w Polsce", Acta Geol. Polon. 28, 1-99.

MESOZOIC STRATIGRAPHIC BREAKS AND PELAGIC STROMATOLITES IN THE BETIC CORDILLERA, SOUTHERN SPAIN

J.A. VERA and A. MARTIN-ALGARRA
Dpto. Estratigrafía y Paleontología. IAGM, Facultad de Ciencias. Universidad de Granada. 18071 Granada (Spain)

ABSTRACT

In the Jurassic and the Cretaceous of the Betic Cordillera (Southern Spain) numerous examples of pelagic stromatolites have been found. In the present study a close relationship between the situation of pelagic stromatolites and stratigraphic discontinuity surfaces is shown to exist. These surfaces can be encrusted by ferruginous, manganiferous or phosphatic stromatolites, or covered by a bed very rich in oncoids with a core made up of ammonites and/or pelagic calcareous sediment. The textural and morphological features of these structures show that they are microbial (possibly bacterial and/or fungal) in origin. They form irregular and laterally discontinuous, decimetric to millimetric horizons which grow in areas of pelagic swell with a very slow sedimentary rate. The discontinuity surfaces are generally related to falls in sea-level that provoked interruptions of sedimentation reflected in omission surfaces, hardgrounds and, more locally, paleokarstic surfaces. The growth of stromatolites and oncoids indicate the base of transgressive cycles, with very slow sedimentation. The discontinuity surfaces with pelagic stromatolites and oncoids are situated in the following: a) Middle Liassic, b) Base of Middle Jurassic, c) Top of Bathonian, d) Dogger-Malm boundary, e) Kimmeridgian, f) Upper Valanginian-Hauterivian, g) Upper Albian and h) Cretaceous-Tertiary boundary. Stromatolite growth accounted preferently on swells, in submarine and moderatelly deep to deep environments, during sea-level rises after lowstands. They mark the beginning of transgressive cycles and, usually, are finally buried under pelagic sediments when sea-level rise progresses. Their growth always accounts in extremely starved environments, and was favoured by submarine hydrothermalism and bottom currents. Under these hard environmental conditions, microbial communities and, in a lesser extent, encrusting foraminifera and serpulids, were the only agents able to survive and accrete pelagic biosedimentary deposits.

INTRODUCTION

Phanerozoic pelagic stromatolites have always been described in close relationship with

J. Bertrand-Sarfati and C. Monty (eds.), Phanerozoic Stromatolites II, 319–344.

zones of very low sedimentation rate, as in Paleozoic examples (Playford & Cockbain, 1969; Playford et al., 1976; Tucker, 1973) or in Jurassic (Jenkyns, 1970, 1971; Bernoulli & Jenkyns, 1974; Seyfried, 1978; Comas et al., 1981; Massari, 1979, 1981, 1983; Massari & Dieni, 1983; Monty, 1986; etc.) or Cretaceous ones (Rioult & Royant, 1975; Massari & Medizza, 1975; Krajewski, 1981a,b,c, 1983) and even as in recent examples from the present-day sea-bottom (Monty, 1973a). In the Betic Cordillera of Southern Spain the Jurassic, and also the Cretaceous, are essentially made up of pelagic sediments, that in some particular sequences show stratigraphically condensed and/or reduced facies bearing many sedimentary gaps (Azéma et al., 1979; García-Hernández et al., 1980). A detailed systematic study of discontinuity surfaces and associated sediments shows that these formed during specific periods of geological history and that pelagic stromatolites are commonly associated with them (Martín-Algarra & Vera, 1982, and this volume; González-Donoso et al., 1983; Martín-Algarra et al., 1983; Vera, 1984a,b; Molina et al., 1985; Martín-Algarra, 1987; Molina, 1987). The purpose of this note is to synthesise some of the data obtained about the stratigraphic position and features of these stromatolites and to propose some hypotheses to explain their spatial and temporal location and genesis.

GEOLOGICAL AND PALEOGEOGRAPHICAL SETTING

The Betic Cordillera (Fig.1) constitutes the westernmost Alpine mediterranean chain. It appears to south and southeast of the Iberian Peninsula and runs, underneath the Mediterranean, to the Balearic Islands, where it outcrops again in Mallorca and Ibiza. It is usual to distinguish two first-order geological ensembles (Fig.1), the External and the Internal Zones, which are separated by a third one, the Flysch Complex of the Campo de Gibraltar (Azéma et al., 1979; García-Hernández et al., 1980).

The external zones (southern iberian margin)

The External Zones are a complex of thrust sheets formed by thick successions of Triassic to Miocene sedimentary rocks which have been detached from their Paleozoic basement because of the tertiary alpine tectonics. During the Mesozoic and the Tertiary these successions formed a several hundreds of kilometres wide continental margin, the **Southern Iberian margin**, which was the westernmost part of the European margin of the Tethys (Fig.2A).

The External Zones are subdivided in two great tectonic ensembles (Fig.1) that derive from two great paleogeographic domains of the Southern Iberian margin (Fig.2A-B): a platform area to the north, the Prebetic, and a pelagic area to the south, the Subbetic.

The *Prebetic* (Fig.1, 2) is the most external domain of the Southern Iberian margin, and was situated nearest to the ancient Iberian continent. During its whole geologic history, it was essentially a pericontinental shelf area, with mostly carbonate, shallow

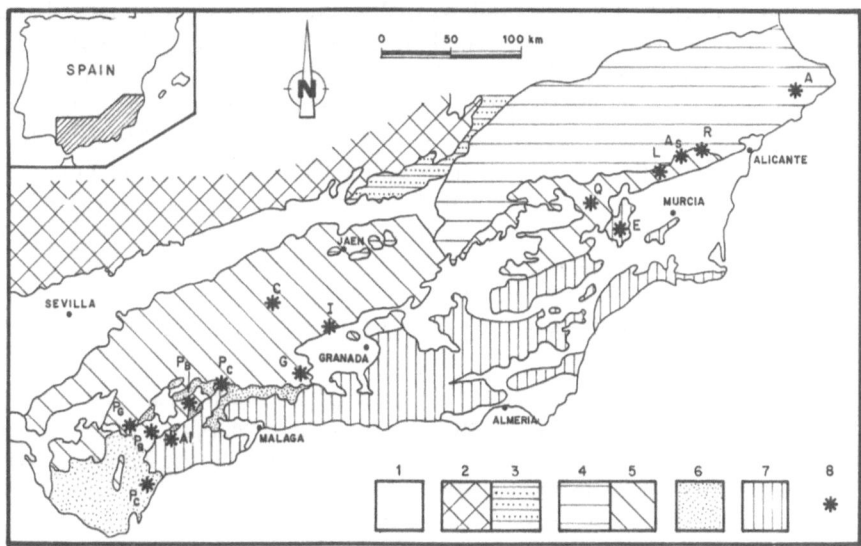

Figure 1 : Geological sketch map of the Betic Cordillera. Key: 1: Neogene and Quaternary. *Iberian Foreland* (2 and 3). 2: Paleozoic basement. 3: Non-folded Mesozoic-Tertiary cover. *External Zones* (4 and 5). 4: Prebetic. 5: Subbetic. 6: *Campo de Gibraltar Complex.* 7: *Internal Zones*. 8: Outcrops with pelagic stromatolites mentioned in the text. A: Sierra Aixorta. As: Sierra del Asiento. R: Sierra de Reclot. L: Sierra de Lugas. Q: Sierra de Quípar. E: Sierra Espuña. I: Illora. C: Sierra de Cabra. G: Sierra Gorda. P: Penibetic (P_G: Grazalema. P_B: Stromatolite Belt. P_C: Canutos Bed). Al: Sierra de la Almola.

marine to coastal, and sometimes continental sedimentation, hemipelagic sediments being rarely found, except at some internal points of the Prebetic platform or during specific periods of the geologic history of pronounced sea-level highstand.

The *Subbetic* (Fig.1, 2) constituted the internal part of the Southern Iberian margin and was an area of mostly pelagic sedimentation. During the Jurassic and the Lower Cretaceous (Fig.2A) the Subbetic had a well-defined topography in troughs and swells individualized during the Middle Liassic by disintegration of a Lower Liassic carbonate platform (García-Hernández et al., 1976, 1980, 1986-87a; Vera, 1981, 1984a,b, 1986, 1988). Two troughs and two swells are distinguished. They are, from the Prebetic towards the interior of the basin, the following (Fig.2A-B): the Intermediate Units Trough, the External Subbetic Swell, the Median Subbetic Trough and the Internal Subbetic Swell (the westernmost part of this latter is also known as Penibetic: Martín-Algarra and Vera, this volume). Thick sequences of mostly carbonate and, in a lesser extent, siliceous and turbiditic deep pelagic sediments, were deposited in the troughs, and an active contemporaneous submarine volcanism also affected these areas. On the contrary, reduced and/or condensed sequences of mostly nodular and other pelagic limestones were deposited in the swells (Fig.2B). Most of these condensed sequences show sedimentary gaps that can be dated by means of fossils. Furthermore, peculiar sedimentological features such as paleokarst, hardground, rockground, boring, burrowing, dissolution and/or different kinds of mineralized crusts, many of them bearing pelagic stromatolites, indicate stratigraphic discontinuities. They can frequently be seen in the Prebetic (Cretaceous-Tertiary boundary of Sierra Aixorta) or in different Subbetic realms, in several points and ages from Jurassic to Cretaceous (Martín-Algarra & Vera, 1982, 1989, and this volume; González-Donoso et al., 1983; Martín-Algarra, 1987; Molina, 1987; García-Hernández et al., 1988; Vera et al., 1984, 1988).

The internal zones (mesomediterranean terrane)

The Internal Zones (Fig.1) are formed by strongly deformed and metamorphosed, mostly Paleozoic and Triassic rocks structured in three nappe complexes (Nevado-Filabrides, Alpujarrides-Rondaides and Malaguides) in which the mesozoic cover has been tectonised together with its paleozoic basement. The Jurassic and Cretaceous sediments are scarce or absent in most nappe complexes, and their original sedimentary features intensely modified by the alpine metamorphism. In fact post-Triassic rocks are only present in the *Malaguides* and the *Rondaides*, a fragmented group of units showing

Figure 2 : A) Palinspastic and paleogeographic reconstruction of the westernmost Tethys during the Lowermost Cretaceous (after Martín-Algarra, 1987). Key of symbols.- a: Continental areas. SOUTHERN IBERIAN MARGIN (b, c, d). b: shelf areas (Prebetic). c: slope, trough and basinal areas (Subbetic). d: swells (Subbetic). OCEANIC TO SEMIOCEANIC BASINS (e). e: Campo de Gibraltar and related flysch basins. MESOMEDITERRANEAN TERRANE (f, g). f: slope and basinal areas (Rondaides). g: shallow water areas (Malaguides: 13) and areas of unknown depositional environment (Nevado-Filabrides: 10; and Alpujarrides: 11). NORTH AFRICAN MARGIN (h, i). h: shelf areas. i: slope and basinal areas. j: position of shelfbreaks and/or swell margins. Key of numbers.- *Iberian Foreland.* 1: Unfolded mesozoic cover. *Southern Iberian margin (2, 3, 4, 5 and 6).* 2: Prebetic. 3: Intermediate Units Trough. 4: External Subbetic Swell. 5: Median Subbetic Trough. 6: Internal Subbetic and Penibetic Swells. *Deep, oceanic to semioceanic basins (7,8 and 9).* 7: Campo de Gibraltar Flysch Basin. 8: North African Flysch Trough. 9: Western dependences of the Ligurian Ocean in the Nevado-Filabrides. *Mesomediterranean Terrane (10, 11, 12, 13 and 14).* 10: Nevado-Filabrides. 11: Alpujarrides. 12: Rondaides (talus slope of the Betic Internal margin). 13: Malaguides (platform of the Betic internal margin). 14: Algerian Kabylias. Letters on this chart correspond to the paleogeographic position of the stromatolite-bearing outcrops described in the text and situated in Figure 1. I-I': Position of cross-sections in Figure 2B. B) Paleogeographic cross-sections showing the Jurassic-Cretaceous evolution of the Betic Cordillera. Situation (I-I') in Figure 2A. Numbers as in Figure 2A. Note that the Campo de Gibraltar Flysch Basin (C.C.G.: 7) was opening during Late Jurassic and Early Cretaceous. Key of symbols: a: Fluviatile and deltaic sandstones. b: Tidal flat and lagoonal carbonates. c: Open-marine carbonate or mixed siliciclastic-carbonate facies. d: Cherty and marly limestones with carbonate turbidite intercalations. e: Radiolarian limestones and radiolarites. f: Nodular and condensed pelagic limestones (Ammonitico rosso and related facies). g: Pelagic marls and marly limestones. h: Siliciclastic turbidites. i: Submarine volcanism. j: Slumps and carbonate breccias. k: Non depositional surfaces. l: Pelagic stromatolites (localities as in Fig.1 and 2A).

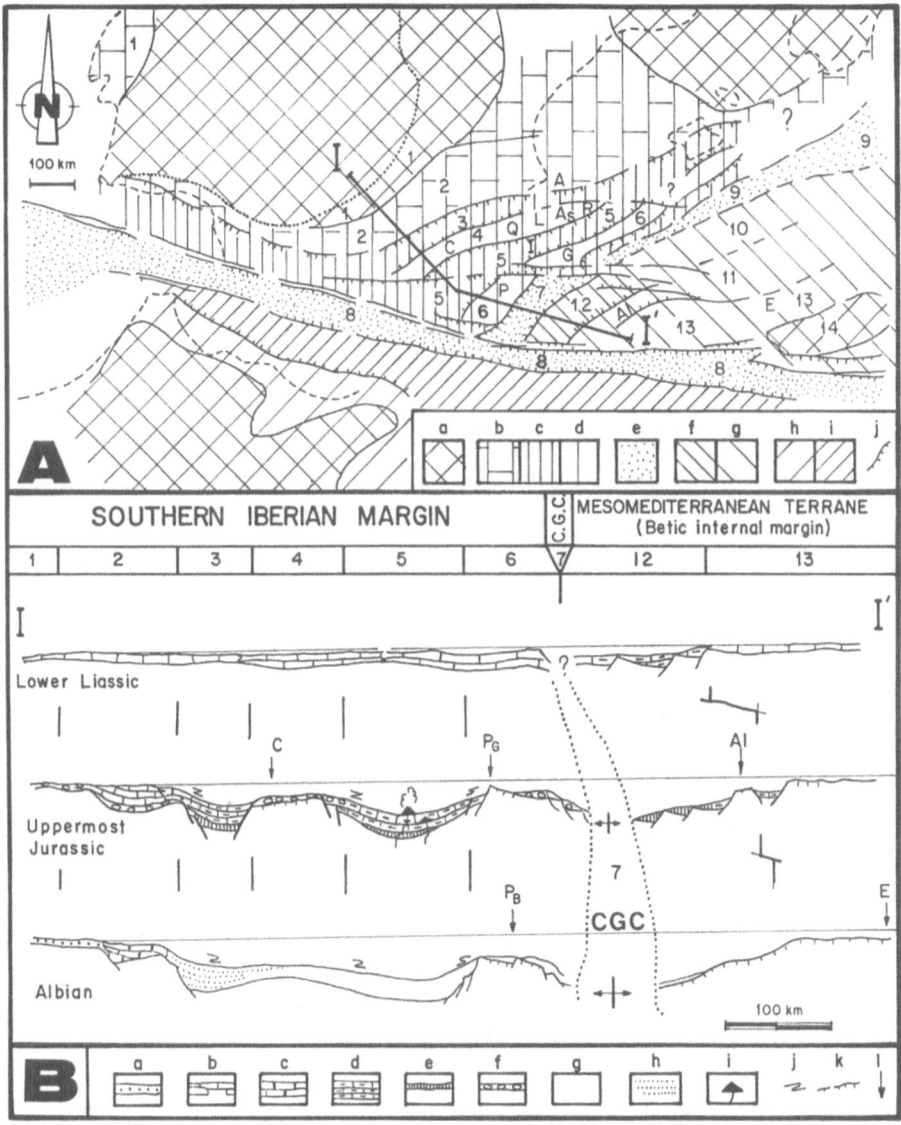

close tectonic, stratigraphic, and facies relationship with the Alpujarrides. This later group of units shows facies, sequences, a tectonic position and a stratigraphic evolution which are similar to those of the Austroalpine domains in the Alps.

The Mesozoic paleogeography of the Internal Zones is not yet completely well understood. Nevertheless, it is clearly established (see Martín-Algarra 1987, for details, and references therein) that during the Jurassic, and especially the Cretaceous and the

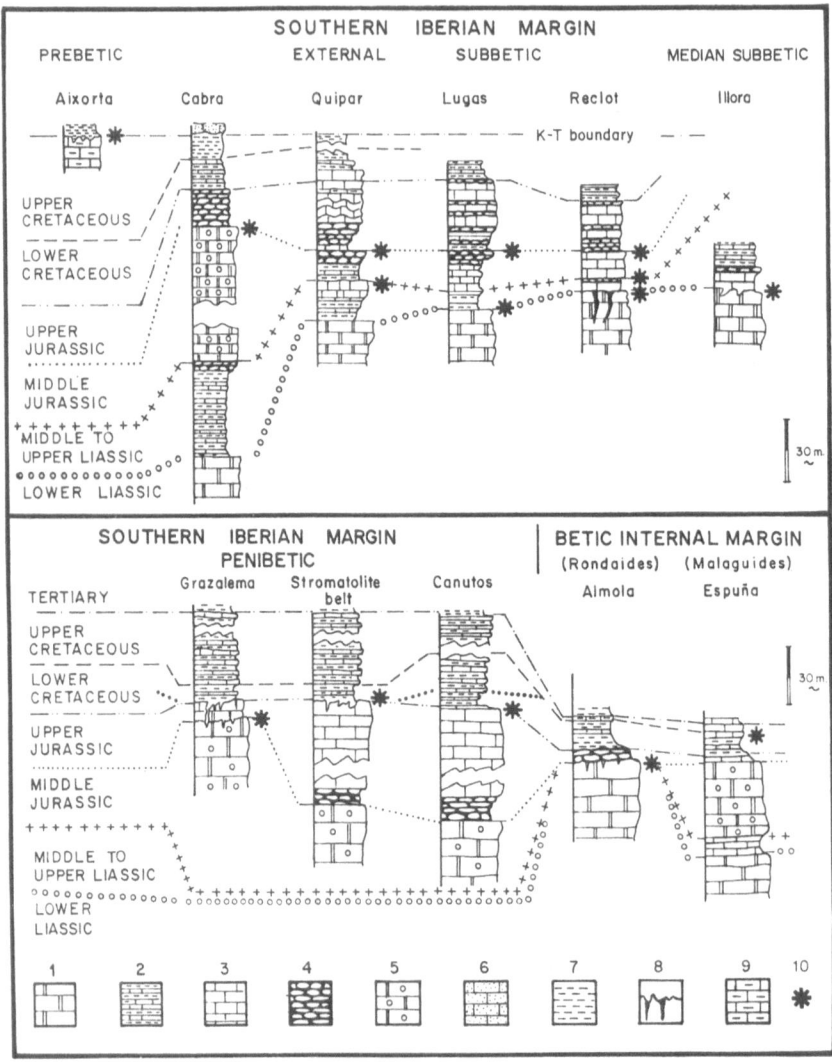

Figure 3 : Jurassic-Cretaceous stratigraphic columns of some of the studied outcrops, with the location of pelagic stromatolites. 1: Shallow water limestones (crinoidal limestones on top) : Lower-Middle Liassic. 2: Pelagic, marly limestones and marls (Late Middle Lias-Cretaceous). 3: Undifferentiated (mostly red) pelagic limestones, sometimes slumped (Late Middle Lias-Late Jurassic). 4: Nodular limestones (Ammonitico Rosso facies: Late Middle Lias-Late Jurassic) 5: Shallow water, oolitic limestones (Dogger) 6: Calcareous turbidites and breccias (Late Jurassic to Paleocene). 7: Marls and marly limestones with planktonic foraminifera (Mid- to Late Cretaceous and Paleogene). 8: Shallow water carbonate deposits (Upper Cretaceous). 9: Erosional surfaces, paleokarsts, neptunian dykes. 10: Pelagic stromatolites.

Paleogene, the Internal Zones of the Betics, together with the Internal Zones of the Rif (Morocco) and Kabylias (Algeria) formed an allochtonous terrane (Fig.2A), called **Mesomediterranean Terrane** or microplate by Durand-Delga & Fontboté (1980). This Terrane was bordered by narrow continental margins and by deep, oceanic or semioceanic basins, that separated it from the Southern Iberian margin of the Iberian Plate and from the North African margin (External Zones of the Maghrebian chains) of the African Plate (Fig.2A-B). In these basins turbidites and deep-sea sediments of the Campo de Gibraltar Complex and its north-african counterparts were deposited (Fig.1,2).

The Mesomediterranean Terrane itself was a hardly subsident, continental-crust area, partially covered by the sea and situated between the Iberian and African plates far more to the east (possibly up to 1000 km) of its present position in the Betic Cordillera (Fig.2A). Towards its Betic-Rif side, it was formed by a hemipelagic platform area, the *Malaguides*, bordered to the west by an abrupt, narrow and extremely starved continental margin of Austroalpine type, the *Rondaides*, which received a scarce and condensed pelagic sedimentation during the Jurassic and the Cretaceous. It was separated from the westernmost part of the Southern Iberian margin (the western Internal Subbetic or Penibetic, see Martín-Algarra & Vera, this volume) by a deep oceanic to semioceanic basin (Fig.2B). Whitin this **Betic internal margin** pelagic stromatolites have been found in two different units: the Upper Jurassic of the Almola unit, a problematic tectonic element of Alpujarride-Rondaide affinity; and the Cretaceous (Albian) of the Malaguides.

In summary, the Betic Cordillera is the result of the convergence and oblique collision of two opposed Mesozoic continental margins, separated by a narrow deep semioceanic basin: the *Southern margin of the Iberian Plate*, where the External Zones (Subbetic and Prebetic) came from; and the *Betic internal margin*, which formed the western part of the Mesomediterranean Terrane. It was later structured as the nappe complexes of Malaguides and Alpujarrides-Rondaides of the Internal Zones. Both margins had a very irregular topography which was controlled by intense synsedimentary block-faulting due to rifting and by an extensional-transtensional tectonics that affected the whole western Tethys during the Jurassic and Cretaceous. This resulted in the formation of swells, slopes and basins. Condensed deposits and stratigraphic discontinuity surfaces bearing pelagic stromatolites are frequently found in close relation with starved areas in both margins, as well as in their alpine mediterranean counterparts.

BETIC PELAGIC STROMATOLITES

In this paper only "mineralized" (ferruginous, manganesiferous or phosphatic) pelagic stromatolites clearly associated with well-defined sedimentary breaks are studied; therefore pelagic calcareous stromatolites associated with pelagic limestones of ammonitico rosso and related facies, that are also present and are locally frequent in the

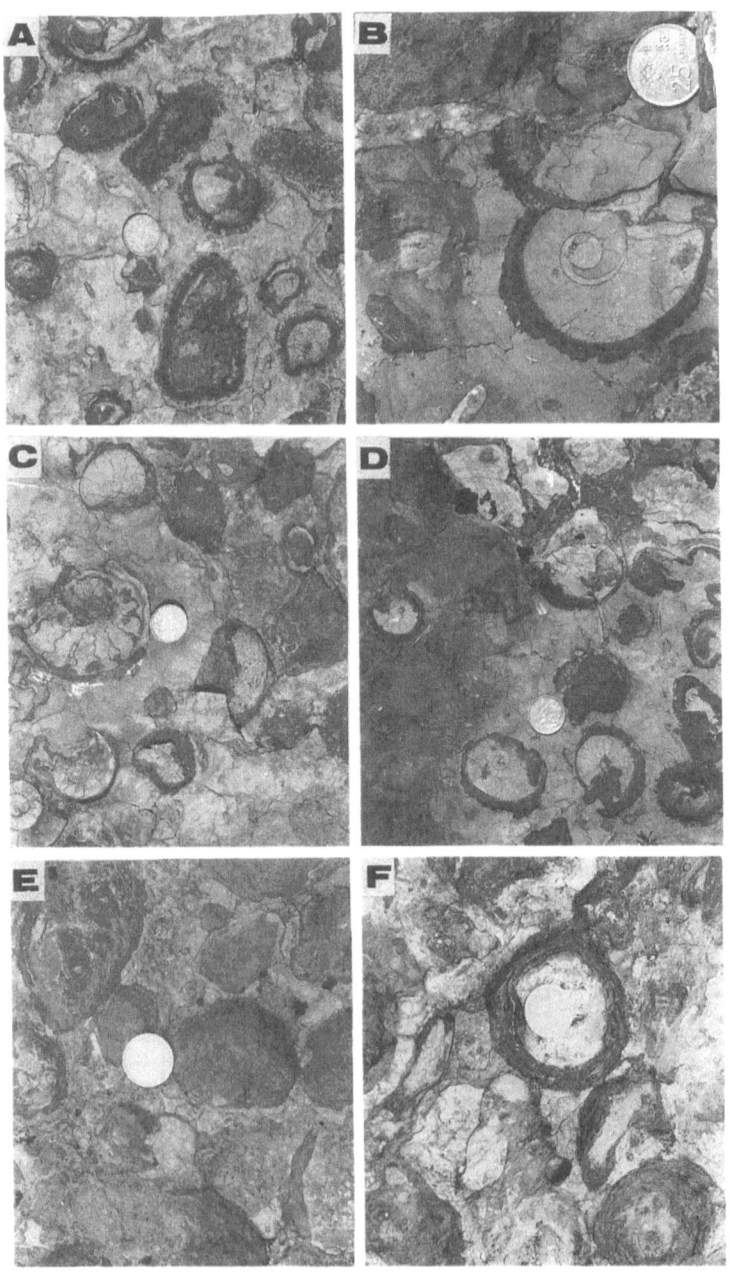

Betic Cordillera will not be considered (Comas et al., 1981, Acosta and García-Hernández, 1988, Martín-Algarra et al., 1990).

The outcrops studied (Fig.3) are most frequently to be found in units which, during the Jurassic and Cretaceous, constituted pelagic swells (Fig.2). In the trough sequences, where sedimentation rate was high, no pelagic stromatolites appear, with the single exception of the very important discontinuty surface of the Middle Liassic, where microstromatolites and more locally well-developed pelagic stromatolites (Illora section: García-Hernández et al., 1976, 1986-87a; Fig.1I,; Fig.3 and 4F) have been recognized. Howewer, the best and most beautiful examples of pelagic stromatolites have been found in swell areas of the Southern Iberian margin (External Subbetic, Internal Subbetic and Penibetic). Stromatolites of Dogger and Malm age, are frequent in the many outcrops of the External Subbetic swell (Fig.1, 2A, 3) Sierra de Cabra (Bathonian-Callovian: C), Sierras de Quípar and Lugas (same age as the former, plus Aalenian occurrences: Q and L, respectively; Fig.4A-E, 7), and Sierras del Asiento and Reclot (several ages throughout the Jurassic sequence: As and R). In the Internal Subbetic, stromatolites are found in the Callovian and Oxfordian of Sierra Gorda (point G) and in the Penibetic, where they appear in many outcrops at several points of the Jurassic-Cretaceous sequence (P, and see Martín-Algarra and Vera, this volume, for details).

The most recent example detected, at the Cretaceous-Tertiary boundary and in the Lowermost Paleocene, comes from Sierra Aixorta (Company et al., 1982), an outcrop pertaining to the most internal Prebetic (Fig.1A; Fig.3, 5, 6).

In the Betic internal margin pelagic stromatolites are found in only a few places. They have been detected in the Sierra de la Almola (Al), on a discontinuity surface situated in the base of the Malm (probably of Upper Oxfordian age) and coating abrasion hardgrounds and fossils within pelagic fossiliferous limestones of Lower Kimmeridgian age (Martín-Algarra et al., 1983; Martín-Algarra, 1987). Phosphatic stromatolites, nodules and crusts have also been detected in discontinuity surfaces, at the base of Upper Albian in sequences pertaining to the Malaguide realm (Sierra Espuña: M; Paquet, 1969).

Figure 4 : Field aspect of several ferro-manganesiferous pelagic oncoids and crusts, included in red pelagic limestones. A to E from Sierra de Quipar: Callovian strata resting on the discontinuity surface of the Upper Bathonian. F from Illora: Middle-Upper Carixian. A) : Plan-view of a slightly abraded oncoidal bed. Note the abundance of arborescent microstromatolites in the sediment-oncoid interface. B) : Close-up of an ammonite-bearing oncoid core. Note the preservation of ammonite shell. C) and D) : Abrasion surface affecting the top of the oncoidal bed. The upper surface of oncoids has been eroded after consolidation and before the renewal of pelagic sedimentation. Ammonites in the cores of oncoids are clearly visible, locally fragmented and strongly corroded before being encrusted by stromatolites. E) : Slighty eroded oncoid surfaces and crusts. F) : Liassic oncoids. Cores are formed by red to pink pelagic limestones, locally with fragmented ammonites of Middle Carixian age.

Composition and mineralogy

The pelagic stromatolites studied are all of a ferro-manganesiferous and/or phosphatic composition. Most frequently, Jurassic examples are rich or very rich in iron-manganese, whereas Cretaceous ones are usually phosphate-rich, but mixed situations appear. Mineralogy is variable in detail, but phosphate is always cryptocrystalline apatite, iron minerals are goethite, hematite, and pyrite, whereas manganese minerals are pyrollusite, psillomelane and other Fe-Mn oxy-hydroxides. The mineralogy is shown by studies using X-R diffraction, transmitted- and reflected-light microscope, SEM and microprobe analysis (preliminary results). Carbonates (mainly calcite) are always present.

Morphological features

Pelagic stromatolites associated to stratigraphic break surfaces show three main morphologies: crustose, oncoidal and microstromatolitic.

Condensation surfaces are frequently covered by, usually ferruginous, smooth to more or less irregular and laterally discontinuous laminated crusts with variable thickness (millimetric to decimetric), which are, in fact, *crustose stromatolites* (Fig.5A-B). They are normally flat, as they encrust planar to slightly erosional hardgrounds. Nevertheless, as their shape is directly controlled by the substrate morphology, these stromatolitic crusts can show very irregular geometries, sometimes labyrinthic. Some of them can be seen attached to walls or roof of microcaves (Fig.6) in the substrate, when this latter has been affected either by dissolution of submarine or subaerial origin (as in

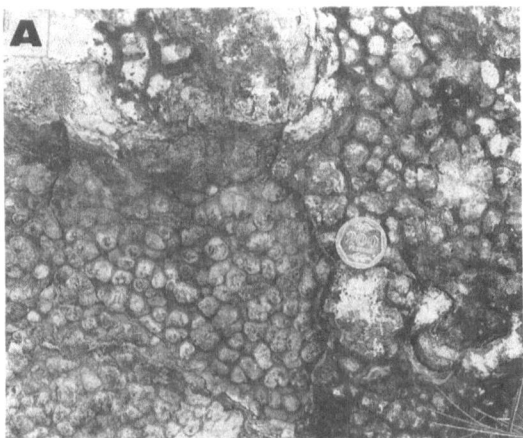

Figure 5 : A) : Surface view of a mammilated stromatolite crusts made of contiguous heads and columns. Sierra Aixorta (Prebetic), Cretaceous-Tertiary boundary. B) : Same as A), slightly eroded. Erosion allows to see the laminated internal structure of stromatolite. Coin = 2 cm.

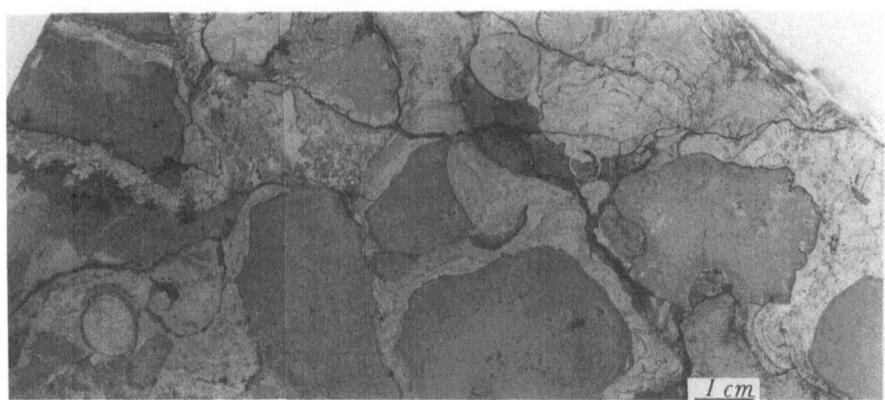

Figure 6 : Polished slab from the break surface at Cretaceous-Tertiary boundary in Sierra Aixorta (Prebetic). Cretaceous sediments (grey) have been strongly burrowed. Burrow walls have been later encrusted by phosphatic pelagic stromatolites in all directions: note stromatolites growing from top to bottom in the lower right corner. Finally, cavities have been filled up with light gray Paleocene calcareous sediment very rich in planktonic foraminifera.

the Penibetic: see Martín-Algarra and Vera, Fig.4, this volume), or by boring and burrowing. Crust surfaces are frequently mammilated (Fig. 5A) resulting from the coalescence of contiguous columns (Fig.5B); they may also be planar, irregular or domal. The best developed examples of crusts can be· seen in the Middle Lias of the Sierra de Reclot, the Upper Bathonian discontinuity in several sites of the External Subbetic (Sierras de Cabra and Reclot, among others), the Upper Albian-Vraconian of the Penibetic and the Cretaceous-Tertiary boundary of Sierra Aixorta in the most internal Prebetic.

In many other cases microbial structures appear as *oncoids* of few millimetres to centimetres in size, isolated or, more usually, grouped in centimetric beds of condensed pelagic limestone, lying directly on or immediately over discontinuity surfaces (Fig.4). In the Jurassic examples, oncoid cores are frequently formed by bioclastic pelagic sediments rich in ammonites (Fig.4A-E, 7A-D) and other microfossils (belemnites, aptychi), showing various degrees of fragmentation, dissolution, lithification and boring (Fig.7). Dense accumulations of oncoids can be seen within centimetric condensed beds of Callovian age (resting on the Upper Bathonian discontinuity surface) in Sierra de Quípar, Lugas and Asiento (Fig.4), in the Uppermost Valanginian-Lower Hauterivian Canutos bed of the Penibetic (see Martín-Algarra & Vera, this volume, Fig.5) and also locally in the Oxfordian of the Sierra de la Almola (Martín-Algarra, 1987).

The walls of neptunian dykes and hardgrounds within pelagic sediments, especially condensed cephalopod limestones and sometimes also nodular limestone (ammonitico rosso and related facies) commonly show millimetric crusts of *arborescent microstromatolites* (Fig.6). They can be seen also sparsely distributed within these

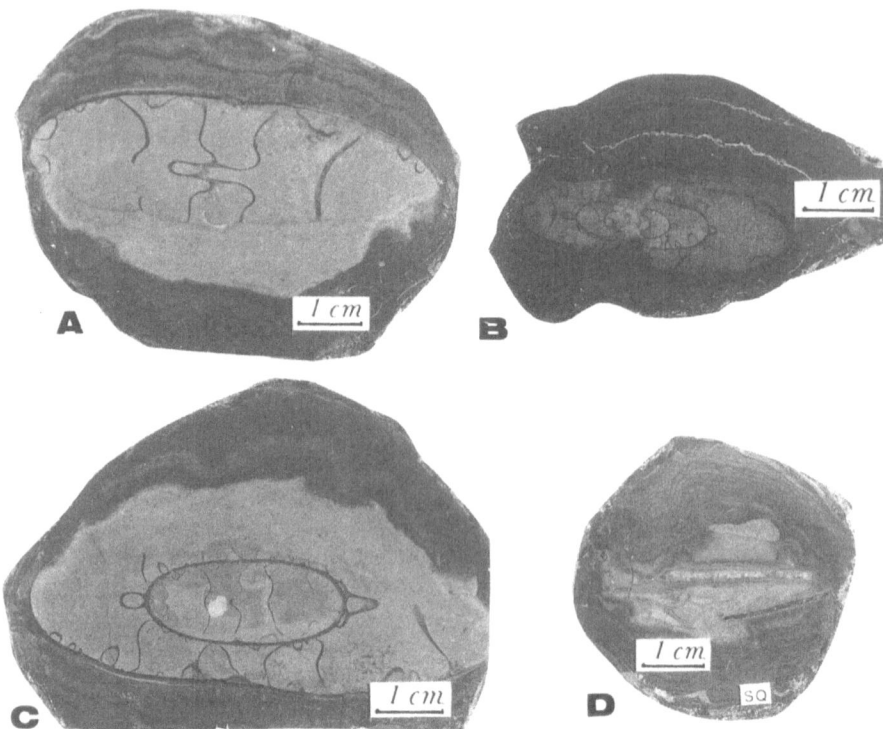

Figure 7 : Jurassic oncoids. A) and C), from the base of Oxfordian in the Sierra de Lugas; B) and D) from the Bathonian-Callovian boundary in Sierra de Quípar. A) B) and C) : Oncoids with a core formed by intensely corroded ammonite shells. Note partial destruction of shell before stromatolite colonization. Ammonites have been extracted from former red pelagic sediments, possibly by current action, and later encrusted by stromatolites. Lamination is hardly visible because of strong remobilization of Fe-Mn during diagenesis. D) : Oncoid showing a bioclastic core made by a mud clast with corroded and mineralized stromatolite fragments. Alternance between carbonate red sediments and ferruginous laminae are visible in the stromatolite.

facies as micronodules and irregular oncoids, usually encrusting more or less corroded bioclastic remains.

In section, to the naked eye, stromatolites show a clear and fine laminated fabric which may be planar or wrinkled, or yet domal leading to the formation of laterally linked hemispheroids, or arborescent microstromatolites (Fig.6, 7). This lamination is visible because of changes in sediment and/or in mineral content, resulting in alternations in colour. It may or may not be associated with filamentous organic remains.

Interpretation

The interpretation of the stromatolite growth is based on field work, macroscopic observation of hand samples and, especially, on microscopic examination. Howewer, no detailed textural description of stromatolite textures will be given here: the reader is referred to Martín-Algarra and Vera (this volume) for this kind of details. We will simply reaffirm that accretion of reported stromatolites results from the interaction of several processes: a) growth of microbial filaments, preferently bacterial and, to a lesser extent, fungal; b) microbially controlled floculation and/or precipitation of minerals like sulphides (pyrite), phosphates (apatite) and iron and manganese oxy-hydroxides (goethite, pirolusite, psilomelane, etc); c) trapping and binding of fine pelagic sediments: biogenic (coccoliths) or bioclastic micrite, sand and silt-sized terrigenous grains (quartz and clay); d) encrustation by foraminifera. The same textures and fabric as found in the Penibetic stromatolites (Martín-Algarra and Vera, this volume) have been detected occurring at different locations and ages. Nevertheless, in Jurassic examples, locally abundant serpulids encrusting the stromatolite growth surfaces have been found. On the other hand, in stromatolites associated with the Cretaceous-Tertiary boundary in the Sierra Aixorta (Prebetic), angular, fine-sand and silt quartz grains as well as planktonic foraminifera abound within stromatolite laminae. They record high sediment supply, a fact which can be easily understood if we take into account the proximity the Iberian continent, when compared with the other sites studied.

STRATIGRAPHIC BREAKS AND PELAGIC STROMATOLITES

In the Betic Cordillera the presence of pelagic stromatolites is restricted to specific discontinuity surfaces. These breaks in sedimentation occupy definite stratigraphic positions, as can be demonstrated when accurate biostratigraphical studies are feasible by the existence of fossils. They present in different stratigraphic sequences of various paleogeographic domains. Coeval events, bearing pelagic stromatolites have also been detected in other Alpine chains. A brief review of stromatolite-bearing Betic stratigraphic breaks and their alpine mediterranean counterparts follows here:

Middle Liassic

This is a regionally very important event because it coincides with the disintegration of the Lower Liassic carbonate platform and the invasion of the Subbetic by pelagic facies. The best example observed of pelagic stromatolite of this age comes from Illora, in the Median Subbetic (Fig.4F). It was previously described by García-Hernández et al. (1976, 1986a). Microbial accretions appear both as oncoids (Fig.3F) having a diameter of several centimetres (5 to 15 cm) and as millimetric, discontinuous and irregular crusts on slightly eroded crinoidal limestones (Middle-Lower Carixian). Their cores are made up of red micritic pelagic limestones and sometimes Middle Carixian ammonite-fragments. Germann (1971) has noted similar ferro-manganese crusts and nodules of the same age in the Tyrol (Northern Calcareous Alps).

Base of Middle Jurassic

Several discontinuity surfaces bearing pelagic ferro-magnanese stromatolites have been detected in the Sierra de Reclot (External Subbetic) and were considered as one single Aalenian discontinuity by Seyfried (1978: his "limonit-kruste II"). In fact a more complex situation has been discovered within the pelagic limestones that make the transition from Lias to Dogger. In this area at least four breaks exist (certainly laterally coalescing), each one with pelagic stromatolites:
- The lower one is a well developed hardground, covered by centimetric stromatolite crusts resting on a discontinuity surface with ammonites of the early upper Toarcian (P. Rivas, pers. com).
- Two minor hardgrounds marked by centimetric, discontinuous ferruginous crusts of stromatolitic origin appear inside yellowish-brown, orange-reddish and beige pelagic limestones of Toarcian(?)-Aalenian age which have so far not been dated precisely.
- The upper break surface coincides with a complex crust developed on a red pelagic fossiliferous limestone bearing many ammonites and clasts partially or wholly covered by stromatolitic ferruginous envelopes. The ammonites indicate the uppermost zone of the Aalenian (*G. concavum* Zone) and the base of the Bajocian (A.Linares and J.Sandoval, pers. com.).
Analogous and age-equivalent examples have been described from Sicily by Jenkyns (1970), Mallorca by Prescott (1988) and fthe North-African chains by Elmi (1981).

Top of Bathonian

This break is one of the most interesting because it is present in various different outcrops and bears, in many of them, abundant and morphologically diversified pelagic stromatolites. It has been very well recorded and preciselly dated in different and distant outcrops of the External Subbetic (Sierras de Cabra, Quípar, Lugas, Asiento and Reclot). There it displays different stratigraphical and sedimentological features that have already been studied by Seyfried (1979, 1981), Delgado et al. (1981) and Molina et al. (1985), but no detailed studies on pelagic ferromanganese stromatolites have been carried out. The latter appear:
- As crusts of variable thickness (1-7 cm), sometimes with colonization levels made of serpulids, visible to the naked eye (Sierra de Cabra).
- As arborescent microstromatolites directly encrusting irregularities on a karstic surface developed on Bathonian oolitic limestones, as is the case of the Fuente de los Frailes outcrop in the Sierra de Cabra (Molina 1987). In other cases they colonize the wall of neptunian dykes cross-cutting nodular pelagic limestones, due to tensional and contractional tectonics.

- As spectacular accumulations of oncoids within the first, ammonite-rich, callovian bed, resting on the discontinuity surface (Fig.4A-E).
Analogous coeval structures have been noted by Brochwicz-Lewinski et al. (1984) in the Carpathians.

Dogger-Malm boundary

In the swell sequences of the Subbetic the base of the Malm coincides systematically with an important stratigraphic gap including the uppermost Callovian and the Lower to Middle Oxfordian, thin sediments of this age having been dated exceptionally in few outcrops (Sequeiros, 1974). In some places where Callovian deposits are found (e.g. the Internal Subbetic of Sierra Gorda), there is no clear lithological change from the Dogger to the Malm, both series being formed by nodular limestones of similar facies. Howewer, accurate examination of these sections shows the presence of slight unconformities (Vera, 1966) or lenticular bedding between the Callovian and Oxfordian strata, making evident the existence of an important discontinuity surface between them. In other sites (External Subbetic in the Sierra de Cabra: Vera et al., 1984; Molina, 1987; and Penibetic: Martín-Algarra, 1987) most of the Callovian-Oxfordian interval is missing and the Dogger shows carbonate platform facies. The stratigraphic gap is clearly shown by obvious sedimentological features, such as paleokarst, hardgrounds or ferromanganese nodules and crusts of pelagic stromatolitic origin.
Pelagic stromatolites have been found in successions of nodular and red pelagic limestones in the Sierras de Reclot and Lugas (External Subbetic; Seyfried, 1981) and in the Sierra Gorda (Internal Subbetic; García-Hernández et al., 1986b), where they mark the Callovian-Oxfordian boundary. They show laterally linked centimetric structures encrusting the discontinuity surface. Pelagic stromatolites (centimetric encrustations and oncoids) have also been recorded at the base of Middle-Upper Oxfordian pelagic fossiliferous limestones (Penibetic, Torcal Fm.: see Martín-Algarra & Vera, this volume). Microbial structures have been noted in the Betic internal margin of the Mesomediterranean Terrane. In the Sierra de la Almola outcrop centimetric, spheroidal to ovoidal oncoids, as well as irregular and discontinuous crusts, sometimes broken, occur within unfossiliferous strata of brown ferro-manganesiferous and strongly condensed pelagic limestones, beneath well-dated ammonite-rich deposits of the lowermost Kimmeridgian (Martín-Algarra et al., 1983; Martín-Algarra, 1987).
Analogous coeval structures (nodules and crusts) have been noted by Bourbon (1980) in the alpine Briançonnais. In the Apennines Farinacci (1967) described arborescent stromatolites associated with Oxfordian sediments.

Kimmeridgian

No clear discontinuity surface has been detected in Kimmeridgian sequences but by places the Lower Kimmeridgian shows clear evidence of condensation. In this case, ferro-manganesiferous pelagic stromatolites may occur, as in the Sierra Gorda (Internal

Subbetic) or in the Sierra del Asiento and Reclot (External Subbetic) where stromatolite crusts and oncolites are found within red pelagic limestones. In the Sierra de la Almola (internal continental margin) well-developed arborescent microstromatolites also appear, encrusting ammonites of Lower Kimmeridgian age.

Analogous structures have been recognised by Germann (1971) and Farinacci et al. (1981, a,b) in the Kimmeridgian of Tyrol (Northern Calcareous Alps) and Northern Apennines, respectively.

Cretaceous

As cretaceous stromatolites of the Penibetic have already been described in detail (Martín-Algarra and Vera, this volume), we should only recall that they are predominantly phosphatic and can appear: a) as oncoids and microcrusts situated in a thin condensed bed (Canutos bed) of uppermost Valanginian-Lower Hauterivian age, overlying a Valanginian discontinuity surface; b) as microstromatolites encrusting discontinuity surfaces of Hauterivian and Aptian age and/or associated with pelagic sediments rich in planktonic foraminifera and nannofossils of Upper Hauterivian-Barremian and of Upper Aptian age; and c) forming an Albian-Vraconian Stromatolite Belt, which was several tens of kilometers long and capped a lower cretaceous paleokarstic surface.

Albian phosphatic nodules of probable stromatolitic origin also exist closely associated with an Albian discontinuity surface in the Malaguide sequences of the Sierra Espuña area, pertaining then to the internal continental margin of the Betics.

Coeval examples have been described in several areas of the Alps (Rioult and Royant, 1975; Gebhardt, 1983; Delamette 1988; Föllmi, 1989) and the Carpathians (Krajewsky, 1981 a,b,c, 1983, etc.). Examples of cretaceous pelagic stromatolites with no Betic counterparts have been described in the base of Campanian in the Southern Alps (Massari, 1979; Massari and Medizza, 1973).

Cretaceous-Tertiary boundary

The only example presently known comes from the Sierra de Aixorta, in the most internal Prebetic, where it was detected by Company et al. (1982). Here, a very irregular, strongly burrowed discontinuity surface (Fig.6) has been carved in Upper Maastrichtian hemi-pelagic yellowish greyish limestones, very rich in large planktonic foraminifera. Discontinuous breccia-levels also appear. Burrowing of *Thalassinoides* type may affect the uppermost metre of Cretaceous limestones to form a complex network. Burrow walls are covered by a thin, millimetric iron crust gradually fading inside the Cretaceous limestones (Fig.6).

Stromatolites grew on the discontinuity surface, forming an irregular dense crust up to 10 cm thick (Fig.5A-B). They are also found as endostromatolites (Monty, 1982) coating the inner wall of burrows in the upper 30-40 cm of the Upper Maastrichtian bioturbated bed (Fig.6). However, they do not occur in the deepest parts of these burrows, which are filled by fine sandy biomicritic calcareous sediment rich in Paleocene planktonic foraminifera. Like their Lower Cretaceous penibetic counterparts, these stromatolites are phosphatic and associated with glauconitic grains, clasts and patinas. Nevertheless they have trapped and binded much more clastic particles than the older stromatolites. Indeed biomicritic, planktonic foraminifera-rich calcareous sediments, with fine angulose quartz and clay particles, are frequently visible between the stromatolite laminae, indicating a much higher sedimentation rate. This fact is easily understood if one considers that the Sierra Aixorta sequence was situated seaward of a carbonate platform with marginal terrigenous influx during Cretaceous-Paleocene times. Coeval and analogous stromatolitic crust have been reported from the Briançonnais Zone, in the Alps (Bourbon, 1980).

PALEOGEOGRAPHIC SIGNIFICANCE

The close association between sedimentary breaks and pelagic stromatolites in Mesozoic times must be understood in terms of: a) the paleogeographic setting of the westernmost Tethys, which favoured breaks in the sedimentation, and b) environmental conditions favouring the microbial activity of stromatolite-building communities. Obviously, the breaks are more easily detectable in swell and shallow platform sequences. In trough sequences they are usually missing because of the more continuous sedimentation. Nevertheless, lithological and facies changes are frequently coeval with discontinuities found in other sequences and domains (see, Martín-Algarra, 1987, fig. 248). This situation shows that the same important events influenced a wide area comprising the two former Betic continental margins as well as the more distant Tethyan regions. But they conditioned different sedimentary results depending on local paleogeographical constraints. These events were tectonic, climatic or eustatic and determined changes in basin morphology or stability, current pattern, terrigenous or biogenic carbonate sediment input, pH, eH and availability in metallic elements, and in the relative position of the coast and the sea-bottom with respect to sea-level.

Role of tectonics.

Syn-sedimentary tectonic activity was related to rifting and later sea-floor spreading during the Jurassic and the Cretaceous in the Tethys. This caused: a) the individualization of Tethyan continental margins (García-Hernandez et al., 1980, Vera, 1981) by the disintegration of the shallow carbonate platform of the Upper Triassic (Betic internal margin) or Lower Liassic (Southern Iberian margin); b) the expansion of pelagic facies throughout the Subbetic realm along the Jurassic and the individualization of swells and troughs caracterized by differencial subsidence; c) the installation of a system of oceanic currents and the general sediment starvation of the margins; d) the existence of contemporaneous active submarine volcanism. These conditions affecting especially swells, favoured condensed and/or reduced

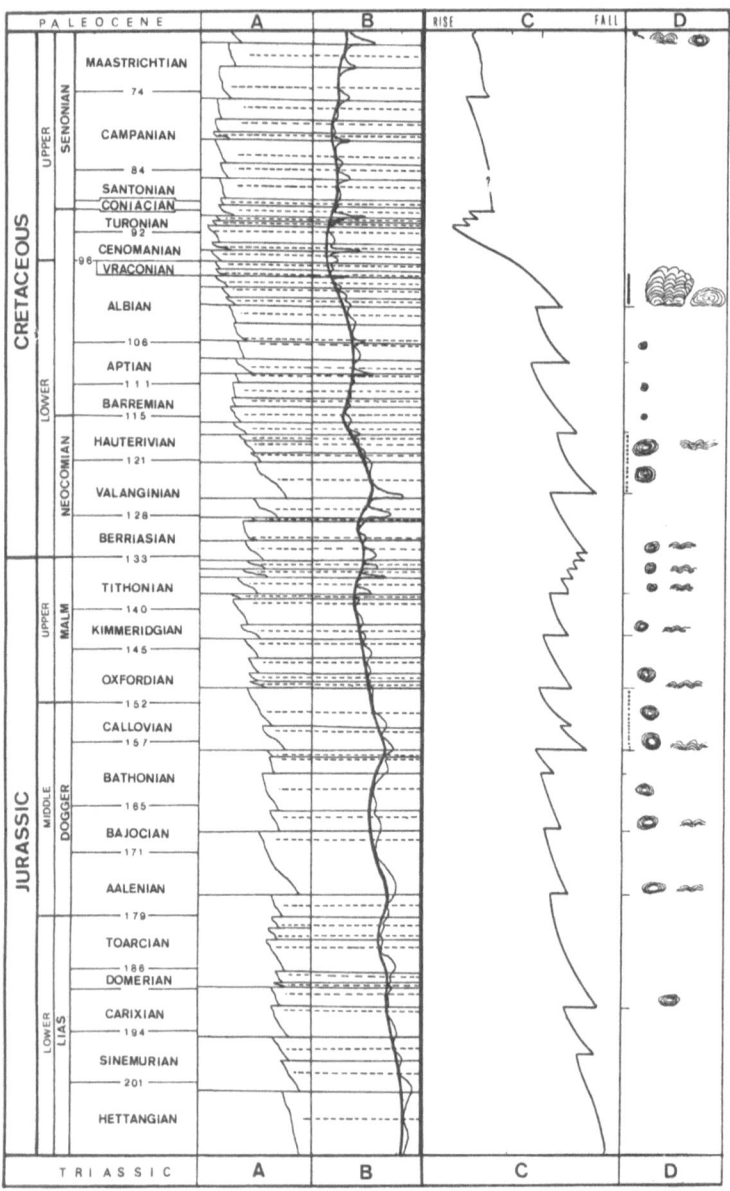

Figure 8 : Jurassic-Cretaceous evolution of the sea-level fluctuations. A) : Sequence stratigraphy after Haq et al. (1987). B) : Eustatic curves after Haq et al. (1987); C) : Fluctuations of the sea-level in the External Zone of the Betic Cordillera (after Vera, 1988). D: Stratigraphic position of pelagic stromatolites in the Betic Cordillera.

sedimentation and, in turn, burrowing and encrustation of the sea bottom by appropiate organisms. Among these communities comprising bacteria, fungi, encrusting foraminifera, and also locally serpulids, were the only agents able to accrete biosedimentary deposits (stromatolites and oncoids) by: a) trapping and binding pelagic sediments (mostly calcareous and biogenic, terrigenous clastics being extremelly scarce in these starved environments), and b) precipitating minerals like Fe-Mn oxides and hydroxides, iron sulphides and phosphates on the sea bottom. During mesozoic oceanic crisis (Monty, 1986) a preliminary enrichment of sea water in metallic elements and/or phosphate existed. This enrichment was induced by current action, decomposition of organic matter (phosphate) or submarine hydrothermalism of volcanic origin. In zones of very low sedimentation rate did exist and, because of the bio-physico-chemical processes associated with bacterial activity, induced mineral deposition in stromatolitic structures. They occurred, possibly in oxidizing environments during the Jurassic and in slightly reducing ones in the Cretaceous because of the general tendency to low-oxygen levels of the Cretaceous sea (Jenkyns, 1980).

Role of eustatism.

The Jurassic pelagic stromatolitic occurrences are systematically associated with the base of transgressive cycles (already noted by Monty, 1973b, when referring to Sicilian Jurassic pelagic stromatolites studied by Jenkyns, 1970, 1971). The facies association and the vertical evolution of stratigraphic sequences illustrate relative sea-level changes which are determined by subsidence and eustasy. The Jurassic stratigraphic breaks of the Betic Cordillera, as can be seen by their position on top of regressive sequences or by their own particular features (such as paleokarstic morphologies on shallow water limestones), seem to have formed most frequently during relative falls in sea-level, partly associated with the raising of the sea-bottom due to tectonic tilting and partly of eustatic origin. These sea-level falls originated sedimentary pauses and abrupt facies changes. They were followed by transgressions much more important than the previous sea-level falls due to the progression of extensional tectonics throughout Jurassic and Cretaceous.

The Cretaceous stratigraphic breaks in the Betic Cordillera are also connected in many cases with eustatic falls (García-Hernández et al., 1982). This has been synthesised in a curve of Mesozoic post-Triassic sea-level evolution for the external continental margin of the Betic Cordillera by Vera (1984b, 1988). The facts revealed by this curve (Fig.8) correspond to the eustatic events recorded by Hallam (1978, 1981, 1984), Vail et al. (1984) and Haq et al. (1987; 1988), as well as the curves proposed for the Apennines and the Alps. Major abrupt falls in sea-level occurred during the Carixian (190 Ma), Upper Bathonian (158 Ma), Valanginian (126 Ma), Aptian (109 Ma) and Lower Albian (100 Ma). Second-order falls in sea-level occurred during the Aalenian (177 Ma), Bajocian (169 Ma), Oxfordian (150,5 Ma), Kimmeridgian (142 Ma), Hauterivian (118 Ma), Turonian (91 Ma), Upper Campanian (75 Ma) and Cretaceous/Tertiary boundary (65 Ma). Pelagic highs were as sensitive to eustatic falls as were the shallow platforms. Resulting sedimentary effects varied according to the site: the shallower pelagic swells (the pelagic platforms) and the neritic carbonate platforms eventually emerged and were karstified, the deeper swells were submitted to submarine erosion (abrasion by currents, bioturbation and bioerosion, submarine dissolution (halmyrolisis) and/or lithification, because of sedimentary still-stand.

In conclusion, the optimal conditions for microbial activity occurred after these sea-level falls and the formation of their associated discontinuity surfaces. Pelagic stromatolite growth on the swells began during periods of rapid sea-level rise, or, in other words, during the development of Transgressive System Tracts (Haq et al., 1987). In this situation only very little sediment, if any, reached the swell bottom, perhaps because of the active current action, and this allowed stromatolite to grow. The microbial activity controlled precipitation of minerals (oxydes, hydroxides, phosphates, etc) and trapped the available sedimentary particles, to form extremely thin, micron-sized, slowly growing mats: their piling up and their overgrowth by encrusting organisms (foraminifera, serpulids, etc) resulted in the accretion of complex stromatolites. Growth periodicity was possibly controlled by poorly known environmental factors that may have been partly climatic. When sedimentation rate increased the growth ceased and the stromatolites were fossilized by pelagic sediments.

The close association of pelagic stromatolites and stratigraphic discontinuity surfaces throughout the Mesozoic, especially on pelagic swells, demonstrate that these surfaces were the most favourable sites for stromatolite growth in the pelagic environment of the Tethys Ocean. With the exception of some outer platform and swell environments where they grew in close association with nodular limestones and associated facies (Jenkyns, 1971; Massari, 1979, 1981, 1983; Comas et al., 1981; Gaillard, 1983; Acosta & García-Hernández, 1988; Martín-Algarra et al., 1990; Dromart et al., this volume) and some particular isolated "pelagic platforms" where pelagic oolitic, stromatolitic and nodular facies developed during the Middle-Late Jurassic and lowermost Cretaceous (Martín-Algarra, 1987), no sites other than break surfaces allowed stromatolites to grow in the pelagic realm. After the lowermost Cretaceous, neither the swells nor the pelagic platforms were colonized any more by stromatolites, except for discontinuity surfaces and very shallow marine, coastal and continental carbonate environments. This fact reflects the Phanerozoic tendency of progressive displacement of stromatolite communities from "normal" marine realms to continental and coastal ones, or to marine areas where environmental conditions were too severe for colonization by organisms other than stromatolite-forming microbes, especially cyanobacteria and bacteria (Monty, 1971, 1973b). This tendency culminated (Monty, 1979) in the disappearance of calcified stromatolites in the marine realm and sediments.

CONCLUSIONS

Finally, from the present study the following conclusions can be drawn:

1.- Stratigraphic break surfaces are frequently found in the Jurassic-Cretaceous pelagic record of the Betic Cordillera, as in the Southern Iberian margin or in the Betic internal margin.

2.- These sedimentary discontinuity surfaces are especially abundant in condensed sequences on pelagic swells, where they constituted very appropriate sites for pelagic stromatolite growth.

3.- Pelagic stromatolites formed on these surfaces are mostly "mineralized" and non-calcareous, preferently ferruginous-manganesiferous and/or phosphatic in composition. They form: crusts on planar to irregular hardgrounds; oncoids with a core of micritic to bioclastic pelagic sediment or ammonites and arborescent microstromatolites on abrasion hardgrounds, shell fragments or walls of neptunian dykes and burrows (endostromatolites).

4.- These stromatolites are possibly bacterial and, to a lesser extent fungal. They were formed by the combination of several processes: accretion of microbial filaments; biogenically controlled precipitation and flocculation of iron and manganese minerals and phosphate; trapping and binding of fine siliciclastic and, especially, fine carbonate sediments (coccoliths); and, finally, encrustation by foraminifera and serpulids.

5.- Significative concentrations of pelagic stromatolites have been found on break surfaces in the following intervals of the Jurassic-Cretaceous record of the Betic Cordillera: Middle Liassic, base of the Middle Jurassic, top of the Bathonian, Dogger-Malm boundary, Kimmeridgian, Upper Valanginian-Lower Hauterivian, Albian-Vraconian and Cretaceous-Tertiary boundary. Small encrustations, mostly microstromatolitic have been found locally in other age intervals.

6.- In all the examples reported above the formation of the break surface itself accounted before stromatolite growth, during periods of sea-level fall and/or lowstand that, in many cases, correlate with tectonic and eustatic events that affected wide areas of alpine mediterranean continental margins. As a consequence of these phenomena, sedimentation stopped and previously deposited sediments were partially eroded, in some cases in subaerial environments. Isolated platforms and pelagic swells were especially sensitive sites for the development of these sedimentary interruptions, while trough areas received more or less continuous sedimentation.

7.- Stromatolite growth occurred later, in submarine and moderatelly deep to deep environments, during sea-level rises after lowstands. They mark, then, the beginning of transgressive cycles and usually are finally buried under pelagic sediments when sea-level rise progresses. Their growth always accounts in extremelly starved environments, with a practically absent or very scarce sediment supply, and probably favoured -at least in some cases- by submarine hydrothermalism and bottom currents. Under these hard environmental conditions, microbial communities and, to a lesser extent, encrusting foraminifera and serpulids, were the only agents able to survive and accrete pelagic biosedimentary deposits.

8.- This kind of stromatolites is different of pelagic calcareous stromatolites associated to some types of nodular limestones of ammonitico rosso facies developped on pelagic swells of the alpine mediteranean regions during the Upper Jurassic-lowermost Cretaceous, and to those associated to Upper Jurassic hemi-pelagic bioconstructions of sponges and other organisms of european epicontinental seas, which are the most recent known examples of calcareous stromatolites within the pelagic realm.

ACKNOWLEDGEMENTS

The authors would like to express their gratitude to Dr. A.Checa, A. Linares, P. Rivas and J. Sandoval, all of the Dpt. Estratigrafia-Paleontologia of the University of Granada, who kindly dated the ammonite fauna and indicated several points where pelagic stromatolites occur; to Dr. N. McLaren for correction of the original manuscript; and to Drs. H.C. Jenkyns (University of Oxford), C.L.V. Monty (University of Nantes) and J. Sarfati (University of Montpellier), who critically read the manuscript. The present study has been financed by CICYT, Research Project no. PB-90-0853.

REFERENCES

Acosta, P. and García-Hernández, M. (1988) "Biohermos de esponjas y estromatolitos en la secuencia transgresiva oxfordense de la Sierra de Cazorla", Geogaceta, 5, 36-39.

Azéma, J., Foucault, A., Fourcade, E., García-Hernández, M., Linares, A., Linares, D., Gonales-Donoso, J.M., López-Garrido, A.C., Rivas, P. and Vera, J.A. (1979) "Las microfacies del Jurásico y Cretácico de las Zonas Externas de las Cordilleras Béticas", Secr. Publ. Univ. Granada, pp. 1-83.

Bernoulli, D. and Jenkyns, H.C. (1974) "Alpine, Mediterranean and central Atlantic Mesozoic facies in relation to the early evolution of the Tethys", in: R.H. Dott and R.W. Shaver (eds.), Modern and ancient geosynclinal sedimentation, Soc. Econ. Paleontologists & Mineralogists, Spec. Publ. no. 19, 129-160.

Bourbon, M. (1980) "Evolution d'un secteur de la marge nord-Tethysienne en milieu pélagique: La Zone Briançonnaise près de Briançon entre le début du Malm et l'Eocène inférieur", Thesis, Univ. L. Pasteur Strasbourg, pp. 1-335.

Brochwicz-Lewinski, W., Gasiewicz, A., Strzelecvki, R., Suffczynski, M., Szatkowski, K., Tarkowski, R. and Zbik, M. (1984) "Anomalia Geochemiczna na pograniczu jury srodkowej i Gornej w poludniowej Polsce", Przeglacl Geologiczny, 12, 647-650.

Comas, M.C., Olóriz, F. and Tavera, J.M. (1981) "The red nodular limestones (Ammonitico Rosso) and associated facies: A key for settling slopes or swell areas in the subbetic Upper Jurassic Submarine Topography (Southern Spain)", in: A. Farinacci and S. Elmi (eds.), Rosso Ammonitico Symposium Proc., Tecnoscienza Roma, pp. 113-136.

Company, M., García-Hernández, M., López-Garrido, A.C., Vera, J.A. and Wilke, H. (1982) "Interpretación genética y paleogeográfica de las turbiditas y materiales redepositados del Senonense superior de la Sierra Aixorta (Prebético interno, Provincia de Alicante)", Cuadernos de Geología Ibérica, 8, 451-465.

Delamette, M. (1988) "Relation between the condensed Albien deposits of the Helvetic domain and the oceanic current-influence continental margin of the northern Tethys", Bull. Soc. Geol. France, (8) IV, 739-745.

Delgado. F., Linares, A., Sandoval, J. and Vera, J.A. (1981) "Contribution à l'étude de l'ammonitico rosso du Dogger dans la Zone Subbeétique", in: A. Farinacci and S. Elmi (eds), Rosso Ammonitico Symposium Proc., Tecnoscienza Roma, pp. 181-197.

Durand-Delga, M. and Fontboté J.M. (1980) "Le cadre structural de la Méditterranée occidentale. In: Géologie des Chaînes Alpines issues de la Téthys", Mem. Bur. Rech. Geol. Min., 15, 67-85.

Elmi, S. (1981) "Sédimentation rythmique et organisation séquentielle dans les Ammonitico-Rosso et les faciès associés du Jurasique de la Mediterranée occidentale. Interprétation des grumeaux et nodules", in: A. Farinacci and S. Elmi (eds.), Rosso Ammonitico Symposium Proc., Tecnoscienza Roma, pp. 251- 299.

Farinacci, A. (1967) "La serie Giurassico-Neocomiana di Monte Lacerone (Sabina). Nuove vedute sull'interpretazione paleogeografica della aree di facies Umbro-Marchigiana", Geol. Romana, 6, 421-480.

Farinacci, A., Malantrucco, G., Mariotti, N. and Nicosia, U. (1981b) "Ammonitico Rosso facies in the framework of the Martani Mountains paleoenvironmental evolution during Jurassic", in: A. Farinacci and S. Elmi (eds.), Rosso Ammonitico Symposium Proc., Tecnoscienza, Roma, pp. 311-334.

Farinacci, A., Mariotti, M., Nicosia, U., Pallini, G. and Schiavinotto, F. (1981b) "Jurassic sediments in the Umbro-Marchean Apennines: An alternative model", in: A. Farinacci and S. Elmi (eds.), Rosso Ammonitico Symposium Proc., Tecnoscienza, Roma, pp. 335-398.

Föllmi, K.B. (1989) "Evolution of the Mid-Cretaceous Triad. Platform carbonates, Phosphatic sediments, and Pelagic Carbonates along the Northern Tethys Margin", Lecture Notes in Earth Sciences 23: 1-153. Springer-Verlag, Berlin.

Gaillard, C. (1983) "Les biohermes à spongiaires et leur environnement dans l'Oxfordien du Jura méridional", Doc. Lab. Géol. Lyon, 90, 1-515.

García-Hernández, M., González-Donoso, J.M., Linares, A., Rivas, P. and Vera, J.A. (1976) "Características ambientales del Lías inferior y medio en la Zona Subbética y su significado en la interpretación general de la Cordillera", in: Reun. Geol. Cord. Bética Mar Alborán, Secr. Publ. Univ. Granada, pp. 125-157.

García-Hernández, M., López-Garrido, A.C., Martín-Algarra, A. and Vera, J.A. (1982) "Cambios eustáticos en el Cretácico de la Cordillera Bética: Comparación de la evolución sedimentaria en un dominio de plataforma (Zona Prebética) y otro de umbral pelágico (Penibético)". Cuadernos de Geología Ibérica, 8, 581-597.

García-Hernández, M., López-Garrido, A.C., Rivas, P., Sanz de Galdeano, C. and Vera, J.A. (1980) "Mesozoic paleogeographic evolution of the external zones of the Betic Cordillera", Geol. Mijnb., 59, 155-168.

García-Hernández, M., Lupiani, E. and Vera, J.A. (1986a) "La sedimentatión liásica en el sector central del Subbético medio: registro de la evolución de un rift intracontinental", Acta Geologica Hispanica, 21-22: 329-337.

García-Hernández, M., Lupiani, E. and Vera, J.A. (1986b) "Discontinuidades estratigráficas del Jurásico de Sierra Gorda (Subbético interno, provincia de Granada)", Acta Geologica Hispanica, 21-22, 339-349.

García-Hernández, M., Martín-Algarra, A., Molina, J.M., Ruiz-Ortiz, P.A. and Vera, J.A. (1988) "Umbrales pelágicos: Metodología de estudio, tipología y significado en el análisis de cuencas", II Congr. Geol. España SGE, Granada, Simposios, pp. 231-240.

Gebhardt, G .(1983) "Stratigraphische kondensation am beispiel mittelkretazischer vorkommen in perialpine raum", Thesis Univ. Tübingen, pp. 1-145.

Germann, K. (1971) "Mangan-Eisen-fuhrende Knollen und Krusten in jurassischen Rotkalken der Nordlichen Kalkalpen", N. Jb. Geol. Palaont. Mh., 197, 133-156.

González-Donoso, J.M., Linares, D., Martín-Algarra, A., Rebollo, M., Serrano, F. and Vera, J.A. (1983) "Discontinuidades estratigráficas durante el Cretácico en el Penibético (Cordillera Bética)", Estudios Geológicos, 39, 71-116.

Hallam, A. (1978) "Eustatic cycles in the Jurassic", Palaeogeogr. Palaeoclimatol. Palaeoecol., 23, 1-32.

Hallam, A. (1981) "A revised sea-level curve for the early Jurassic", J. Geol. Soc. London, 138, 735-743.

Hallam, A. (1984) "Pre-quaternary sea-level changes", Ann. Rev. Earth Planet. Sci., 12, 205-243.

Haq, B.U. Hardenbol, J. and Vail, P.R. (1987) "Chronology of Fluctuating Sea Level Since the Triassic", Science, 235, 1156-1167.

Haq, B,U,, Hardenbol, J. and Vail, P.R. (1988) "Mesozoic and Cenozoic chronostratigraphy and cycles of sea-level change", in: C.K. Wilgus, B.S. Hastings, C.G.St.C. Kendall, H.W. Posamentier, C.A. Ross and

J.C. van Wagoner (eds.), Sea-level changes: an integrated approach, Soc. Econ. Paleontologists & Mineralogists, Spec. Publ. no. 42: 71-108.

Jenkyns, H.C. (1970) "Fossil Manganese Nodules from the West Sicilian Jurassic", Eclog. Geol. Helv., 63, 741-774.

Jenkyns, H.C. (1971) "The genesis of condensed sequences in the Tethyan Jurassic", Lethaia, 4, 327-352.

Jenkyns, H.C. (1980) "Cretaceous anoxic events: from continents to oceans". J. Geol. Soc. London, 137, 171-188.

Krajewski, K.P. (1981a) "Phosphate microstromatolites in the High- Tatra Albian limestones in the Polish Tatra Mountains", Bull. Acad. Pol. Sci. Ser. Sci. Terre, 29, 175-183.

Krajewski, K.P. (1981b) "Phosphate pizolite structures from condensed limestones of the High-Tatric Albian (Tatra Mts.)", Ann. Soc. Geol. Pol., 51, 339-352.

Krajewski, K.P. (1981c) "Pelagiczne stromatolity z wapieni albu wierchowego Tatr", Kwart geol, 25, 731-759.

Krajewski, K.P. (1983) "Albian pelagic phosphate-rich macrooncoids from the Tatra Mts. Poland", in: T.M. Peryt (ed), Coated Grains, pp. 344-357, Springer-Verlag, Berlin.

Martín-Algarra, A. (1987) "Evolución geológica alpina del contacto entre las Zonas Externas y las Zonas Internas de la Cordillera Bética (sector centro-occidental)", Thesis, Univ. Granada, pp. 1-1171

Martín-Algarra, A., Acosta, P., García-Hernández, M. and Checa, A. (1990) "Oxfordian bioconstruction of stromatolites and sponges in the Sierra de Cazorla (Prebetic Zone, Southern Spain)", 13th Int. Sedim. Cong. (Sediments 1990), Nottingham, Abstracts of Papers, pp. 329-330.

Martín-Algarra, A., Checa, A., Oloriz, F. and Vera, J.A. (1983) "Un modelo de sedimentación pelágica en cavidades kársticas: la Almola (Cordillera Betica)", X Congr. Nac. Sedimentología, Mahón, 3, 22-25.

Martín-Algarra, A. and Vera, J.A. (1982) "El Cretácico del Penibético, de las unidades del Campo de Gibraltar, las Zonas Internas y las unidades implicadas en el contacto entre las Zonas Internas y las Zonas Externas", in: A.García (ed.), El Cretácico de España, Univ. Complutense, Madrid, pp. 603-638.

Martín-Algarra, A. and Vera, J.A. (1989) "La serie estratigráfica del Penibético (Cordillera Bética)", in: Libro Homenaje a Rafael Soler, Asoc. Geol. Geof. Petrol., Madrid, pp. 67-76.

Martín-Algarra, A. and Vera, J.A. (this volume) "Mesozoic pelagic phosphate Stromatolites from the Penibetic (Betic Cordillera, Southern Spain)".

Massari, F. (1979) "Oncoliti e stromatoliti pelagiche nell Rosso Ammonitico veneto", Mem. Sc. Geol., 32, 1-21

Massari, F. (1981) "Cryptalgal fabrics in the Rosso Ammonitico sequences of the Venetian Alps", in: A. Farinacci and S. Elmi (eds), Rosso Ammonitico Symposium Proc., Tecnoscienza, Roma, pp. 435-469.

Massari, F. (1983) "Oncoids and Stromatolites in the Rosso Ammonitico Sequences (Middle-Upper Jurassic) of the Venetian Alps, Italy", in: T.M. Peryt (ed.), Coated grains, Springer-Verlag, Berlín, pp. 358-366.

Massari, F. and Dieni, I. (1983) "Pelagic oncoids and ooids in the Middle-Upper Jurassic of Eastern Sardinia", in: T.M. Peryt (ed), Coated grains, Springer-Verlag, Berlin, pp. 367-376.

Massari, F. and Medizza, F. (1973) "Stratigrafia e paleogeografia del Campaniano-Maastrichtiano nelle Alpi Meridionali (con particolori riguardo agli hardgrounds della Scaglia Rossa Veneta)", Mem. Ist. Geol. Min. Univ. Padova, 28, 1-62.

Molina, J.M. (1987) "Análisis de facies del Mesozoico en el Subbético Externo (provincia de Córdoba y sur de Jaén)", Thesis, Univ. Granada, pp. 1-512.

Molina, J.M., Ruiz-Ortiz P.A. and Vera J.A. (1985) "Sedimentación marina somera entre sedimentos pelágicos en el Dogger del Subbético Externo", Trabajos de Geología, Oviedo, 15, 127-146

Monty, C.L.V. (1971) "An autoecological approach of intertidal and deep-water stromatolites", Ann. Soc. Geol. Belgique, 94, 265-276.

Monty, C.L.V. (1973a) "Les nodules de manganèse sont des stromatolites océaniques", C.R. Acad. Sci. Paris, D-276: 3285-3288.

Monty, C.L.V. (1973b) "Precambian background and Phanerozoic history of stromatolite communities, an overwiew", Ann. Soc. Geol. Belgique, 96, 585-624.

Monty, C.L.V. (1979) "Scientific reports of the belgian expedition on the Australian Great Barrier Reefs. Sedimentology: 2. Monospecific stromatolites from the great barrier Reef Tract and their paleontological significance", Ann. Soc. Geol. Belgique, 101, 163-171.

Monty, C.L.V. (1982) "Cavity or fissure dwelling stromatolites (endostromatolites) from Belgian Devonian mud mounds (Extended Abstract)", Ann. Soc. Geol. Belg., 105, 343-344.

Monty, C.L.V. (1986) "Interactions événements géologiques-Stromatolites", Bull. Cent. Rech. Explor. Prod. Elf Aquitaine, 10, 537-553.

Paquet, J. (1969) "Etude géologique de l'Ouest de la province de Murcie (Espagne)", Mem. Soc. Geol. France, 111: 1-270.

Playford, P.E. and Cockbain, A.E. (1969) "Algal stromatolites: deep-water forms in the Devonian of Western Australia", Science, 165, 1008-1010.

Playford, P.E., Cockbain, A.E., Druce, E.C. and Wray, J.L. (1976) "Devonian Stromatolites from the Canning Basin, Western Australia", in: M.R. Walter (ed.), Stromatolites, Devel. in Sedimentology 20, Elsevier, Amsterdam, pp. 543-563.

Prescott, D.M. (1988) "The geochemistry and paleoenvironmental significance of iron pisoliths and ferromanganese crusts from the Jurassic of Mallorca, Spain", Eclogae Geol. Helv., 81, 387-414.

Rioult, M. and Royant, G. (1975) "La croute stromatolique in Manara horizon repère de l'Aptien-Albien dans la serie briançonnaise du Monte Armelta (Alpes Ligures)",Congr.Inter. Sediment., Nice 10, 123-128.

Sequeiros, L. (1974) "Paleobiogeografía del Calloviense y Oxfordense en el sector central de la Zona Subbética", Thesis Univ. Granada, 65, 1-635.

Seyfried, H. (1978) "Der Subbetische Jura von Murcia (Sudost-Spanien)", Geol. Jb. (B), 29, 1-21.

Seyfried, H. (1979) "Ensayo sobre el significado paleogeográfico de los sedimentos del Jurásico de las Cordilleras Béticas orientales", Cuad. Geol. Univ. Granada, 11: 317-348.

Seyfried, H. (1981) "Genesis of "regressive" and "transgressive" pelagic sequences in the Tethyan Jurassic", in: A. Farinacci and S. Elmi (eds.), Rosso Ammonitico Symp., Tecnosc. Roma, pp. 547-579.

Tucker, M.E. (1973) "Ferromanganese nodules from the Devonian of the Montagne Noire (S. France and West Germany)". Geol. Rundschau, 62, 137-153.

Vail, P.R. Hardenbol, J. and Todd, R.G. (1984) "Jurassic Unconformities, Chronostratigraphy, and Sea-Level Changes from Seismic Stratigraphy and Biostratigraphy", in: Amer. Assoc. Petrol. Geol., mem. 36, 129-144.

Vera, J.A. (1966) "Estudio geológico de la Zona Subbética en la transversal de Loja y sectores adyacentes" Thesis Univ. Granada (Mem. Inst. Geol. Min. Esp. 72, pp. 1-192, 1969).

Vera, J.A. (1981) "Correlación entre las Cordilleras Béticas y otras cordilleras alpinas durante el Mesozoico", in: PICG Real Acad. Cienc. Exact. Fis. Nat. Madrid, 2, 125-160.

Vera, J.A. (1984a) "Discontinuidades estratigráficas en materiales pelágicos: caracterización, génesis e interpretación". I Congr. Nac. Geol., Segovia, 3: 109-122.

Vera, J.A. (1984b) "Aspectos sedimentológicos en la evolución de los dominios alpinos mediterráneos durante el Mesozoico", in: A. Obrador (ed), Libro Homenaje a L. Sánchez de la Torre, Grup. Esp. Sed., Publ. Geología, Barcelona, 20: 25-54.

Vera, J.A. (1986) "Las Zonas Externas de las Cordilleras Béticas", in: Geología de España, Inst. Geol. Min. España, Madrid, 2: 218-251.

Vera, J.A. (1988) "Evolución de los sistemas de depósito en el margen ibérico de las Cordilleras Béticas", Rev. Soc. Geol. España, 1, 373-391.

Vera, J.A., Molina, J.M. and Ruiz-Ortiz, P.A. (1984) "Discontinuidades estratigráficas, diques neptúnicos y brechas sinsedimentarias en la Sierra de Cabra (Mesozoico, Subbético Externo)", in: A. Obrador (ed), Libro Homenaje a L. Sánchez de la Torre, Grup. Esp. Sed., Publ. Geología, Barcelona, 20, 141-162.

Vera, J.A., Ruiz-Ortiz, P.A., García-Hernández, M. and Molina, J.M. (1988) "Paleokarst and related Pelagic Sediments in the Jurassic of the Subbetic Zone, Southern Spain", in: N.P. James and P.W. Choquette (eds.), Paleokarst, Springer New York, pp. 364-384.

MESOZOIC PELAGIC PHOSPHATE STROMATOLITES FROM THE PENIBETIC (BETIC CORDILLERA, SOUTHERN SPAIN)

A.MARTIN-ALGARRA and J.A.VERA
Departamento Estratigrafía-Paleontología and I.A.G.M.- Facultad de Ciencias.- Universidad. 18071.- Granada ; Spain

ABSTRACT

In the Penibetic (Southern Spain), a Mesozoic pelagic swell of the Western Betic Cordillera, several examples of pelagic phosphate stromatolites have been detected. They lie on three regionally important stratigraphic discontinuity surfaces of different ages respectively: the Dogger-Malm boundary, the base of the Upper Valanginian and the Lower Albian. The stromatolites are usually several centimetres in size and their morphology varies between ovoidal, subspheroidal or flattened nodules (oncoids) and laminated crusts of planar, cupola-like or irregular shape. Microscopic examination reveals that stromatolite building is essentially the result of the textural interaction of four types of elements: 1) microbial filaments of probable bacterial origin; 2) phosphates and minor amounts of other minerals (goethite, pyrite, etc.); 3) pelagic sediment rich in planktonic foraminifera and coccoliths; and, 4) encrusting foraminifera. These elements appear interrelated in various ways giving rise to several mesoscopic morphologies and microscopic stromatolite fabrics, sometimes arranged in cycles. Stromatolite accretion occurred as a result of the piling-up of microbial, especially bacterial filament bundles and, due to their life processes, these bacteria precipitated pyrite and floculated jelly-phosphate substances around the filaments. Small pelagic particles, especially coccoliths, were trapped and bound during stromatolite growth. Encrusting foraminifera consolidated the structure. The growth rhythms or cycles reveal that there have been periodic variations of different duration in the speed and mode of stromatolite growth (variations which are especially obvious in Albian-Vraconian examples). These periodic changes that affected a pelagic realm could have been caused in the last instance by climatic factors which have modified environmental parameters (currents, temperature, planktonic productivity, etc.) and/or determined the existence of complex microbial ecological successions. The formation of Penibetic phosphatic stromatolites occurs in not very deep (several hundred meters at most) zones of very low sedimentation rate, affected by oceanic (upwelling or geostrophic) currents, near the O_2 minimum layer. In some examples (Albian-Vraconian) the contemporaneous existence of a low-temperature submarine hydrothermalism cannot be excluded.

J. Bertrand-Sarfati and C. Monty (eds.), Phanerozoic Stromatolites II, 345–391.
© 1994 *Kluwer Academic Publishers.*

INTRODUCTION.

The existence of phosphatic and/or ferruginous-manganiferous stromatolites in the Mesozoic pelagic, stratigraphic record of the European Alpine regions is well known. They have frequently been considered as not stromatolites but accretionary or concretionary nodular structures or crusts caused by chemical precipitation at the sea bottom. These structures have been detected in Mesozoic rocks from numerous plateaus or swells in the Alps (Heim, 1946, 1958; Rioult and Royant, 1975; Massari, 1975, 1979, 1981, 1983; Massari and Medizza, 1973; Föllmi, 1989), Sardinia (Massari and Dieni, 1983), the Apennines (Farinacci, 1967), the Carpathians (Krajewski, 1981a,b,c, 1983, 1984), Sicily (Jenkyns, 1967, 1970a,b) and in the Betic Cordillera (Chauve et al., 1968, González-Donoso et al., 1983; García-Cervigón et al., 1986; Martín-Algarra, 1987; Vera and Martín-Algarra, this volume). In the most recent studies these structures are extremely useful in improving our knowledge of the sedimentary processes and the palaeogeographical setting of the environments and the series in which they appear. This, in turn, has provoked an increasing interest in their detailed study (Krumbein, 1983). This interest has been further stimulated by the fact that these microbial accretions are regularly found lying on stratigraphic gaps or discontinuity surfaces frequently closely related to sea-level changes (Vail et al., 1977; García-Hernández et al., 1982; Vera 1984a,b; Haq et al., 1987; Vera and Martín-Algarra, this volume).

New data on these stromatolites are presented herein, as well as hypotheses about their growth pattern and their palaeogeographic setting. Some preliminary results have already been published in earlier articles (González-Donoso et al., 1983; García-Cervigón et al., 1986-87) and especially in Martín-Algarra (1987).

GEOLOGICAL SETTING.

The Penibetic is the innermost (south-western) tectonic and palaeogeographic realm within the External Zones of the Betic Cordillera (Fig.1). It consists of an essentially carbonate Mesozoic stratigraphic sequence, with important stratigraphic gaps, especially as regards the Lower Cretaceous (González-Donoso et al., 1983; Martín-Algarra, 1987). Pelagic phosphate stromatolites appear at several levels in the three lithostratigraphic groups of the Penibetic (Fig.2). In chronological order they are: Hidalga, Líbar and Espartina Groups (Martín-Algarra, 1987; Martín-Algarra and Vera, 1989).

The Hidalga Group is composed of Triassic rocks and contains no pelagic stromatolites. These are present, however, in the Líbar Group, and notably in the contact between this group and the Espartina Group, an example of stratigraphic discontinuity surface extremely important regionally, resulting in a systematic absence of most of the Lower Cretaceous in the Penibetic. We will now describe briefly some of the salient stratigraphic features of the Líbar and Espartina Groups.

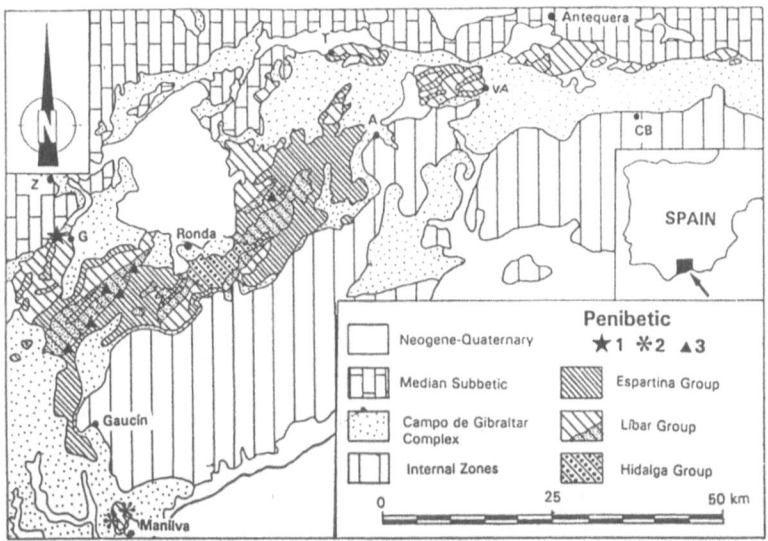

Figure 1 : Geological sketch map of the Western Betic Cordillera (Southern Spain). Pelagic phosphate stromatolites: 1.- Jurassic stromatolites (Oxfordian). 2.- Cretaceous pre-karst stromatolites (Canutos beds: Uppermost Valanginian-Lower Hauterivian). 3.- Cretaceous post-karst stromatolites (Stromatolite Belt: Albian-Vraconian). Stippled of the Libar Group correspond to the Dolomite Belt. Localities: CB.-Casabermeja, VA.-Valle de Abdalajís, A.-Ardales, T.-Teba, G.-Grazalema, Z.-Zahara.

The Líbar Group

This group is made exclusively of carbonate rocks ranging in age from the Liassic to the Lower Valanginian. It can be subdivided into two limestone formations: the lower, made of shallow-water facies of Lias-Dogger age (Endrinal Fm.), and the upper, constituted of pelagic facies of Malm-Lower Valanginian Age (Torcal Fm.). Its base has been systematically dolomitized, probably by thermal processes (Martín-Algarra, 1987): this is especially well-developed along a regional trend where most of the Líbar Group has been dolomitized up to its uppermost part, term here the Dolomite Belt (Fig.2).

The Endrinal Fm. was deposited in a carbonate bank isolated from the Iberian Continent by the deep trough of the Median Subbetic. This "isolated platform" (cf. Read, 1982, 1985) was individualized by rifting from the Liassic onwards (García-Hernández et al., 1976, 1980, 1986; Azéma et al., 1979).

The contact between the Endrinal Fm. and the Torcal Fm. is marked by a discontinuity surface becoming progressively more erosional and morphologically irregular towards the outer northwestern sectors (Fig.2). There, it has locally the features of a palaeokarst (Fig.3A,B), bearing the oldest pelagic phosphate stromatolites discussed in the present

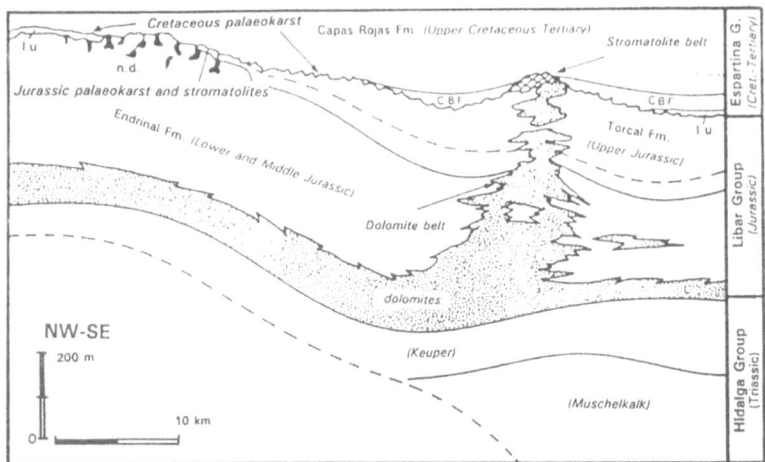

Figure 2 : Simplified stratigraphical scheme of the Penibetic. Note superposition of Jurassic and Cretaceous palaeokarsts and neptunian dykes in the external part (NW) and correlative thinning of pelagic formations. Key: n.d.- neptunian dykes. l.u.- lower stratigrafic units of the Espartina Group such as the Canutos Bed and others. CBF.- Capas Blancas Fm.

study (Fig.3D). This surface formed at the same time as the replacement of shallow neritic facies of the Endrinal Fm. by the pelagic facies of the Torcal Fm. during the Dogger-Malm transition. This stratigraphic break is connected with a tectonic tilting of blocks which brought about the emergence and karstification of the northwestern sectors of the Penibetic and their later sinking at the beginning of the Malm. Similar synchronic phenomena have been described in other swells of the same basin by Vera et al. (1988).

Figure 3 : A) Field view at Grazalema of exhumed Cretaceous paleokarst on the limestones of the Líbar Group. Bedding to the left. In the wall, the contact between the Endrinal Fm. (E) and the Torcal Fm. (T) is visible (arrows). Jurassic stromatolites and oncoids (e) lye on this contact. Note neptunian dyke filling of Torcal Fm. (T, arrowed) within the Endrinal Fm. (E) in the lower right part of the photo (close-up in B and C). Note well-preserved Cretaceous lapies (karren) morphologies, perpendicular to bedding (then originally subvertical), still partly filled with pelagic sediments (s) very rich in planktonic foraminifera of Cretaceous age (Upper Aptian, at this point). B) Close-up of the lower right part of a, showing a fisure of karstic origin in the Endrinal Fm (E) filled up with fossiliferous pelagic limestones, rich in dwarf ammonites of Upper Oxfordian age of the Torcal Fm. (T). C) Close-up of B; note preserved, formerly aragonitic, ammonite shells. D) Detail of the lower left part of A, showing the planar abrasion discontinuity surface between Torcal Fm. (T) and Endrinal Fm. (E) bearing pelagic stromatolites (black crusts) and oncoids (e). E) Probably autoclastic collapse breccias of Endrinal Fm. grainstones bound by Upper Aptian pelagic micritic sediment (s).

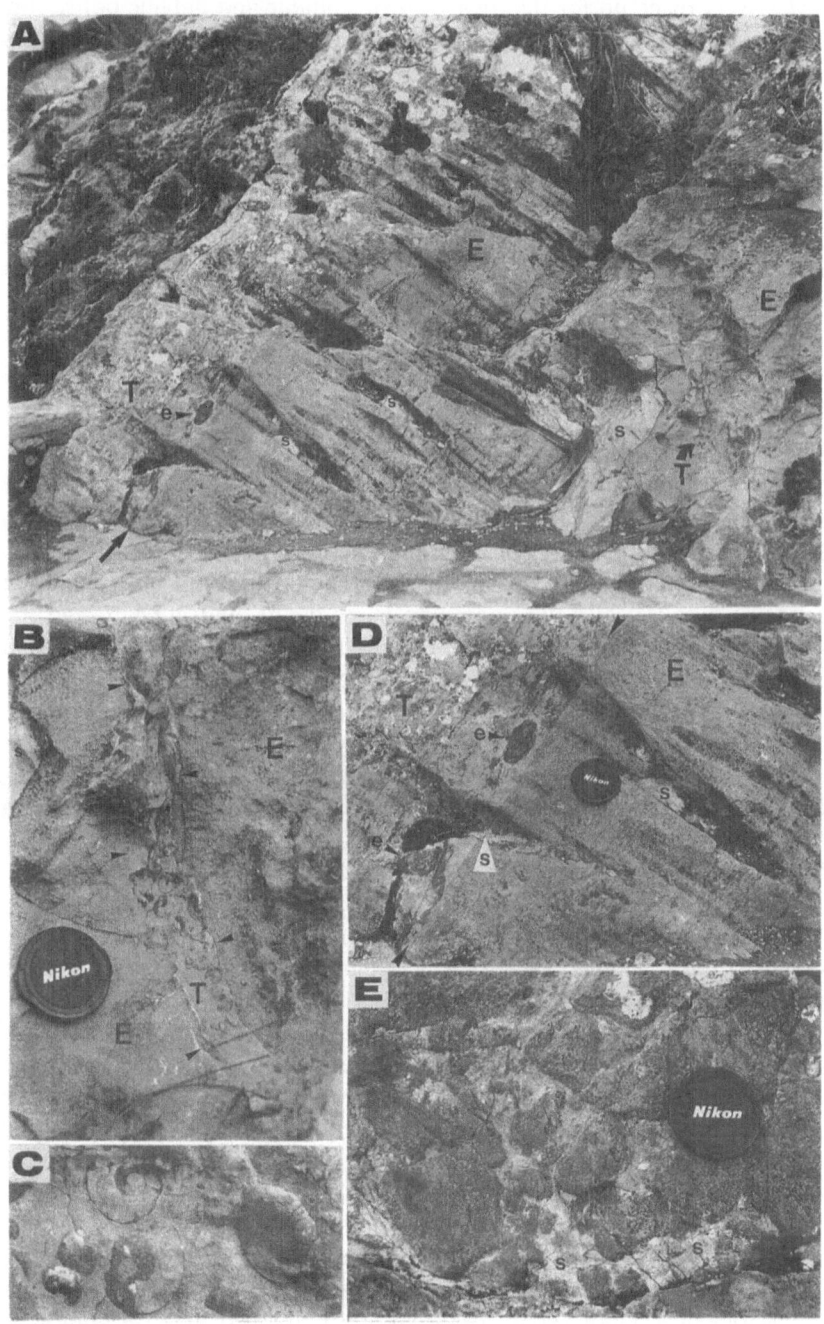

The Torcal Fm. is made up for the most part of nodular and pelagic oolitic limestones (sensu Jenkyns, 1972, 1980b) and, to a lesser degree, of other facies. These rocks contain abundant ammonites (Fig.3B,C) and other macrofossils (belemnites, solitary corals, brachiopods, bivalves, crinoids, etc.), as well as many pelagic microfossils (planktonic foraminifera and crinoids, calcispherulids and calpionellids, etc.). The existence of these fossils allows us to establish several biozones within the Upper Jurassic and Neocomian. The Torcal Fm. was deposited on a very shallow pelagic swell, termed a "pelagic platform" by Martín-Algarra (1987), which had some peculiar features. Together with many pelagic macro- and micro-organisms, other organisms proliferated (probably algae and cyanobacteria) on this swell, and these in turn produced pelagic ooids, oncoids and stromatolites (see Martín-Algarra, 1987).

The Espartina Group.

Breaks in the Lower Cretaceous and Lower Units.

The upper limit of the Líbar Group is established by the appearance of a very clear erosional surface in which a large stratigraphic gap affecting most of the Lower Cretaceous can be detected (Fig.2, 3A). A close study of this contact has shown that this discontinuity surface is really very complex in its detail, and is composed of various minor sedimentary gaps, one superimposed on the other and separated by thin (millimetric to decimetric) pelagic episodes of strongly condensed and laterally discontinuous sediments of the Lower Cretaceous. Despite their extreme thinness, these sediments can be dated, since they are very rich in planktonic foraminifera, nannoplankton and locally even ammonites, so that these gaps were dated as: intra-Valanginian, intra-Hauterivian, intra-Aptian and intra-Albian (Martín-Algarra and Vera, 1982; González-Donoso et al., 1983; Martín-Algarra, 1987).

The sedimentary breaks mentioned above occurred in a submarine environment, except for that of Hauterivian age, which was caused by a generalized emergence of the Penibetic. This emergence, like that which occurred at the Dogger-Malm limit, was probably due to the combined effects of a fall in sea-level (García-Hernández et al., 1982) and a tectonic tilting that caused a new raising and emergence of the northwestern sectors of the Penibetic: because of this, a karstic paleorelief was formed, which, in the Ubrique-Grazalema area, affects the uppermost hundred metres of Líbar Group limestones (Fig.2). Thus, during the most part of the Lower Cretaceous the top of the Líbar Group was the rock ground of a relatively shallow pelagic plateau, on which there was scarcely any sedimentation. Together with other factors, this favoured in turn the growth of pelagic phosphate stromatolites, both before and after the emergence and karstification.

The pelagic phosphate stromatolites (Fig.4, 5A) formed prior to emergence are situated in a carbonate layer some centimetres thick, dated as Uppermost Valanginian-Lower Hauterivian (González-Donoso et al., 1983) and formally termed the "Canutos Bed" by

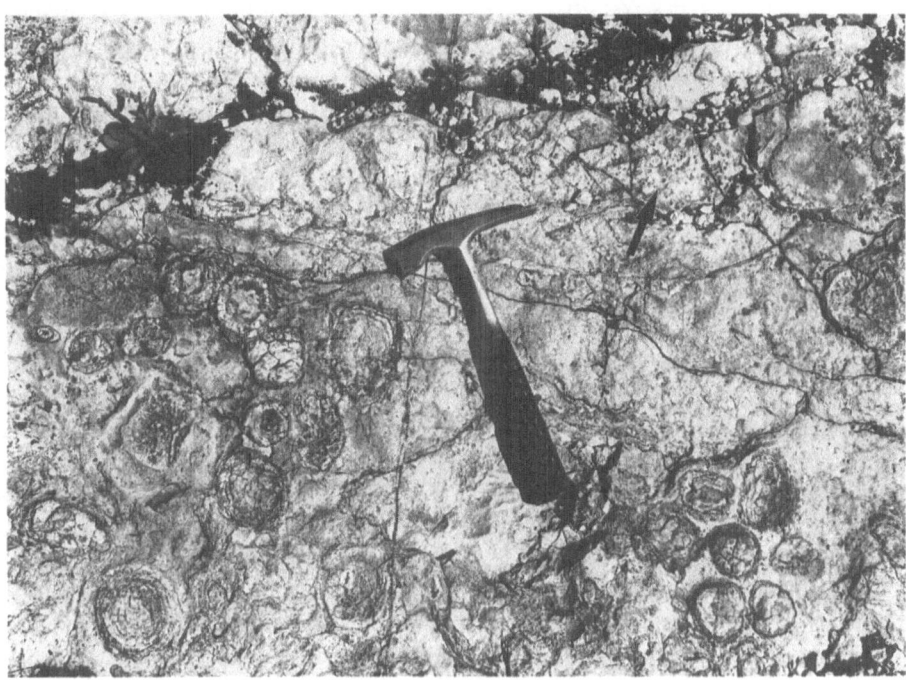

Figure 4 : Pre-karst Canutos Bed (Uppermost Valanginian-Hauterivian), Los Canutos de Manilva (plan view).

Martín-Algarra (1987); it crops out only in the most internal sectors of the Penibetic. Some small pelagic phosphate microstromatolite crusts formed
after the karstification are Upper Hauterivian and Upper Aptian, but most of them are found during the Albian-Vraconian, coinciding with the renewal of the sedimentation (Fig.5B). The growth of the latter stromatolites on the palaeokarst only occurred in an area of minimal sedimentation that coincided approximately with the position of the Dolomite Belt of the Líbar Group in earlier times, and which we have termed the Stromatolite Belt (Fig.2).

The bedded formations of the Espartina Group.

From the Middle-Upper Albian up to the Tertiary, marly and limy-marly sediments, rich in planktonic foraminifera and nannofossils of the Espartina Group progressively buried the palaeokarst and the pelagic stromatolites. Nevertheless, it should be noted that, while in some areas of the Penibetic the deposition of the Albian-Vraconian stratified sediments (Capas Blancas Fm.) took place, in the Stromatolite Belt the greatest known proliferation of pelagic phosphate stromatolites occurred at the same time.

The Capas Blancas Fm. (similar to the Apenninic *Scaglia Bianca*) is made up of well-bedded marly limestones and marls, generally in decimetric strata and showing a rhythmic pattern. Although light coloured on outcrop, they are grey, greenish or black on clean fractures; these dark colours result from their abundance of organic matter, locally concentrated in thin layers of asphalt or bituminous shales. Within the Capas Blancas Fm. limonitized pyrite nodules and, particularly, chert nodules frequently appear; the latter become more and more abundant toward the top of the formation, allowing us to differentiate two members: the lower member is made up of limestones and marls, and the upper one of cherty platy limestones. Both are rich in planktonic foraminifera with some ammonites in the lower one, allowing very accurate dating (González-Donoso et al., 1982, 1983). The Cherty limestones Boquerón Member disappears at the northwestern limit of the Stromatolite Belt, due to a stratigraphic break beneath the overlying Capas Rojas Fm. In the outcrops which coincide with the Stromatolite Belt (e.g. Puerto del Viento), the Member shows local slump folds, thus demonstrating former depositional paleoslopes in the neighbourdhood, and tectonic instability at the time of deposition. On the other hand, to the SE of the Stromatolite Belt, complete sedimentary continuity has been found, and there is a quick but gradual transition from the Cherty limestones Boquerón Member to the Capas Rojas Fm.

The Capas Rojas Fm. (similar to the Apenninic *Scaglia Rossa*) is made up of marly-limy rhythmites with abundant planktonic foraminifera and coccoliths, as in the Capas Blancas Fm. The essential difference between them is in the salmon-red colour, indicating an important change in redox depositional conditions. The Capas Blancas Fm. was deposited in anoxic marine waters which invaded the Penibetic during the Mid-Cretaceous Oceanic Anoxic Event (OAE) (Schlanger and Jenkyns, 1976; Jenkyns, 1980a, 1985; Lloyd, 1982), while the deposition of the Capas Rojas Fm. occurred in well-oxygenated waters. Two members can be differentiated: the lower one is limy, Coniacian-Lower Campanian in age; the upper member is marly, and ranges from the Upper Campanian to the Upper Eocene-Oligocene (Martín-Algarra and Martínez-Gallego, 1984).

The sedimentary features of the bedded formations of the Espartina Group show that throughout the Upper Cretaceous and the Tertiary the Penibetic deepened and evolved to a drowned isolated platform (Read, 1982, 1985). Preliminary studies of the marly-limestone rhythms of the Capas Blancas Fm. and the Capas Rojas Fm. favour the hypothesis of their climatic control in the Milankovitch band (Martín-Algarra, 1987), as proposed by several studies for similar sediments of other alpine areas (Fischer, 1976, 1981, 1982; de Boer, 1983; de Boer and Wonders, 1984; Arthur et al., 1986). The deepening tendency of the Penibetic is confirmed by the increase in the marl content from the Upper Cretaceous onwards, and by the appearance of flysch-like detrital facies during the Tertiary, before the definitive stop of the sedimentation. Regional palaeogeographic studies show that although there is a clear deepening trend during the last stage of the post-Albian sedimentary history of the Penibetic, it still constituted a relative elevation within the whole southwestern Betic continental margin, when compared to deeper basins adjacent. In the more external Median Subbetic trough of the

External Zones, or the more internal Betic Flysch Basin, contemporary deposits were of deeper facies, more clayey and sedimentation probably occurred near or below the CCD (Martín-Algarra, 1987; López-Galindo and Martín-Algarra, 1992).

Stromatolite beds and associated sediments.

The pelagic stromatolites of the Penibetic have been found in three different stratigraphic positions, with varying development in each case. They have similar morphological, compositional, textural and genetic features, and show only minor differences, depending on their relative stratigraphic positions and on the detailed sedimentological context. We will first describe the Jurassic examples and later the pre- and post-karstification Cretaceous occurrences as in each case they show very different field relations with the surrounding rocks.

Stromatolites from the Dogger-Malm boundary.

Stromatolites from this age have only been found in the Grazalema area (Fig.1). These structures are visible in the wall of one of the big prominences of the Cretaceous palaeokarst, precisely on the discontinuity surface between the Endrinal Fm. and the Torcal Fm (Fig.3A). In this area, the Endrinal Fm. is composed of tabular cross-bedded calcarenites and calcirudites. Texturally, they are oolitic and bioclastic grainstones, with benthic foraminifera of the Dogger. Locally, the textural elements show incipient micritization, and may have been leached along karstic discontinuity surfaces that make the contact between the Endrinal Fm. and the Torcal Fm. This contact is an erosional surface that laterally crosscuts the Endrinal limestones, revealing irregular cavities and fissures that are filled up with pelagic sediments of the Torcal Fm. (neptunian dykes: Fig.3B,C). The Torcal Fm. limestones show the facies of condensed, fossiliferous pelagic limestones (Schlager, 1974). These limestones are of light brown colour and contain biomicritic microfacies, rich in *Protoglobigerina, Saccocoma*, calpionellids and macrofossil fragments (brachiopods, belemnites, corals, aptychi and above all ammonites). Like some other fossils, ammonites preserve, either partially or totally, their formerly aragonitic, neomorphosed shell (Fig.3C), intensely corroded by endolithic organisms, possibly algae and fungi. The existence of the ammonites allows us to date the base of the Torcal Fm., where the stromatolites appear at this point, as of Middle-Upper Oxfordian age. In this area, the thickness of the Torcal Fm. is greatly reduced (hardly reaching 5 m: Fig.2), although its upper part can be dated, thanks to the presence of several calpionellid species and genera, as having been deposited during uppermost part of the Lower Berriasian (González-Donoso et al., 1983).

Jurassic (Oxfordian) microbial structures form centimetric nodules (oncoids) just on the contact between the Endrinal Fm. and the Torcal Fm., and thin (millimetric to centimetric) crusts on abrasion hardgrounds (Seyfried, 1979; Martín-Algarra, 1987), which truncate the upper part of the macrofossils and mark fine stratification surfaces

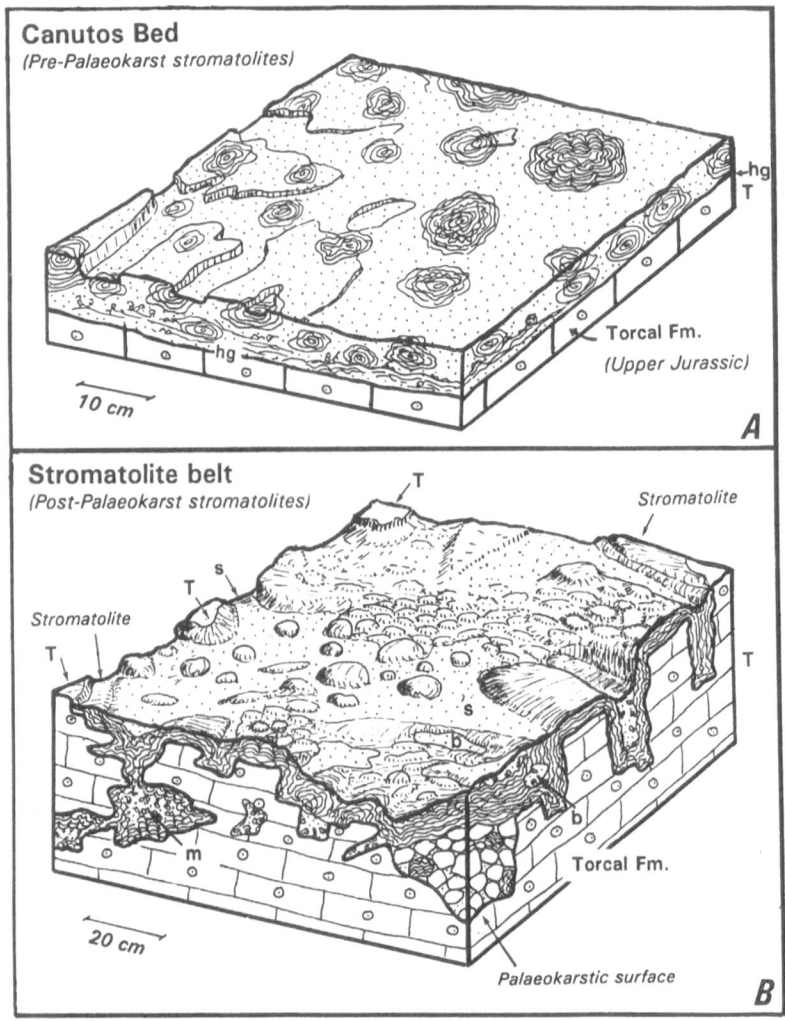

Figure 5 : A) Field aspect of the Canutos Bed. The contact with the Torcal Fm. (T) is a hardground (hg). B) Schematic block diagram showing field structural relationships between palaeokarstic substrate (Torcal Fm.: T), pelagic sediment (s: stippled) and post-karst stromatolites of the Stromatolite Belt. Note the existence of oncoids and crusts, the reencrustation of oncoids by crusts, the cavity fillings by stromatolites and microstromatolites (m), and the local boring (b), also partially filled with endo-microstromatolites. Pelagic fossilizing sediment is stippled. Based on a real example near Benaoján.

within the condensed fossiliferous pelagic limestones of the base of the Torcal Fm. A Microscopic examination of these nodules and crusts reveals an organic origin (microbial remains and encrustig foraminifera) and the existence of associated micro-stromatolites.

Pre-karstification Cretaceous stromatolites.

They appear in the "Canutos Bed" (Martín-Algarra, 1987), a discontinuous and centimetric bed of light-coloured pelagic limestones (yellowish, greenish or beige), limited at base and top by discontinuity surfaces (Fig.4, 5A). The lower boundary is a planar abrasion surface cutting the upper part of the Líbar Group. In the Los Canutos de Manilva the upper limit is an erosional and biocorrosional surface; in the Chorro-Valle de Abdalajís it shows centimetric hollows (kamenitzas) and highs, a morphology identical to features formed by karstification in the northernmost areas, although smaller in size (Fig.5A). This "Canutos Bed" limestone has a micritic or biomicritic crinoid-rich microfacies, and contains calcareus nannoplankton of the Uppermost Valanginian-Lower Hauterivian (González-Donoso et al., 1983).
The microbial growths within this bed comprise centimetric oncoids, spheroidal to flattened, bearing little interstitial sediment, though some laminar structures (flat to undulose stromatolites) trapping abundant sediments are also present (Fig.4, 5A).

Post-karstification Cretaceous stromatolites.

After the Hauterivian karstification the Cretaceous palaeokarst of the Penibetic was again covered by sea-water. The rockground formed in this way became a most suitable place for the formation of pelagic phosphate stromatolites before it was buried by pelagic sediments (Fig.5B). Nevertheless, from the Upper Hauterivian to the Upper Aptian, only sporadic and isolated microstromatolite growths have been detected. These structures developed mostly during Albian-Vraconian times, and only in one specific region: along the Stromatolite Belt (Fig.2). There, the stromatolites appear as laminated crusts whose thickness may exceed one decimetre, of dark brown, greenish or reddish colour (Fig.5B, 6A-C). Besides these crusts, centimetric oncoids ovoidal, spheroidal or flattened shape abound (Fig.7A). The crusts cover the karstic dissolution cavities partially or totally, especially at the tops of prominences (Fig.5B,6A), but also cover their subvertical walls and even the inside of microcaves (cavity dwelling or endostromatolites of Monty, 1982, 1986) (Fig.5B, 6A,B). They often appear fragmented and coated with later laminae which also include ground clasts, glauconite grains and pelagic biomicritic sediment rich in planktonic foraminifera (Fig.6c). On the basis of these data it can confidently be asserted that the growth of these structures generally occurred during the Upper Albian and Lower Vraconian. The pelagic sediment is often burrowed and shows signs of synsedimentary lithification and plasticdeformation (Fig.6c). Its lithological features, microfossil content and age are exactly the same as those of the Capas Blancas Fm., but the colour can be locally red.

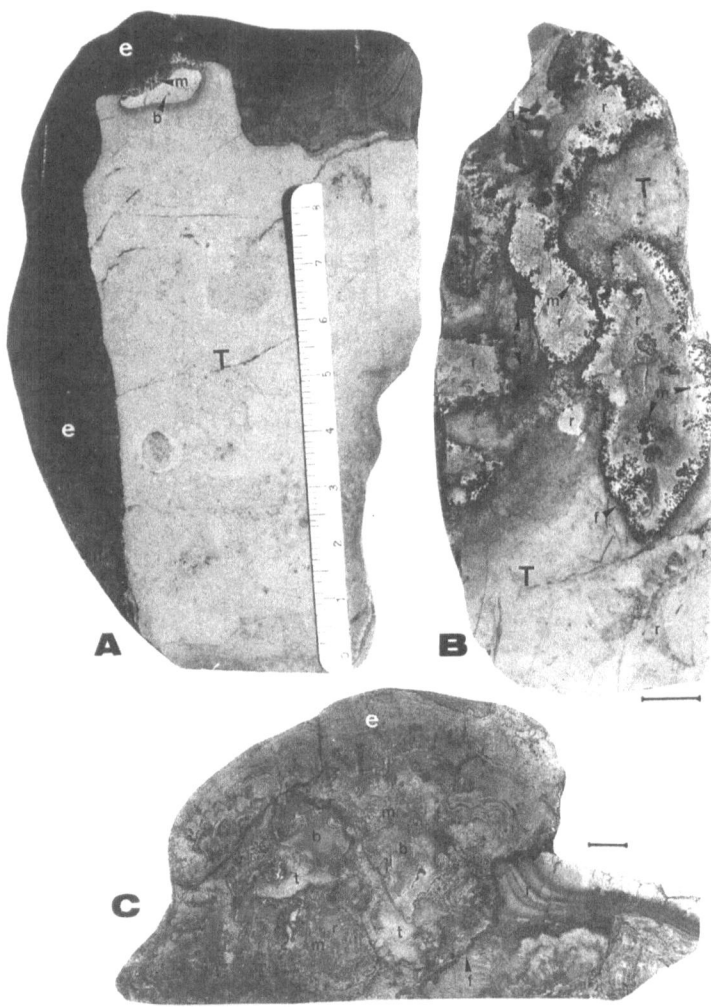

Figure 6) Cretaceous post-karstic stromatolites. a from Hacho de Montejaque (scale is in cm), b and c from the way from Montejaque to Cortijo de Líbar, (bars: 1 cm). a) A palaeokarstic relief on "pelagic oolitic" limestones of the Torcal Fm. (T), is encrusted by a dense stromatolitic coating (e). The boring (b) was formed after the first encrustation, is bordered by a glauconitized aureole, and is filled with endostromatolites (m) growing towards the centre of the cavity, that has been finally filled with pelagic sediment. b) Borings in the Torcal Fm. (T) are internally covered by ferrugenized crusts (f) and arborescent microstromatolites (m); borings have been later filled with pelagic red sediment (r) with glauconite clasts (g). c) Stromatolite head (e) growing on a complex substrate made of arborescent microstromatolites (m), red sediment rich in planktonic foraminifera, and clasts of the Torcal Fm. (t). Later on, the stromatolite was bored (b), the walls of borings were encrusted with ferruginous endo-microstromatolites (f) and voids were finally filled up with foraminifera-rich red pelagic sediment (r), locally laminated.

PELAGIC PHOSPHATE STROMATOLITES

Material and Methods

We have studied and sampled fifteen stratigraphic sections bearing phosphate stromatolites from the Penibetic. Inspection of the outcrop has been completed by the study of several hundred samples, polished slabs and thin sections, under binocular magnifying glass and transmitted-light microscope in order to establish their basic petrographic features (composition and texture). Some selected samples were studied under the SEM some of them after having undergone an acid attack of several minutes with diluted nitric or hydrochloric acid or with H_2O_2; they were then subjected to ultrasonic cleaning and later washed, to eliminate reaction precipitates on the sample; they were finally dried at room temperature for 24 hours.

The study of the chemical composition (phosphate content) was first of all carried out using the Shapiro's (1952) rapid technique based on colorimetry; some samples were later studied using more conventional analytical techniques (spectrometry). A mineralogical analysis by X-ray diffraction (García-Cervigón et al., 1986) followed, and some specially interesting and attractive specimens were studied under the light-reflected microscope, in order to identify opaque mineral phases and to establish their textural relationships with the microbial structures. E.D.X. and microprobe data confirm the observations, hypotheses and conclusions that will be presented below.

Composition and mineralogy

The composition of the different types of phosphate stromatolites from the Penibetic is fairly uniform. They are made up of carbonates, phosphates, iron oxides and hydroxides, quartz and, to a lesser extent, glauconite, sulphates (baryte), sulphides and clay minerals (Table 1). By using X-Ray diffraction techniques (García-Cervigón et al., 1986) we were able to identify the phosphate as cryptocristalline apatite, associated with calcite, quartz and goethite (sometimes in considerable amounts); we also found hematite, glauconite and pyrite. All these opaque minerals can be also identified under reflected-light microscope. Data from EDX microanalysis of microstromatolites agree fairly closely with the mineralogical and chemical data obtained using other methods; if we order the elements identified from those <u>most</u> frequently to those <u>least</u> frequently found, we have: Ca, P, Si (locally very frequent), S, Fe, and, more sporadically, Mg, Cu, Mn and others. Chemical analysis (Table 2) shows a mean content of apatite of about 25%, although in some cases this can exceed 45%.

Macroscopic and mesoscopic morphologies and rhythms.

The phosphate microbial accretions from the Penibetic provide good examples of all possible macroscopically structural transitions between two extreme morphological types. The first type is that made up of thin (millimetric) to thick (decimetric) laminated

Sample	Locality	Apatite	Calcite	Quartz	Goethite
PB-6	Berrueco	v.a	p	p	p
PB-7	Berrueco	a	p	p	p
PB-8	Berrueco	a	v.a	p	p
Mo-8	Montejaque	v.a	p	a	p
CA-4	Canutos	v.a	p	a	-
CC-11	Castillo	a	p	p	-
HM-4	Hacho Mont.	a	a	p	-
B-6	Boquerón	a	v.a.	p	-
OR-1	Ortegícar	a	p	p	p
F-3	Torcal	a	v.a.	p	-
BJ-2	Benaoján	a	p	p	p
CH-10	Chorro	p	v.a.	p	-

v.a.- very abundant; a.- abundant;   p.- present

Table 1 : Mineralogical composition analysed with X-ray diffraction

Sample	Locality	% PO4	% apatite
PB-6a	Berrueco	13.54	22.10
PB-6c	Berrueco	14.32	23.38
MO-2	Montejaque	10.78	17.61
CA-4	Canutos	23.25	37.99
HM-4a	Hacho Mont.	13.40	21.88
HM-4c	Hacho Mont.	1.96	3.20
B-6a	Boquerón	10.70	17.48
B-6b	Boquerón	28.06	45.83
BJ-2	Benaoján	27.42	44.78
OR-1	Ortegicar	26.65	45.53
F-3b	Torcal	12.12	19.79

Table 2 : Chemical composition of studied samples

crusts, or stromatolites s.s. of planar, domal or irregular shape according to the morphology of their substrate (Fig.6A,C). The second one comprises nodules or oncoids, of ovoidal, subspheroidal or flattened shape, macroscopically very similar to the Mn nodules of the ocean-floor (Fig.4, 5, 7A,B). A detailed examination of thinner, millimetric, ferruginous and phosphatic crusts shows that they are microstromatolitic. All these types are laterally and vertically related, especially in the best developed post-karst examples. The stromatolite surface may be planar or mamillated to botryoidal, with a low to moderate synoptic relief. This surface can be encrusted by foraminifera. Many oncoids show a dissymetric development of upper and lower outer laminae: outer laminae tend to envelope the structure or to coalesce with other adjoining oncoids to finally form laterally continuous crusts (Fig.6C, 7A). Another common macroscopic feature of crusts and oncoids is the existence of retractional fissures filled up with pelagic sediments and, less frequently, the repeated brecciation of the stromatolites. They may also be associated with sedimentary breccias that are locally overgrown by later stromatolite laminae (Fig.6C).

In section, several types of lamination can be found:

- *Flat lamination*, strictly *planar and laterally continuous* or more or less *undulose and/or discontinuous*.
- *Hemispheroidal*, formed by *isolated or laterally linked cupolas* (Fig.7B).
- *Columnar to pseudocolumnar*, of *cylindrical or club-shaped* morphology (Fig.7A). The synoptic relief is generally low to moderate and the stacking of hemispheroids shows a generally high degree of inheritance, much more variable in detail (Fig.7A,B).
- *Dendroid or arborescent to spongy (microstromatolitic)*, in some cases clearly delimited (Fig.6B), or gradually disappearing into the adjacent sediment (Fig.7A).

Flat to undulose, hemispheroidal and microcolumnar stromatolites can have a dense or spongy internal structure generally of dark colours, or they can show a rhythmic succession of well-defined laterally continuous, dark and light-coloured laminae of consistent to regularly variable thickness (Fig.7A,B). Rhythms of at least three sizes can be distinguished (Fig.8). The only visible to the naked eye are the two bigger ones, (Fig.7A); the smaller must be seen under the microscope. The thickness of the first order rhythms ranges from some millimetres up to one centimetre, and three parts can be distinguished within each one (Fig.7A, 8A):

- The lower part is formed by planar to undulose laminae that tend to smooth out the irregularities of the substrate.
- Outwards, in the intermediate part of the cycle, which is generally the thicker one, the lamination change to cylindrical, turbinate or club-shaped columns or pseudocolumns, thus producing a "cauliflower-like" structure.
- Finally, in the upper part, the columns may be capped by a new continuous lamina; in other cases, however, they are truncated by an erosional surface, or capped by a millimetric horizon of microstromatolites associated with pelagic sediment, rich in planktonic foraminifera (Fig.9C).

The thickness of the second order rhythms is comprised between some hundred microns and one millimetre (Fig.8D). They are formed by an alternation of calcareous sediments and phosphates arranged in well-defined laminae, that are especially well visible in the intermediate parts of first-order cycles. However, these rhymths, like the third-order ones (micrometre-sized), are better studied under the microscope, and they correspond to the dense laminated, alternating and repetitive fabrics described below. The third order rhythms are also present in the inside of the microstromatolites (Fig.8C,D).

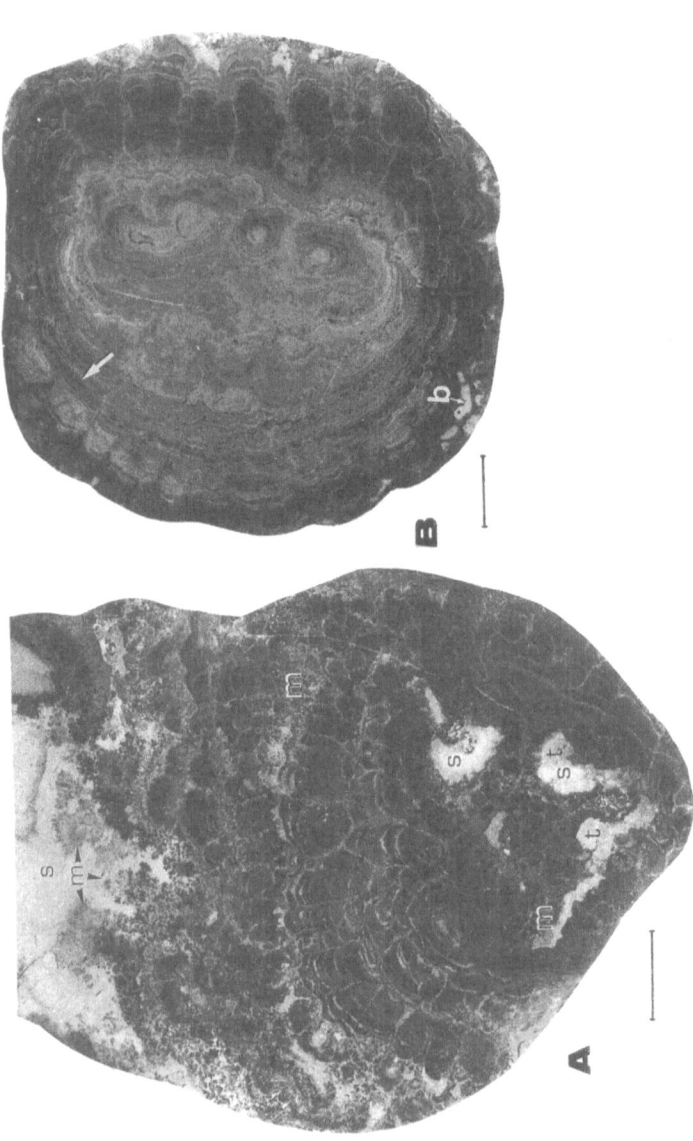

Figure 7 : Morphologies of stromatolites and oncoids. Scale bar is 1 cm long in all cases. A) Vertical section through a post-karst turbinate to subsphaerical stromatolite. Stromatolite Belt, Hacho de Montejaque. The substrate is an oncoid, whose core is made of mud clasts of pelagic white biomicrites (s) including microstromatolites and limestone clasts of the Torcal Fm. (t). A laterally continuous dense laminated crust covers this core; it is followed outward by columnar to pseudocolumnar stromatolites with alternating lamination; then grade to arborescent microstromatolite levels (m) rich in clear pelagic sediments (s). B) Section through a pre-karst oncoid from the Canutos Bed, cut parallel to bedding. Los Canutos de Manilva. The core is an assymmetric oncolite made of alternating sediment-rich laminae and microbial plus phosphate-rich laminae; outwards carbonate sediment content decreases, and laminae become darker and closely packed; structure changes from undulose to crenulated patterns, then to columns or pseudocolumns. Note the dissymmetric development of laminae on the left- and right-hand side; note also the unconformity (arrow); b: indicates a boring filled with pelagic micritic sediment.

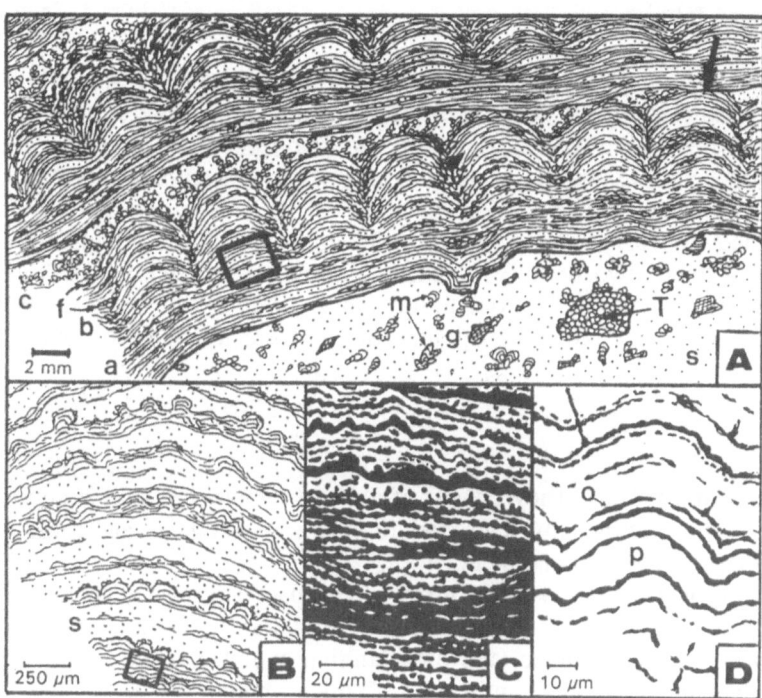

Figure 8 : Rhythms in stromatolite lamination. A) First order rhythms. The substratum is formed by a clast of pelagic biomicritic sediment (s: stippled) with fragments of oolitic limestones of the Torcal Fm. (t), glauconite (g) and arborescent microstromatolites (m). Two, and part of a third, first order rhythms exist locally lined by a discordance (arrow). Note the upward succession from planar dense laminated, repetitive to alternating fabrics (a), to domal and pseudocolumnar stromatolites with alternating fabrics (b), then to arborescent microstromatolites and pelagic sediments. Sediments (s) stippled. Note the abundance of encrusting foraminifera (f), that give locally a spongy texture to the stromatolite (upper left corner). (See fig.7A). B) Second order rhythms, formed by the alternance of phosphate (p) plus microbial (o) and sediment-rich (s) laminae. Note the domal microstromatolitic texture of phosphate plus microbial laminae. C) Third order rhythms, formed by bundles of organic filaments (black) within phosphate (white). Note the existence of small discordances between some larger rhythms, the existence of prostrate and erect filaments and their saussage shape. D) Third order rhythms of the smallest possible size: alternances of phospahte laminae (white) and prostrate isolated filaments (black).

Textures and fabrics.

Microscopic examination reveals that stromatolite building is essentially the result of the textural interaction of four types of elements: phosphates, microbial filaments, pelagic sediments and encrustation by foraminifera (Fig.9). These elements are interrelated in various ways, giving rise to several stromatolite fabrics that show all kinds of mutual lateral and vertical transitions (Fig.9A). Some of these fabrics have been defined by Krajewsky (1983), in pelagic phosphate stromatolites from the Carpathian Albian: two densely laminated fabrics, respectively repetitive and alternating, and two loose microstromatolitic fabrics, which are respectively domal and arborescent; the fifth one is a spongy fabric formed by encrusting foraminiferal constructions.

Fabrics

Dense laminated fabrics. They consist of a dense superposition of composite laminae and microlaminae formed by microbial filaments, phosphates and sediments, with a repetitive or alternating arrangement.
Repetitive fabric (Fig.9F) is made up of a compact aggregate, up to some hundred microns thick, of micrometre-sized, more or less laterally continuous, clear (yellowish to orange) phosphatic laminae, and dark laminae which include abundant remains of prostrate, planar to undulose microbial filament bundles (see below). This fabric contains only scarce sediments and it is typically found in planar zones of stromatolites and oncolites (Fig.6, 7).
Alternating fabric is composed of alternating laminae several hundred microns to slightly more than one millimetre thick, made up of fine-grained calcareous sediments, and of phosphate plus microbial remains. The sedimentary laminae are coccolith-rich micrites or sometimes microsparites of (sedimentary or diagenetic?) uncertain origin. In detail, phosphatic laminae show a dense repetitive texture and a planar or microdomal shape (see below) (Fig.9B).

Microstromatolites. Microstromatolites are millimetric to some tens of microns in size, domal to branching, finely laminated bodies (Fig.10). They correspond to forms defined as "*Frutexites*" by Hofmann and Grotzinger (1985). Laminae are very well-defined, are laterally continuous an show a consistent to slightly variable thickness (micrometre-sized: Fig.10A-D, 11A-D). Phosphates and opaque minerals (iron sulphides, oxides and hydroxides) constitute practically the only components of microstromatolites, with little or no interstitial carbonate sediment (Fig.10A,C,D; 11A,B, 12, 13). Microstromatolites show all possible morphological transitions from isolated or cumulated hemispheroids to small domes, to cylindrical or turbinate columns or pseudocolumns and, especially, to branching arborescent contructions. Two basic types can be distinguished (Krajewsky, 1983): domal and arborescent.

Isolated or laterally linked domal microstromatolites normally appear in close connection with dense laminated fabrics, especially with the alternating-one and they grow starting on the biggest and most continuous phosphatic laminae (Fig.9B). They show a low to moderate synoptic relief and good inheritance of laminae.

Arborescent microstromatolites are much more frequent (Fig.9C,G, 10). They show extremely variable irregular morphologies and have a high to very high synoptic relief, with enveloping outer laminae (Fig.8A,C, 11A, 12A,B,C). They can appear closely connected with the general lamination, but frequently they show a more irregular growth pattern and their distribution is independent of the main lamination. They can encrustate small clasts of pelagic limestones from the Torcal Fm., bioclasts or glauconite grains that are included in pelagic biomicritic sediments (Fig.11C,D; 13A). Another frequent disposition of arborescent microstromatolites is as microcavity dwellers (micro-endostromatolites of Monty, 1982, 1986): they encrust the walls of organic borings and retractional fissures crosscutting former microbial envelopes (Fig.6A,B).

Evidences of plastic deformation of semiconsolidated microstromatolite-rich and planktonic foraminifera-rich sediments, repeated breaking-up of clasts of the underlying rockground, brecciation of former stromatolitic crusts, and re-encrustation by later stromatolites are abundant in the examples studied (Fig.6C, 7A). These sediments, removed after their deposition by current action and by burrowing, are presently found as mud clasts, rimmed by iron oxy-hydroxides or glauconite, that have been trapped and bound by later stromatolitic laminae, and constitute the core of many oncoids. As a consequence of the embodiment of these originally soft, microstromatolite-rich mud clasts by later stromatolite laminae, a local growth pattern of microcolumnar to arborescent microstromatolites, opposite to that of the main stromatolite, can result (Fig.8A, 14A). In this case the mineralized, sometimes slightly corrosional rims of mud clasts also crosscut the microstromatolites includes in them, a fact that allows to distinguish this arrangement from endostromatolites (Fig.6A,C, 14B). This latter observation is important because microstromatolites included in mud clasts are predate the main lamination, whereas endostromatolites postdate it.

Spongy fabric built by encrusting foraminifera. Contrary to what has been claimed for the examples discussed in the literature, this fabric plays a very important, sometimes crucial, part in the building of stromatolic structure (Fig.,9A,D,E,G). Under the microscope the foraminifera assemblages give the stromatolite a reticulated, spongy appearance as it shows a marked porosity: the lenticular, planoconvex voids of the chambers of foraminifera are filled with microsparite and/or phosphate, and their walls are externally encrusted by microbial dark laminae and associated phosphates (Fig.9D,G).

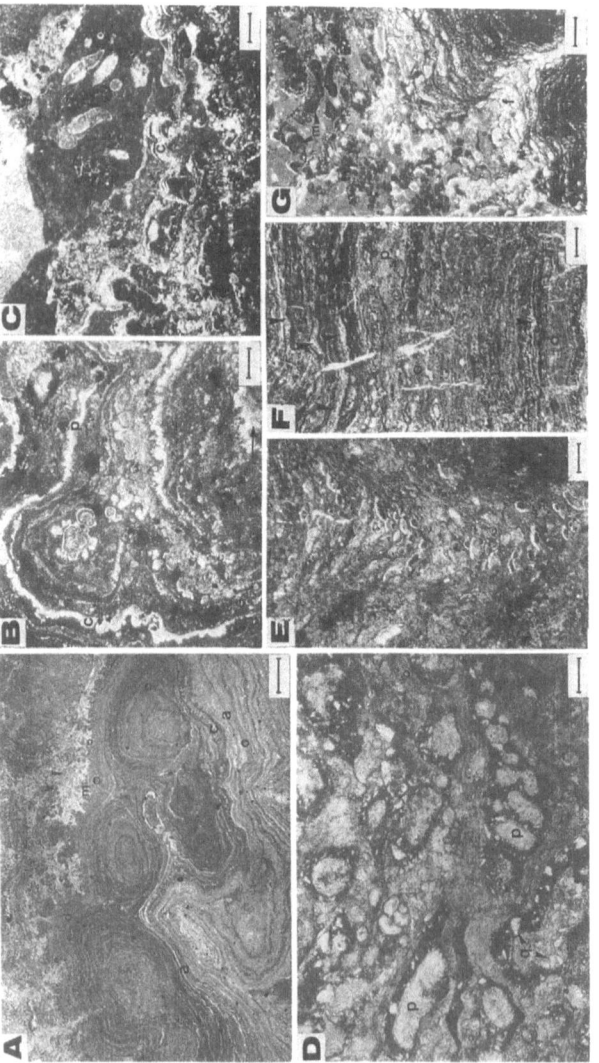

Figure 9 : Microscopic fabrics (transmitted light). A and B from the pre-karst Canutos Bed, Los Canutos de Manilva. C, D, E, F and G from the post-karst Stromatolite Belt, Hacho de Montejaque. A) Tangential section of an oncoid. General relationship pattern between stromatolite fabrics: repetitive (d) and alternating (a) dense laminations; domal (c) and arborescent (m) microstromatolites, and encrusting foraminifera (f). Black spots are oxidized pyrite. Scale=1 mm long. B) Alternating lamination formed by sediment-rich micritic laminae (black), and phosphate-rich (white) laminae (p); some of these latter are formed by laterally linked domal microstromatolites (c). Note oxidized pyrite spots and fenestrae with microsparite geopetal filling (arrow). Same sample as A), Scale= 250 μm. C) Sediment lamina containing planktonic foraminifera, covering domal microstromatolites to the right (c) and partly microsparitized (r) arborescent one to the left. Scale= 250 μm. D) Encrusting agglutinating foraminifera: note trapped quartz (q) in their test, phosphate (p) infillings of chambers, and overlying microbial laminae (o). Same locality and age as C) Scale= 150 μm. E) Encrusting foraminifera colony between stromatolite pseudocolumns. Scale= 250 μm. F) Repetitive dense laminated fabric: see dense stacking of dark microbial laminae between clear phosphatic ones (p), and laminar fenestrae (arrow). Scale= 250 μm. G.) Hemispheroids coated by encrusting foraminifera (f) and arborescent microstromatolites (m), and covered by sediments. Scale=250 μm.

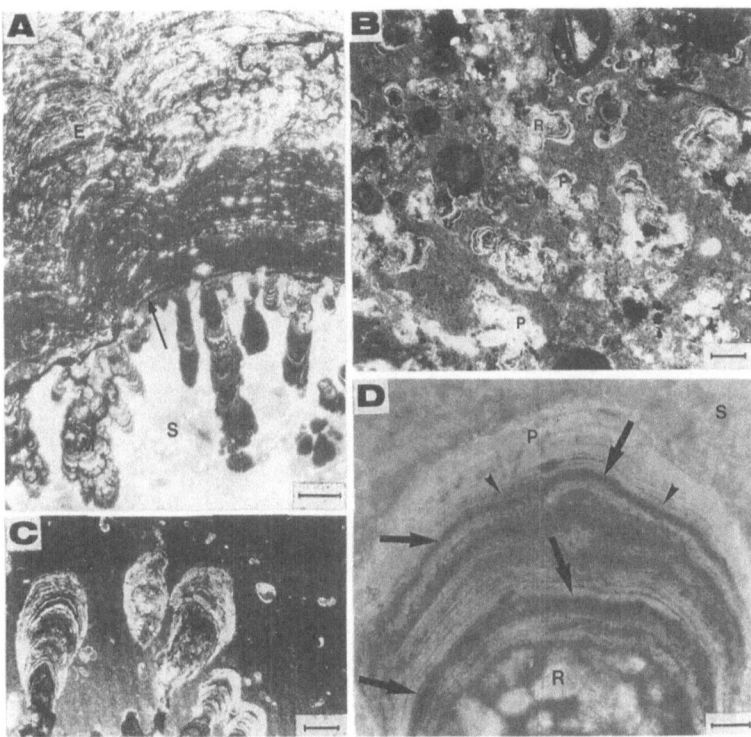

Figure 10 : Arborescent microstromatolites (transmitted-light) from the Stromatolite Belt. A, C and D, Peñón del Berrueco, B, Hacho de Montejaque. A) Microstromatolites (M) growing in an opposite sense to the principal stromatolite (E). Note the discontinuity stained by iron oxydes (arrow) between the sediment-rich (S) core and the pseudocolumnar stromatolite. Black irregular spots in the upper half of photo are artifacts. Scale= 1 mm. B) Texture of a mud clast from the core of one oncoid: pelagic micrite including abundant microstromatolites. Note the presence of isolated pelagic ooids removed from the Torcal Fm. (upper part of the photo), and the micro-sparitized parts (R) of phosphatic (P) microstromatolites. Little black spots are oxidized pyrite and bigger ones are glauconite clasts. Scale=250 µm. C) Contact between arborescent microstromatolites and overlying sediments (wackestone with planktonic foraminifera). Note the high synoptic relief and enveloping outer laminae. Scale=200 µm. D) Close-up of stromatolite lamination of an arborescent microstromatolite. Key= S: pelagic sediment; R: recrystallized central part of the column; P: phosphate. Alternating clear (phosphate) and dark (microbial) laminae are evident. Biggest arrows point to organic filaments bundles, but more or less isolated filaments are also visible. Little arrows point to erect filaments. Scale= 25 µm.

Textural elements

Phosphates. Phosphates appear under the microscope as a yellowish transparent and amorphous mass (Fig.9B, 10C,D, 11A-D). However, X-ray diffraction reveals their

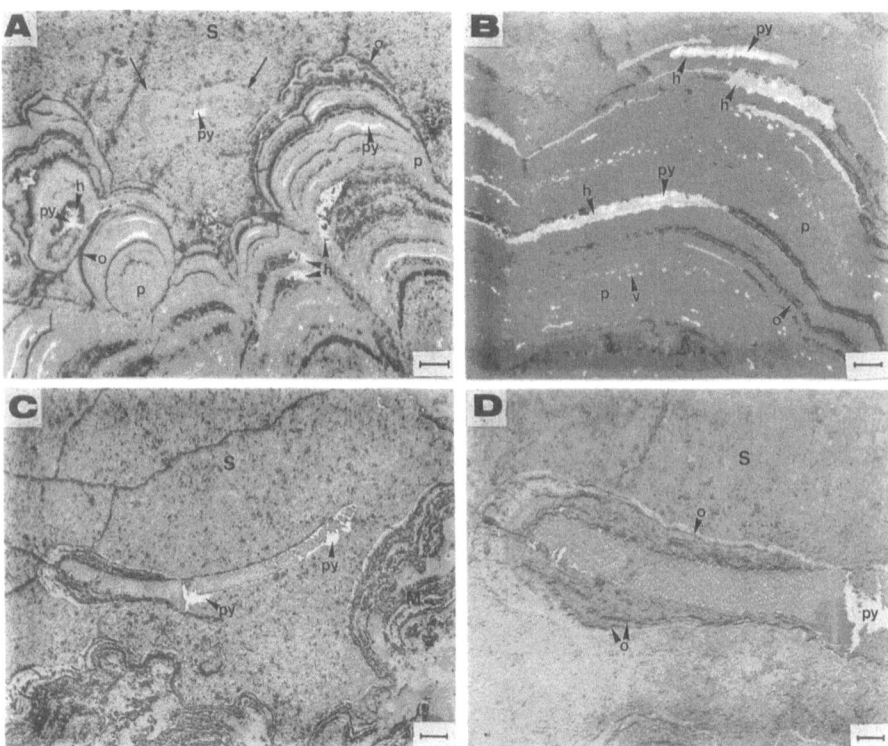

Figure 11 : Stromatolite textures of the post-karst Stromatolite Belt (reflected-light). A,B, same sample as Fig. 6A; C, D, Peñón del Berrueco. A) Domal to pseudocolumnar microstromatolites showing a high degree of inheritance. Note lateral continuity and regular variation of the thickness of laminae, and the enveloping outer laminae. Lamination is defined by alternances of clear phosphatic laminae (p) and black organic (o) laminae, with subordinate pyrite (py) partly oxidized to hematite (h) and/or goethite. Microstromatolites are covered by micritic pelagic sediments (s) with planctonic foraminifera (arrows). Bar= 30 μm. B) Close-up of a, showing details of the alternance between phosphate-rich (p) and organic-rich laminae (o), with pyrite (py) and hematite (h). Voids (v) in phosphate may be empty moulds of bundles of filaments. Bar= 10 μm. C) Contact between microstromatolites (M) and carbonate pelagic sediment (S), shell-fragment with crystalline texture, partly mineralized in pyrite (py), encrusted by phosphatic microstromatolite. Bar= 30 μm. D) Close-up of the encrusted shell fragment of c) Bar=10 μm .

cryptocrystalline nature and under the SEM the phosphate laminae appear as loosely packed aggregates of micrometre-sized (or smaller) crystals of apatite (Fig.12A,B,C). Phosphates are closely related to dark laminae or fill empty chambers of encrusting foraminifera or yet fenestrae parallel to the lamination (Fig.9D; 12D,E,F). These fenestrae formed by the retraction of the stromatolite during lithification (sheet craks). Some of these fenestrae are also filled by a microsparitic sediment (Fig.12B; see also

Krajewski 1981b, 1983). Phosphatic, originally colloidal, substances appear to have moved frequently during diagenesis, partially obliterating the original laminated structure. Phosphates form cupolas with a perforated surface (Fig.13A,B) that seems to have been caused by leaching of erect filaments. In some places coccoliths may be bound to the surface or wrapped in phosphate laminae. In some samples the coccoliths are abundant and, in those previously attacked with acids, they can be seen under the SEM as internal moulds, which sometimes can be classified, a fact that allows accurate dating of the stromatolite growth (Fig.13C,D,E,F).

Microbial filaments. The filamentous texture of the dark laminae alternating with phosphatic laminae is clearly to be seen under the transmitted-light microscope (Fig.10D, 15, 16). Their original organic nature (microbial mats) can be confirmed by study under a reflected-light microscope (Fig.11A,B), since these filaments are well isolated within the phosphatic mass and are delimited by alignments of carbonaceous matter closely connected with pyrite, locally altered into hematite and goethite more o less completely (Fig.11A,B). Framboidal pyrite, transformed to a greater or lesser extent, also appears in phosphatic fenestrae within the stromatolite laminations.
At least two kinds of filaments have been identified. The largest-ones (tens of micrometres thick) are associated to microsparitic laminae or fenestrae. The smaller ones are most widely and best represented, show analogous features both inside the microstromatolites and in the laminated structure in general, and are the essential component of microbial laminae. They are smaller than one micron in diameter and are generally clustered: they form prostrate mats or films of variable thickness (from a few to more than a hundred microns) parallel to the general lamination (Fig.10D, 11A,B,D, 12d,E,F, 15A,B,C,E). Sometimes, erect, fan-shaped dendritic bundles of filaments can be identified (Fig.15D,E, 16). The filaments are micrometre-sized boudinated and segmented tubes formed by strings of linked or closely situated spheroidal to sausage-shaped forms similar in morphology and size to coccoidal, bacillary and especially filamentous bacteriae. In the samples studied under the SEM the filamentous nature of the dark laminae is nonetheless clearly to be seen. The microbial filaments systematically appear as elongated empty voids in the phosphate, both in samples previously attacked or not with acids (Fig.12D,E,F). This fact shows the process of leaching and decomposition of the original organic matter after deposition. In most cases, the SEM study shows that these elongated empty voids have irregular limits formed by apatite crystals, but in some points true internal moulds of small, more or less cylindrical tubes can be seen (Fig.12D,E,F; 13D), sometimes segmented (Fig.12E): the tubes have a morphology and size similar to that of many filamentous or rod-like bacteria and their interpretation as bacterial filaments formed by a row of cells seems reasonable. Other non-filamentous, more or less spheroidal forms, frequently associated to framboidal pyrite, can be detected here and there in the stromatolitic texture (Fig.15D): they also are probable remains of microbial colonies of coccoidal bacteriae, which can precipitate early diagenetic pyrite inside of the stromatolite. Moreover, in one sample the surface of a fragment of microstromatolite (Fig.17) is formed by a dense

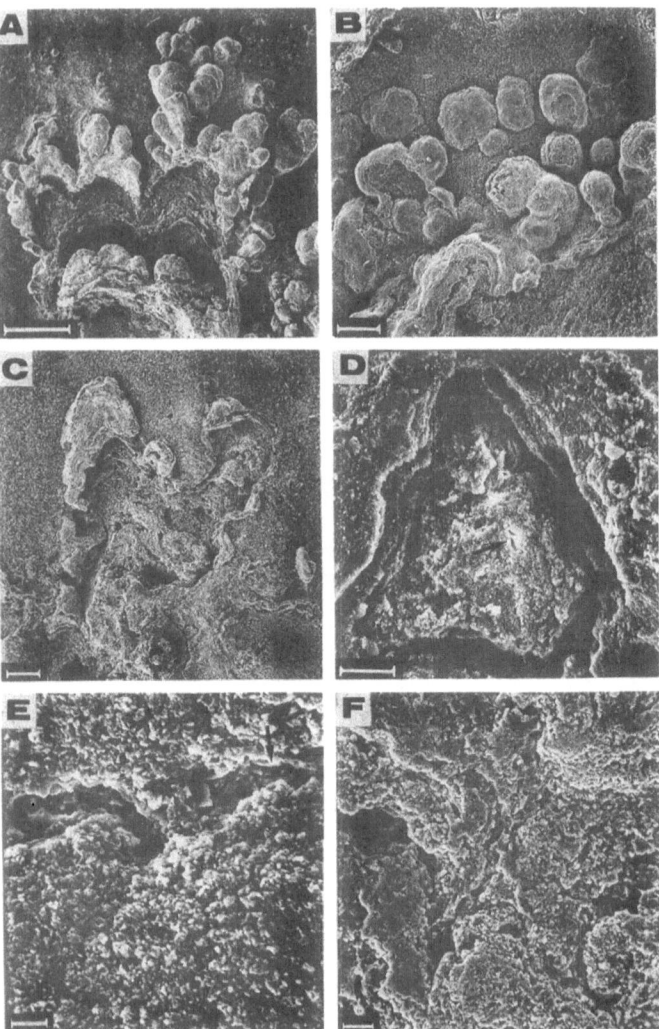

Figure 12 : SEM photos of post-karst arborescent microstromatolites (A,B,C) and organic filaments moulds (D,E,F). A,C,D,E,F) Peñón del Berrueco, b)Hacho de Montejaque. A) Typical arborescent discrete microstromatolite. Bar= 200 µm. B) Oblique view of microstromatolite hemispheres, showing "onion" concentric structure. Bar= 100 µm. C) Note the finely laminated internal texture of the upper part of the microstromatolite, and the recrystallized core. Bar= 100 µm. D) Alternation of phosphate-rich and microbial laminae. Arrow, moulds of filaments. Bar= 20 µm. E) Filament moulds in phosphate. See tubular and segmented disposition of arrowed filament. Bar= 5 µm. F) Organic-rich laminae with abundant filament moulds in phosphate (arrow). Note porous texture of the phosphate laminae. Bar= 10 µm.

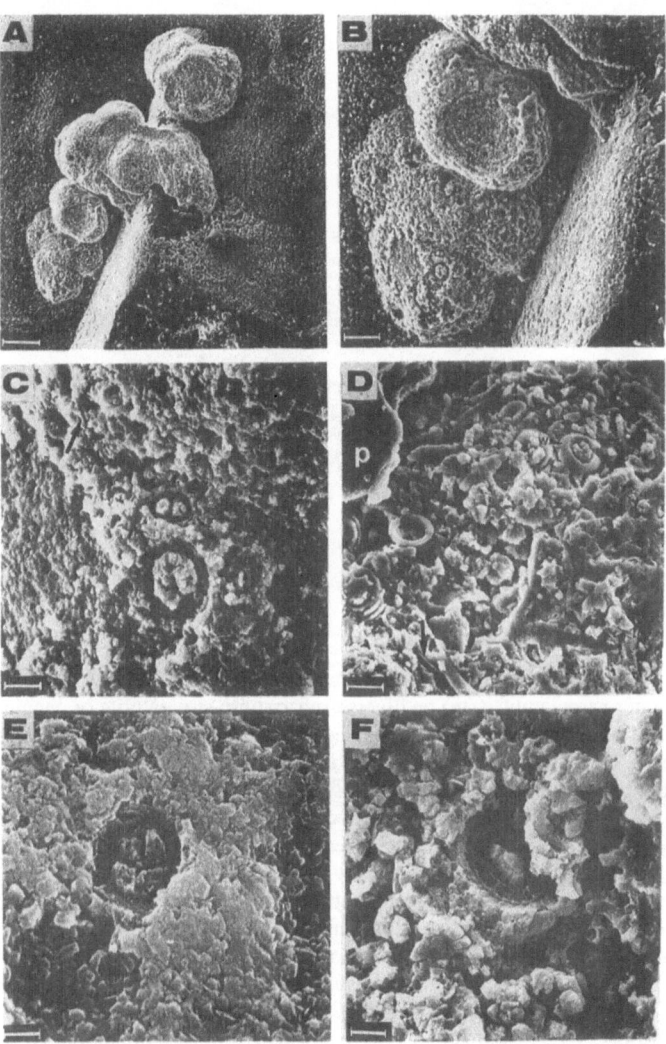

Figure 13 : Phosphate microstromatolites trapping coccoliths (SEM photos). A, B, C) Peñón del Berrueco, D, E, F) Hacho de Montejaque. A) Microstromatolites encrusting a glauconite clast. Bar= 50 μm. B) Detail of A, concentric lamination of microstromatolite pustules, perforated surface of phosphate laminae and external moulds of carbonate grains and coccoliths. Bar= 20 μm. C) Close-up of B, coccoliths (*Zygodiscus cf. elegans* and unidentifiable forms) and organic filaments moulds (arrow). Bar= 5 μm. D) Sedimentary lamina made of coccolith-rich carbonate trapped between organic filaments moulds (arrow), phosphate laminae (p). Coccoliths identified are *Zygodiscus elegans* (z), *Loxolythus aff. armilla* (l), *Watznaueria barnesae* (W). Bar= 5μm. E) External coccolith-moulds on crystaline phosphate. *Prediscosphaera cf. spinosa*. Bar= 1μm. F) Externalcoccolith-moulds on crystalline phosphate. *Watznaueria barnesae*. Bar= 2μm.

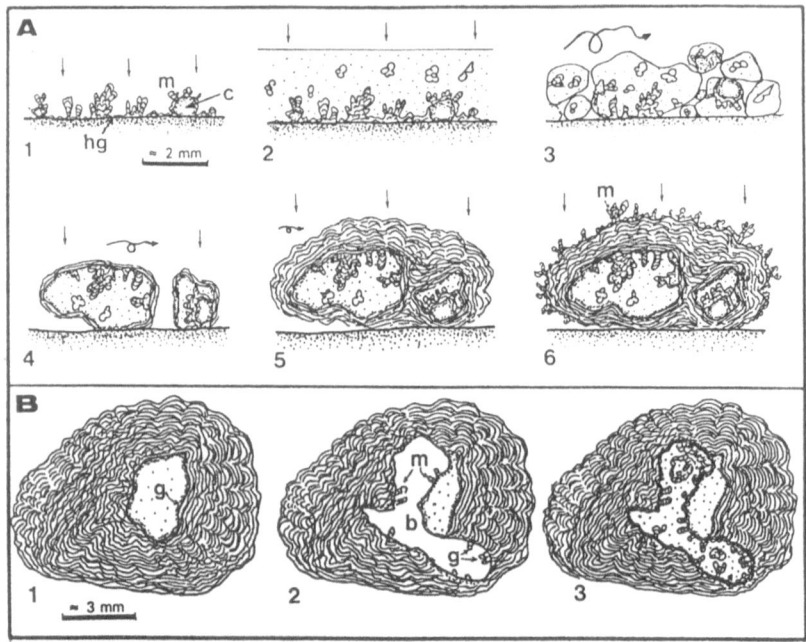

Figure 14 : A) Origin of mud clasts with microstromatolites in the nucleus of some oncoids. 1: Growth of microstromatolites (m) on diverse substratum: stromatolite or oncoid (see number 6), more or less glauconitized or ferrugenized hardground (hg), or clast removed from the underlying rock (c). Small vertical arrows indicate a low sediment supply. 2: Deposition of pelagic sediment rich in planctonic foraminifera and burial of microstromatolites. Long vertical arrows indicate higher sediment supply. 3. Sediment disturbance by burrowing and/or by current, and formation of clasts including arborescent microstromatolites. 4: Encrustation of isolated clasts by planar microbial laminae. Frequent rolling under moderate current, and low sediment supply (vertical arrows). 5: Disymmetric upwards development of microbial encrustation by stacking hemispheroids. Sporadic rolling under low energy currents and low to moderate sediment supply. B) Origin of endostromatolites within oncoids: 1: Oncoid grown on a core formed by a clast (g: glauconitized or ferrugenized rim). 2: Burrowing of the crust or bed where the oncoid is included, and boring (b) that get into its core. Later glauconitization (g) of boring walls and growth of microstromatolites occurs. 3: End of microstromatolite growth and filling by pelagic sediments (stippled). Repeated boring and re-encrustation can occurt, so that the distinction between the sediment filling and the original sediment core can be difficult.

mass of spheroidal to ellipsoidal objects arranged in strings, interpreted as colonies of phosphate precipitating bacteria.

Associated pelagic sediments. In oncoids cores, sediments are abundant. They are frequently micrite-supported microbreccias containing small stromatolite fragments, rockground clasts (especially those of oolitic pelagic limestones of the Torcal Fm.) and

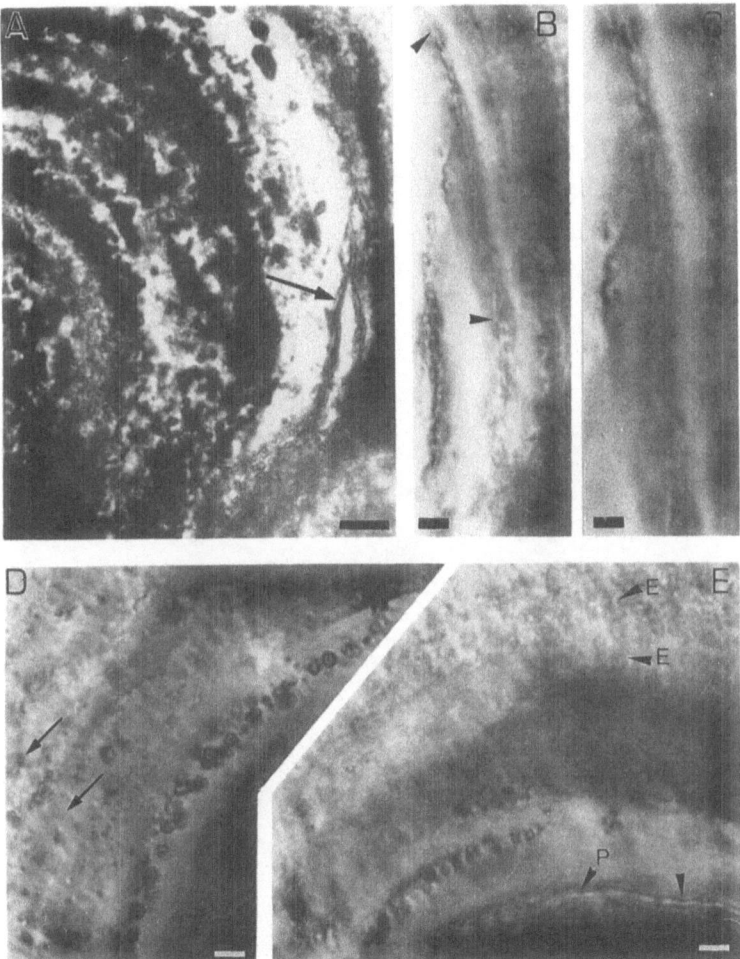

Figure 15 : Microbial, possibly bacterial filaments from microstromatolites. Transmitted-light. A) Dark laminae are bundles of filaments (arrow, prostrate filament) plus opaque minerals. Clear laminae are phosphatic. Scale= 10 μm. B) Successive prostrate filaments within the phosphate. Dark aureole around the filaments, may be a ghost of a bacterial sheath. Scale= 2 μm. C) Close-up of the filament (arrow on b). Scale= 1 μm. D) Fan-shaped erect filaments (sausage-like filament arrow). E) Same as d, with prostrate (P) and erect (E) filaments. D,E , Scale=2.5 μm.

fragments of glauconitic and ferruginous crusts, that have been colonized in many cases by microstromatolites (Fig.7A, 13A,B). However, the <u>sediments</u> within the stromatolitic laminae are normally scarce, although they are found everywhere, as it is shown both by microscopic study and by the carbonate content of all samples. These sediments are made up partly of detrital fine-grained particles, especially quartz and clay, and above

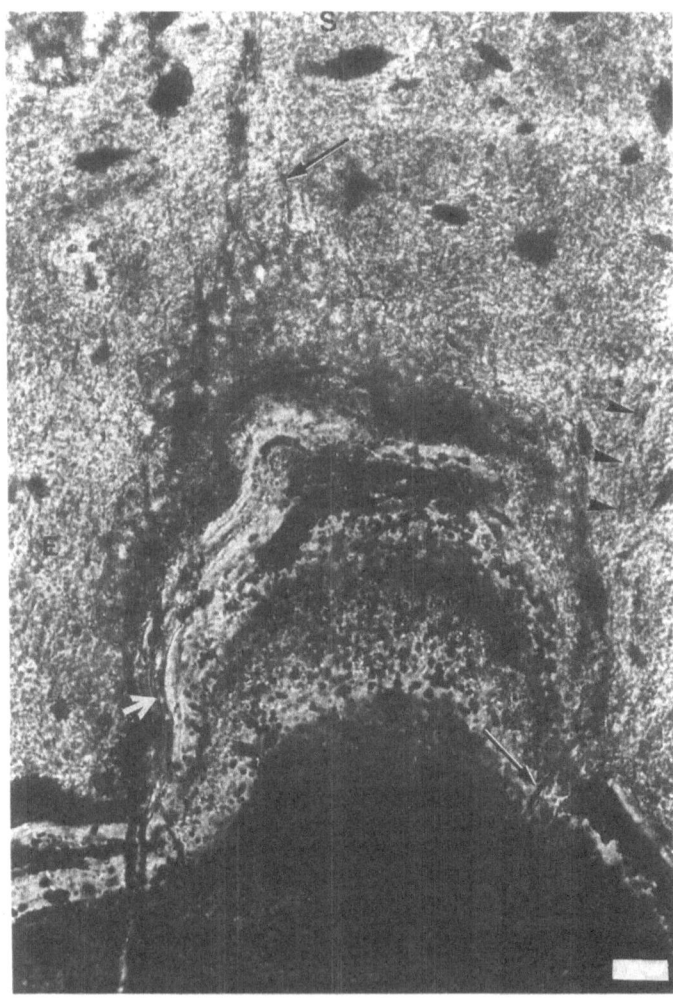

Figure 16 : Prostrate (white arrow) and long erect (black arrows) filaments within arborescent microstromatolite. Note the fan-shaped disposition of erect filaments between prostrate mats and also outside of them (E), trapping clastic particles (S). Scale bar is 20 μm.

all of micrites bearing planktonic foraminifera and coccoliths, and microsparites. They appear in laminae several micronmetres thick, alternating with phosphate-rich and microbial laminae (Fig.9B), or as millimetric levels associated with microstromatolites which culminate rhythmic growth cycles (Fig.7A). The pelagic sediments also appear filling borings and burrows, or fractures and microjoints formed during early diagenesis.

Encrusting foraminifera. Encrusting foraminifera form lenticular planoconvex accretions that grew leaning against each other; they appear isolated in the stromatolite laminae, or form colonies that are frequently found between adjacent hemispheroids (Fig.9E). The wall structure, often visible, enables us to approach their taxonomy :
- Some of them, the most abundant and the biggest (up to 800 μm per chamber), have a clearly agglutinating wall, formed by particles of silt size, preferably quartz, bound by a calcareous cement (Fig.9D). They belong to the Tolypanminninae sub-family of the Ammodiscidae family and several genera have been detected.
- Other, generally smaller organisms (100μm per chamber) seem to have porcelanaceous walls and are similar to certain forms of the Fischerinidae and Nubeculariidae families. Similar forms have been identified in manganese nodules (Wendt, 1974).

Diagenesis

Stromatolitic textures and structures are frequently partially or totally obliterated by diagenetic processes. We should distinguish between early synsedimentary diagenesis (occurring during or immediately after deposition), from late diagenetic processes.

Early diagenesis comprises various phases of cementation, mineralization, lithification, appearance of fenestrae and, often, the breaking-up of stromatolites. Dehydration of gels joined to early lithification caused collapse of the structure and the consequent appearance of sheet cracks, transversal and parallel to the lamination. These are usually filled by microsparite and/or pelagic sediments, and may not affect the whole structure, illustrating their synsedimentary origin. Cementation and lithification must have been related with microsparite precipitation in the original pores of the structure. Other associated mineralization processes included the filling of the empty chambers of encrusting foraminifera and of the fenestral voids between laminae by phosphatic gels, as well as the formation of framboidal pyrite nodules. The biocorrosion and organic boring which is clearly to be seen in many samples and the action of water currents must have been the cause of the partial breaking-up and brecciation frequently affecting stromatolites (Fig.6C). In several post-karst samples, taken from the Hacho de Montejaque (Upper Albian-Lower Vraconian), fan shaped (or drusiform) aggregates of barite have been observed between layers of the stromatolite lamination (Fig.18). In the example shown in this figure, the stromatolitic lamination settles on the sharply truncated apex of most of crystals (arrows). Nevertheless, in protected areas of these fan-shaped aggregates (lower left part in Fig.18), long crystals with well-preserved pointed apices have been found: they have been fossilized by microsparitic sediment and trapped by stromatolitic laminae. These observations demonstrate that these aggregates, probably of inorganic or not entirely organic origin, must have been formed during periods in which the growth of the structure was interrupted; they were also partly destroyed, perhaps by current action and/or boring, before microbial mats bound them into the stromatolite. Furthermore, early diagenetic barite has been detected as fracture-fillings partly crosscutting the stromatolite lamination.

During latest stages of diagenesis, the following phenomena seem to have occurred: a) Migration of the phosphates which, in some cases, has almost completely obliterated all depositional features. b)Recrystallization by microsparitization which sometimes affects either microstromatolites or the whole structure (Fig.10B). c) Oxidization of pyrite to hematite and goethite (Fig.11). Later tectonic deformation occurred, as well as frag- mentation of some stromatolites, plus the secondary acquisition of a peculiar laminar structure, similar to some kinds of zebra-like structures.

DISCUSSION AND INTERPRETATION.

Role of organisms in the stromatolite growth.

The organo-sedimentary nature of the phosphate stromatolites of the Penibetic is obvious and irrefutable in the light of the micro- and macroscopic observation that has been carried out. These stromatolites have been built basically by the activity of filamentous micro-organisms and, to a lesser extent, of encrusting foraminifera. The

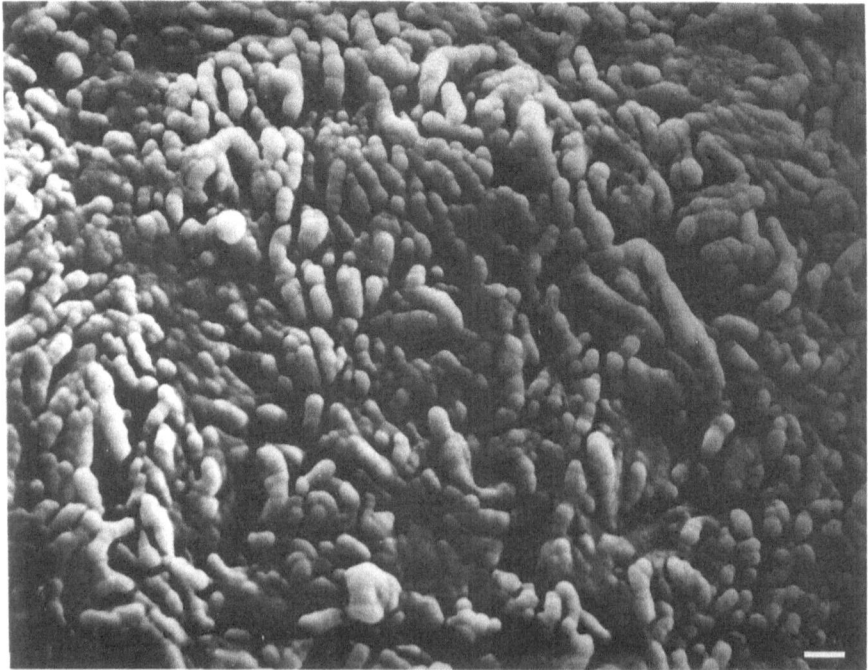

Figure 17 : Surface of a bacterial mat from a microstromatolite. Peñón del Berrueco. Lowermost Vraconian. Abundant filaments formed by strings of spheroidal to elliptical bodies. Bar=2µm.

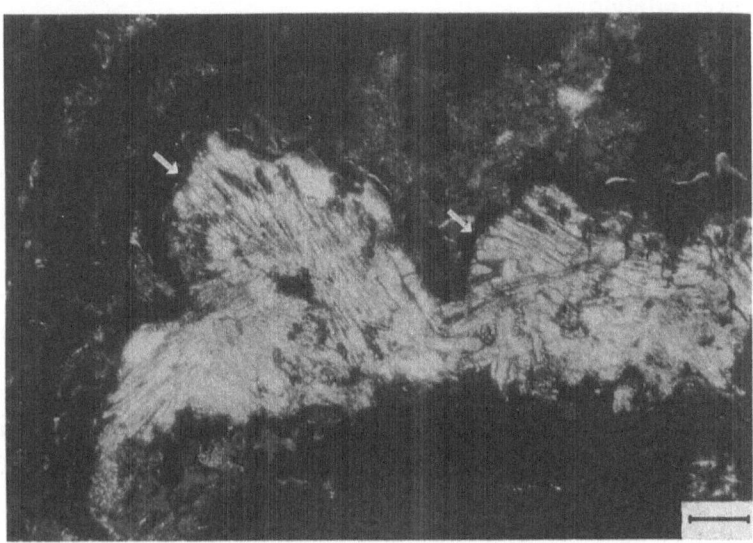

Figure 18 : Drusiform aggregate of barite, trapped in a post-karstic stromatolite from the Hacho de Montejaque. Truncation and corrosion of crystal ends at points arrowed is clear, as is their preservation on the lower left. Note that the upper laminations adapted to the surface of the aggregate. Scale bar is 250 µm.

morphology and the dimensions of the filaments, their grouping into irregular bundles, their segmented appearance in some examples, and their close relationship to carbonaceous matter and pyrite allow us confidently to affirm their microbial, probably bacterial origin. Bacteria were possibly chemolito- or chemoorganotrophic bacteria, capable of precipitating sulphides and phosphates at the sea bottom. Similar filamentous bacteria forming extensive mats and able to precipitate several kinds of minerals are known from the present-day sea-bottom (Jannasch, 1984a,b; Jannasch and Mottl, 1985). As far as the encrusting foraminifera are concerned, they contributed to consolidation of the stromatolite, making it resistant to current action and burrowing. A question not yet completely solved is: why did these organisms live in an essentially bacterial environment, where hard physico-chemical conditions possibly existed, allowing sulphides and phosphate to precipitate? A suggestive hypothesis could be that foraminifera and bacteria were symbiotically related: encrusting foraminifera should need bacteria to live in the stromatolitic environment, as food or for digestion, respiration or any other life process. A symbiotic association between bacteria and several groups of invertebrates has been described from deep-sea environments (Jannasch, 1984 a,b; Jannasch and Mottl, 1985).

Phosphate precipitation.

Textural data demonstrate that stromatolite and especially microstromatolite accretion occurred basically as a result of the piling up of bacterial filament bundles. Most of them are prostrate and then subparallel to the main lamination, but the accretion of the structure, and particularly of the microstromatolites, occurred by means of the emission of erect filaments that spread over the phosphate and sedimentary overlying lamina to form a new film, in the way that we illustrate in figure 19. Because of bacterial chemosynthetic life processes, pyrite-precipitated and colloidal, phosphate-rich, possibly organic substances flocculated, forming a mucilaginous envelope around filaments. The common occurrence of coccoliths wrapped in phosphate, shows that: a) the precipitation took place on the sea-bottom, and b) precursory phosphate-rich substances must have formed a colloidal deposit around bacterial filaments, onto or inside which sedimentary particles could adhere. These phosphate-rich substances were not originally apatite; this mineral was possibly formed after deposition, perhaps after

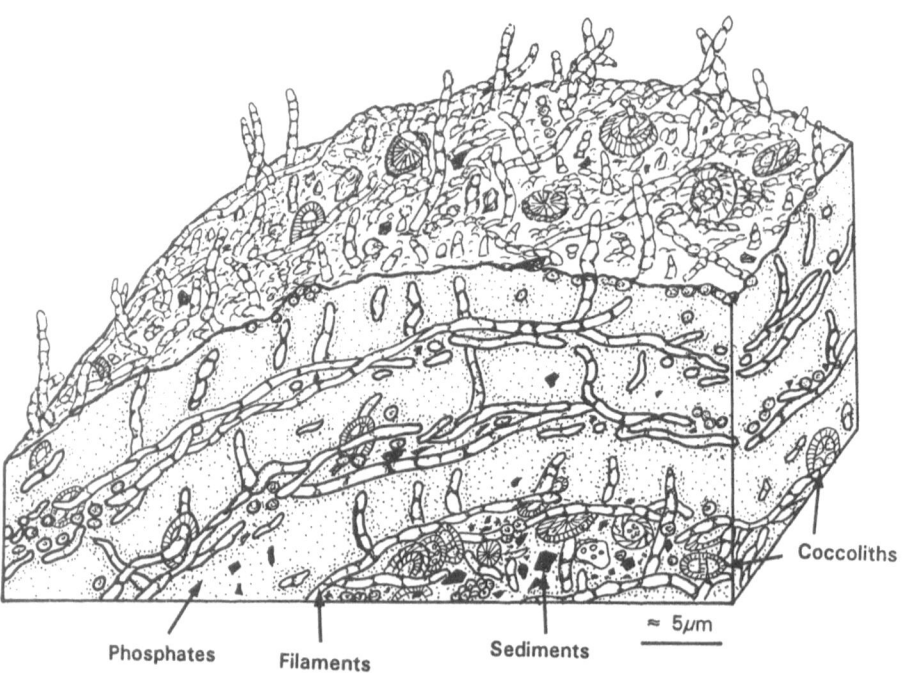

Figure 19 : Hypothetical sketch showing the proposed growth pattern of bacterial filaments, floculation of phosphates, and trapping and binding of "detrital" pelagic particles, and coccoliths.

or during decay of a precursory phosphate-rich organic substance, and later on recrystallized.

The fact of organic participation in phosphate fixation and precipitation from sea water is known, and organic activity plays an important role both in the preliminary concentration of phosphorus in sediments and interstitial waters, as well as in the formation of phosphate deposits (Slansky, 1982). As far as the mineralizing role of microorganisms, particularly bacteria, which make up the stromatolites is concerned, we must point out that their participation in phosphate and sulphur precipitation is already well-know (Mendelson, 1976; Trudinger and Mendelsohn, 1976; Malone and Towe, 1970; Hölm, 1987) and that the formation of apatite by bacterial action from calcite and phosphatic organic matter has been demonstrated experimentally (Lucas and Prévot, 1981). Again, the close textural connection between phosphate precipitates and filamentous structures of microbial origin has been emphasized in several recent studies of phosphate deposits (Soudry 1979, 1987; Soudry and Champetier, 1983; Soudry and Southgate, 1989). The biochemical details of the phosphate precipitation process are not yet well know, but we do know phosphatization is usually associated with or preceded by glauconitization (Odin and Letolle, 1980); this fact is also detected in the examples from the Penibetic, where the presence of patinas and glauconite grains is commonly found both in sediments which appear encapsulated in stromatolites and in those which themselves include stromatolites.

Carbonate and clastic sediments and growth-rate of stromatolites.

The trapping and binding of the sediments is due to the accretional dynamics of the bacterial filaments themselves, and in addition was helped by the existence of phosphate-rich colloidal precipitates. Nevertheless, the role played by the encrusting and agglutinating foraminifera should not be underestimated. They contributed to the consolidation of the structure and its growth, incorporating into their walls appreciable quantities of fine terrigenous grains (quartz, clays), so much so that these stromatolites can sometimes be called "zoogenic" (Catalov, 1983). The presence of planktonic foraminifera and coccoliths among the sediments trapped between stromatolitic laminae allows us to date the stromatolites themselves and the adjacent sediments and, thus, to estimate their growth-rate. For example, in an outcrop near Ortegícar, there is a decimetric stromatolite encrusting on a breccia dated as Late Albian (*R. ticinencis* Zone) which is fossilized by late Early Vraconian marls (*P.buxtorfi* Zone of González-Donoso et al., 1982) allowing us to date the crust itself as lowermost Vraconian (*P.praebuxtorfi* Zone); this conclusion agrees well with microfossil associations found in stromatolites coming from other localities such as Benaojan or Hacho de Montejaque (see González-Donoso et al., 1983 for micropaleontological details and location): then, in these localities, stromatolites grow during a time interval that can be reasonably estimated as several tens to hundreds of thousand of years (Harland et al., 1989). If we consider stromatolite growth as a more or less continuous phenomenon (a hypothesis that obviously is an oversimplification) these and other examples indicate that the reasonable average growth-rate of Penibetic (at least the Albian-Vraconian examples) was very slow: about tenths of a millimetre per 1000 years, or even slower.

The significance of growth rhythms.

As it was described before, there are at least three orders of rhythms in the stromatolites of the Penibetic. It seems clear, then, that there have been at least three periodic variations of differing duration in the speed and mode of stromatolite growth, variations which are especially obvious in Cretaceous examples (although they are not exclusively found there) and above all in those of the Albian-Vraconian. Anyway, several question arise: a) What periodicity did each rhythm have? b) What sedimentary processes caused these periodic changes in stromatolite growth? And, finally, c) How do we understand the different kinds of cyclicity in terms of regional palaeogeographic evolution?.

Periodicity

At the present time it is not possible to establish a precise estimate of the length of each rhythm. Nevertheless, as it was previously argued, a reasonable average time-span is several tens to several hundreds of thousand years for the growth of centimetric to decimetric stromatolitic crusts. Accounting for the fact that the stromatolite growth was probably not uniform (because different fabrics and elements could have grown at different rates, and because there exist erosional surfaces between cycles), if our previous estimates are correct, the first order rhythms could have been formed over several thousands to tens of thousand of years; second order rhythms could have developed over several hundred to thousand years; and third order rhythms could be yearly episodes.

Processes

Different processes can be invoked to understand changes in stromatolite lamination. A possible mechanism could be related with microbial ecological successions, in a way similar to that found in modern stromatolites (see Monty 1967, 1976, for example). The existence of complex microbial communities in our stromatolites seems certain, because different filamentous and non-filamentous organic remains have been observed, although they cannot be exactly identified (a common situation when fossil stromatolites are studied: Monty, 1976; Krumbein, 1983). Then, the possibility of a control of changes in micrometric lamination (our third order rhythms) by succesive growth of different microbial communities seems reasonable: the growth of one microbial community would originate ecological conditions allowing the later ones to proliferate. Other mechanisms can be invoked to explain rhythmic changes in stromatolite growth, which are compatible with the hypothesis proposed above. The upward succesion of fabrics could have originated by periodic changes in environmental parameters, so that third order rhythms could result from yearly seasonal environmental variations. In the case of second order rhythms, environmental changes could be related to periodic variations in the pelagic rain, controlled by the productivity of the surface waters. On their hand, first order rhythms can be controlled by an upward decrease of water agitation and a concomitant increase in sediment input to the stromatolite, followed by a new reactivation of water agitation. In fact, the repetitive dense laminated fabric that characterises the lower parts of these rhythms, can have originated under conditions of greater energy than the other fabrics: it is the poorest-one in trapped interstitial sediment, it grows on erosional surfaces and is typically found trapping and binding reworked aggregates of pelagic muds with microstromatolites and limestone clasts of the underlying rockground. Intermittent or continuous water movement could favour an initial detachment and reworking of previously deposited and semi-lithified sediments (Fig.14); their frequent rolling on the substrate prevented sedimentation and growth of cupolas and, then, repetitive dense laminated fabrics would develop. No high energy or strong currents are necessary; in fact the common occurrence of discoidal, flattened forms with assymmetric development of laminae does not agree with a very agitated environment and rather suggests quiet waters (Jones and Wilkinson, 1978); the existence of frequent erosional surfaces limiting cycles of differing size demonstrates, however, that rolling occurred from time to time. In conclusion, only weak bottom currents would act in the stromatolite-forming settings of the Penibetic. Intermittent decrease in water agitation and/or contemporaneous increase in sediment input onto microbial mats could have favoured individualization of cupolas and columns. Finally, the rhythm concludes with the development of delicate arborescent microstromatolites under very quiet energy conditions, an interpretation also supported by the common occurrence of microstromatolites as cavity dwellers, encrusting walls of boring and cracks of the structure, in intricate areas, clearly well-protected from current action.

Regional palaeogeographic significance.
The last question concerning growth rhythms is if they can be better undestood in terms of local environmental conditions or perhaps of more important and extended, palaeogeographically controlled, circumstances. Sedimentary and facies changes in the pelagic realm usually have a wider environmental significance than a strictly local one. Indeed, long-term sedimentary stability and uniformity are more frequent in open marine than in shallow-water settings (Fischer, 1976, 1981, 1982; etc.). Then, if growth rhythms of our stromatolites are controlled by cyclic environmental variations of regional or wider significance (temperature, composition, salinity, sea-water currents, or biogenic planktonic productivity, the basic source of the carbonate sediments which reached the sea-bottom), these variations should have controlled the sedimentation in areas situated outside of strictly stromatolitic settings, and caused contemporaneous rhythmic changes in their deposits. Rhythmic deposits, coeval with the formation of Jurassic and Cretaceous pre-karstic stromatolites of the Penibetic, are frequently found in the Median Subbetic and other realms of the External Zones of the Betic Cordillera such as Oxfordian radiolaritic-marly alternances (Ruiz-Ortiz et al., 1989; O'Dogherty et al., 1989) or Neocomian marly-limy rhythmites (Azéma et al., 1979; García-Hernández

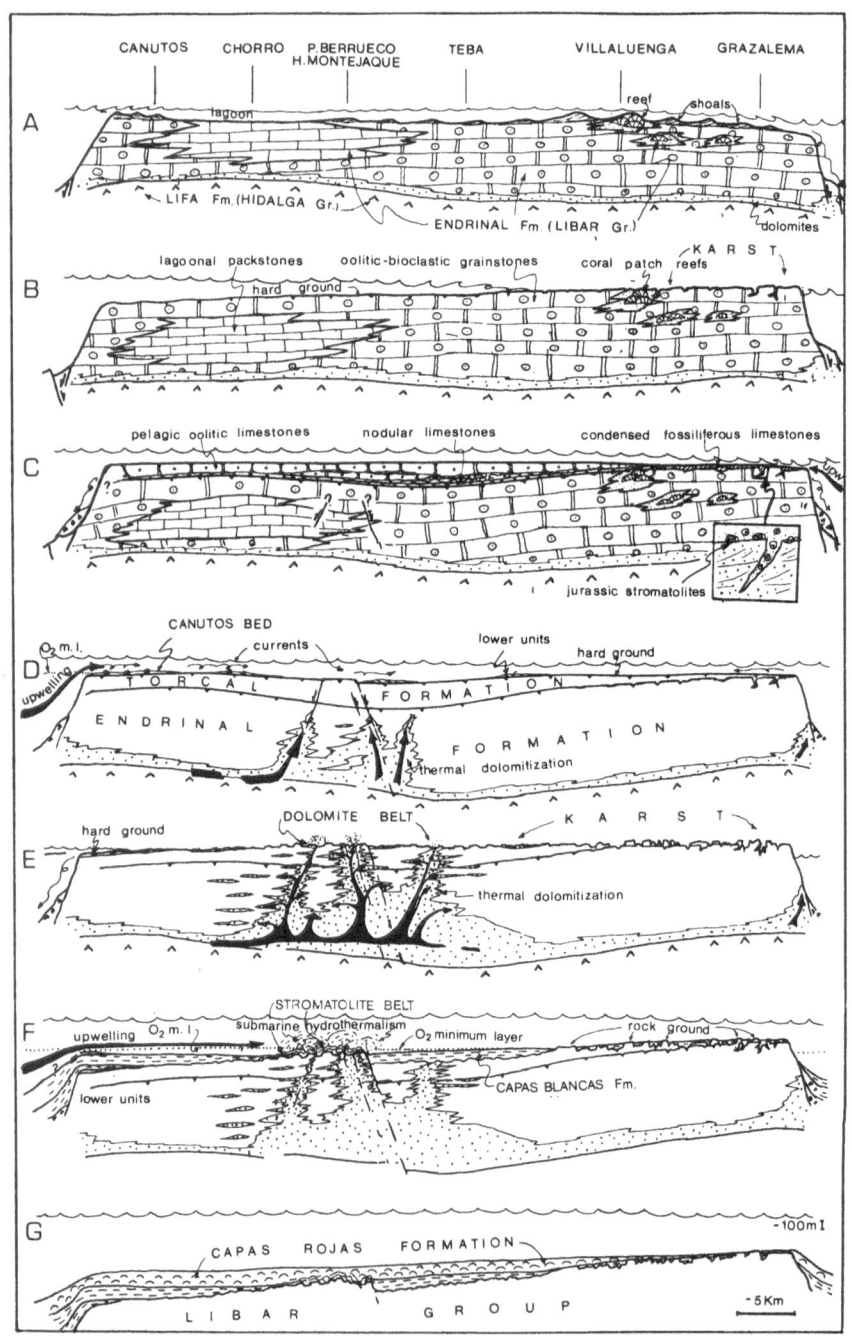

et al., 1980). Within the Penibetic itself, the Capas Blancas Fm., which is also made up of marly-limy rhythmites, is contemporaneous and is laterally connected with the post-karst Albian-Vraconian Stromatolite Belt; the fact that the climatic origin of these rhythms was probably astronomically controlled, has been commented on above and their periodicity could be approximately of the same order of magnitude as estimated periodicities of the coetaneous stromatolites. An extremely suggestive hypothesis would be, then, that stromatolite growth cycles (at least the first order ones) were thus the expression on a small scale of posibly astronomically controlled marly-limy (or radiolaritic) rhythms which are detected in coeval sediments both in the Penibetic and outside of it.

The position of the Albo-Vraconian Stromatolite Belt and its possible connection with thermal submarine surges.

The phosphate stromatolites of the Jurassic and the pre-karst Canutos Bed of the Lower Cretaceous were formed in outlying areas of the Penibetic pelagic swell, on its external and internal boundaries respectively. On the other hand, the post-karstic phosphate pelagic stromatolites of the Albian-Vraconian make up a Stromatolite Belt. This was located in the central zone of the swell, where the Líbar Group limestones were previously extensively dolomitized up to the palaeokarst surface locally. If we take into account the upward decrease in dolomitization, the morphology of dolomitized masses and the fact that no dolomitized rocks have been detected in the Espartina Group, we

Figure 20 : Palaeogeographic evolution of the Penibetic and setting of pelagic stromatolites. Northwest to the right. Scale indicative, similar in all cases. A) Lias-Dogger: isolated platform (bank) stage and deposition of shallow water (neritic) carbonate facies. B) Dogger-Malm transition: stratigraphic break originated by tectonic tilting of the bank, emergence of external parts. C) Malm to Lower Valanginian: pelagic platform stage (Torcal Fm.). At the beginning of this stage do pelagic phosphate stromatolites occur in the Grazalema area. During the rest of the stage there was pelagic-non-phosphatic, stromatolite growth, not studied in this article. D) Uppermost (?) Valanginian-Lower Hauterivian: after the intravalanginian discontinuity caused by the renewal of tilting of the Torcal Fm. stopped and there were only local lag deposits on the plateau (lower units). The Canutos Bed was formed in the southeastern part of the bank. Thermal dolomitization progressed rapidly upwards along normal faults, possibly listric. E) Hauterivian: emergence and karstification, especially developed in the most external part. Dolomitization reached the karst surface. F) Uppermost Hauterivian to early Upper Cretaceous: drowned platform or pelagic plateau stage. Renewed sedimentation of pelagic marls (Capas Blancas Fm.) occurs on palaeokarstic rockground below the topmost surface of oxygen minimum layer. There is localised phosphate stromatolite colonization on highs (the Stromatolite Belt) in the area where previous thermal dolomitization was active, the latter having been replaced by submarine low-temperature hydrothermalism. Active upwelling on the bank margins and interior. G) Senonian-Paleogene: drowning of the plateau and deposition of Capas Rojas Fm.

can deduce, then, that dolomitization moved from the underground upwards, probably following faults which were active during the tilting that caused the Hauterivian emergence and karstification. The dolomitizing fluids were probably driven by thermal convection processes (Kohout convection: Simms, 1984) similar to those that have been detected in recent carbonate platforms (Mullins et al., 1985). These fluids have risen to the surface of the palaeokarst, since dolomitic masses have been detected up to the top of the Líbar Group; higher in the section they then lost their force, or ceased completely, after the drowning of the Penibetic below the sea-waters. It is posible, however, that low temperature hydrothermal activity did not completely disappear; this possibility, together with the other oceanographic, paleotectonic and palaeogeographic factors could explain the close spatial relationship which exists between the Dolomite Belt and the Albo-Vraconian Stromatolite Belt. This relationship would result from the fact that the localised development of submarine microenvironments characterised by particular temperature and salinity differing somewhat from those found else where in the surrounding sea-floor (Fig.20). This would in turn have favoured the proliferation of bacteria, biogenetically induced mineral precipitation and the construction of stromatolites. It is well-know that bacterial activity and stromatolite formation can occur in hydrothermal waters, even at high temperatures, in subaerial environments (see e.g. Walter et al., 1976); in submarine environments the presence of hydrothermal exhalations may be one of the factors which determined the organically-induced precipitation of ferruginous-manganiferous minerals (Burns and Burns, 1979; Murray, 1979; Jannasch, 1984a,b; Jannasch and Mottl, 1985; Hölm, 1987) and the formation of manganese nodules and crusts which, according to Monty (1973a, 1977), are oceanic stromatolites. Submarine hydrothermal activity, however, although it is possible, and indeed probable, is not a prerequisite for the explanation of the formation of the Stromatolite Belt; also it is not necessarily required for the other examples we have studied, where no clear geological evidence for the existence of such hydrothermalism has been found. Of course, this is still possible, given that the examples under consideration were located near the limits of the swell, which were important synsedimentary fracture zones. In the case of the most recent example other factors seem to have intervened, since the Stromatolite Belt must have been formed on a relative submarine high within the Penibetic on which there was hardly any sedimentation. This high separated two areas of the Penibetic where limestone-marl rhythmites were deposited (Capas Blancas Fm.). This Belt also formed the outer (NW) boundary of the outcrops belonging to the Boqueron cherty limestones Member which shows synsedimentary folds in its inmediate surrounding and which is missing, due to a stratigraphic gap, in the more external regions. Finally, the thickness of the Espartina Group in general is minimal on the Stromatolite Belt. All these phenomena, interpreted in the palaeogeographical context, can be clearly seen in Fig.20.

Bathimetry and palaeo-oceanographic considerations.

Our phosphate stromatolites were formed on sedimentary discontinuity surfaces on a shallow pelagic swell which was repeatedly exposed to emergence for short periods of time. The colonization of the discontinuity surfaces by stromatolites marked the beginning of a new sedimentation, locally of very brief duration (as in the case of the pre-karst Cretaceous Canutos Bed), which in all cases is closely connected with the beginning of transgressive regional cycles resulting from eustatic rises (Vera and Martín-Algarra, this volume). The features of the stromatolites, the sediment facies and the fossils associated with them, together with other regional and paleontological data, clearly show that they grew in an open marine environment of warm, alkaline, and slightly reducing waters. This setting was characterised by a very low sedimentation rate and was affected by currents, at least sporadically. During the Cretaceous, this environment laid next to areas of deeper, poorly oxygenated or even anoxic waters, where marly limestone rhythmites with euxinic facies were deposited. According to the results of recent studies, such conditions are precisely similar to those most propitious for the deposition of phosphate and the formation of large deposits of submarine phosphates (Cook and McElhinny, 1979; Slansky, 1980; Bentor, 1980; Sheldon, 1980, 1981; Arthur and Jenkyns, 1981; Baturin, 1982; Soudry and Champetier, 1983; Soudry, 1979, 1987; etc.). These deposits were usually formed in the most open areas of the platforms, at relatively shallow depths (at a maximum of a few hundred metres), located near the boundary of the oxygen minimum layer, and intensely affected by upwelling. It thus appears probable that some of the conditions, under which the phosphate stromatolites of the Penibetic were formed, were quite similar to those which gave rise to great phosphorus deposits.

Palaeogeography may hinder a complete comparison between Alpine Mesozoic stromatolitic phosphates and Tertiary to Recent great phosphorous deposits. The second are more frequently found in intermediate latitudes, where strong upwelling brought cold bottom waters rich in phosphorous from deeper levels of the ocean, favouring a very high organic productivity. Nevertheless, the Penibetic was situated in an "equatorial" or nearly equatorial context, as a part of the Western Tethys, and there during the Mesozoic possibly no cold bottom water of polar provenance existed, since the North- and South-Atlantic oceans were still closed. Then, if upwelling did exist in the Western Tethys, it was probably driven by winds or geostrophic currents. This hypothesis resembles the oceanographic situation found in the Australian east-coast continental margin (O'Brien and Veeh, 1980), where a bacterial origin of phosphorites has also been proposed (O'Brien et al., 1981), and it has been applied by Delamette (1985) in order to explain mid-Cretaceous alpine phosphorites, at least partly of stromatolitic origin. The palaeogeographic and palaeoclimatological data favour this hypothesis and do not support the idea of a local climatic deterioration in the Southern Iberian continental margin due to cold water ascent, as in present-day coast affected by upwelling. The existence of a geostrophic current coming from the eastern to the western Tethys is demonstrated by palaeobiogeographical data (Rat, 1982, 1984).

The oxygen depletion of deep waters as a consequence of the rise in sea-level during the Upper Valanginian, Upper Aptian and, above all, the Upper Albian-Vraconian (Vail et al., 1977; Hancock and Kauffman, 1979; Jenkyns, 1980a, 1985; García-Hernández et al., 1982) must have favoured the phosphate enrichment of contemporary anoxic sediments (Arthur and Jenkyns, 1981; etc). Their action prevented sedimentation, or made it more difficult, and encouraged stratigraphic condensation, thus permitting the growth of micro-organisms. The most propitious areas for phosphatization and glauconitization processes are the sea bottoms situated at depths near the limits of the O_2 minimum layer. During the Mid-Cretaceous this layer expanded and their boundaries ranged from to 100 to 600 m (Arthur and Jenkyns, 1981). Therefore, it seems reasonable to conclude that the greatest depth at which phosphate pelagic stromatolites of the Penibetic formed must have been of the order of a few hundred metres.

The position of the phosphate stromatolites of the Penibetic, varied through time from outer (northern) sectors during Jurassic to the inner (southern) and then central areas of the pelagic swell during the Cretaceous. This clearly shows the interaction between the tectonic tilting which caused the raising of the most external sectors, and the progress of the eustatic rise in sea-level and thus of the O_2 minimum layer throughout the Lower Cretaceous (Martín-Algarra et al., 1992). Together with this rise, the great Oceanic Anoxic Event of the mid-Cretaceous took place and created the conditions which allowed the deposition of limestones and phosphate crusts on a great number of pelagic swells of the Mediterranean Alpine regions (Arthur and Jenkyns 1981; Lloyd 1982; Jenkyns 1980a, 1985). The fact that well-developed phosphate pelagic stromatolites with similar features and the same age (Albo-Vraconian) as those of the Stromatolite Belt of the Penibetic have also been found in the Alps (Rioult and Royant, 1975; Bourbon, 1980; Delamette, 1985, 1988; Föllmi, 1989) and the Carpathians (Krajewsky, 1981a,b,c, 1983, 1984), suggests that they could result from the same general sedimentological, oceanographical and palaeogeographical conditions.

CONCLUSIONS

The following general conclusions can be drawn from the present study:

1) The sea-bottom of the Penibetic, a Mesozoic swell related to the external Betic continental margin and situated in the westernmost part of the Tethys, was repeatedly colonized by stromatolite building microbial communities, formed essentially by filamentous bacteria and encrusting foraminifera.

2) The ground colonization occurred at times of very low sedimentation rate, immediately after regionally very important sedimentary breaks, during the base of the Upper Jurassic (Oxfordian), the Uppermost Valanginian-Lower Hauterivian, and especially during the Upper Albian-Vraconian.

3) The colonized substratum was, in every case, a hard- or rock-ground, affected, at least sporadically, by currents; in these conditions scarce pelagic, essentially biomicritic, calcareous sediments were deposited; later they were semiconsolidated or lithified at a very early postdepositional stage.

4) Microbial, possibly bacterial filaments mediated the precipitation of sulphides (pyrite) and especially of phosphate, directly at the sea-bottom. Phosphates, and to a lesser extend sulphides, formed some sort of mucilaginous envelope around organic filaments trapping pelagic detrital particles and coccoliths.

5) As in manganese nodules encrusting foraminifera can play an essential role in stromatolite building and consolidation.

6) Stromatolite growth shows clear pulsations marked by particular fabrics. Their hypothetical relationships with periodical ecological, climatically induced changes have been proposed. Stromatolitic lamination and cycles would have thus be the expression, on a small scale, of climatically controlled marly- limestone rhythms, which are present in contemporary sediments of the adjacent parts of the pelagic plateau or also in deeper adjacent basins.

7) Stromatolite formation occurred in zones affected by current action (upwelling or most probably geostrophic in origin) and near the O_2 minimum layer, which favoured phosphate precipitation. The precise depth of formation is not known, but it seems probable that it is not deeper than some hundred meters.

8) There is some evidence favouring the contemporary existence, of active, low-temperature submarine hydrothermalism in the area of maximun growth of phosphate stromatolites of the Penibetic: this could have favoured bacterial activity and mineral deposition, and proliferation of stromatolite-building submarine microbial communities.

9) Our examples clearly illustrate the displacement of stromatolite communities from shallow-water and epicontinental seas, where they had been dominant throughout the Phanerozoic, to the pelagic realm. They were replaced by better adapted and more competitive organisms, and found the appropiate niche in the pelagic realm, where environmental conditions were harder. In this case, greater water depth, poor light, the influence of bottom current, sporadic boring and possibly hydrothermalism, hindered the settlement of most organisms except bacteria and encrusting foraminifera, which were perhaps symbiotically related.

ACKNOWLEDGEMENTS

The authors would like to express their gratitude to Dr. E. Barea (Department of Inorganic Chemistry, University of Granada) for his help in the preparation of the samples studied under SEM; also to Dr. P. Fenoll (Department of Mineralogy, University of Granada) for her assistance in the microscopic study of the minerals and the taking of the light-reflected microscope photographs; to R. Aguado (Department of Stratigraphy and Paleontology, University of Granada) for the classification of the coccoliths detected under the SEM; to Drs. E. Montoya and M.A. Ribadeneyra (Dpt. of Microbiology, University of Granada) for their valuable comments on the bacterial remains observed under the SEM. The critical reading of the earlier versions and suggestions made by Drs. M. García-Hernández, (University of Granada), C.L.V. Monty (University of Nantes), H.C. Jenkyns (University of Oxford), J. Sarfati (University of Montpellier) and H. Hofmann (University of Montréal) have contributed very much to improve this paper. Also we thank Dr. N. McLaren for the translation of the first manuscript into English, and the technicians of the Servicios Tecnicos (University of Granada) and E.U.I.T.I. El Sario (University of Navarra) for their care in obtaining and processing the photographs under the SEM. The present study has been financed by the CAICYT, Research Projects no. 0971-87 and 0853-90.

The authors would like to express their gratitude to Dr. E. Barea (Department of Inorganic Chemistry, University of Granada) for his help in the preparation of the samples studied under SEM; also to Dr. P. Fenoll (Department of Mineralogy, University of Granada) for her assistance in the microscopic study of the minerals and the taking of the light-reflected microscope photographs; to R. Aguado (Department of Stratigraphy and Paleontology, University of Granada) for the classification of the coccoliths detected under the SEM; to Drs. E. Montoya and M.A. Ribadeneyra (Dpt. of Microbiology, University of Granada) for their valuable comments on the bacterial remains observed under the SEM. The critical reading of the earlier versions and suggestions made by Drs. M. García-Hernández, (University of Granada), C.L.V. Monty (University of Nantes), H.C. Jenkyns (University of Oxford), J. Sarfati (University of Montpellier) and H. Hofmann (University of Montréal) have contributed very much to improve this paper. Also we thank Dr. N. McLaren for the translation of the first manuscript into English, and the technicians of the Servicios Tecnicos (University of Granada) and E.U.I.T.I. El Sario (University of Navarra) for their care in obtaining and processing the photographs under the SEM. The present study has been financed by the CAICYT, Research Projects no. 0971-87 and 0853-90.

REFERENCES

Arthur, M.A., Bottjer, D.J., Dean, W.E., Fischer, A.G., Hattin, D.E., Kauffman, E.G., Prat, L.M., Scholle and P.A. (1986) "Rhythmic bedding in Upper Cretaceous pelagic carbonate sequences: varying sedimentary response to climatic forcing", Geology, 14, 153-156.

Arthur, M.A. and Jenkyns, H.C. (1981) "Phosphorites and paleoceanography", Oceanologica Acta, Proc 26th Inter Geol. Congr. Paris (1980) n. sp., pp. 83-96.

Azéma, J., Foucault, A., Fourcade, E., García-Hernández, M., Linares, A., Linares, D., López-Garrido, A.C., Rivas, P. and Vera, J.A. (1979) "Las microfacies del Jurásico y Cretácico de las Zonas Externas de las Cordilleras Béticas", Secr. Publ. Univ. Granada, pp. 1-83.

Baturin, G.N. (1982) "Phosphorites on the seafloor. Origin, composition and distribution", Devel. in Sedimentology 33, Elsevier, Amsterdam, pp. 1-343.

Bentor, J.K. (1980) "Phosphorites: The unsolved problems", in: J.K. Bentor (ed.), Soc. Econ. Paleontologists & Mineralogists, Spec. publ., 29, 3-18.

Bourbon, M. (1980) "Evolution d'un secteur de la marge nord-Tethysienne en milieu pélagique: La Zone Briançonaise près de Briançon entre le début du Malm et l'Eocène inférieur", Thesis, Univ. L. Pasteur, Strasbourg, pp. 1-335.

Burns, R.G. and Burns, V.M. (1979) "Manganese Oxides", in: Marine Minerals (Chapter 1), Min. Soc. Amer. Shourt course Notes, 6, 1-46.

Catalov, G.A. (1983) "Triassic oncoids from central Balkanides (Bulgaria)", in: T.M. Peryt (eds.), Coated Grains, Springer-Verlag, Heidelberg, pp. 398-408

Chauve, P., Didon, J. and Peyre, Y. (1968) "Le Crétacé inférieur du Pénibétique (Zone de Ronda-Torcal, Cordillères Bétiques, Espagne)", Bull. Soc. Geol. France (7), 10, 56-64.

Cook, P.J. and McElhinny, M.V. (1979) "A reevaluation of the Spatial and Temporal Distribution of Sedimentary Phosphate Deposits in the light of Plate Tectonic", Econ. Geol., 74, 315-330.

De Boer, P.L. (1983) "Aspects of middle cretaceous pelagic sedimentation in southern Europe; production and organic matter, stable isotopes and astronomical influences", Geol. Ultraiectina, 31, 1-112.

De Boer, P.L. and Wonders, A.A.H. (1984) "Astronomically induced rhythmic bedding in Cretaceous Pelagic sediments near Moria (Italy)". In: A. Berger, J. Imbrie, J. Hays, G. Kukla and B. Saltzman (eds.), Milankovitch and Climate, part 1, Reidel Publ. Co, Dordrecht, Boston, Lancaster, pp. 177-190.

Delamette, M. (1985) "Phosphates et paléocéanographie: l'exemple des phosphorites du Cretacé moyen delphino-helvétique", C. R. Ac. Sc. Paris, 300, II, 20, 1025-1028.

Delamette, M. (1988) "Relation between the condensed Albien deposits of the Helvetic domain and the oceanic current-influenced continental margin of the northern Tethys", Bull. Soc. Geol. France (8) IV, 739-745.

Farinacci, A. (1967) "La serie Giurassico-Neocomiana di Monte Lacerone (Sabina). Nuove vedute sull'interpretazione paleogeographica della aree di facies Umbro-Marchigiana". Geol. Romana, 6, 421-480.

Fischer, A.G. (1976) "Pelagic sediments as clues to the Earth behavior", Mem. Soc. Geol. It., 15, 1-8.

Fischer, A.G. (1981) "Climatic oscillations in the biosphere", in: Biotic crisis in ecological and evolutionary time, Academic Press, New York, pp. 103-131.

Fischer, A.G. (1982) "Long-term climatic oscillations recorded in Stratigraphy", in: Studies in Geophysics: Climate in Earth History, National Academy Press, Washington, 9, 97-104.

Föllmi, K. (1989) "Evolution of the Mid-Cretaceous Triad. Platform carbonates, Phosphatic sediments, and Pelagic Carbonates Along the Northen Tethys Margin", Lecture Notes in Earth Sciences, 23, pp. 1-153, Springer-Verlag, Berlin.

García-Cervigón, A., Martín-Algarra, A., Montealegre, L. and Vera, J.A. (1986) "Estromatolitos pelágicos fosfatados relacionados con discontinuidades estratigráficas en el Cretácico del Penibético (provincia de Málaga)", Acta Geológica Hispánica, 21-22, 361-372.

García-Hernández, M., González-Donoso, J.M., Linares, A., Rivas, P. and Vera, J.A. (1976) "Características ambientales del Lías inferior y medio en la Zona Subbética y su significado en la interpretación general de la Cordillera", in: Reun. Geol. Cord. Bética Mar Alborán, Secr. Publ. Univ. Granada, pp. 125-157

García-Hernández, M., López-Garrido, A.C., Martín-Algarra, A. and Vera, J.A. (1982) "Cambios eustáticos en el Cretácico de la Cordillera Bética: Comparación de la evolución sedimentaria en un dominio de plataforma (Zona Prebética) y otro de umbral pelágico (Penibético)", Cuadernos de Geología Iberica, 8, 581-597.

García-Hernández, M., López-Garrido, A.C., Rivas, P., Sanz de Galdeano, C. and Vera J.A. (1980) "Mesozoic paleogeographic evolution of the external zones of the Betic Cordillera", Geol. Mijnb., 59, 155-168.

García-Hernández, M., Lupiani, E. and Vera, J.A. (1986) "La sedimentación liásica en el sector central del Subbético medio: registro de la evolución de un rift intracontinental", Acta Geológica Hispánica, 21-22, 329-337.

González-Donoso, J.M., Linares, D., Martín-Algarra, A., Rebollo, M., Serrano, F. and Vera, J.A. (1983) "Discontinuidades estratigráficas durante el Cretácico en el Penibético (Cordillera Bética)", Estudios Geológicos, 39, 71-116.

González-Donoso, J.M., Linares, D., Rebollo, M. and Serrano, F. (1982) "Bioestratigrafía del Albense medio-Turonense medio basada en foraminíferos planctónicos, Penibético, Cordilleras Béticas", Cuadernos de Geología Ibérica, 8, 739-758.

Hancock, J.M. and Kauffmann, E.G. (1979) "The great transgressions of the late Cretaceous", Jour. Geol. Soc. London, 136, 175-186.

Haq, B.U., Hardendol, J. and Vail, P.R. (1987) "Chronology of sea-level changes since the Triassic", Science, 235, 1156-1167.

Harland, W.B., Armstrong,R.L., Cox, A.V., Craig, L.E., Smith, A.G. and Smith, D.G. (1989) "A Geologic Time Scale 1989", Cambridge Univ. Press, Cambridge, pp. 263.

Heim, A. (1946) "Problemas de erosión submarina y sedimentación pelágica del presente y del pasado", Rev. Museo de la Plata, 4-22, 125-178.

Heim, A. (1958) "Oceanic sedimentation and submarine discontinuities", Eclog. Geol. Helv., 51, 642-649.

Hofmann, H.J. and Grotzinger, J.P. (1985) "Shell-facies microbiotas from the Odjick and Rocknest formations (Epworth Group; 1.89 Ga), Northwestern Canada", Can. J. Earth Sci., 22, 1781-1792.

Hölm, N.G. (1987) "Biogenic influences on the geochemistry of certain ferruginous sediments of hidrothermal origin", Chemical Geology, 63, 45-57.

Jannasch, H.W. (1984a) "Microbial processes at deep sea hydrothermal vents", In: P.A. Rona, K. Borstrom, L. Laubier and K.L. Smith (Eds.), Hydrothermal processes seafloor spreading centers, Plenun Publ. Co, 677-709.

Jannasch, H.W. (1984b) "Chemosynthesis: The Nutritional Basis for life at deep-sea vents", Oceanus, 27, 73-78.

Jannasch, H.W. and Mottl, M.J. (1985) "Geomicrobiology of Deep-sea Hydrothermal vents", Science, 229, 717-725.

Jenkyns, H.C. (1967) "Fossil Manganese nodules from Sicily", Nature, 216, 673-674.

Jenkyns, H.C. (1970a) "Submarine volcanism and Toarcian Iron Pisolites of Western Sicily", Eclog. Geol. Helv., 63, 549-572.

Jenkyns, H.C. (1970b) "Fossil Manganese Nodules from the West Sicilian Jurassic", Eclog. Geol. Helv., 63, 741-774.

Jenkyns, H.C. (1972) "Pelagic "oolites" from the Tethyan Jurassic", J. Geol., 80, 21-33.

Jenkyns, H.C. (1980a) "Cretaceous anoxic events: from continents to oceans", J. Geol. Soc. London, 137, 171-188.

Jenkyns, H.C. (1980b) "Tethys: past and present", Proc. Geol. Ass., 91, 107-118.

Jenkyns, H.C. (1985) "The early Toarcian and Cenomanian-Turonian anoxic events in Europe: comparisons and contrasts", Geol. Rundsch., 74, 505-518

Jones, F.G. and Wilkinson, B.H. (1978) "Structure and growth of lacustrine pisoliths from recent Michigan Marl Lakes", J. Sedim. Petrol., 48, 1103-1110.

Krajewski, K.P. (1981a) "Phosphate microstromatolites in the High-Tatra Albian limestones in the Polish Tatra Mountains", Bull. Acad. Pol. Sci. Ser. Sci. Terre, 29, 175-183.

Krajewski, K.P. (1981b) "Phosphate pizolite structures from condensed limestones of the High-Tatric Albian (Tatra Mts.)", Ann. Soc. Geol. Pol., 51, 339-352.

Krajewski, K.P. (1981c) "Pelagiczne stromatolity z wapieni albu wierchowego Tatr", Kwart. geol., 25, 731-759.

Krajewski, K.P. (1983) "Albian pelagic phosphate-rich macrooncoids from the Tatra Mts. Poland", in: T.M. Peryt (ed.), Coated Grains, Springer-Verlag, Berlin-Heidelberg, pp. 344-357

Krajewski, K.P. (1984) "Early diagenetic phosphate cements in the Albian condensed glauconitic limestone of the Trata Mountains, Western Carpathians", Sedimentology, 31, 443-470.

Krumbein, W.E. (1983) "Stromatolites - the challenge of a term in space and time", Precambrian Res., 2, 493-531.

Lloyd, C.R. (1982) "The mid-Cretaceous earth: paleogeography, ocean circulation and temperature: atmospheric circulation", Jour. Geol., 90, 393-413.

López-Galindo, A. and Martín-Algarra, A. (1992) "Palaeogeography and Mineralogy of Mid-Cretaceous Flysches in the Gibraltar arc area", Cretaceous Research, 13, 421-443.

Lucas, G. and Prevot, L. (1981) "Synthèse d`apatite à partir de matière organique phosphorée (ARN) et de calcite par voie bacterienne", C. R. Acad. Sci. Paris, 292, 1203-1205.

Malone, P.G. and Towe, K.M. (1970) "Microbial carbonate and phosphate precipitates from sea-water cultures", Mar Geol., 9, 301-309.

Martín-Algarra, A. (1987) "Evolución geológica alpina del contacto entre las Zonas Externas y las Zonas Internas de la Cordillera Bética (sector centro-occidental)", Thesis, Univ. Granada, pp. 1-1171.

Martín-Algarra, A. and Martínez-Gallego, J. (1984) "El Paleógeno del Penibético (Cordillera Bética)", Mediterránea, Ser. Geol., 3, 41-64.

Martín-Algarra, A., Ruiz-Ortiz, P.A. and Vera, J.A. (1992) "Factors controlling Cretaceous turbidite deposition in the Betic Cordillera", Rev. Soc. Geol. España, 4, 53-80.

Martín-Algarra, A. and Vera, J.A. (1982) "El Cretácico del Penibético, las unidades del Campo de Gibraltar, las Zonas Internas y las unidades implicadas en el contacto entre la Zonas Internas y Externas", in: El Cretácico en España, A. García (ed.), Univ. Complutense, Madrid, pp. 603-638.

Martín-Algarra, A. and Vera, J.A. (1989) "La serie estratigráfica del Penibético (Cordillera Bética)", in: Libro Homenaje a Rafael Soler, Asoc. Geol. Geof. Esp. Petrol., 67-76.

Massari, F. (1975) "The hardgrounds of the "Scaglia Rossa veneta" (Southern Alps, Italy)". Publ. IX Cong. Int. Sediment. Nice, 4, 243-247.

Massari, F. (1979) "Oncoliti e stromatoliti pelagiche nell Rosso Ammonitico veneto", Mem. Sc. Geol., 32, 1-21.

Massari, F. (1981) "Cryptalgal fabrics in the Rosso Ammonitico sequences of the Venetian Alps", in: A. Farinacci and S. Elmi, (eds.), Rosso Ammonitico Symposium Proc., Tecnoscienza Roma, pp. 435-469.

Massari, F. (1983) "Oncoids and Stromatolites in the Rosso Ammonitico Sequences (Middle-Upper Jurassic) of the Venetian Alps, Italy", in: T.M. Peryt (ed.), Coated grains, Springer-Verlag Berlin Heidelberg, pp. 358-366.

Massari, F. and Dieni, I. (1983) "Pelagic oncoids and ooids in the Middle-Upper Jurassic of Eastern Sardinia", in: T.M. Peryt (ed.), Coated grains, Springer-Verlag, Berlin Heidelberg, pp. 367-376.

Massari, F. and Medizza, F. (1973) "Stratigrafia e paleogeografia del Campaniano- Maastrichtiano nelle Alpi Meridionali (con particolari riguardo agli hardgrounds della Scaglia Rossa Veneta)", Mem. Ist. Geol. Min. Univ. Padova, 28, 1-62.

Mendelsohn, F. (1976) "Mineral Deposits Associated with stromatolites", in: M.R. Walter (ed.), Stromatolites, Devel. in Sedimentology, 20, Elsevier Amsterdam, pp. 645-662

Monty, C.L.V. (1967) "Distribution and Structure of recent stromatolitic algal mats, Eastern Andros Island, Bahamas", Ann. Soc. Geol. Belg., 90, 55-99.

Monty, C.L.V. (1973a) "Les nodules de manganèse sont des stromatolites océaniques", C. R. Acad. Sci. Paris, D-276, 3285-3288.

Monty, C.L.V. (1973b) "Precambrian background and Phanerozoic history of stromatolite communities, an overview". Ann. Soc. Geol. Belgique, 96, 585-624.

Monty, C.L.V. (1976) "The origin and development of criptalgal fabrics", in: M.R. Walter (ed.), Stromatolites, Devel. in Sedimentology, 20, Elsevier Amsterdam, pp. 193-249.

Monty, C.L.V. (1977) "Evolving concepts on the nature and the ecological significance of stromatolites: a rewiew". in: E. Flugel E (ed.), Fossil Algae, Springer-Verlag, Berlin-Heidelberg, pp. 15-35.

Monty, C.L.V. (1982) "Cavity or fissure dwelling stromatolites (endostromatolites) from Belgian Devonian mud mounds (Extended abstract)", Ann. Soc. Geol. Belg., 105, 343-344.

Monty, C.L.V. (1986) "Interactions événements géologiques-Stromatolites", Bull. Rech. Explor. Prod. Elf Aquitaine, 10, 537-553.

Mullins H.T., Wise (jr.) S.W., Gardulski, A.F., Hinchey, E.J., Masters, P.M. and Siegel, D.I. (1985) "Shallow subsurface diagenesis of Pleistocene periplatform ooze: northern Bahamas", Sedimentology, 32, 473-494.

Murray, J.W. (1979) "Iron oxides", in: Marine Minerals, Chapter 2, Min. Soc. Amer. Short Course Notes, 6, 47-98.

O'Brien, G.W. and Veeh, H.H. (1980) "Holocene phosphorites on the East Australian continental margin", Nature, 288, 690-692

O'Brien, G.W., Harris, J.R., Milnes, A.R. and Veeh, H.H. (1981) "Bacterial origin of East Australian Continental margin phosphorites", Nature, 294, 442-444.

O'Dogherty, L., Sandoval, J., Martín-Algarra, A. and Baumgartner, P. (1989) "Las facies con radiolarios del Jurásico subbético", Rev. Soc. Mex. Paleont., 2, 70-77.

Odin, G.S. and Letolle, R. (1980) "Glauconitization and phosphatization environments: a tentative comparison", in: J.K. Bentor (ed.), Soc. Econ. Paleontologist & Mineralogists, Spec. publ., 29, 227-237.

Rat, P. (1982) "Factores condicionantes en el Cretácico de España", Cuadernos de Geología Ibérica, 8, 1059-1076.

Rat, P. (1984) "La tectonique de plaques confrontée à la dynamique externe", Bull. Soc. Geol. France (7), 26, 377-399.

Read, J.F. (1982) "Carbonate platforms of passive (extensional) continental margins: types, characteristics and evolution", Tectonophysics, 81, 195-212.

Read, J.F. (1985) "Carbonate Platform Facies Models", Amer. Assoc. Petr. Geol. Bull., 69, 1-21.

Rioult, M. and Royant, G. (1975) "La "croute stromatolitique in Manara" horizon repère de l'Aptien-Albien dans la serie briançonnaise du Monte Armelta (Alpes Ligures)", Proc. Congr. Inter. Sediment., Nice, 10, 123-128.

Ruiz-Ortiz, P.A., Bustillo, M.A. and Molina, J.M. (1989) "Radiolarite sequences of the Subbetic, Betic Cordillera, Southern Spain", in: J.R. Heim and J. Obradovic (eds.), Siliceous deposits of the Tethys and Pacific regions, Springer-Verlag, New York, 227-253.

Schlager, W. (1974) "Preservation of cephalopods skeletons and carbonate dissolution on ancient Tethyan sea-floors", in: K.J. Hsü and H.C. Jenkyns (eds.), Pelagic Sediments; on land and under the sea, Inter. Assoc. Sediment., Spec. publ., Blackwell, 1, 49-70.

Schlanger, S.O. and Jenkyns, H.C. (1976) "Cretaceous oceanic anoxic events: causes and consequences", Geol. Mijnb., 55, 179-184

Seyfried, H. (1979) "Ensayo sobre el significado paleogeográfico de los sedimentos del Jurásico de las Cordilleras Béticas orientales", Cuad. Geol. Univ. Granada, 10, 317-348.

Shapiro, L. (1952) "Simple field method for the determination of phosphate in phosphatic rocks", Amer. Mineral., 37, 391-342.

Sheldon, R.P. (1980) "Episodicity of phosphate deposition and deep ocean circulation-an hypothesis", in: J.K. Bentor (ed.), Soc. Econ. Paleontologists & Mineralogists, Spec. publ., 29, 239-247.

Sheldon, R.P. (1981) "Ancient marine phosphorites", Ann. Rev. Earth Planet. Sci., 9, 251-284.

Simms, M. (1984) "Dolomitization by Thermal convection in carbonate platforms (abstract)", Amer. Assoc. Petr. Geol. Bull., 68, 528.

Slansky, M. (1980) "Géologie des phosphates sédimentaires", Mem. Bur. Rech. Geol. Miner., 114, pp. 1-92

Slansky, M. (1982) "Importance du rôle des organismes et de la matière organique dans la sedimentation phosphatée", in: Livre Jubilaire G. Lucas, Geologie Sedimentaire, Mem. Geol. Univ. Dijon (7), pp. 215-224.

Soudry, D. (1979) "Intervention de schyzophytes dans la phosphomicritization de débris oseaux", C.R. Acad. Sci. Paris, 288D, 666-671.

Soudry, D. (1987) "Ultra-fine structures and genesis of the Campanian Negev high phosphorites (southern Israel)", Sedimentology, 34, 641-660.

Soudry, D. and Champetier, Y. (1983) "Microbial processes in the Negev phosphorites (southern Israel)", Sedimentology, 30, 411-423.

Soudry, D. and Southgate, P.N. (1989) "Ultrastructure of a Middle Cambrian Primary Nonpelletal Phosphorite and its Early Transformation into Phosphate Vadoids: Georgina Basin, Australia", J. Sedim. Petrol., 59, 53-64.

Trudinger, P.A. and Mendelsohn, F. (1976) "Biological Processes and Mineral Deposition", in: M.R. Walter (ed.), Stromatolites, Devel. in Sedimentology, 20, Elsevier, Amsterdam, pp. 663-672.

Vail, P.R., Mitchum, R.M. and Thompson III, S. (1977) "Seismic stratigraphy and global changes of sea level. Part.4. Global cycles of relative changes of sea level", in: C.E. Payton (ed.), Seismic Stratigraphy, Amer. Assoc. Petrol. Geol., mem., 26, 83-97.

Vera, J.A. (1984a) "Discontinuidades estratigráficas en materiales pelágicos: caracterización, génesis e interpretación", I Congr. Esp. Geol., Segovia, 1, 109-122.

Vera, J.A. (1984b) "Aspectos sedimentológicos en la evolución de los dominios alpinos mediterráneos durante el Mesozoico", in: A.Obrador (ed.), Libro Homenaje a L.Sánchez de la Torre, Grup. Esp. Sed., Publ. Geología, Barcelona, 20, 25-54.

Vera, J.A. and Martín-Algarra, A. (1993) "Mesozoic stratigraphic breaks and pelagic stromatolites in the Betic Cordillera, Southern Spain" (in this volume)

Vera, J.A., Ruiz-Ortiz, P.A., García-Hernández, M. and Molina, J.M. (1988) "Paleokarst and related Pelagic Sediments in the Jurassic of the Subbetic Zone, Southern Spain", in: N.P. James and P.W. Choquette (eds.), Paleokarst, Springer, New York, 364-384.

Walter, M.R., Bauld, Y. and Brock, T.O. (1976) "Microbiology and morphogenesis of columnar stromatolites (Conophyton, Vacerilla) from hot springs in Yellowstone National Park", in: M.R. Walter (ed.), Stromatolites, Devel. in Sedimentology 20, Elsevier, Amsterdam, pp. 273-310.

Wendt, J. (1974) "Encrusting organism in deep-sea manganese nodules", in: K.J. Hsu and H.C. Jenkyns (eds.), Pelagic Sediments; on land and under the sea", Inter. Assoc. Sedim., Spec. publ., Blackwell, 1, 437-447.

PART IV

Paleozoic Stromatolites and Thrombolites

SILICICLASTIC-CARBONATE STROMATOLITE DOMES, IN THE EARLY CARBONIFEROUS OF THE AJJERS BASIN (EASTERN SAHARA, ALGERIA)

J. BERTRAND-SARFATI
Institut des Sciences de l'Evolution, URA-CNRS, 327 ; Sciences et Techniques du Languedoc, 34095 Montpellier Cedex 5, France.

ABSTRACT

Stromatolite domes crop out widely in the upper Visean of eastern Sahara, Algeria and Lybia. Their growth was conditionned by the sedimentary regime prevailing in the eastern Sahara during Early Carboniferous times, as well as the specific features of sedimentation over cratonic areas. The stromatolites of the two major units are discrete juxtaposed domes with high total relief. The same kind of laminae, probably built by cocoid cyanobacteria and bacteria, are found in all domes. Some of them are characterized by a high percentage of siliciclastic material. Stromatolites grew in very quiet subaqueous environments protected from open marine water. The stromatolites of the lower unit, are strictly the same from base to top of the sequence (except for an increase in size). The upper unit is a typical shallowing-upward para-sequence, starting with subaqueous large domes passing to stromatolite heads in intertidal settings. In similar deposits in Lybia, Whitbread and Kelling (1982) interpreted the stromatolite "Collenia interval" as resulting from a widespread delta abandonment. In terms of sequence stratigraphy the two sequences containing the stromatolites are High Stand System Tracts following a sea level rise. However some differences observed in the stromatolites indicate distinct conditions of growth: the stromatolites of the lower unit have developed in lakes fed by fresh-water streams, while those of the second unit involved a marine lagoon protected by oolitic bars, filled during sea-level rises and shallowing-upward progressively. The very large extension and the thickness of the stromatolite layers imply a deposition over large, flat, stable areas, with low subsidence rate, a characteristic feature of the West African Craton since the Proterozoic.

INTRODUCTION

Stromatolites partly siliciclastic, in juxtaposed domes crop out widely over the eastern Sahara, Algeria and Lybia. The two units bearing stromatolites will be described and their environmental conditions discussed. According to paleontological data (brachiopod faunas) they are Late Visean in age. They are totally absent from coeval Carboniferous deposits westward of the area. Such widespread and well developed

J. Bertrand-Sarfati and C. Monty (eds.), Phanerozoic Stromatolites II, 395–419.
© 1994 *Kluwer Academic Publishers.*

marine stromatolite beds are relatively rare in late Paleozoic as well as in more recent formations except for some Cenozoic lacustrine deposits. The stromatolite beds are directly related to the evolution of a large deltaic environment prevailing during the Early Carboniferous in the Fezzan-Ajjers basin. However their extension is also controled by the geodynamic features of the West-African-Saharan craton which guides the sedimentation since the Proterozoic.

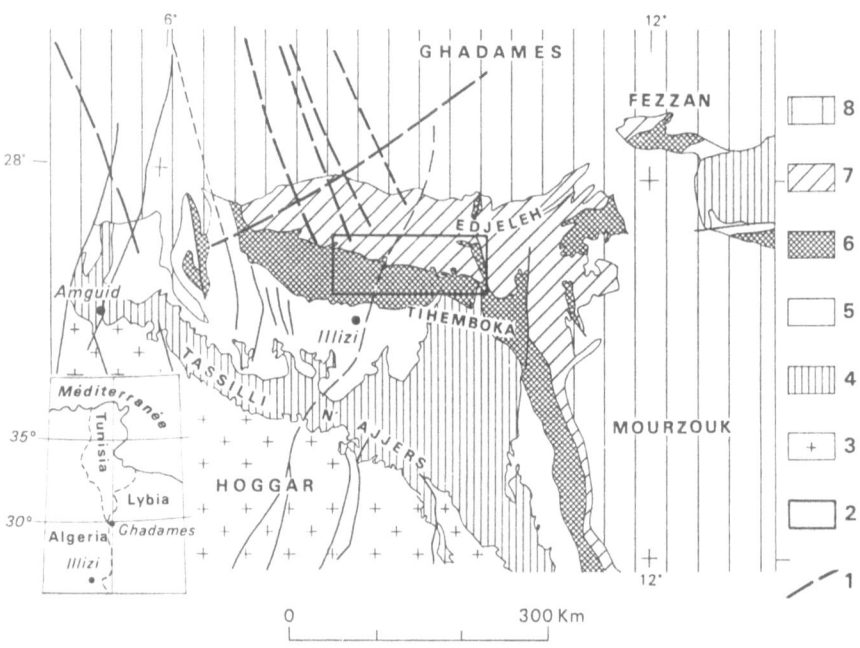

Figure 1 : Schematic geological map of the Ajjers-Ghadames-Mourzouk basin (from the 1/5.000.000° geological map, Fabre et al. 1978). Some of the gravimetric anomalies of the basement are represented, according to the new geophysical map of Algeria (Takherist 1990). Insert : map of Eastern Africa showing the situation of the study area. Legend : 1= basement faults; 2= area of study; 3= basement; 4= Ordovician-Silurian; 5= Devonian; 6= Lower Carboniferous; 7= Upper Carboniferous; 8= Post-Paleozoic.

The Ajjers basin lies at the Algerian-Lybian border (Fig.1). It is limited to the west by the Amguid uplift which is guided by NS Pan-African faults. New geophysical data (Takherist 1990) emphasize the existence of gravimetric anomalies (NW-SE) in the Ajjers basin, as well as a SW-NE fracture towards the north of the area studied. Eastward, the basin grades to the Ghadames-Fezzan basin in Lybia. In the Hoggar, the Pan-African orogeny was completed about 600 Ma (Black et al. 1979). Erosion and peneplanation took place before the Lower Ordovician, leading to the formation of a flat discontinuity surface : the Infra-Tassilian surface. It is covered by an almost

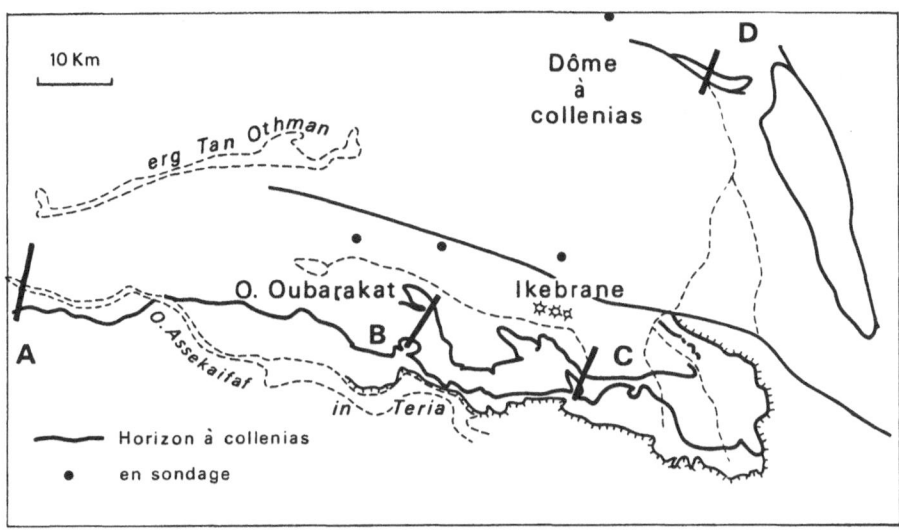

Figure 2 : Schematic geological map of the study area (from the 1/500.000° map of the CREPS 1960). Measured sections : A= West of Oued Assekaïfaf ; B= South of Oued Oubarakat ; C= South of Ikebrane ; D: Dôme-à-collenias.

continuous blanket of mature sandstones : the Tassilis (Fig.1). The Ordovician sedimentation came to an end with a widespread glacial event, the ice-cap being centered on the Hoggar. This was followed by silurian marine sediments, mainly shales. During the Devonian, the Ajjers basin was subdivided by the north-south uplift of the Tihemboka area which was active until the Upper Devonian resting unconformably over more ancient devonian sediments.

In the Ajjers basin as well as in the Lybian Ghadames basin, the Carboniferous started with a marine transgression of Tournaisian age (Fabre, 1988). The Visean comprises mainly calcareous shales followed by the sandstones and shales of the Issendjel formation, lower "Grès à champignons" (Fabre, 1976 ; 1983). The Upper Visean is composed of a sequence containing thin carbonate beds representing the lower member of the Assekaïfaf Formation,   according to Legrand-Blain (1985). Numerous stromatolite layers have been recorded and are organized in two main units. Between these two units a fauna of Brachiopods (*Lybis hericinus, Neospirifer "gwinneri"*, Legrand-Blain, 1978) and Foraminifers (Massa and Vachard, 1979) testify for an Upper Visean age. The stromatolite layers are overlain by the Assekaïfaf carbonates belonging to the Serpukovian (*Syringothyris, Neospirifer "gwinneri"* and in the western sections, *Saharopteria gevini*, Legrand-Blain, 1985). These carbonates are in turn, overlain by continental sandstones with a well known flora of *Lepidodendron* etc.. (Lejal-Nicol, 1972). The contact is generally disconformable in the study area, especially in the

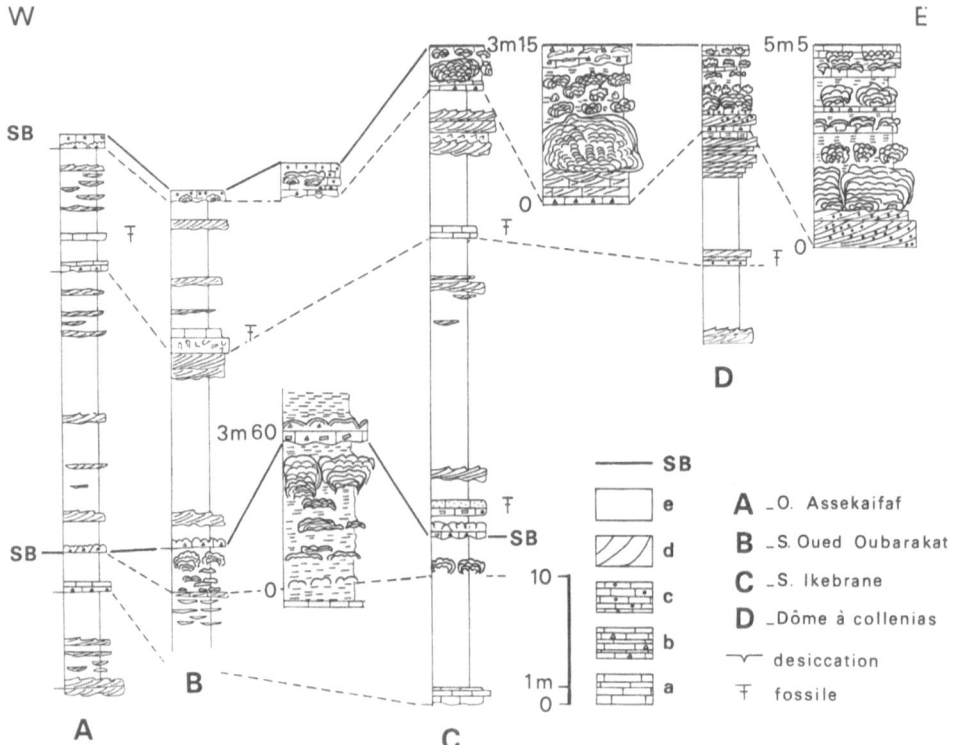

Figure 3 : Measured sections from west to east : a= West of Assekaïfaf; the stromatolites are reduced to a centimetric coating ; b= South of Oued Oubarakat; type section for the stromatolites of the lower unit ; c= South of Ikebrane and d= Dôme-à-collenias ; types sections for the stromatolites of the upper unit. a= carbonates; b= carbonate breccias; c= oolitic carbonates; d= sandstones; e= shales.

oriental part where steep channels deeply eroded the underlaying beds. In the uppermost Carboniferous continental sequences, stromatolites (oncolites and flat domes up to 0,60 m) have been recorded in small fresh-water ponds (Bertrand-Sarfati and Fabre, 1972 ; 1974). The Ajjers basin is one of the most famous eastern Saharan oil fields.

Two distinct units of stromatolites have been traced over long distances, in the area shown in Fig. 2. The lower unit can be followed east-west along strike, for over a hundred kilometers without any major variations. About 50 Km to the north and to the north-east, in the Dôme-à-collenias area, this unit has been recognized in subsurface. The upper unit seems to thin rapidly toward the west, but crops out extensively in the Dôme-à-collenias area. In Lybia, the stromatolite units have been recorded in the Ghadames-Fezzan basin, 250 km toward the northeast (Freulon, 1963 ; 1964 ; Massa, 1988; Whitbread and Kelling, 1982) and in the western part of the Mourzouk-Djado basin, over 600 km (Massa et al., 1974).

DESCRIPTION OF SECTIONS

Three complete sections have been measured, from west to east, close to the Lybian border (Fig.3). The fourth section in Dome-à-collenias, comprises only the upper stromatolite unit. The lithostratigraphic succession will be described rapidly with a special attention to the stromatolite units.

According to sequence stratigraphy, the sections have been divided in three sequences. The sequence boundary are indicated by carbonate beds with brachiopodes shells or by emersion surfaces followed by brecciated carbonates (Fig.3).

1) *First sequence* : The basal Sequence Boundary is materialized in two of the sections by a carbonate bed with brachiopodes shells representing a marine condensed interval, overlaying yellow-gray-green shales. The sedimentation of the following strata is terrigenous mainly clays with very minor silts. Among them are found the first stromatolite unit.

The Lower unit with stromatolites is well defined in the section South of Oued Oubarakat where it presents its maximum thickness and diversity (Fig.3b). Stromatolites are generally reddish and yellow in color and contain large amounts of siliciclastic material.

Very small domes and biscuits grow directly above the shales ; about 2-5 cm thick and

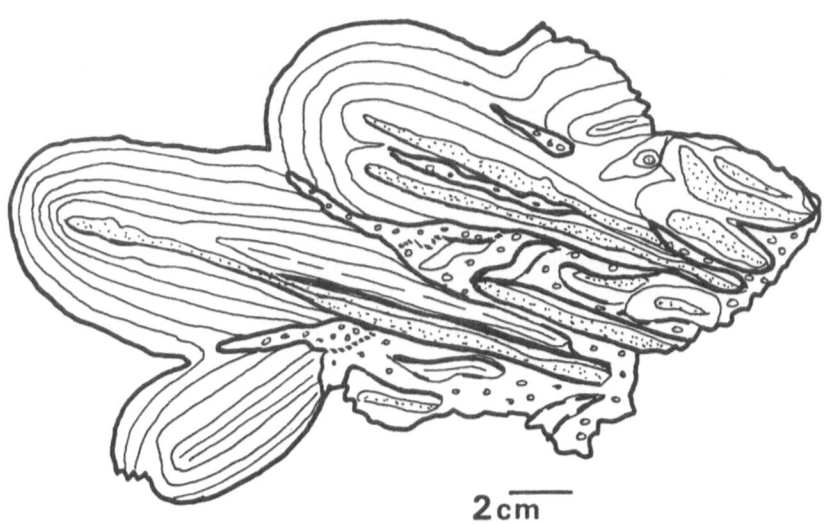

2 cm

Figure 4 : Accretion in the domes of the lower unit : flat mud-chips displaying features of pedogenetic alteration coated and imbricated along the dome edges.

Figure 5 :   Small dome on mud polygone. The surface of the block is scalloped (erosion) and thin layers of silt are intercalated in the mud. The lateral sides are clear-cut and erosional. The basal stromatolitic layer smoothens the surface and coats it entirely. The first coating (laminae with blisters) presents a peculiar shape with a lateral growth (arrow) 4 times thicker than the upward growth. The second coating is different (siliciclastic-micritic and to a lesser extant, micritic-peloidal) with an equal thickness of growth leading to upward bulbous columns. The flakes in the depression left by the stromatolite growth are not coated. Scale-bar= 2.5 cm.

10-15 cm in diameter (Fig.7a), they are closely packed in an almost continuous layer. Firstly, they remain relatively flat but reach 30-50 cm diameter, still growing randomly within the shales. The first laminae establish over reworked mud-chips lying on the shale surface. These mud-chips (Fig.4) suffered subaerial exposure as shown by their pedogenetic alteration. They are episodically brought in from adjacent areas and contribute to the accretion of the bioherm. Intermediate-size domes are often built over an eroded mud surface (Fig.5). They grow first asymmetrically, then laminae type changes and they grow in an upward direction building bulbous columns. The largest domes (the uppermost ones) comprise closely juxtaposed and overlaping hemispherical bioherms, up to 0,8 to 1 m height and 0,5-O,8 m in diameter (Fig.6, 7b).

The outer laminae of the buildup, generally dip almost vertically around ancient ones. The originality of this feature is that the more recent laminae dip down in the shales further than the initial stromatolite biscuit. This gives the buildup the aspect of a tooth-root, especially if the mud has been removed from the centre (Fig.7c,e). Similar structures are displayed in Green River lacustrine stromatolites domes (personal observation, 1978) and with a smaller size in some lacustrine stromatolites in Limagne (Bertrand-Sarfati et al., 1966). Laminae never envelop the lower surface of the buildup.The stromatolite is composed of pseudocolumnar laminae, with very high

degree of inheritance (Hofmann, 1969). Frequent horizontal fenestrae separate the laminae in packages, like onion-skins (Fig.7c). The horizontal section of the domes as well as the columns on the upper surface are subcircular. No elongation or ridges can be seen (Fig.7d). The stromatolite heads are buried by shales (O.5-1m). Westward, the stromatolites persist with a great morphological continuity and the same vertical succession over at least 30-40 km. Further west (Fig.3a) they are replaced by shales.
Toward the east, they are represented by large domes similar to those of the section 3b, in a unique layer. They have not been studied further east in Lybia. Except in size, the stromatolites display no variation from base to top.

2) The second sequence : The Sequence Boundary is an emersion surface followed by a transgressive red carbonate breccia, reworking mud-chips and pebbles with desiccation features and pedogenetic alteration. It contains broken shells and is overlained by carbonate mudstones with polygonal, low relief domal structures (10-15 cm diameter). These are essentially thrombolitic in texture ; the uppermost part only is laminated. Numerous fenestrae in the thrombolite have a gypsum crystal outline and are usually empty now. This carbonate bed is found in all sections and may grade into bioclastic

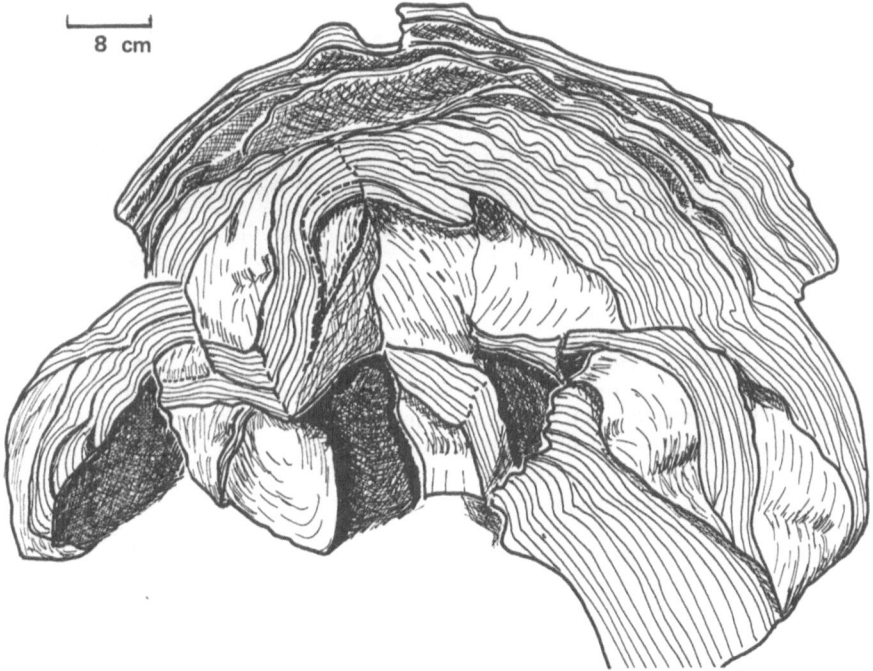

8 cm

Figure 6 : Large dome at the top of the lower unit, outer laminae dipping vertically, deeper than the initial stromatolite coating.

Figure 7 : Domes of the lower unit in section 3b : a) small dome at the base of the lower stromatolite unit ; Scale bar = 2 cm ; b) three overlaping domes, the laminae of the third one envelope the others (arrow) ; the upper surfaces of the three domes display the same hemispherical bumps as in d) ; c) large dome with onion-skin laminae due to concordant fenestrae, now empty ; d) upper surface of a dome, hemispherical bumps are the termination of the pseudocolumnar laminae ; domes as well as pseudocolumns are subcircular in section ; e) cake-like dome, the outer laminae cap entirely the lateral surface, stopping clearly at the base (arrow). all scale bar = 20 cm.

carbonates of various thickness. The sequence continues with shales containing rare, thin channeled sandstones with fauna on the upper surface. In section 3b, thicker storm deposited sandstones (up to 1m) occur with reworked mud-chips at their base.

3) Third sequence : The Sequence Boundary is a carbonate condensed interval which contains unbroken shells (Legrand-Blain, 1978 ; 1985) in the section South of Ikebrane (Fig.3c). It is reduced toward the west to a sandy grainstone with only broken shells. It is overlained by shales, siltstones ands sandstones with planar cross-stratifications. Shells are found along the set surfaces. Siltstone and sandstone layers in thickening-coarsening upward sequences display borings and burrows.

The Upper unit with stromatolites is well represented in the two westernmost sections : South of Ikebrane and Dôme-à-collenias where the succession is thicker (Fig.3c,d). The stromatolite para-sequence overly a detrital carbonate bed comprising reworked mud chips breccia (essentially desiccated mud pebbles), grading to a bioclastic-oolitic carbonate sand. In Dôme-à-collenias, the oolitic and bioclastic cross-bedded sand, with oscillation ripples on the surface reach a metre thick. The constitutive grains, well sorted, are either strongly abraded bioclasts or reworked ooids with a bioclast nucleus. They are included in a grainstone with an entirely oxidized matrix (Chauvel and Massa, 1981). South of Ikebrane, the stromatolites started as flat laminae growing straight on the rippled surface of the bioclastic carbonate. They rapidly form discrete domes (20 cm) coalescing in large domes (1-1.5 m thick and up to 2-2.5 m diameter). These do not present the tooth-root appearance of the domes of the lower unit. On the contrary, the outer laminae envelop the whole dome surface (Fig.8), coating also the flat base of the

Figure 8 : Dome of the upper unit : flat-base sub-hemispherical dome, the outer laminae envelope entirely the dome; crest is visible on the uppermost part (arrow); type section 3c. Scale-bar=20 cm.

domes which are eroded as discrete cake-like individuals. At places a kind of crest oriented east-west (Fig.8) appears on the upper surface of the domes (2-4 cm of relief and 20-30 cm lengh). Domes then may be slightly elongated. Laminae in the domes are pseudo columnar with a very high inheritance and simply undulated in the dome envelope. Sometimes the pseudocolumns are bent northward asymmetricallly in respect to the crest. Above these large domes, smaller juxtaposed domes (0.1 to 0.8 m) are separated by thin veneer of shales. They are subspherical with small pseudocolumns sometimes slightly radiating (Fig.9). The last stromatolitic layer found in the section 3c, South of Ikebrane, comprises layers of bioclastic and conglomeratic carbonate with isolated stromatolite heads.   The small flat stromatolite domes are growing over desiccated flakes and mud polygones. They are covered by a bioclastic and oolitic sand with vertical borings.

In Dôme-à-collenias the stromatolites are in three para-sequences. a) The first stromatolite heads start directly over the oolite surface, they rapidly increase in size by incorporating coated pebbles (oolite, siltstones, bioclastic carbonate) and coalescencing with adjacent domes. The large hemispherical domes (1m high and 2-2.5 m diameter) are very similar to those already described : plunging and enveloping outer laminae ; low crest and slight elongation. They are overlained by stromatolite heads (30-50 cm height and 20-30 cm in diameter) which often presents two distinct episodes of growth:

Figure 9 :   Small dome of the upper unit comprising vertical pseudocolumns in the centre ramifying and anastomosing in the lateral portions of the dome ; Section 3c; scale bar = 5 cm.

Figure 10 : Ovoid oncolite from the section 3b, South of Oued Oubarakat ; the nucleus is complex : a mud polygone (desiccation cracks and fenestrae) is thinly coated along its edges. It is overlain by a new oncolite made of clastic carbonate coated by a discontinuous thin veneer of stromatolitic laminae. It has probably been overturned. Between the two oncolites a thin layer of clastic sediment is trapped. The stromatolite head is growing around the complex nucleus. The growth is of equal thickness on side and top but thinner in the overhanging parts. Laminae are thin with a high degree of inheritance however numerous fenestrae and sheet cracks are visible in the lateral coating. Scale bar = 2,5 cm.

one central hemispherical smooth dome, overlaped by pseudocolumnar to columnar "curly" cap. The para-sequence ends with shales but subaerial exposure is materialized by karstic dissolutions recorded in the large domes of the upper unit and filled by endostromatolites (Fig.11e ; Monty, 1986). b) the second para-sequence begins with a bioclastic-oolitic carbonate bed with mud-chips (desiccated) breccias. Shell fragments are entirely dissolved and replaced by sparry calcite. The stromatolites are small flat low domes (40-50 cm high) with numerous large fenestrae (horizontal as well as globular-angular) filled with calcite cement. Coarse lithoclasts are found in the interspaces of pseudocolumns and ooids are sticked within the distorted laminae. The stromatolites also contain traped *Syringothyris* brachiopods debris and cylindrical borings.

c) The last para-sequence starts over bioclastic carbonate bed (with echinoderm plates) with desiccated and subaerially altered carbonate chips. Discrete stromatolite heads (40-50 cm high and 20-40 cm diameter) are built by pseudocolumnar and rare columnar

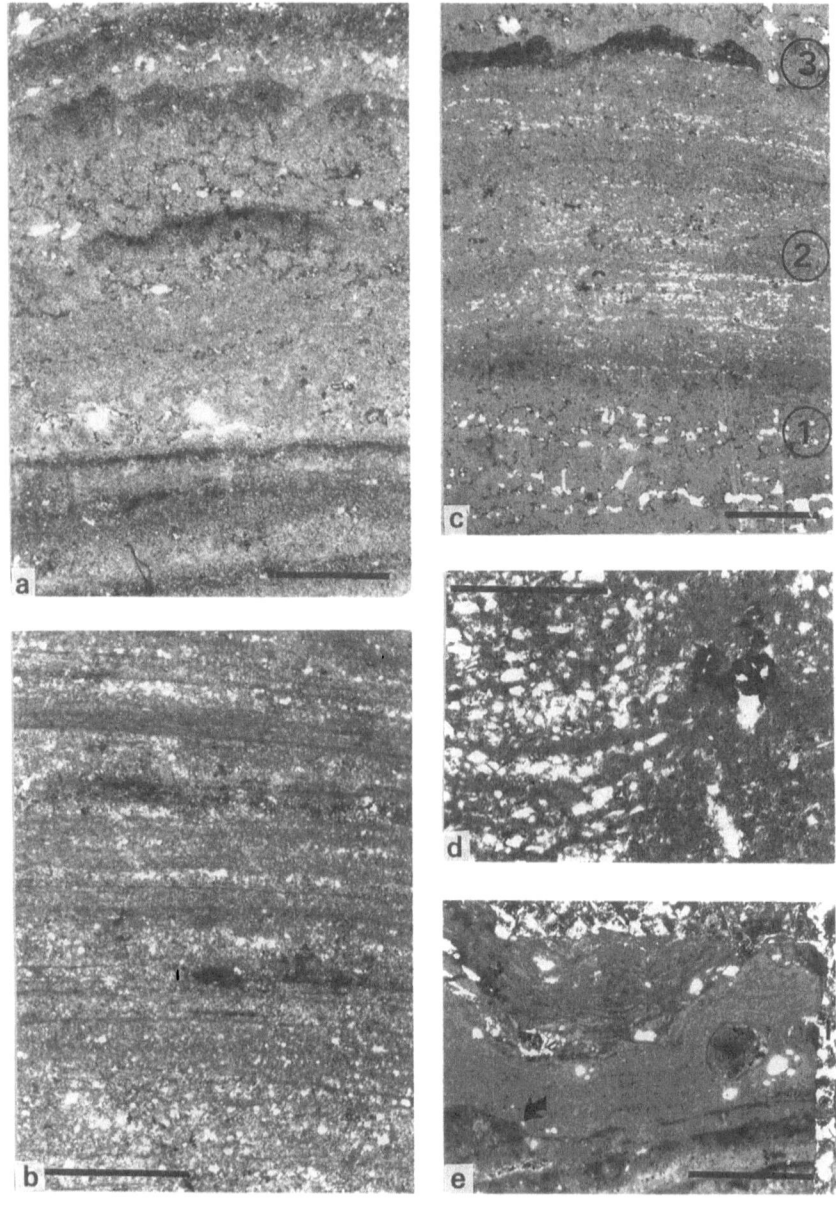

laminae. They form a pavement of small low domes with eroded central part, embedded in a bioclastic-oolitic discontinuous layer.

d) The upper limit of the stromatolite unit in Dôme-à-collenias displays only a change in the lithofacies from a very shallow intertidal carbonate to an open marine shale bed containing a very rich fauna of brachiopods (Legrand-Blain, 1980 ; 1985).

Toward the west the stromatolite upper unit is represented by oncolites capped by small bioherms (Fig.3b). The oncolites (5-15 cm) are almost spherical with an asymmetrical growth, the underneath coating being very thin (Fig.10). They are overgrown by coalescing laminae forming low relief isolated bioherms (10-20 cm high). The oncolites grow on carbonate breccia with mud polygones and their nucleus is often a desiccated mud polygone. Numerous coarse ooids or bioclasts are trapped within the laminae or in the desiccation laminar fenestrae or reworked as oncolite nucleus. They are buried by bioclatic-oolitic sand filling the depressions between the domes. There the limit of the sequence displays clearly defined subaerially exposed surface.

Higher up the sequences are siliciclastic with minor carbonate beds but never contain any other stromatolite layers.

STROMATOLITE MICROSTRUCTURES AND LAMINAE ORGANIZATION

The presence of stromatolites in a dominantly terrigenous sequence as well as the relatively high content of siliciclastic material found within the stromatolite laminae constitute a very original feature. Terrigenous material, clay and silt size quartz is present in almost all the laminae in amounts between 10 to 50 % . They correspond to what Riding (1993) termed "siliciclastic-carbonate stromatolites". In all the stromatolites, the laminae are moderately convex, with pseudocolumnar, very close to contiguous packing (Hofmann, 1969). The laminae degree of inheritance is generally high except in low flat-laminated domes of the upper parts of the para-sequences in Dôme-à-collenias. Discontinuities are abundant but they often remain parallel to laminae. The basic feature of the stromatolite laminae is a siliciclastic-micritic mat, interpreted as built by non filamentous micro-organisms, probably bacteria. Many of

Figure 11 : laminae sequence is from base to top : a) alternation of siliciclastic-micritic and red oxidized laminae, then higher up, micritic-peloidal alternating with red oxidized laminae ; upper unit (Dome-à-collenias) ; b) alternation of micritic films and spar-microspar laminae with rare silt-size quartz and small fenestrae ; base of the big domes of the upper unit (S of Ikebrane) ; c) (1) repetitive micritic-peloidal laminae with oxidized filaments (?) and angular empty fenestrae, followed by (2) repetitive composite siliciclastic-micritic laminae (macrolaminae, more than 10 % of silt-size quartz) (3) upper limit of the sequence is a red oxidized microbial lamina, with a granular texture ; large domes, upper unit (Dôme-à-collenias) ; d) detail of a siliciclastic-micritic laminae with, to the right, a micritic clot, displaying branched fenestrae in its centre; the relief is emphasized by the silt-sized quartz flat-lying over the microbial laminae ; domes of the lower unit (S of Oued Oubarakat) ; e) endostromatolite in karstic cavities in the upper part of the big domes (upper unit). Arrow pointed toward the eroded laminae ; to the right, a late fissure filled by siderite cuts the stromatolite and the endostromatolite (S of Ikebrane). All scale-bar= 0.75mm, e= 2 mm.

these laminae contain detrital material, clay, silt-sized quartz, carbonate mud and more rarely ooids. Others of these micritic laminae have suffered diagenetic alteration. A few laminae, essentially micritic, reflect biogenic features, filaments, blisters, obviously microbial. A peloidal-micritic fabric widespread in all domes is probably diagenetic.

The most frequent microstructure : thick micritic laminae

Micritic-siliciclastic laminae.

The thin micritic laminae (Fig.11a,c) with confuse boundaries (10 to 15 µ) are generally flat to moderately convex, constant in thickness and contain high percentage (> 10 %) of trapped clay and silt-sized grains (mainly quartz and corroded micas). The silt-sized elements are flat lying on the micritic mat surface. The sedimentary content of the lamina may be up to 30% of the total volume. Sometimes, along one lamina, the grains concentrate in small areas outlining micritic masses with a certain relief and deprived of clastics (Fig.11d). The central part of these micritic masses is riddled by fenestrae frequently branching. The laminae are repetitive forming a macrolamina (3 to 10 laminae) which is often terminated by an oxidized grumelous mat.

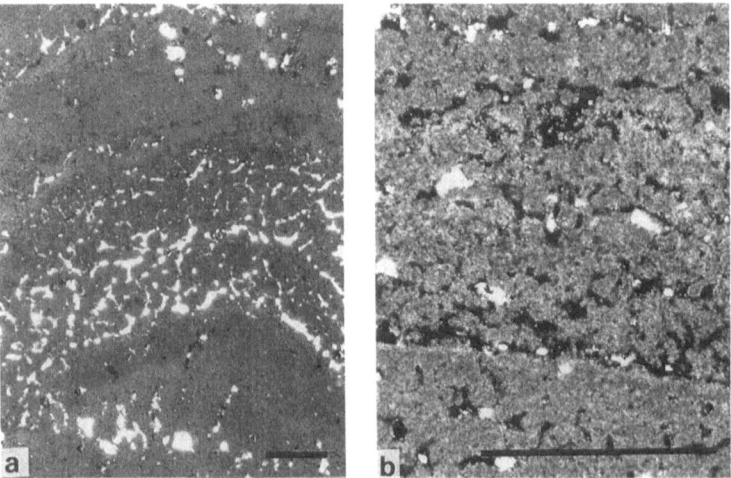

Figure 12 : Micritic-peloidal laminae, biggest domes of the upper unit (Dome-à-collenias) : a) repetitive micritic laminae with contorted cracks defining peloids, most of the fenestrae are empty, their abundance differ from one lamina to the other. The cracks do not progress from the upper surface (differing from boring or desication) ; b) micritic-peloidal lamina; the fenestrae stopped at the lamina basal interface, they are filled by sparite with iron oxide. All scale bar = 0,5 mm.

The micritic mats are interpreted as microbial, probably bacterial mucilaginous mats on which sticked siliciclastic elements lying flat over the smooth surface of the mat as illustrated by Monty (1967 ; 1976). The low relief of the micritic isolated patches and branched fenestrae evidenced the coherent-resistant nature of the microbial buildup. The silt and clay content of the water are not sufficient to prevent the growth of the microbial community, however they are probably introduced or reworked intermitently, may be by low energy wave action.

Micritic-peloidal laminae with numerous fenestrae.

Generally they comprise thick micritic laminae (up to 5 mm), riddled with fissures cemented or not by a sparry calcite, sometimes entirely oxidized (Fig.11a,c). These thin and regular fissures display a contorted network leading to an organization of the micrite in peloids (Fig.12b).The fissures may also resemble large filaments. The laminae contain a few clay and silt-sized quartz (less than 10%). The cracks may be enlarged by dissolution processes (Fig.12a), in this case, large voids cut through the

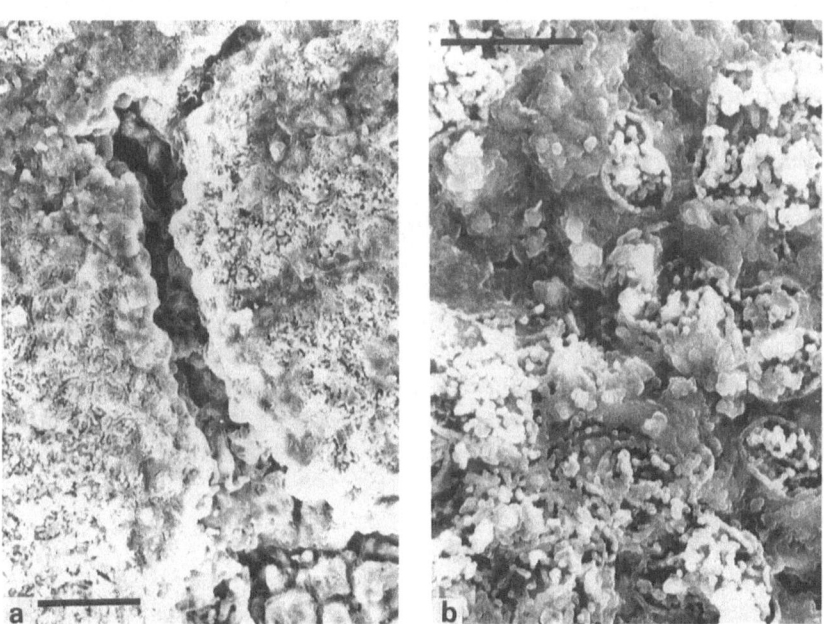

Figure 13 : SEM photographs of the micritic-peloidal lamina (domes in the lower unit) : a) adjacent peloids separated by an empty "fissure"and lined by sparite cement and composed of rounded clots; scale bar = 35 µ. b) detail of the peloid core showing the spheres (diametre < 10 µ) partially filled and lined by smaller round bodies isolated or aggregated in rods (about 1 µ) probably bacteria ; Scale bar = 10 µ

laminae. However these fissures are never filled with sediment as in a post-depositional process, but only by cement. The base and top of the lamina, lacking cracks, differenciate them from borings. They do not change significantly in thickness in the vertical part of the lamina. These laminae form about 80% of the large dômes in the upper unit, but are found occasionnaly in the lower unit.

This peloidal feature is not a desiccation process, because the fissures appear after the growth of laminae and are restricted to the central part of a bunch of laminae, differing also from boring by endoliths. Dewatering processes, syneresis cracks within the water-rich laminae can lead to the appearance of peloids. Early compaction of the mats led to degassing and enlargement of some fenestrae. SEM studies of this microstructure confirm the fact that peloids are separated by depression outlined by crystal of cement but remaining partially empty (Fig.13a). Closer views of the inside of the peloids reveal numerous spheres (10 μ, Fig.13b) filled and outlined by intertwined rods and elongate agregates of bacteria-like bodies (>1 μ). This can be interpreted as coccoid cyanobacteria having be colonized by bacteria, probably after their death (Chafetz and Buzinski 1992). The microstructure itself can be interpreted as the diagenetic evolution of a mat composed of mucilaginous gel embedding coccoid microbes. This occurs probably before the total decay and colonization by bacteria which induces the precipitation of micritic calcite (no bacteria are found in the peloid interspaces).

Such lithification process forming peloids have been described in living bacterial mucilaginous mud (Camoin, 1989).

Micritic films.

Thin micritic, very flat and continuous films (about 20 μ) are found in the basal laminae of the stromatolites of the upper unit. Very constant in thickness, they are grouped by two or more, separated by clear micritic laminae (Fig.11b). Others are isolated, in thick micritic laminae or alternating with the discrete colonies described beneath (Fig.16b,e).The micritic films are usualy interpreted as coherent cyanobacterial mat. Not very frequent, they may be compared to the films built by *Schizothrix* in other stromatolites (also in marine settings ; Monty, 1982) or the *Microcoleus* mats in salinas. It is worth noting that they only appear in the basal laminae of large domes from the upper unit and in the oncolites.

Micritic laminae with blister-like features.

Generally they are thick laminae (reaching 1 to 3 mm), starting as micritic layer and differenciated into closely packed blisters of micrite with a darker envelope (Fig. 14a,c). In section the blisters are circular. They may appear in relief in the upper part of laminae. They are found occasionally in successive sequences, in some domes of the lower unit. There can be clastic material (always less than 10 %) trapped between or on top of the blisters, never within. When the amount of silt grains increases, the blister size diminishes. Another type of microbial colony resembling blisters comprises

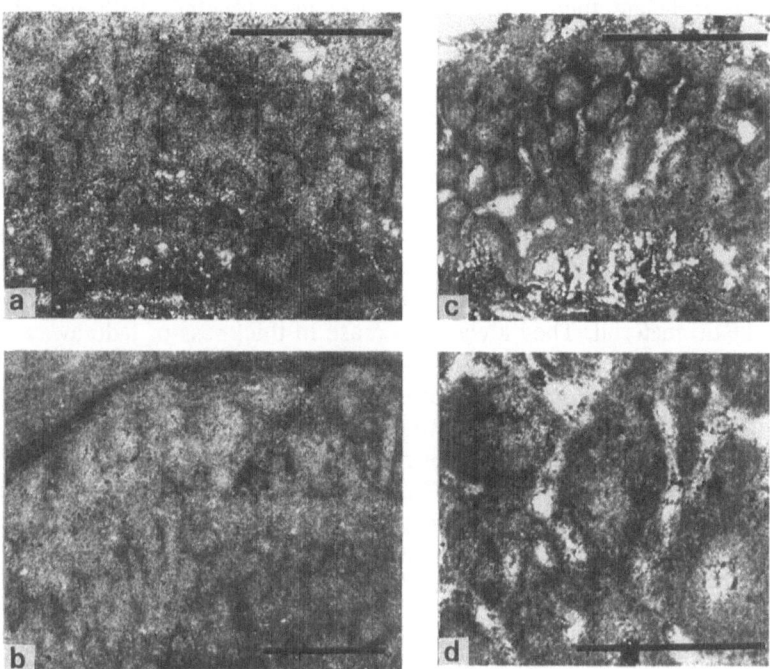

Figure 14 : Laminae with blister-like features : a) micritic bushes outlined by darker micrite ; rare silt-sized quartz (< 10 %) are trapped between the bushes ; larger domes of the lower unit (South of Oued Oubarakat) ; b) elongated blister-like colonies in a lamina outlined by a micritic layer ; the decreases in lamina thickness along the flank of the pseudocolumns is clearly visible ; small isolated biscuits of the lower unit (South of Oued Oubarakat) ; c) elongated vesicles with a darker limit ; voids between them resemble filaments ; some of the vesicles are empty ; small domes of the upper unit (Dôme-à-collenias); d) sections of the vesicles in b) like microbial lumps and clots lined by sparite cement (Monty, 1976); All scale-bar= 0.75 mm, e= 0.40 mm.

laminae with juxtaposed vesicles. These are subspherical to ellipsoidal. A darker micritic outer limit and a microsparitic centre, sometimes sparitic and actually empty (Fig.14d). These vesicles are separated by voids, sometimes outlined by a fringe of sparitic cement, with a fairly constant diameter, that may resemble filaments (Fig.14b). These blisters and vesicles may be interpreted as microbial colonies probably non-filamentous comparable to *Renalcis*. In interpreting "saccate *Renalcis*", Pratt (1984) proposed that precipitation of micrite was occurring either within the mucilaginous sheaths of colonies of cyanobacteria while the "clotted" *Renalcis* may be due to precipitation throughout the colonies and the sheaths. However, the laminated habit of the blisters differenciate them from known *Renalcis* together with the absence of branching.

Other laminae microstructures with biogenic remains

Laminae with red oxidized bushes.

They are very frequent, present in all stromatolites as thin oxidized films or mats (50-100 µ) which often present irregular bumps or bushes with a grumelous structure, composed in fact, of juxtaposed small vesicles (Fig.11c). They may contain small amounts of clastic material. They always decrease in thickness or fade away along the plunging parts of the laminae. Numerous fenestrae, probably due to biological activity, are found within these laminae. They are isolated on top of several other laminae defining sequences (Fig 11c), or alternating with micrite laminae (Fig.11a).

The grumelous-filamentous habit of these bushes is certainly microbial ; they may be compared to some mats in Suguta lake (Casanova, 1986), built by very thin bacterial filaments. The fact these are often oxidized may indicate an enrichment in iron which corresponds to a specific bacterial community or to organic matter favouring the iron fixation.

Laminae with thin erect micritic filaments.

Unbranched, erect dark filaments (Fig.15a) are found in grumulous micrite laminae comprising micro-peloids, perhaps microbially generated (25-50 µ). More rarely they may be prostrate and interwoven at the laminae base. Occasionally a kind of spiraled internal structure can be seen (Fig.15c). These laminae with erect filaments are generally isolated. They never present the diagenetic fenestrae (cracks) seen previously in the peloidal-micritc type. They only occur in the small domes from the upper unit. Sometimes filaments are coated by cloudy grumelous micrite that can be dense (Fig.15b) or fibrous (Fig.15d). Numerous voids are found between the coated filaments also composing a peloidal texture, resembling the micritic-peloidal microstructure described above.

Erect filaments are rare in the Carboniferous stromatolite record. Their spiraled structure is comparable to *Microcoleus* or *Schizothrix* filaments twisted within a sheath. They are not as densely packed as usual in the laminae built by filamentous cyanobacteria. We consider them as part of the lamina builder community, together with other micro-organismes, probably bacteria. The hairy micritic coating found around some of the filaments is obviously bacterial.

Laminae with angular filaments.

Always isolated, they include a tube of sparite with a central darker line. The tubes vary from 40 to 80 µ and the darker central filament, about 10 µ is comparable in size to the

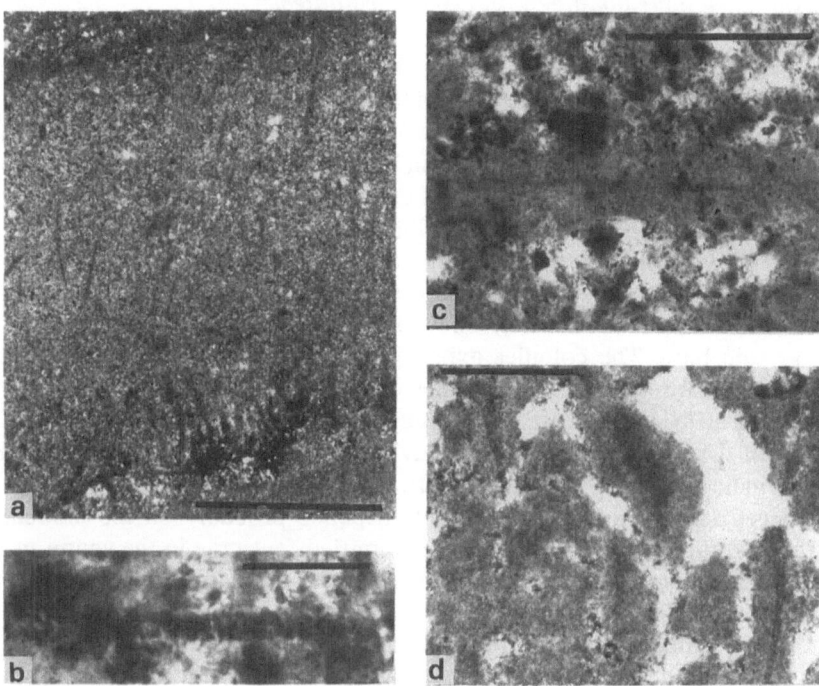

Figure 15 : Micritic filaments, small domes of the second sequence of Dôme-à-collenias : a) granular-micritic lamina with scattered erect micritic filaments ; scale-bar=0.75 mm; b) detail of a) : the filament shows a spiraled structure ; the micritic mesh forming the matrix of the lamina seems also bacterial in origin (filaments ?) ; scale-bar= 50 µ ; c) dense micritic coating, probably bacterial, around a thin dark filament ; scale-bar= 50 µ ; d) micritic filaments slightly enlarging are encased within a filamentous micritic coating, large voids actually empty are left between the "peloid-like" structure ; scale-bar= 150 µ.

thin filaments described above. These filaments are generally preserved at the base of micritic laminae (Fig.16,c). When entirely sparitic, they may be confused with the cracks of the peloidal-micritic laminae.

They resemble the thin filaments previously described, differently preserved as empty molds with a microsparitic coating, (Monty, 1976) or large sheath with partly preserved organic matter in the central tube. In all cases they participate to the growth of the mat together with another bacterial community responsible for the general shape of the laminae.

Siliciclastic-carbonate laminae with discrete colonies.

These discrete colonies vary from very well defined hemispheres to diffuse patches of micrite, all displaying ramified fenestrae in their center (Fig.16a,d). The most definite in size and shape are lobate to hemispherical micritic bumps (up to 0,6-1,5 mm) with positive relief. No sediment is found within these colonies ; very coarse material (ooids and quartz) is trapped between them and is embedded in carbonate mud matrix. Some of these bumps display molds of short thin filaments (12 to 25 μ ; Fig.16d). The ramified fenestrae found in their central part are filled by sparite-microsparite cement, generally oxidized. These colonies are found in the very first laminae, growing upon the oolite, in the upper unit of Dôme-à-collenias. They are associated with a few films and micritic laminae at the base. The colonies expend through thick macrolamina (5 mm high). They grade to high relief micritic colonies encased in a wackestones with very coarse oolitic material (Fig.16e). These composite laminae are repeated over the 10-15 first centimetres of the buildup. The colonies having a positive relief are obviously microbial and rapidly indurated. They contain some filaments molds or angular network of cracks. The habit in discrete colonies versus mats depends probably on the amount of very coarse material supplied with the mud. The ramified fenestrae in the centre may be due to early or late cementation of the voids left when the living organisms move toward the periphery. Their branching is not clearly explained (evaporite rosettes ?). A "second generation" of inhabitants of the microbial mat may also be involved in the actualstructures of the branched fenestrae of these colonies. These structure in discrete bodies can be compared to some *Renalcis* with a saccate micritic habit, implying the same process of calcification of bacterial or cyanobacterial colonies (Pratt, 1984). However, they differ from *Renalcis* by their isolated growth, flat base and laminated occurrence.

DISCUSSION : PALEOENVIRONMENTS OF GROWTH

The stromatolite units appear only in the Ajjers-Ghadames basin where sandstones and shales represent the most abundant facies with very subordinate carbonate-coquinoid beds. Recently, the Mrar Formation in Lybia has been interpreted by Whitbread and Kelling (1982, p. 1105) as a complex deltaic system, ranging from continental to tidal margino-marine. The stromatolite depositional settings succeeded to a sheltered marine deltaic environment. The area, slightly subsident, is characterized by interdistributary bays protected from wave influence by prograding sand bars. The stromatolite "collenia interval" is interpreted as intertidal in origin, resulting from a sea level rise. The different units (3 recorded in Lybia) were reflecting minor fluctuations of the sea-level. In the area studied, the Hassi Issendjel Formation which crops out beneath the stromatolite units, appears to have many common features with the Mrar Formation. It comprises sandstone with continental plants, shallow marine siltstones and shales with only few interbedded marine shell beds. The stromatolites can also be interpreted as

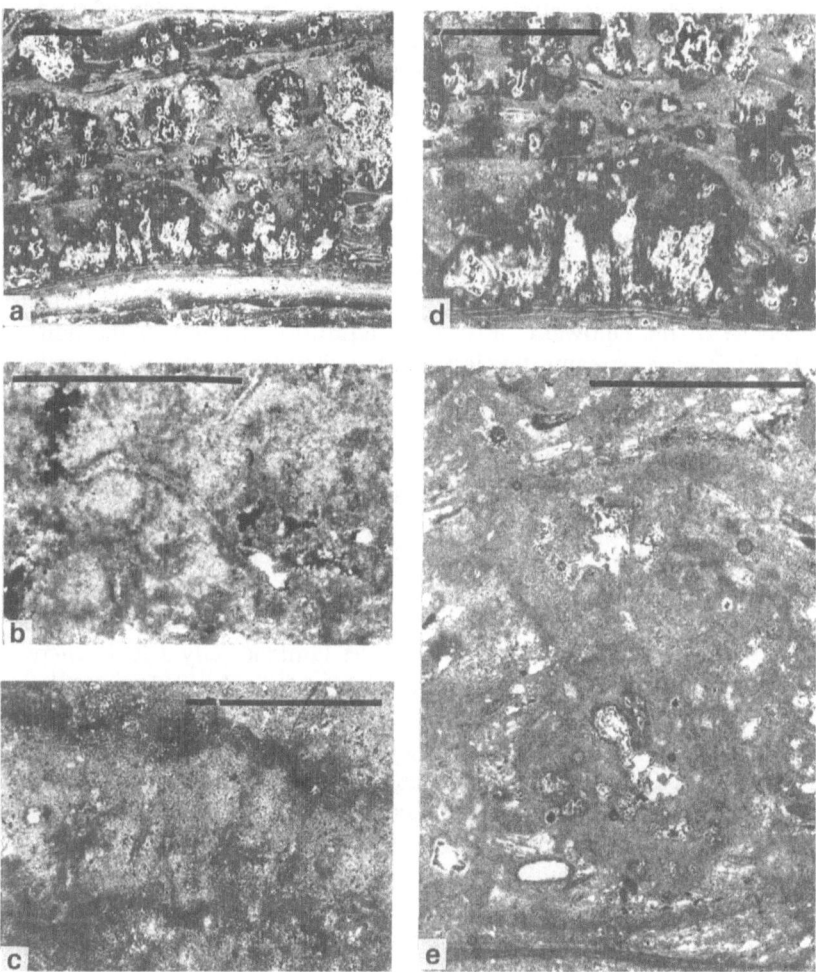

Figure 16 : : a) Thick composite micritic laminae comprising hemispherical colonies with branched fenestrae and large clasts embedded in the micrite between the colonies (to the right) ; b) Contorted filaments scattered in a micritic lamina, a thin darker line is visible in the centre. This may be a filament and its sheath ; c) Angular filaments with a dark coating and a clear sparitic infill also included in a micritc lamina ; d) Detail of a) isolated sub-hemispherical micritic colony growing over micritic films, the relief of the colony is emphasized by the sediment infill (rare quartz) on which smaller colonies are growing ; molds of filaments are visible in the colony ; e) Large vase-shaped micritic colony with central fenestrae ; the relief is outlined by the large clasts (less than 10 % quartz) deposited around the colony. Dome-a-collenias, a) b), e) base of upper unit and b), c) domes of the upper unit. All scale-bar= 1mm, b=o.75 mm.

growing in an abandonment delta plain, however the specific study of the stromatolite sequences allows us to propose more precise conditions of growth for the two units.
The stromatolite beds extend largely (more than 70 km east-west and 30-40 northward, only in the Algerian part of the basin) ; stromatolites are not indicative of a shoreline but developed over the whole surface as discontinuous patches.

Interpretation of the lower stromatolite sequence. The terrigenous sequence in which the stromatolites are found, correspond to a High Stand System Tract. It follows a carbonate shell-bed which represents a Transgressive System Tract corresponding to the maximum flooding during a sea level rise. According to Posamentier and Vail (1988), because of the landward shift of the equilibrium point of the delta, shallow lakes fed by fresh-water streams developed in the abandonned plain. The sediment input is therefore very low : clays and silts, very minor sands. The lack of bioclasts, shells and ooids, together with the lack of elongation of the stromatolites suggest a very quiet depositional environment, protected from the direct influence of marine water. An intertidal origin for these stromatolites is excluded. They are interpreted as deposited under water, growing when the sediment supply stopped, allowing the microbial mat to settle and proliferate. However these siliciclastic-carbonate stromatolites (Martin et al., 1993) were growing in very muddy waters, an unusual environment for photosynthetic cyanobacteria. It is the reason why we interpret these stromatolites as mainly built by bacteria. No desiccation features are found in the laminae only a few syneresis cracks. The rare clasts are redeposited carbonate mud-chips brought in from the lake borders, where the mud was subaerially exposed. These chips are used by stromatolites as a hard substrate for their growth. There is no shallowing-upward sequence of the stromatolites. In the area of Oued Oubarakat, the stromatolites remain in the same conditions from base to top however duration favourable to growth was increasing. The presence of only one series of bioherms in the area of Ikebrane indicates that the body of water was prograding eastward. The repetitive and monotonous laminae reflect poorly contrasted climate. The Sequence Boundary is a subaerially exposed surface, distinct from the stromatolites. It is followed by a new sea level rise : the carbonate contains reworked mud polygones. Polygonal mud-cracks are coated by thin stromatolite and thrombolite.

Interpretation of the Upper unit setting. The environmental setting is not very much different from that of the lower unit. The terrigenous sequence including the stromatolite unit, begins with a carbonate condensed interval (Maximum Flooding). The sequence is also a High Stand System Tract corresponding to a sea level rise in a delta plain environment. However this environment can be interpreted as an intertidal lagoon protected from open marine influence by oolitic-bioclastic sand bars. The size of the large domes, absence of elongation, laminae high degree of inheritance and existence of syneresis cracks testify for a stromatolite growth in subaqueous quiet conditions, under probably 1 to 2 metres of water for the first para-sequence (large domes). Between the pseudocolumns of the domes, ooids and bioclasts are trapped while no sediment is found between adjacent domes, as evidenced by the continuous peripheral coating (wall). From base to top of the unit, the domes diminished in size, the variability of the laminae and the trapping of bioclasts and ooids were increasing. Karstification of the stromatolite surfaces, increasing sheet cracks, desiccation polygones and fenestrae characterize shallowing-upward para-sequences. At Ikebrane the stromatolites comprise only one para-sequence starting by a bioclastic carbonate and clearly limited on top of the stromatolites by a subaerially exposed surface marking the Sequence Boundary. Further west stromatolites are reduced to an oncolitic layer in an environment situated clearly landward. In Dome-à-collenias there are three prograding para-sequences : reflecting marine fluctuations. The Sequence Boundary is not as well defined by a subaerial erosional surface. Despite the hot and arid conditions revealed by numerous desiccation features, relatively few evaporites are recorded, this indicating probable dissolution during humid periods. The repetitive succesion of the laminae involves the absence of seasonal alternation as that visible in modern lacustrine settings (Casanova, 1986) or in Miocene lacustrine deposits (Bertrand-Sarfati et al., 1966 ; 1991). Paleomagnetic data indicate that the area was in tropical latitudes during Early Carboniferous (Irving and Irving, 1982 and Feinberg et al., 1990).

CONCLUSIONS

The Upper Visean stromatolite units croping out in the Ajjers basin coincide with the abandonment plain of a marine-deltaic system. However the two units were not deposited in strictly similar conditions. The large extant of the stromatolite units and some specific characters of the microstructure constitute original features.
1) The two units are deposited within a High Stand System Tract. The lower unit was probably coastal lakes in an abandonned delta plain, protected from marine waters. The lakes were fed by fresh-water streams and sediment supply is exclusively terrigenous. The upper unit is deposited in an intertidal lagoon isolated from open sea by low oolitic-bioclastic sand-bars.
2) The stromatolite gross morphology is always a dome, with slight variations in the size, the external coating, and the pseudocolumnar or columnar attitude. This latter may be controled by the duration of the optimum conditions of growth.
3) The large domes in both units have been built in subaqueous quiet to very quiet environments. They spread over the bottom of the entire area.
4) The lower unit was entirely deposited in the same environment while the upper one contains successive prograding shallowing-upward para-sequences, ending with subaerial exposures.
5) The laminae display almost the same types of microstructures in all the domes ; differences are in the relative proportions of each types. The lack or scarcity of filamentous structures seems to indicate that most of the laminae were built by coccoid cyanobacterial and bacterial communities.
6) The siliciclastic-micritic laminae which are predominant in domes of the lower unit and frequent elsewhere imply a continuous input of terrigenous material with which the

microbial growth has to cope. Such siliciclastic-carbonate stromatolites are not reported very often. The micritic-peloidal laminae with syneresis cracks in the large domes of the upper units are also a very specific features of the Carboniferous stromatolites. All these may reflect a buildup essentially dominated by bacterial associations.

7) The local environment in eastern Sahara is deltaic-marine; stromatolites are located in abandonment plain, growing during sea level rises, climate may have been tropical (paleomagnetic data) with low seasonal diversity. The large extent of the stromatolite layers involves the deposition over large, flat, stable areas, with a low subsidence rate. These settings have few counterparts to-day and belong to the cratonic sedimentation areas, recorded over the West-African-Saharan Craton since Proterozoic times.

AKNOWLEDGEMENTS

This study has been supported by the Algerian National Petroleum Company SONATRACH, during two field seasons (1973-1990) and by the CNRS, for laboratory work. I thank J. Casanova and G. Camoin for their critic lecture of the manuscript.

REFERENCES

Bertrand-Sarfati, J. and Fabre, J. (1972) "Les stromatolites des formations lacustres post-moscoviennes du Sahara septentrional (Algérie)", 24th Internat. Geol. Congress 7, 458-470.
Bertrand-Sarfati, J. and Fabre J. (1974) "Les stromatolites nodulaires de l'Autunien lacustre du bassin d'Abadla-Bechar (Sahara occidental Algérien)" Bull. Soc. d'Hist. Naturelle Afrique du Nord, Alger 40, 179-206.
Bertrand-Sarfati, J., Freytet, P. and Plaziat, J. C. (1966) "Les calcaires concrétionnés de la limite Oligo-miocène des environs de Saint Pourçain sur Sioule : (Limagne d'Allier) : rôle des Algues dans leur édification, analogie avec les stromatolites et rapport avec la sédimentation" Bull. Soc. Geol. Fr. VII, 652-664.
Bertrand-Sarfati, J., Freytet, P. and Plaziat, J. C. (1993). "Microstructure and biogenic remains in non-marine stromatolites (Tertiary, France). Comparison with some Proterozoic microstructures", in J. Bertrand-Sarfati and C.L.V.Monty (eds), Phanerozoïc Stromatolites II, Kluwer Academic publishers, Dordrecht,
Black, R., Caby, R., Moussine-Pouchkine, A., Bayer, R., Bertrand, J.M.L., Bouiller, A.M., Fabre, J. and Lesquer, A. (1979) "Evidence for late Precambrian plate tectonics in West Africa" Nature 278, 223-227.
Camoin, G. (1989) "Les plate-formes carbonatées du Turonien et du Senonien de Méditerranée Centrale (Tunisie, Algérie, Sicile)", Thèse Université Aix-Marseille.
Casanova, J. (1986) Les stromatolites continentaux : paléoécologie, paléohydrologie, paléoclimatologie : application au rift Gregory, Thèse Aix-Marseille.
Chafetz, H. S. and Buczynski, C. (1992) "Bacterially induced lithification of microbial mats" Palaios 7, 277-293.
Chauvel, J. J. and Massa D. (1981) Paléozoïque de Lybie occidentale, constantes géologiques et pétrologiques, signification des niveaux ferrugineux oolothiques Compagnie Fr. des Pétroles, Paris.
CREPS. (1960) Carte géologique de la région de Fort Polignac, au 1/500.000.
Fabre, J. (1976) Introduction à la géologie du Sahara algérien et des régions voisines, Société Nationale d'Edition (SNED), Alger.
Fabre, J. (1983) Afrique de l'Ouest. Introduction géologique et termes stratigraphiques, Lexique stratigraphique international. Pergamon Press, Oxford.
Fabre, J. (1988) "Les séries Paléozoïques d'Afrique : une approche", Journ of African Earth Sci. 7, 1-40.

Fabre, J. et al. (1978) Carte géologique du Nord-Ouest de l'Afrique au 1/500.000.

Feinberg, H., Aifa, T., Pozzi, J.P., Khattach, D. and Boulin, G. (1990) "Courbes de dérive apparente des pôles magnétiques de l'Afrique et de la Meseta marocaine pendant le Paléozoïque", C.R. Acad. Sc. Paris 310, 913-918.

Freulon, J. M. (1963) "Existence d'un niveau à stromatolites (Collenia) dans le Carbonifère marin du bassin de Fort-Polignac (Shahara oriental)", C. R. somm. Société Géologique Fr., 233-234.

Freulon, J. M. (1964) Etude géologique des séries Primaires du Sahara Central (Tassili n'Ajjer et Fezzan) Publication C.N.R.S., C.R.Z.A. Géologie. Paris.

Freytet, P. and Plaziat, J.C. (1982) Continental carbonate sedimentation and pedogenesis. Late Cretaceous and Early Tertiary of Southern France, Schweizerbart'sche Verlags. Stuttgart.

Hofmann, H. J. (1969) "Attributes of stromatolites", Geol. Survey Canada, Pap. 69-39, 43 pp.

Irving, E. and Irving, G.A. (1982) "Apparent polar wander paths Carboniferous through Cenozoic and the assembly of Gondwana", Geophysical survey 5, 141-188.

Legrand-Blain, M. (1976-1978) "Lithostratigraphie du Carbonifère marin du bassin d'Illizi (Sahara algérien occcidental). Les formations d'Assekaifaf et de l'Oued Oubarakat", Bull. Soc. d'Histoire Nat. Afrique du Nord, Alger 67, 103-118.

Legrand-Blain, M. (1980) "Le Carbonifère marin du bassin d'Illizi (Sahara algérien oriental). Mise au point stratigraphique", C. R. Somm. Société Géologique Fr. 3, 81-83.

Legrand-Blain, M. (1985) Dynamique des Brachiopodes carbonifères sur la plateforme carbonatée du Sahara algérien, paléoenvironnements, paléogéographie, évolution, Thèse Université de Bordeaux.

Lejal-Nicol, A. (1972) "Contribution à l'étude des Lycophytes paléozoïques du bassin de Fort-Polignac (Illizi)", Bull. d'Histoire Nat. Afrique du Nord, Alger 63, 49-79.

Martin, J. M., Braga, J. C.and Riding, R. (1993) "Siliciclastic stromatolites and thrombolites, Late Miocene, SE Spain", Journal of sedimentary petrology 63, 131-138.

Massa, D. (1988) "Paléozoïque de Lybie occidentale stratigraphie et Paléogéographie", Thèse Université de Nice.

Massa, D., Termier, H. and Termier, G.. (1974) "Le Carbonifère de Lybie occidentale, stratigraphie, paléontologie", Notes et Mem. C.Fr.Pétroles, Paris 11, 139-206.

Massa, D. and Vachard, D. . (1979) "Le Carbonifère de Lybie occidentale : biostratigraphie et micropaléontologie. Position dans le domaine tethysien d'Afrique du Nord", Rev. Inst. Fr. du Pétrole, Paris 34, 3-65.

Monty, C. L. V. (1967) "Distribution and structure of recent stromatolitic algal mats, Eastern Andros Island, Bahamas", Ann. Soc. Geologique. Belge 90, 55-100.

Monty, C. L. V. (1972) "Recent algal stromatolithic deposits, Andros Island, Bahamas, Preliminary report", Geol. Rundschau 61, 742-786.

Monty, C. L. V. (1976) "The origin and development of cryptalgal fabric", in M.R. Walter (eds), Stromatolites, Elsevier, Amsterdam, pp. 193-205.

Monty, C. L. V. (1986) "Interactions énements géologiques-stromatolithes", Bull. centre Rech. Explor-Prod. Elf-Aquitaine 10, 537-553.

Posamentier, H. W. and Vail, P. R. (1988) "Eustatic controls on clastic deposition II- sequence and Systems Tract models", in C.K. Wilgus et al. (eds), Sea-level changes : an integrated approach. Society of Economic paleontologists and mineralogists 42, 125-154.

Pratt, B. (1984) "*Epiphyton* and *Renalcis*, diagenetic microfossils from calcification of Cocoid Blue-green Algae" Journal of Sedimentary Petrology 54, 948-971.

Takherist, D. (1990) Structure crustale, subsidence mésozoïque et flux de chaleur dans les bassins nord-sahariens (Algérie) : apport de la gravimétrie et des données de puits, Thèse Université de Montpellier II.

Whitbread, T. and Kelling, G. (1982) "Mrar formation of Western Lybia. Evolution of an Early Carboniferous Delta system", Amer. Ass. Petroleum Geologists Bull. 66, 1091-1107.

THROMBOLITIC-STROMATOLITIC CYCLES OF THE CAMBRO-ORDOVICIAN BOUNDARY SEQUENCE, PRECORDILLERA ORIENTAL BASIN, WESTERN ARGENTINA

C. ARMELLA
Museo Argentino de Ciencias Naturales "Bernardino Rivadavia", Av. Angel Gallardo 470 ; 1405 Buenos Aires ; Argentina

ABSTRACT

The La Flecha Formation (Cambrian-Ordovician boundary) from the Carbonate Basin of Argentina (South America) has a clear shallowing-upward cyclic pattern. Each cycle is formed by a subtidal unit and ends with a supratidal one. This sequence has a biosedimentary style with a great development of thrombolites and stromatolites. An ideal cycle is described in which the facies occur and the most usual vertical associations suggest that this sequence may be a protected ramp or an island model, which was accreted vertically and migrated laterally over a wide and protected area.
The thrombolitic megastructure of Argentina are interpreted in terms of Cambrian-Ordovician global changes.

INTRODUCTION

The Lower Paleozoic Precordillera Oriental carbonate platform of Argentina was fully developed during the Cambrian and Ordovician periods. Its palaeogeographic distribution is controlled by the old Pampeano-Brasileño Craton perimeter (Fig.1a). Precordillera platform outcrops are the best exposed in Western Argentina. In other areas (Central and Southern Argentina) platformal outcrops are poorly studied.
The section of the Precordillera Oriental platform containing the Cambrian-Ordovician boundary (Fig.1b) is characterized by a monotonous biosedimentary pattern in the Zonda Formation (Upper Middle Cambrian), and develops largelly into regressive cycles in the La Flecha Formation (Upper Cambrian-Lower Ordovician). Northwise of the Precordillera Basin (Precordillera Central) this sequence is known as San Roque Formation.
The incipient cyclic behavior of the sequence appears at the top of Zonda Formation together with the first thrombolites of the geologic record from Precordillera. Therefore the transition from the Zonda Formation to the La Flecha Formation is gradual.

J. Bertrand-Sarfati and C. Monty (eds.), Phanerozoic Stromatolites II, 421–441.
© 1994 *Kluwer Academic Publishers.*

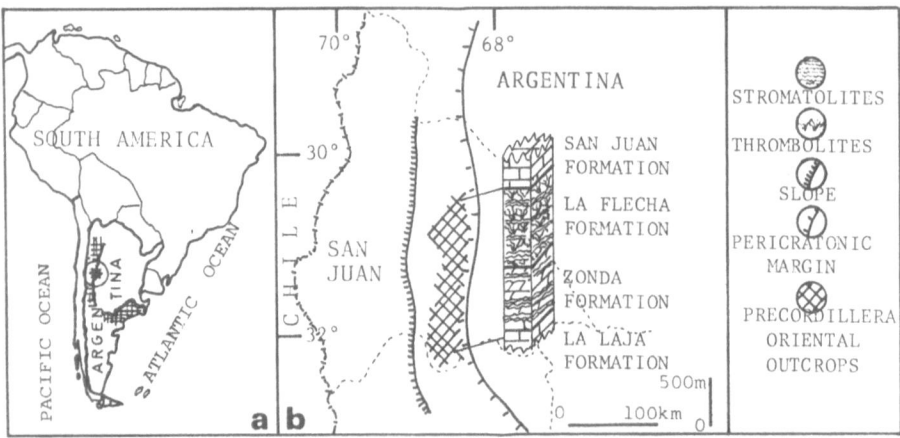

Figure 1: Location of the studied area a) Pericratonic Latest Cambrian-Earliest Ordovician basin developed in Argentina; b) Precordillera Oriental outcrops and stratigraphical relationships of the La Flecha Formation.

The development of the microbiolite cycles took place on top of an extended stable carbonate ramp (Zonda Formation) with wide and continuous facies under peritidal conditions and a general regressive tendency.

Although aproximately 1000 meters thick, this sequence does not contain any fauna, an Upper Cambrian to Tremadocian age is indicated by the occurrence of Middle Upper Cambrian trilobites (*Bollaspidella* Zone) in the underlying La Laja Formation, and lower Arenigian conodonts (*Oepicodus evae*) in the overlying San Juan Formation (Bordonaro, 1983). Therefore, the Cambrian-Ordovician boundary cannot be identified in the sequence. Cyanobacteria are the main palaeoecological element within the sequence.

SUMMARY OF MICROBIAL STRUCTURES

The microbial structures comprise thrombolites and stromatolites. Other structures are present with *Renalcis, Epiphyton* and *Nuia.*

Thrombolites

Thrombolites are well developed in the La Flecha Formation, especially within buildups which have a wrinkled upper contact and a rugged upper surface.
Thrombolites are generally dominant in the cycles and show an irregular rough-weathering surface in the field. Walter and Heys (1985) considered that thrombolites are disturbed stromatolites in which the activity of metazoan has disrupted the

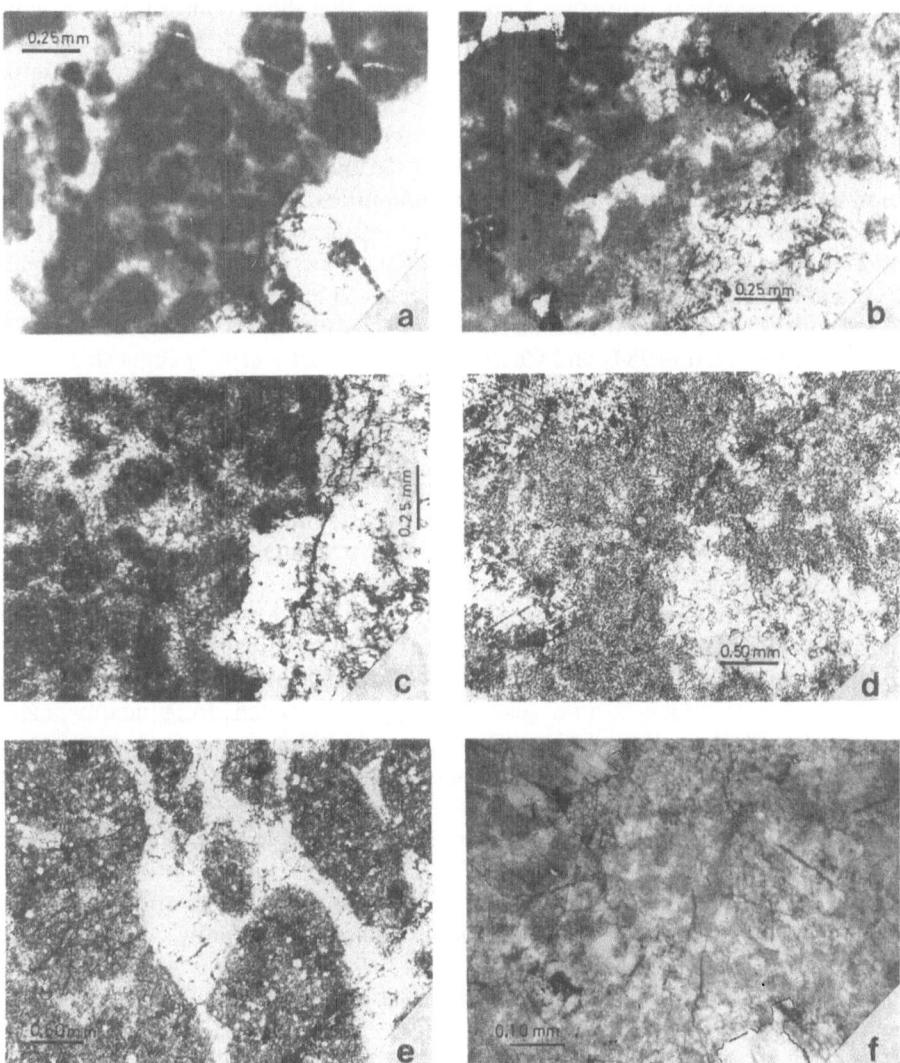

Figure 2: Microstructure of thrombolites. a & b) Part of micritic lobate thromboids with a) grumous-peloidal b) and grumous microstructure. These thromboids show dark micritic peripheries and sparitic cement. c) Peloidal micritic microstructure of a portion of an amoeboid thromboid. A neat contact is observed with the silica blocky cement. d) Partially interconnected lobate and amoeboid thromboids with massive and grumous microstructure and blocky cement of sparite and dolosparite. e) Amoeboid and rounded thromboids with grumous microstructure. The cement is a mosaic of fine sparite. f) Peloidal-grumous microstructure with superimposed dolosparitic crystals

lamination. In this paper it is not considered possible that the origin of the thrombolitic fabrics have been produced by boring or grazing organisms as no fauna has been found neither within thromboids or in the inter-thromboids sediments, except few (less than 1%) isolated, scattered bioclasts of indeterminated shells fragments.

Bioturbation effects have been found within the upper levels of the La Flecha Formation related to the appearance of abundant Ordovician metazoan of the San Juan Formation that competed with the microbial communities. Therefore, the origin of these fabrics is considered to be strictly microbial. According to Kennard and James (1986), the clotted fabric is attributed to the calcification of coccoid microbial communities. Riding (1991) considers that any deposit that appears to be microbial, with unlaminated and clotted fabrics, can be termed a thrombolite, so, there are a variety of origins for these microbialites. Pratt (1984) and Chafetz and Buczynsky (1992) suggested that the clotted mesostructure may represent calcification by coccoid microbial communities, therefore, *Epiphyton* and *Renalcis* are also due to bacterial precipitation of calcium carbonate.

The thromboids (macroscopic clots, see Kennard, 1989) are predominantly composed by micrite, but black siliceous thromboids also occur. The grumous microstructure is the most abundant (Fig.2b,d,e). However the microstructure may change within the thromboid, from massive micritic to grumous-peloidal (as a spongy fabric) towards the periphery. The grumous-peloidal is made up of "microclots" of 100 to 150 mm (Fig.2a,f). Less frequently there are thromboids with peloidal microstructure. It is formed by micritic peloids, of rounded shape, 100 to 250 µm diameter with thin dark microsparitic matrix (Fig.2c) and generally appears related to grumous-peloidal thromboids. The first inter-thromboids cement phase is a thin isopachous rim of equant crystals. There are no evidences that thromboids or micritic peloids have been removed, either by burrowing or dissolution.

Some thromboids form a network or wall-like structures, 0.5 to 1 centimeter thick, with inter-thromboid spaces, 1 or 2 cm wide.

Different kinds of arrangement are determined by the growth vector attitude and the degree of interconnection of thromboids (isolated, partially interconnected, anastomosed and coalescent (Fig.3).

Six main arrangements are observed (Armella, 1989, a ,b, 1990):

Encephalic: (Fig.4a) Isotropic and uniform growth of thromboids. They are mainly amoeboid or lobate in shape. They may be either in contact or present a spacing resembling a brain pattern. The density of the thromboids varies from one thrombolite to the other, or within a particular biostrome or bioherm (Fig.5a,b).

Horizontal: (Fig.4b) Horizontal layers are formed by prostrate, lobate-pendant or prostrate-amoeboid thromboids (Fig.6b). Projections which join different levels have been observed. Laterally coalescence of lobate thromboids generally gives kidney-shaped chains.

ISOLATED   PARTIALLY INTERCONNECTED   ANASTOMOSED   COALESCED

Figure 3 : Degree of interconnection of thromboids. Approximate scale: 1:2

Concentric: (Fig.4c) Concentric dendritic thromboids, concave towards the centre, like a cabbage. These thromboids are commonly siliceous.

Vertical: (Fig.4d) Vertical digitated thromboids with two or more, gently anastomosed branches, parallel to the main column, are linked by bridges of nearly 1 cm wide (Fig8a,b,c).

Radial: (Fig.4e) Nonbranched or branching digitated thromboids (alpha or beta style of branching) with a divergent radial growth pattern (Fig.9a,c).

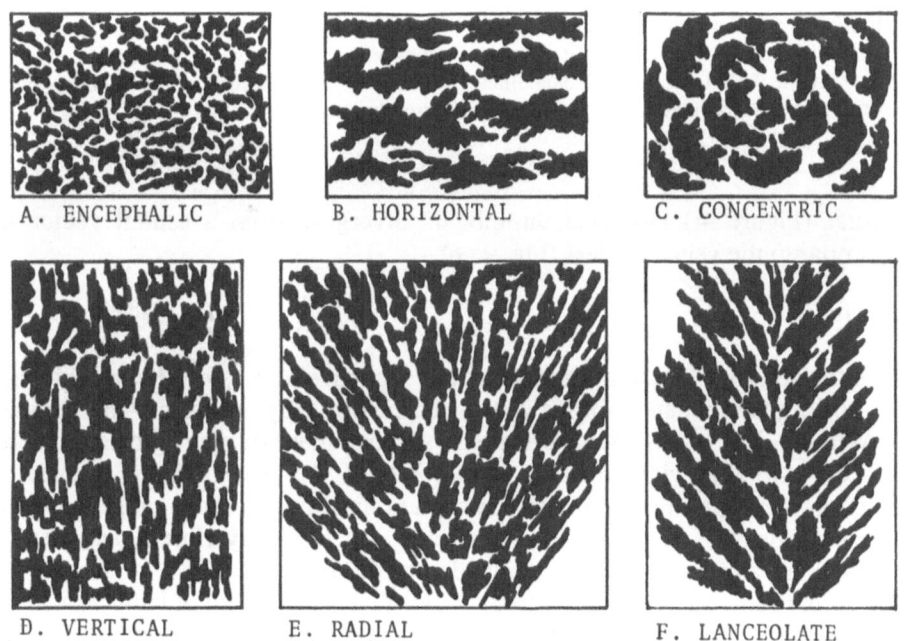

A. ENCEPHALIC   B. HORIZONTAL   C. CONCENTRIC

D. VERTICAL   E. RADIAL   F. LANCEOLATE

Figure 4 : Arrangement of thromboids. Approximate scale: 1:5

Figure 5 : a) Stratiform (tabular) thrombolite (Facies V) with encephalic mesostructure "e". Dolomitic mudstone (Facies VIII) has been deposited over an erosional surface on top of the thrombolite "d". Low relief stromatolites (Facies VII) grew straight upon the mudstone. b) Encephalic mesostructure with high density of thromboids.

Lanceolate: (Figure 4f) Digitated thromboids diverging from a central vector in a pattern similar to the veins of a leaf (Fig. 9a,b).

Stromatolites

Stromatolites are very frequent throughout the sequence and are always abiophoric (deprived of organic remains). In the Zonda Formation they are mainly flat laminated forms. Within the La Flecha Formation stromatolites display greater diversity (Fig5a,6a,d,11a,b), especially towards the south of the basin. They include the following morphologies: planar and low synoptic relief stromatolites, subhemispherical domes or mounds and oncolites. They show either open or very close linkage, smooth or wavy laminae, and generally have a high degree of inheritance.
The stratiform to low synoptic relief stromatolites dispaly a pattern of laminae accretion repetitive, forming sets of flat, undulatory, cumulate and rarely pseudocolumnar laminations of 2 mm thick. Each set is 1 to 3 cm thick and they are integrated by the alternance of massive dolosparitic laminae (of fine crystalline dolostone) and fenestral micritic laminae (with dark micritic boundaries and peloidal or grumous-peloidal microstructures). The isolated domes with a rectangular profile are not so common. They are less than 0.2 m height and more than 0.4 m long.

Figure 6 : a) Example of cycle of the lower section of the La Flecha Formation. The bottom of the photography shows stratiform and low relief stromatolites (Facies VII) and nodular forms with chert interbedded (Facies VI) as the lower cycle facies. Mound shaped thrombolite (Facies V) grew over these stromatolites without a discontinuity surface. It has encephalic mesostructure "e" composed by thromboids of black chert and it is covered by black thrombolitic chert "c". Domal stromatolites (Facies VII) with interbedded thrombolitic chert developed over the thrombolite. The upper cycle boundary surface is at the top of these stromatolites. b) Horizontal mesostructure composed by prostrate and pendant thromboids. An encephalic "e" tendency is observed towards the top of the thrombolite. c) Encephalic subspherical thrombolites. The arrow indicates the cycle top. (Facies IV). d) Flat laminated stromatolite with chert interbedded and an oncolite with concentric lamination, over an encephalic thrombolite "t".

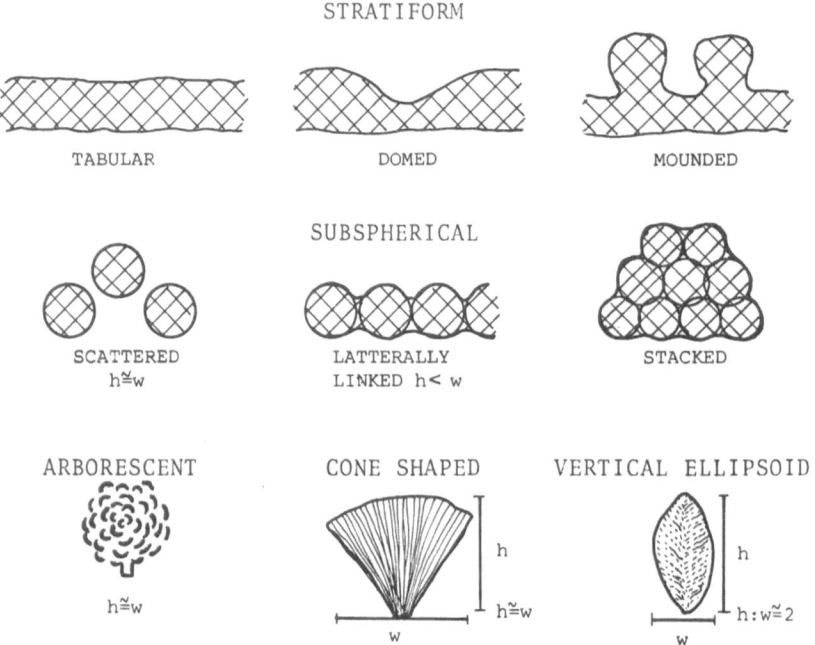

Figure 7 : Megastructures of thrombolites.

The subhemispherical domes are very frequent too, reaching more than 0.5 m of height, with a circular plan-outline of 0.5 m diameter (Fig.11a). Laminae are smooth and gradually pass toward the top to wavy and crinkled external laminae with mudcracks and black chert crinkled cover. This chert cap may be 0.2 m thick. The pattern is formed by the alternation of fenestral micritic laminae (0.1m thick) and chert.

Isolated cherty oncolites have been observed within stratiform stromatolites (Fig.6d). They are ellipsoidal or subspheroidal (0.1 m diameter), and usually have concentric lamination (1 cm thick) with the best development of laminae in an upward direction. This asymetrical pattern suggests that these oncolites grew in situ.

The stromatolites are often more dolomitized than the thrombolites and show fenestral fabric. They are associated with mudcracks and storm layers.

Other Structures

Some buildups with low density (nearly 30%) of rounded, lobate, saccate, grape-like and amoeboid thromboids have a mottled fabric. They comprise *Renalcis* colonies, with chambered and saccate microstructure. *Renalcis* locally occurs in association with minor fan-shaped growth of *Epiphyton* (0.5 mm high).

The oncolites (0.5 mm in size) made of *Nuia* (Chlorophyta), occur together with ooids and build aggregate grains within inter-thrombolitic grainstones (Fig.10b).

THROMBOLITE MEGASTRUCTURES

The megastructures which can be observed in outcrops have been classified according to the indications for stromatolites descriptions given by Preiss (1976), Grey (1989), Grey and others (1990) and Armella (1990, Fig.7). They are described in relation to their frequency of occurrence.

Stratiform

Tabular (Fig.8c), domed (Fig.6a) and mound thrombolites are recognised. The average thickness is 1.5m and the domes frequently have 0.5m of synoptic relief. Mound of stratiform thrombolite have a thickness of 2 m, with diameter of one meter. Planar layered wackestone occurs surrounding the domes or the mounds.
The main mesostructures are encephalic and vertical, and they can exhibit either massive or partially bedded growth layers. The density of the thromboids network tends to decrease from the bottom to the top of each buildup.

Subspherical

Subspherical thrombolites have circular to subcircular plan outlines, and range between 0.5 to 1.5 meters in diameter (Fig.6c). The spheroids may be scattered in carbonate wackestone or laterally linked forming beds of thrombolites. They can also show a stacked mesostructure, resembling an inverted bunch of grapes of nearly 2 m high.
Scattered, laterally-linked and stacked subspherical bioherms invariably have an encephalic arrangement in concentrical layers.
The spheroids are often almost symmetrical, and the asymmetrical ones have their larger axis horizontal .
It is considered that these thrombolites grew in situ due to the lack of deformation and the intergrowth of the peripheric thromboids with the flanking sediment. Some scattered spheroids with eroded peripheric thromboids, show evidences indicating that they have been affected by a short transport.

Cone Shaped

These thrombolites are 0.5 to 2 m height and have a height/width ratio of 1. They are subcircular in plan outline and fan shaped in section (Fig. 9a,c). The main mesostructure is radial but encephalic thrombolites can also occur. Large cone shaped thrombolites frequently exhibit lateral gradation from radial to either vertical or encephalic arrangements. Some bioherms show several growth stages, marked by an upward decrease in the density of the thromboids network (Fig.9c).

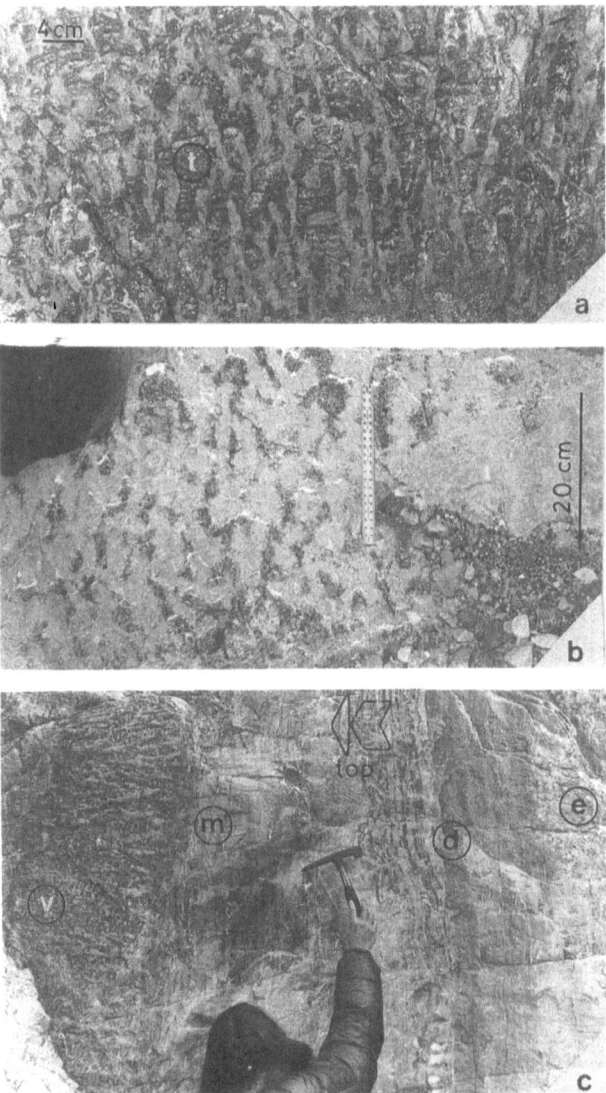

Figure 8 : a) Vertical mesostructure with anastomosed digitated thromboids.b) Vertical mesostructure with low density of thromboids. c) Example of cycles : Lower cycle : Tabular stratiform thrombolite with encephalic arrangement "e" (Facies V) and dolomitic mudstone "d" (Facies VIII) with upper erosional surface. The next cycle begins with flat laminated and domed stromatolites (Facies VII) followed by a layer of mudstone "m" (Facies I). Tabular stratiform thrombolite with vertical mesostructure (Facies V) developed over the mudstone. This cycle ends with an erosional surface.

The tops of the thrombolites are gently convex and are sometimes laterally linked by a black or brownish silica cover. The inter-thrombolite inter-space is filled by a striped wackestone carbonate.

Vertical Ellipsoids

This megastructure resembles a laurel leaf (Fig.9a,b), but it is relatively rare. The average heigh is nearly 1 meter and the height/width ratio is more than 2. They are generally closely spaced and partially linked, with striped wackestone or packstone in the inter-thrombolitic space. Sometimes they show a gradational contact with this sedimentary filling.
The lanceolate mesostructure is typical of these thrombolites, but encephalic one may also occur.

Arborescent

This morphology is less frequent and it is composed of a subspherical upper part with a concentric mesostructure over a massive "trunk" of 0.1 m wide. These are less than 0.5 m high and have a height/width ratio of nearly 1. They occur scattered and isolated on a micritic substratum. They have only been observed at the top of the stratotype section of the La Flecha Formation.

THEORETICAL CYCLE

The La Flecha Formation provides an excellent example of cyclic biosedimentation. Cycles began with well developed thrombolites at the base and pass to stromatolites at the top. An ideal cycle consists of eight successive facies, but different facies combinations result in different kinds of cycle. The facies are generally transitional and commonly exhibit local changes and/or lateral variations which are not considered in this general description. The overall facies are interpreted, on the basis of lithological and palaeoecological analysis, as having be deposited in shallow subtidal to supratidal environments. The normal sequence is represented by:

Facies 1.- Black and Fetid Mudstone : Beds of one or more meters height, with high micritic content. It has grumous or grumous-peloidal microstructure and a rugged aspect that suggests a microbial origin. Towards the upper levels of the La Flecha Formation, in transition to the San Juan Formation (Ordovician) the thickness of this facies increases considerably within the cycles (reaching 4 m or more) and show layered arrangement with wavy boundaries surfaces as an effect of bioturbation (these are the first evidences of traces of an ordovician fauna).

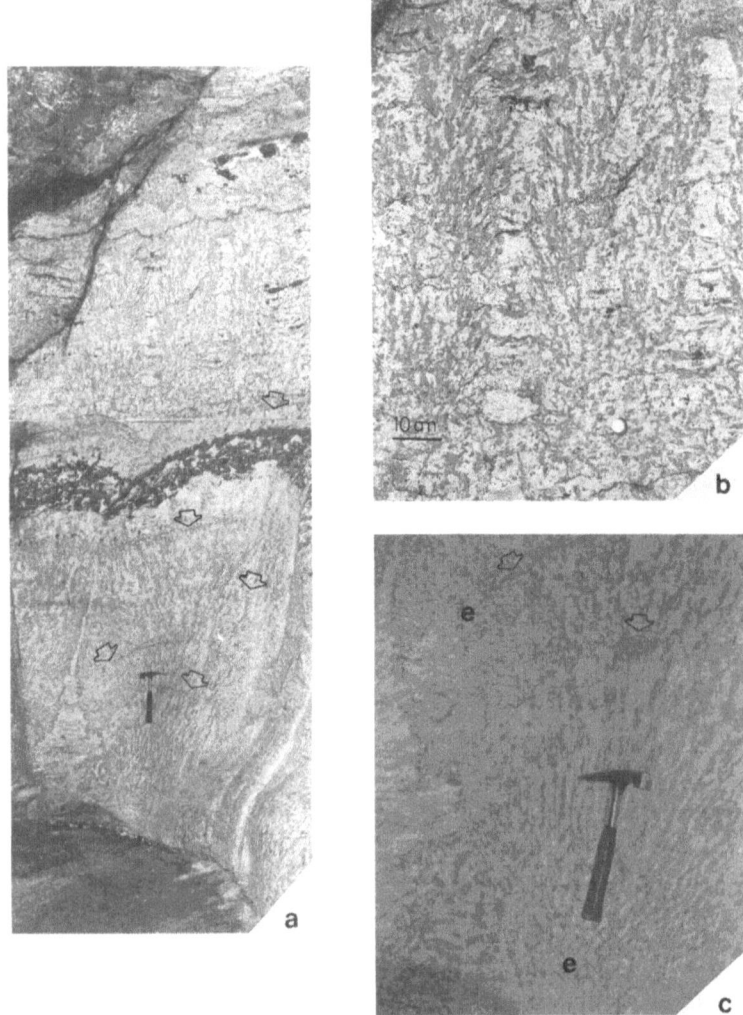

Figure 9 : Facies III: a) Coalescent cone shaped and vertical ellipsoidal thrombolites with close spacing. These superimposed biostromes are separated by a thrombolite black chert layer (parasequense) indicative of a discontinuity of growth more important than the ordinary stops (arrowed). The cone shaped thrombolites have radial mesoestructure but laterally pass to encephalic one. On top of the upper biostrome there is an erosional surface. Mounded thrombolite with encephalic mesoestructure (Facies V) developed over this discontinuity. b) Detail of vertical ellipsoidal megastructure and lanceolate arrangement. c) Detail of cone shaped megastructure. The thrombolite has mainly radial mesostructure which laterally changes to encephalic "e". This latest is found in the cone apex and in the upper part of the thrombolite. Several growth steps have been observed within the cones (arrows, also in a).

This facies grades laterally and vertically to the Facies II, III or V and the environment of formation may have been subtidal with low energy, under wave base, far from the erosive marine currents.

Facies II.- Arborescent Thrombolites : This facies is less frequent and only appears in the upper cycles of the La Flecha Formation.
Arborescent forms may have grown on a soft substrate under lower energy conditions (subtidal, under wave base) and usually show their concentric thromboids "inmerse" in the surrounded massive sediment (Facies I).

Facies III.- Cone Shaped and Vertical Ellipsoidal Thrombolites : These thrombolites generally are closely or very closely spaced and vertical in attitude (Fig.9). In some cycles they are overlying stratiform thrombolites, but their association to subspherical thrombolites have never been observed.
Buildups are flanked by striped carbonate wackestone, mainly composed of microbial peloids, intraclasts, ooids and aggregated grains (comprising *Nuia* and ooids, Fig.10b). These particles indicate an in situ origin by disruption of the thrombolites before induration. It is suggested that these delicate megastructures have grown in a low energy environment (subtidal) and the divergent growth pattern has been favoured by a very low sedimentation rate.

Facies IV.- Subspherical Thrombolites : The subspheroids closely spaced (Fig.6c) make biostromes. Each thrombolite individual has nearly 1 meter in diameter. Banded mudstone are found underlaying and overlaying the closely spaced thrombolites beds. This mudstone is composed of layers of massive and grumous micrite. The subspheroids may also be scattered within a massive mudstone. Then they frequently are sandwiched between the biostromes of closely spaced subspheroids and the mound-stratiform thrombolites of Facies V.
Subspherical thrombolites have grown in a shallow subtidal environment, but buildups of reworked thrombolites have been observed within thin beds of storm deposits (wackestones) which contain thrombolitic intraclasts.

Facies V.-Stratiform to Mound Shaped Thrombolites : These thrombolites form extended biostromes of nearly 1.50 m average height (Fig.6a,8c) though some have more than 4 meters. They frequently show in the basal parts horizontal mesostructure (Fig.6b) in layers of 5 cm in height, which grades to encephalic mesostructure and may end vertically towards the top.
The base has a clear contact with the carbonate sediment comprising frequent storm layers (floatstone or rudstone) while at the top the network of thromboids is less dense and the encrease in sediment filling makes the contact diffuse and interdigitated.

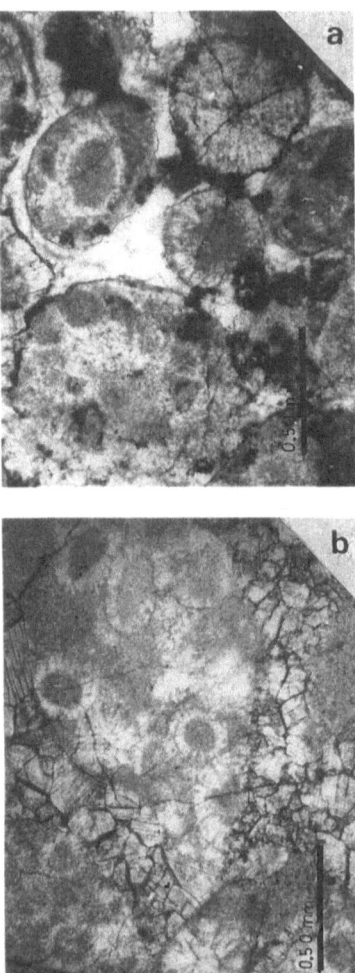

Figure 10 : a) Inter-thrombolitic grainstone composed by aggregated grains, ooids and mud-intraclasts (plasticlasts) with sparitic blocky cement. b) Inter-thrombolitic grainstone composed by aggregated peloids, ooids and *Nuia*, with sparitic blocky cement.

Biostromes may grow one on top of the other or intercalated by thin layers of carbonate packstone or grainstone with algal peloids, intraclast and aggregated grains (ooids, Fig.10a). Thrombolitic intraclast are frequently found within the sediment near the contact thrombolite/sediment. These stratiform thrombolites are flanked by layered mudstone composed by sparitic and micritic stripes of 1.5 cm in height.
Each cycle often shows in the basal parts tabular-stratiform thrombolites, grading towards the top into domes and rarely into mounded thrombolites.

The microfacial analysis of the sediments (mudstone, packstone, wackestone and grainstone) suggests that the stratiform thrombolites developed under a subtidal environment. The presence of high energy particles such as ooids in the aggregated grain may indicate a periodical increase in water turbulence.

Facies VI.- Subhemispherical Domal Stromatolites : These domes occur as closely packed, although isolated ones are also observed. Dark micritic mudstone fills the empty spaces between the domes, and detrital contribution is absent.
This facies is best developed southwards of the depositional basin, it grades laterally and vertically to Facies VII (Fig.6a,11a) and laterally to Facies VIII. The main difference between this facies and Facies VII is morphological, Facies VI has bigger domal stromatolites with moderate relief (diameter similar to hight) and less fenestral porosity than the planar stromatolites.
Pratt and James (1986) suggest for these kind of stromatolites a lower intertidal palaeoenvironment protected from strongly erosive tidal currents. However the presence of mudcracks indicates periodic subaerial exposure.

Facies VII.- Planar and Laterally Linked Low Synoptic Relief Domal Stromatolites :
They occur in biostromes of 0.2 to 0.5 m thick but southwards they can reach about 2 m thickness. Black or brownish chert layers often occur interbedded within the stromatolites or as a discontinuous dark cover overlying the stromatolite. Mudcracked beds and storm deposits (0.1 m thick) are also interbedded with the stromatolites.
Flat laminated stromatolites (Fig.6d) are the more frequent type but pseudocolumns may occur (Fig.11b). Micritic laminae are frequently alternating with thin layers (1 to 5 mm thick) composed of clast of laminae .
The lateral extension of these stromatolites, the association with mudcracks, some tepee structures, fenestral fabric and the alternation with graded storm layers suggest that this facies was formed on upper intertidal-supratidal flats.

Facies VIII.- Dolomitic Mudstone : It is interbedded with Facies VI and VII. The mudstones form sterile beds of 0.5 m average thickness, rarely they reach 2 m in thickness (southwards). They are massive but occasionaly they may show a rough layered pattern. Each layer (0.15 m thick) is bounded by rusted discontinuity surfaces with mudcracks due to periods of subaerial exposure. Storm layers comprising packstone with mud-intraclasts (plasticlasts) are often found interbedded with the dolomitic mudstone.
The microfacial analysis indicates an intertidal-supratidal environment.

PALAEOENVIRONMENTAL SCHEME

The La Flecha Formation is a clear example of small scale cyclic sequence. The cycles are markedly built and dominated by microbialites.

Figure 11: a) Subhemispherical stromatolites with very close spacing and moderate synoptic relief. The laminae begins as smooth and ends as wrinkled (Facies VI). The top of the stromatolite dome is erosional. Stratiform stromatolites (Facies VII) developed over domal ones. Stratiform thrombolite (Facies V) with erosional discontinuity surface underly the stromatolites of Facies VI. b) Subhemispherical domes with low synoptic relief stromatolite and pseudocolumnar to rarely columnar lamination. Micritic and black cherty laminae are interbedded (Facies VII).

These cycles are repeated forming a thick sequence (nearly 1000m) of biosedimentary pattern which was developed on an extensive, protected, gently sloping marine carbonate ramp which had little or no terrigenous sediment input. In the southern portion of the basin more than a hundred cycles have been recorded. The cycles are thought to result from relatively small eustatic sea level fluctuations of unknown origin.

Each cycle begins probably under wave-base in subtidal conditions with the deposition of carbonate mud (Facies I) then the thrombolites grew on top of this substratum building different megastructures under shallow subtidal conditions. Thrombolites from Facies III or Facies IV may build low mounds, as a local feature, that reduce the water circulation. Other important factor of this reduction is non-physiographic (Enos, 1983) due to a broad expanse of shallow water across a cratonic area in which tidal and waves energy are dampened out , as it ocurrs within epeiric seas.

Wide and extend planar stromatolites (Facies VII) and domal stromatolites (Facies VI) developed under shallower waters over the thrombolitic buildups and they gradate laterally and vertically to the dolomitic mud (Facies VIII) in the low intertidal environment submitted to subaerial exposure.

Therefore this facial sequence represents shallowing upwards cycles developed on a low slope carbonate ramp under low and continous subsiding conditions.

The preservation of excellent lamination (stromatolites) and clotted fabrics (thrombolites) was favoured by the absence of macrobentos. The lack of boring or grazing organisms suggests waters not favourable for their development.

Encephalic mesostructures indicates an uniform (isotropic) growth of cyanobacteria, frequently in shallow subtidal facies. Increasingly anisotropic mesostructures (notably vertical and radial types) results from more rapid and punctual growth of cynobacterial communities, probably a response to phototropism within more shallow subtidal environments. This is a typical biogenic buildup (Flügel, 1982) constructed by stromatolites and thrombolites which do not inherit or mimic the shape of the original substrate.

The sequential analysis of shallowing cycles indicates a carbonate ramp that overgoes periodic eustatic changes. These shallowing cycles developed over a carbonate ramp with continuous subsidence (Fig.12a). The vertical accretion was effective until the high tide level was reached (Fig.12b,c, see cross sections). The thickness of each cycle was controlled by the water depth (Fig.12b). The succession of cycles was, therefore, caused by gentle water sea level increments (Fig.12c). Low energy tidal currents were an important agent of sediment transport on carbonate and arround the buildups, but the major sedimentary supply was effectuated during storms. Vertical accretion of each cycle was due to microbial growth.

The global tendency of this sequence is transgressive as the upper levels of the La Flecha Formation the subtidal facies (Facies I; II; III; IV and V) are dominant and the

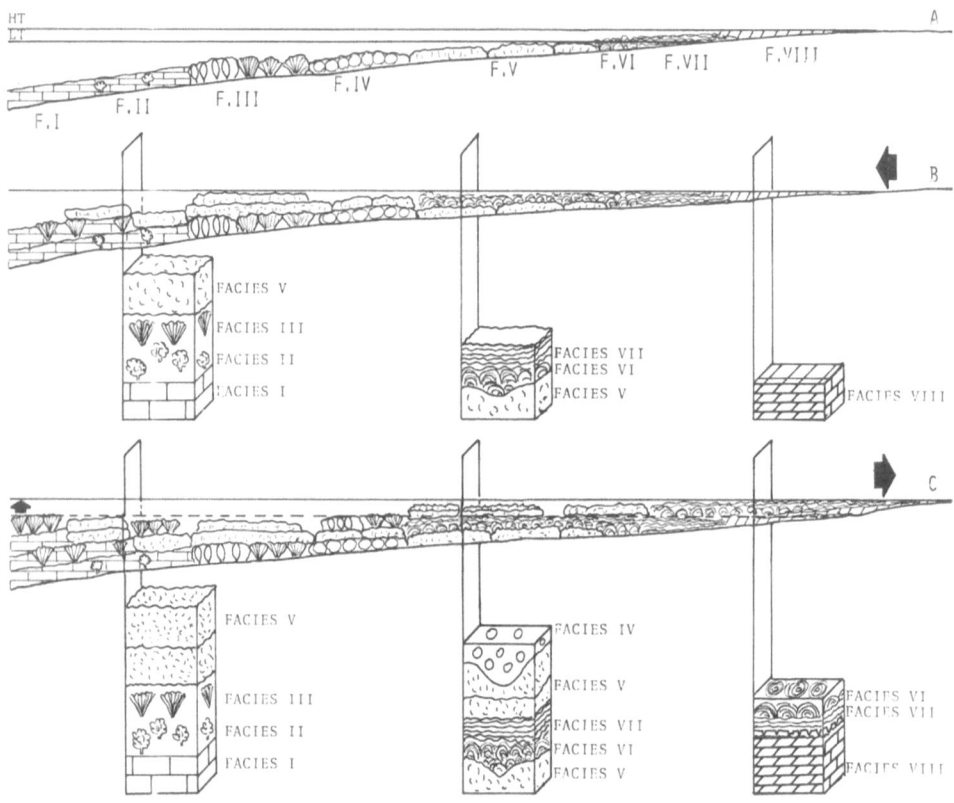

Figure 12: Sketch illustrating the vertical accretion of cycles as an effect of sea-level fluctuations over a carbonate ramp. a) Lateral variations of facies according to environment. b) Shallowing-upward sequences and progradation of facies belt developed when the sea-level is stable or when it begins to fall. c) Back stepping of facies and vertical accretion of cycles occur when the sea-level rises.

intertidal-supratidal facies (Facies VI; VII and VIII) are often absent (stromatolites are lacking in the upper levels). This transgressive tendency is a prelude to the great arenigian transgression (San Juan Formation). During the Arenigian, stromatolites disappears with the appearance of an abundant ordovician fauna.

In the northern portion of the ramp, thick cycles are developed (up to 20 meters), mainly stratiform thrombolites (Facies V) and arborescent thrombolites scattered within mud (Facies II) together with few domal stromatolites (Facies VI). Thrombolitic constructions are thicker in the upper cycles in the south of the basin and the number of thinner cycles also increases. These cycles were generated within shallower water than the northern ones, where minor sea level fluctuations induce significant changes in the depositional environment. Each "megacycle" in the northen portion of the basin is equivalent to three or more cycles in the southern basin. This contrast between northen and southern parts of the basin is most significant in the upper 300 m of La Flecha Formation.

Pratt and James (1986) suggest an island tidal flat model for the shallowing upward cycles of St. George's Group (Lower Ordovician) of Western Newfoundland. In this scheme the peritidal sediment are deposited over and around the islands that are scattered over an extended platform, particularly during periods in which the sedimentation rate was similar to the water sea level increment.

In this paper it is suggested that the palaeoenvironmental model of the Latest Cambrian-Earliest Ordovician pericratonic basin of Precordillera (La Flecha Formation) can be considered as a carbonate ramp with facies grading from subtidal to inter-supratidal. In the south it could be a shallower, probably flatter and perhaps less subsiding area, where the tidal island model applied explaining the greater number of cycles and their lateral variability.

The cycle lateral continuity was not observed mainly because the area was affected by tectonic activity and there are no intermediate outcrops between the different localities studied which are separated by a long distance.

CONCLUSIONS

No evidence of grazing and boring organisms was found therefore the clotted fabric of the thrombolites is considered to be of microbial origin.

Stromatolites are very frequent at the bottom of the sequence of the La Flecha Formation and they grow in size and diversity towards the south of the basin, suggesting shallower water conditions or lower amplitude of eustatic sea level changes.

The analysis of an ideal cycle, indicates that the thrombolitic growth took place over carbonate mud flat, under shallow subtidal conditions. On tops of these thrombolites, the stromatolites grew with shallower water, in intertidal to supratidal conditions.

The cycle developed vertically until the stromatolites reached the sea level. The cyclic fluctuation of the sea level produces the cyclic biosedimentation pattern.

The thickness of the cycles markedly increases towards the north of the depositional basin (5 to 20 meters thick) while southwards they are thinner (2 to 8 meters thick) and more abundant. Sets of these thin cycles probably correlate with "megacycles" in the northern portion of the basin. Therefore each cycle reflects local sea level fluctuation and indicates that the ramp could not be exposed and submerged at the same time, but reflect progradation effect.

The microbialites grew over a very flat and very large pericratonic ramp with flat islands towards the south of the depositional basin, slow and continuous subsidence and periodic sea level fluctuation.

There are no evidence of barriers that may have restricted the water circulation. The reduction of water movement was produced by the presence of thrombolitic buildups and by tidal. Wave energy was progressively dampened out as it occurs in extended shallow water areas within epeiric seas.

Each cycle, individually, is regressive but towards the top of the sequence of the La Flecha Formation the deeper environmental facies are more important (20 to 30 meters thick) indicating a global transgressive tendency. This transgressive tendency announces the very important arenigian transgression represented by the overlying unit (San Juan Formation), together with the sudden development of the ordovician fauna, the disappearence of thrombolites and the decline of stromatolites.

ACKNOWLEDGEMENTS

I am very thankfull for the critical reading and constructive comments by J. Bertand-Sarfati, R.V. Burne and C. Monty. I also thank B. Baldis, O. Bordonaro, N.G. Cabaleri, L. Castro and V. Outes and to the authorities of the Museo Argentino de Ciencias Naturales "Bernardino Rivadavia" for their contribution. This work was partially financed by CONICET (Consejo Nacional de Ciencia y Tecnica).

REFERENCES

Armella, C. (1989a) "Estratigrafía de las Formaciones del límite Cambro-Ordovícico en la Precordillera Oriental". PhD Thesis, Universidad de Buenos Aires, Argentina (Unpublished).

Armella, C. (1989b) "Mesostructure, megastructure and palaeoenvironmental meaning of Cambro-Ordovician thrombolites from Argentina", in J.M.Kennard, and R.V.Burne (Eds), Stromatolite Newsletter, 14, 12-15.

Armella, C. (1990) "Guía práctica para la clasificación descriptiva de trombolitos". XI Cong.Geol.Argentino, II, 195-198.

Baldis, B.A., Bordonaro, O.L., and Beresi, M.S. (1981) "Estromatolitos, trombolitos y formas afines en el limite Cámbrico-Ordovícico del oeste argentino". II Congr. Latinoam. de Paleont. Brazil, 1 19-31.

Bordonaro, O.L. (1983) "El Cámbrico de la Sierra Chica de Zonda, San Juan". PhD Thesis,Universidad Nacional de San Juan,Argentina (Unpublished)

Chafetz, M and C. Buczynsky, C. (1992) "Bacterially induced lithification of microbial mats". Palaios, 7, 277-293.

Enos, E. (1983) "Shelf", in P. Scholle, D. Bebout, and C. Moore,(Eds), Carbonate Depositional Environments, AAPG Memoir 33, 267-297, Tulsa.

Flügel, E. (1982) "Microfacies analysis of limestones". Springer-Verlag, Berlin, 633pp.

Grey, K. (1989) "Handbook for the study of stromatolites and associated structures", in J.M.Kennard and R.V.Burne, (Eds), Stromatolite Newsletter, 14, 82-140.

Grey, K., Awramik, S.M., Bertrand-Sarfati, J., Hofmann, H.J., Pratt, B.R and others, (1990) "Handbook for the study of stromatolites and associated structures". Third draft (Unpublished).

James, N.P. (1984) "Shallowing-upward sequence in carbonates",in R.G.Walker (Ed), Facies Models. Geosci. Can. Repr. Ser.1, 213-228.

Kennard, J.M. (1989) "Cambro-Ordovician thrombolites,western Newfoundland", in J.M.Kennard and R.V.Burne, (Eds), Stromatolite Newsletter, 14, 41-45.

Kennard, J.M. and James, N.P. (1986) "Thrombolites and stromatolites: Two distinct types of microbial structures". Palaios 1, 492-503.

Pratt, B.R. (1984) "Epiphyton and Renalcis : diagenetic microfossils from calcification of coccoid blue-green algae". Journal of Sedimentary Petrology 54, 948-971.

Pratt, B.R. and James, N.P. (1986) "The St. George Group (Lower Ordovician) of Western Newfoundland: tidal flat island model for carbonate sedimentation in shallow epeiric seas". Sedimentology 33, 313-343.

Preiss, W.V. (1976) "Basic field and laboratory methods for the study of stromatolites", in M.R.Walter (Ed), Stromatolites, Developments in Sedimentology, 20. Elsevier, Amsterdam, 360-370.

Riding, R. (1991) "Classification of microbial carbonates", in R.Riding (Ed.) Calcareous Algae and Stromatolites, Springer-Verlag, Berlin. 21-51.

Walter, M.R. and Heys, G.R. (1985) "Links between the rise of Metazoa and the decline of stromatolites". Precambrian Res. 29, 149-174.

THROMBOLITES AND STROMATOLITES WITHIN SHALE-CARBONATE CYCLES, MIDDLE-LATE CAMBRIAN SHANNON FORMATION, AMADEUS BASIN, CENTRAL AUSTRALIA

J. M. KENNARD
Australian Geological Survey Organisation ; GPO Box 378, Canberra City, ACT, 2600 Australia

ABSTRACT

Thrombolites and stromatolites within shallowing-upward carbonate cycles in the Cambrian Shannon Formation, central Australia, display a systematic zonation of mega- meso- and microscopic structures. Subtidal to intertidal thrombolite-stromatolite zonations are interpreted to record ecologic successions of benthic microbial communities in response to sea-level fluctuations and shoaling sedimentation. Following each sea-level rise, distinct microbial communities colonised laterally adjacent environments differentiated by water depth, turbulence, frequency of exposure, abundance of metazoans, salinity and supply of detrital sediment. Systematic analysis of the thrombolites and stromatolites suggests that their microstructure is primarily controlled by the internal organisation and morphologic composition (coccoid or filamentous) of the successive microbial communities, that their mesostructure is controlled by a complex interaction of biological and environmental factors, and that their megastructure is primarily controlled by environmental factors.

INTRODUCTION

Thrombolites (Aitken, 1967; Kennard and James, 1986b) and stromatolites are widespread within shallowing-upward carbonate cycles in the Middle to Late Cambrian Shannon Formation in the Amadeus Basin, central Australia. Within each shallowing-upward cycle, these microbial structures display a systematic zonation of megascopic, mesoscopic, and microscopic structures, and this zonation can be interpreted as an ecologic succession of distinct microbial communities. Systematic analysis of the thrombolites and stromatolites thus enables an evaluation of the relative importance of biological and environmental factors in controlling their megascopic, mesoscopic, and microscopic morphology.

J. Bertrand-Sarfati and C. Monty (eds.), Phanerozoic Stromatolites II, 443–471.

GEOLOGICAL SETTING

The Amadeus Basin, the remnant of a broad intracratonic depression, now 800 km long and 300 km wide, contains a thick succession of shallow marine and terrestrial deposits of Late Proterozoic and Early Palaeozoic age (Wells et al., 1970; Kennard et al., 1986; Lindsay and Korsch, 1989; Korsch and Kennard, 1991). Cambro-Ordovician strata, and some Late Proterozoic strata, extended beyond the present limits of the basin, the present southern and northern margins being defined by tectonic uplift and erosion in Latest Proterozoic and Late Devonian-Early Carboniferous times, respectively (Lindsay and Korsch, 1989; Korsch and Lindsay, 1989). During Middle-Late Cambrian time, sedimentation was characterised by a west to east transition from: 1) coarse fluvial and deltaic siliciclastic facies, to 2) mixed fluvial and shallow marine siliciclastic facies, 3) shallow marine fine siliciclastic and minor carbonate facies, and 4) mixed shallow marine carbonate and fine siliciclastic facies (Kennard and Lindsay, 1991; Fig.1). This

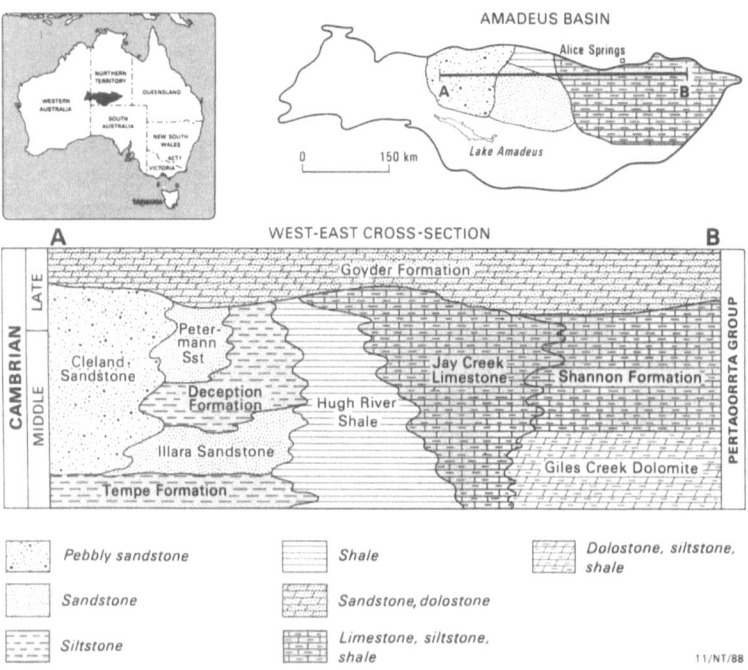

Figure 1. Location, distribution and relationship of Middle and Late Cambrian facies in the Amadeus Basin, central Australia. Modified after Shergold (1986), and Wells et al. (1970).

eastern carbonate-siliciclastic sequence, represented in part by the Shannon Formation and the equivalent Jay Creek Limestone, contains the thrombolites and stromatolites analysed in this study.

SEDIMENTOLOGICAL SETTING AND FACIES ASSOCIATION

The Shannon Formation is characterised by the alternation of poorly exposed siliciclastic mudrocks and thin carbonate cycles of great lateral continuity. Individual cycles can be traced in outcrop for tens of kilometres, and correlated over distances of up to 120 km. The formation is readily subdivided into two units: a lower mudrock-rich unit, and an upper carbonate-rich unit (Fig.2). In both units, the carbonate cycles are sharply overlain by siliciclastic mudrocks, and their basal contact may be either sharp and erosional, or gradational over several decimetres. Mudrock intervals are typically 2--10 m thick in the lower unit, and 1--5 m thick in the upper unit. The

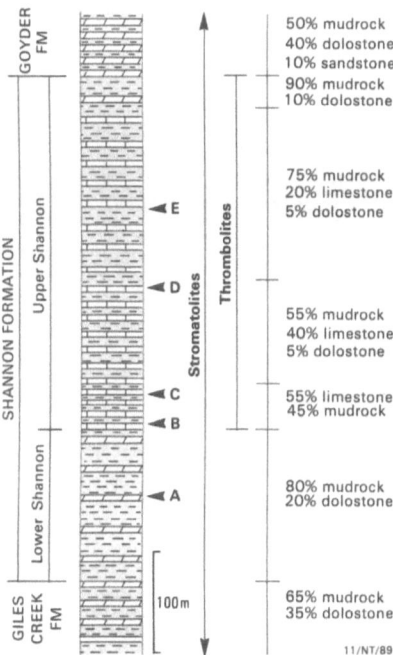

Figure 2 : Generalized lithological section of the Shannon Formation on the southern limb of the Ooraminna Anticline, northeast Amadeus Basin (Locality B in Fig.1). Carbonate cycles A to E are shown in detail in Fig.3.

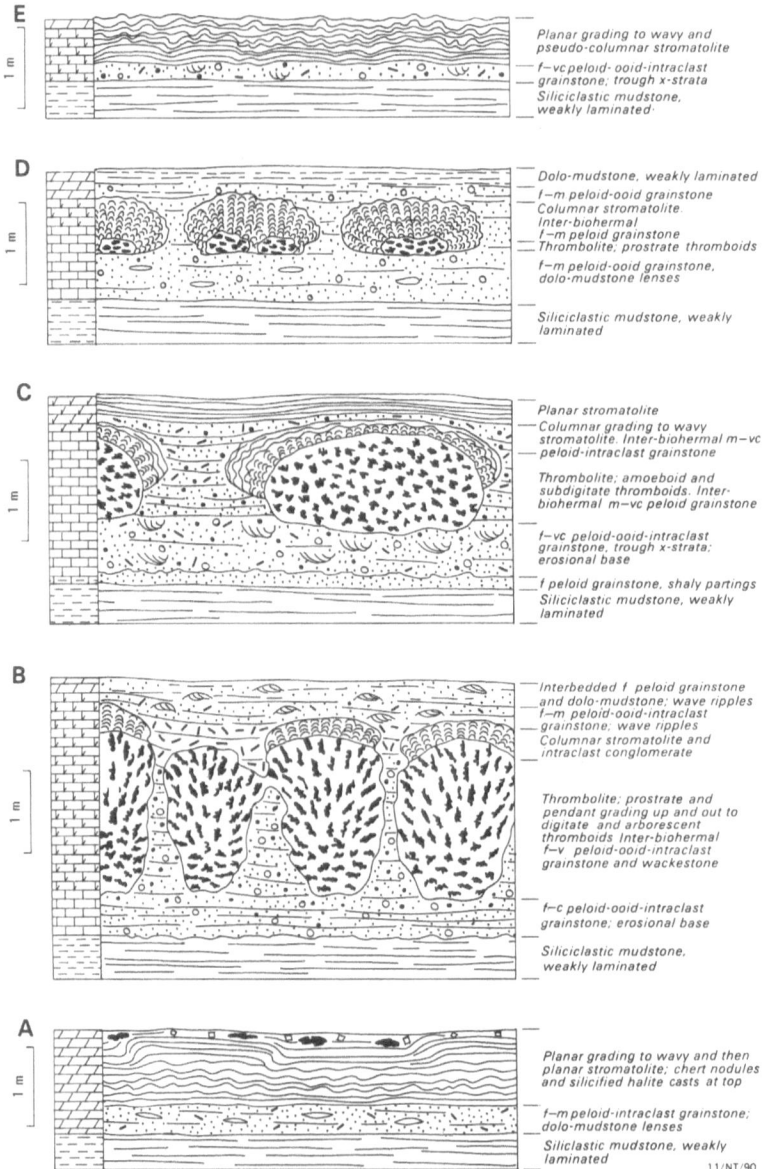

Figure 3 : Detailed sections of repesentative stromatolitic and thrombolitic carbonate cycles in the Shannon Formation. A) Lower unit. B-E) Upper unit.

mudrocks are red-brown, micaceous and weakly laminated, and locally contain desiccation cracks and
runzel marks. Very fine sandy laminae and lenses, and thin interbeds of current-rippled and wave-rippled peloid grainstone, dolo-mudstone, and in a few instances, gypsum, occur within the mudrocks. Mudrock intervals in the south are locally cut by tidal sandstone channels.
Carbonate cycles in the lower unit comprise dolostone, are typically 0.2--2 m thick, and are dominated by stromatolites of low synoptic relief. They commonly exhibit the following shallowing-upward sequence (Fig.3A, 4): 1) Fine to coarse peloid or ooid

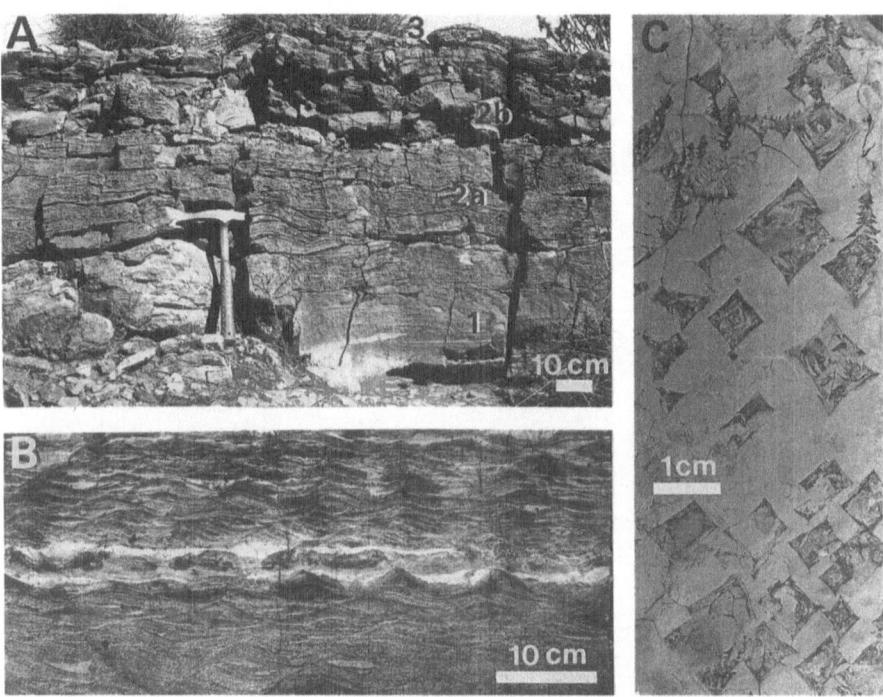

Figure 4. Carbonate cycles in the lower unit of the Shannon Formation. A) Stromatolite-dominated cycle; recessive mudrock is overlain by (1) grainstone with dolo-mudstone lenses, (2a) wavy and (2b) hemispherical stromatolite, and (3) undulose stromatolite. B) Wave-rippled ooid grainstone with lenticular and wavy dolo-mudstone drapes. C) Cubic and pagoda halite pseudomorphs within dolo-mudstone at top of cycle.

448   J. M. KENNARD

Figure 5 : Carbonate cycles in the upper unit of the Shannon Formation. A, B) Thrombolite-dominated cycles; recessive mudrock is overlain by (1) grainstone, (2) thrombolite bioherms, (3) inter-biohermal grainstone, (4a) columnar and (4b) wavy stromatolite, (5) intraclast conglomerate, (6) wave-rippled grainstone, and (7) dolo-mudstone and undulose stromatolite. C) Columnar stromatolite capping thrombolite bioherms shown in A.

grainstone, commonly wave-rippled with lenticular and wavy dolo-mudstone drapes; 2) Wavy-laminated and linked hemispherical stromatolite biostromes; 3) Undulose or planar-laminated stromatolite biostromes, and less frequently; 4) Thinly interbedded dolo-mudstone, peloid grainstone and flake conglomerate. Silicified evaporite pseudomorphs, including cubic, pagoda and reticulate ridge halite (terminology of Southgate, 1982), former ?halite ooids, and cauliflower chert nodules after anhydrite, commonly occur at the top of these cycles (Fig.4B). Thrombolites are not present in this lower unit.

Carbonate cycles in the upper unit consist of limestone and subordinate dolostone, are typically 1–3 m thick, and are characterised by thrombolites of high synoptic relief (generally 30–100 cm) capped by stromatolites of moderate to low synoptic relief (less than 30 cm). They commonly exhibit the following shallowing-upward sequence (Fig.3, 5): 1) Fine to very coarse grained, locally trough cross-stratified, peloid-ooid-intraclast grainstone, which forms a foundation for; 2) Large thrombolite bioherms, or alternatively; 2A) Small thrombolite bioherms and poorly defined thrombolitic masses enclosed within ribbon and nodular bedded, peloid-bioclast packstone and wackestone; 3) Inter-biohermal fine to very coarse grained, peloid-intraclast grainstone with abundant thrombolite fragments; 4) Columnar stromatolites grading upward to wavy-laminated stromatolites (these cap the thrombolite bioherms and are flanked by); 5) Intraclast conglomerate; 6) Fine to medium grained peloid-ooid grainstone with symmetrical wave ripples; 7) Undulose or planar-laminated stromatolite biostromes and dolo-mudstone with desiccation cracks.

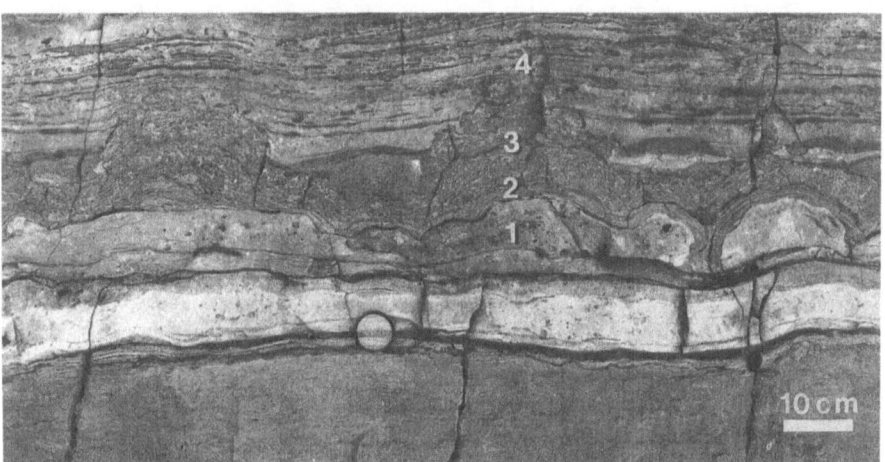

Figure 6 : Thin transgressive sequence at the base of a carbonate cycle in the upper unit of the Shannon Formation; (1) karst eroded dolostone, (2) stromatolite, (3) thrombolite, and (4) grainstone.

TABLE 1. Summary of the Mega-, Meso- and Micro-structure of thrombolites and stromatolites in the Shannon Formation.

	MEGASTRUCTURE	MESOSTRUCTURE	MICROSTRUCTURE
THROMBOLITES			
Type 1	Club, spherical & ellipsoidal bioherms	Prostrate, amoeboid, digitate & arborescent thromboids; spar-filled shelter cavities; inter-framework peloid, ooid, skeletal grainstone & packstone	Variegated lobate & massive; saccate & diffuse spherulitic lobes
Type 2	Irregular masses & small bioherms	Larger lobate, saccate & amoeboid thromboids; spar-filled burrows; inter-framework silty peloid, skeletal wackestone	As for Type 1
STROMATOLITES			
Columnar	Domed & discoidal bioherms	Bumpy tuberous columns; intercolumn intraclast, peloid grainstone	Vermiform or alternating vermiform & massive; locally grumous
Hemispherical	Domed bioherms & biostromes	Hemispherical stromatoids	Banded micro- & crypto-crystalline
Wavy	Domed & tabular biostromes	Wavy sinusoidal stromatoids	Banded micro- & crypto-crystalline
Planar	Tabular biostromes	Planar stromatoids	Banded micro- & crypto-crystalline
Undulose	Tabular biostromes	Undulose stromatoids; evaporite pseudomorphs	Irregular streaky microcrystalline

The cycles are frequently incomplete (Fig.6), however, and any of the above lithotypes may be abruptly overlain by siliciclastic mudrock or, less frequently, truncated by a karst erosion surface. A thin transgressive (deepening upward) sequence of karsted carbonate encrusted by stromatolite is present at the base of some cycles.

Throughout the upper unit there is a progressive upward trend (Fig.3) to thinner carbonate cycles, a decrease in the relative proportion of thrombolites to stromatolites, a decrease in the synoptic relief of both thrombolites and stromatolites, and an increase in the proportion of dolostone, ooid grainstone and quartz sand.

DESCRIPTION OF THROMBOLITES AND STROMATOLITES

The following descriptions comply with the three-tiered analytical scheme (incorporating megastructure, mesostructure and microstructure) and descriptive terms presented by Kennard and James (1986a, 1986b) and Kennard (1989). Individual millimetre to centimetre-sized clots within thrombolites are designated **thromboids** (cryptalgal clots *sensu* Aitken, 1967; mesoclots *sensu* Kennard and James, 1986a, 1986b), and superimposed layers or laminae within stromatolites are designated **stromatoids**. The descriptions are summarised in Table 1.

THROMBOLITES

Megastructure

The thrombolites display a spectrum of bed forms ranging from large club, spherical and ellipsoidal shaped bioherms (Fig.3b,c,d, 5a,b), measuring up to 2 m thick, 4 m long, and with 50–100 cm synoptic relief, to irregular decimetre and centimetre-sized masses of little synoptic relief that are enclosed within ribbon and nodular bedded packstone and wackestone. Thrombolite biostromes (20–40 cm thick) are scarce.

Mesostructure

The thrombolites comprise a framework of dark thromboids, inter-framework detrital sediment (commonly selectively dolomitized) and cement-filled shelter cavities. Two main mesostructural forms are recognised.

Type 1. This is the predominant type which forms large bioherms of moderate to high relief. The thromboids have a variable prostrate, amoeboid, digitate or arborescent shape, and range from several millimetres to 2 cm in size (Fig.7). They commonly exhibit an upward transition from prostrate and pendant forms at the base of bioherms (Fig.7B), to digitate and arborescent forms at the crest and margins of the bioherms (Fig.7C). Most are internally variegated with three distinct components discernible in hand samples (Fig.7B, D): 1) very dark grey to black, millimetre-sized, lobate bodies or

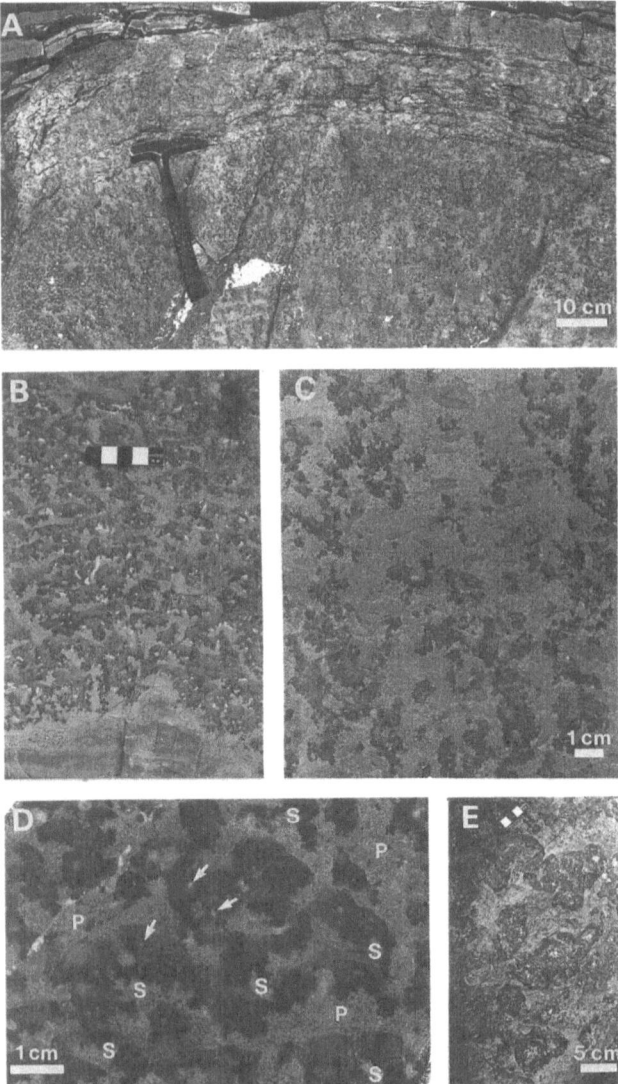

Figure 7 : Mesostructure of thrombolites. A) Digitate and arborescent thromboids at crest of bioherm. Note capping columnar stromatolite (above hammer head). B) Prostrate and pendant thromboids (variegated black and medium grey), spar-filled shelter cavities (white) and inter-framework sediment (light grey). Natural exposure surface, base of bioherm. C) Arborescent thromboids and inter-framework sediment with numerous intraclast. Polished slab, crest of bioherm. D) Amoeboid thromboids comprised of clusters of dark lobate bodies and medium grey micrite. Thromboids are disrupted by burrows (arrows) and support small shelter cavities occluded by white sparry calcite (s). Inter-framework sediment comprises dolomitic packstone (p). E) Lobate and saccate thromboids, Type 2 thrombolite. Natural exposure surface.

clusters of coalesced lobate bodies, 2) irregular patches of grey to brown micrite between the dark lobate bodies, and 3) patches of white spar cement that occlude shelter cavities and burrows within the thromboids. Inter-framework sediment consists of burrowed, poorly sorted, peloid grainstone and packstone, and commonly contains scattered intraclasts, ooids, minor quartz silt, and fragments of trilobites, molluscs and pelmatozoans.

Type 2. This type forms irregular masses and poorly defined bioherms of low synoptic relief within ribbon and nodular bedded packstone and wackestone. The thromboids have lobate, saccate or amoeboid shapes, and are significantly larger than those in the preceding type, typically 4–10 cm in size (Fig.7E). They exhibit a similar three phase internal variegation to Type 1 thromboids, except that the dark lobate bodies are less distinct, slightly smaller and more densely clustered, and cement-filled burrows are more abundant. Inter-framework sediment comprises burrowed silty and bioclastic peloid wackestone.

Microstructure

The thromboids comprise individual, coalesced and multiple lobate bodies of xenotopic microcrystalline calcite (black coloured in hand sample), separated by irregular patches of massive or silty cryptocrystalline calcite (grey-brown in hand sample) (Fig.8A, B); that is, they have a variegated lobate and massive microstructure (Kennard, 1989). Individual lobes are 500–2000 µm in diameter in Type 1 thromboids, and generally 100--500 µm in Type 2 thromboids. They are locally partially enclosed within a thin rim (5–15 µm) of cryptocrystalline calcite (Fig.8B, C, D), in which case they are said to have a saccate lobate microstructure (Kennard, 1989). The crystals within the lobes have a characteristic turbid, amber-coloured appearance due to numerous sub-micron sized inclusions, and they commonly display sweeping extinction (Fig.8C, D). Groups of adjacent crystals locally define spherulites, 100--300 µm in diameter, which have a poorly defined radial-fibrous texture as indicated by irregular sweeping extinction when viewed under cross-polarised light (Fig.9A, B). This type of microstructure, designated spherulitic lobate microstructure, is widespread and clearly defined within thrombolites in western Newfoundland and the Virginian Appalachians (Kennard, 1989).
In Type 2 thromboids, lobate microstructures are either microcrystalline or cryptocrystalline, individual lobes are smaller, and spherulitic microstructures are generally less obvious (Fig.9C).
Lobate microstructures are interpreted to result from the incipient calcification of coccoid microbial colonies (Kennard and James, 1986a; Kennard, 1989). Saccate forms are analogous to the incipiently permineralised colonies of coccoid cyanobacteria[1] within aragonitic stromatolites at Laguna Mormona (now known as Laguna Figueroa), Baja California (Horodyski and Vonder Haar, 1975). Gelatinous sheaths around these colonies are resistant to bio-degradation, and are selectively permineralised to form a

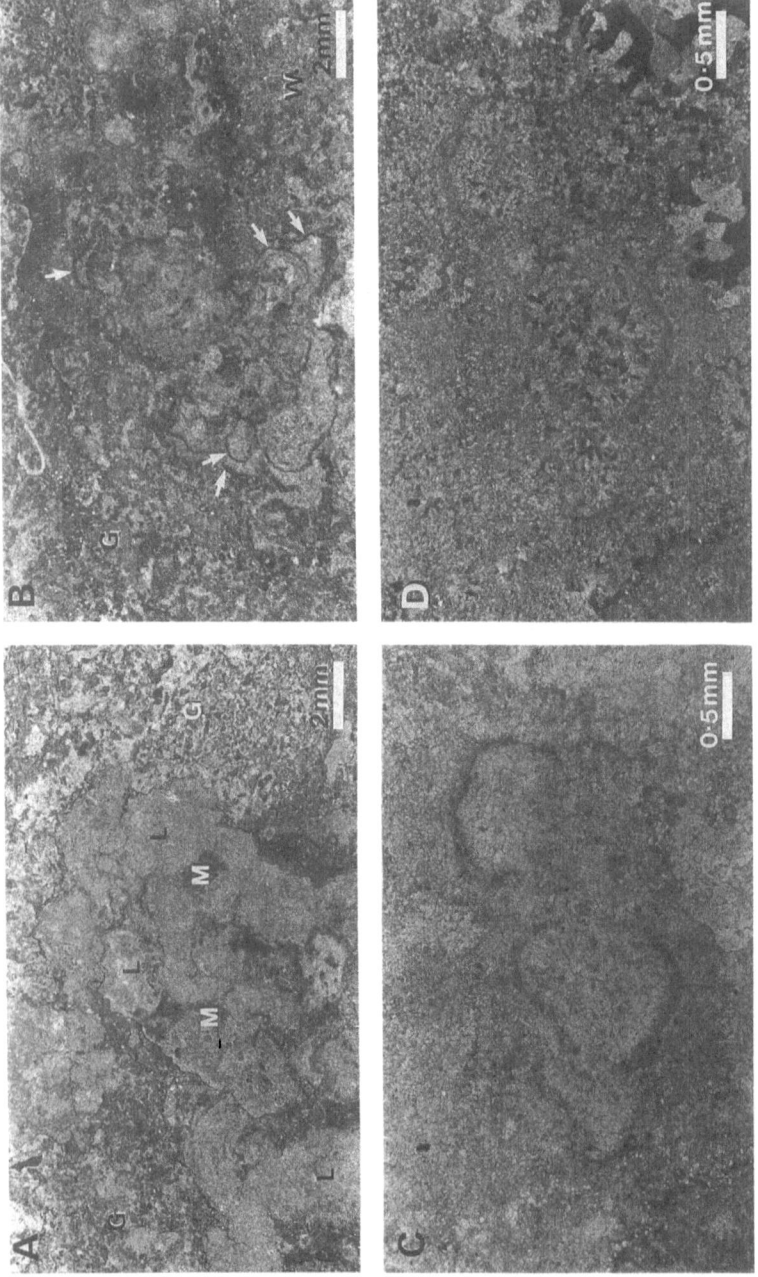

Figure 8. Microstructure of thrombolites. **A)** Thromboid comprised of variegated microcrystalline lobate (L) and cryptocrystalline massive (M) microstructures. Inter-framework peloid grainstone (G). **B)** Thromboid with multiple saccate lobate microstructure (cryptocrystalline rims of lobes are arrowed). Inter-framework peloid-skeletal grainstone (G) and wackestone (W). Note trilobite fragment top left. **C, D)** Saccate lobate microstructure, plane-polarized and cross-polarized light, respectively. Note undulose extiction and poorly defined spherulites within turbid microcrystalline lobes.

cryptocrystalline rim around a void of degraded intracellular material. Similar saccate lobate microstructures are described by Golubic (1983) within the pustular coccoid (*Entophysalis*) mats in Hamelin Pool, Shark Bay, Western Australia. Spherulitic lobate forms also occur within pustular coccoid mats at Shark Bay, Western Australia (Monty, 1976, fig. 27), and within probable bacterial mats at Mono Lake, California (Monty, 1976, fig. 29). They are also similar to the spherulitic microstructures within Pleistocene stromatolites in the precursor Dead Sea, Israel (Buchbinder, 1981). The precipitation of these spherulites is probably triggered by bacterial activity within degraded coccoid communities (communication by Monty, 1979, cited in Buchbinder, 1981, p. 192). Similar spherulitic bacterial precipitates have been produced experimentally in sea water by Lalou (1957), Oppenheimer (1961), McCallum and Guhathakurta (1970), Deelman (1975), Krumbein (1974), Krumbein and Cohen (1977), and Krumbein et al. (1977).

Thus depending on the relative extents of degradation, calcification and bacterial precipitation, the coccoid colonies are variously preserved as massive lobate, saccate lobate or spherulitic lobate microstructures.

The patches of micrite and silty micrite within the thromboids probably represent precipitates within organic mucilage excreted by or associated with the coccoid colonies (Monty, personal communication 1989), together with minor detritus that accumulated between, and then subsequently overgrown by, successive coccoid colonies. Apart from minor quartz silt, there is no other indication of trapped and bound detrital particles within the thromboids. This contrasts with an abundance of peloids, bioclasts and terrigenous silt within the inter-framework sediment between the thromboids. Locally ragged and embayed margins of the thromboids are probably a consequence of partial degradation of the coccoid colonies and or metazoan grazing and burrowing.

Thus the framework of thromboids was constructed by the calcification, rather than the sediment-trapping and binding activities, of a coccoid-dominated microbial community.

The only indication of former filamentous microbes within the thrombolites is a single observation of tufted, *Girvanella*-like tubules within a thrombolite fragment incorporated in inter-biohermal grainstone (Fig.9D). Detrital peloids were clearly trapped and bound between the individual filaments of this tuft, whereas the ubiquitous calcified coccoid colonies (now represented by spherulitic and saccate lobate microstructures) within the thrombolites evidently had little or no ability to trap and bind detrital particles other than minor amounts of silt.

---

[1] These colonies were originally identified as *Entophysallis* by Horodyski and Vonder Haar (1975), but were subsequently re-identified as *Pleurocapsa* by Stolz (1983).

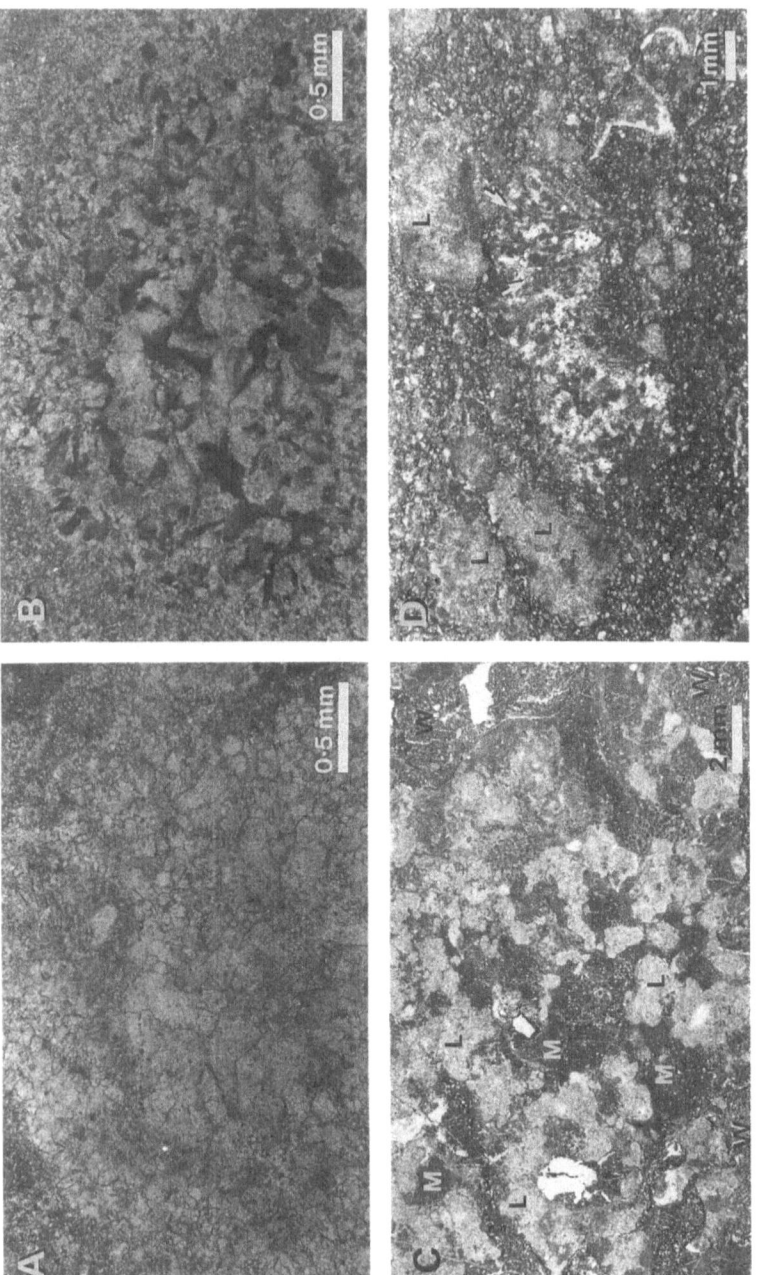

Figure 9. Microstructure of thrombolites. **A, B)** Turbid microcrystalline calcite within weakly spherulitic lobate thromboid, (pl-pol and cross-pol). **C)** Thromboid with variegated microcrystalline lobate (L) and silty cryptocrystalline massive (M) microstructure; inter-framework silty peloid wackestone (W). Type 2 thrombolite. **D)** Radiating tuft of poorly preserved calcified *Girvanella*-like filaments (arrowed) within inter-biohermal grainstone. Note diffuse silt-sized peloids trapped between filaments, reworked fragments of thromboids with turbid microcrystalline lobate microstructure (L), and metazoan fragments (lower right).

STROMATOLITES

Megastructure

The stromatolites generally have greater lateral continuity and lower synoptic relief (less than 30 cm) than the associated thrombolites. They range from domed, discoidal and elongate bioherms several decimetres to one metre thick, to mounded, gently domed and tabular biostromes a few decimetres or less thick (Fig.3, 4A). Within the lower unit, all stromatolites are biostromal and typically have less than 10 cm synoptic relief.

Mesostructure

Five stromatolite mesostructures are present: columnar, hemispherical, wavy, undulose, and planar. Columnar stromatolites are apparently restricted to the upper unit of the formation, and undulose forms are most abundant in the lower unit; the remaining types are equally abundant within both units. Vertical transitions between different forms are widespread.

Columnar. These forms (Fig.5C, 10A, B) were described in detail by Walter (1972) who classified them as *Madiganites mawsoni*. The columns are 5–30 mm wide, and up to several centimetres high. They are characterised by a bumpy tuberous shape, sub-parallel to slightly divergent branching, a discontinuous wall, infrequent lateral linkage, vertical impersistence and coalescence. They are radially disposed within bioherms, generally erect within biostromes, and commonly cap thrombolite bioherms. The stromatoids have an irregular, commonly asymmetrical, gently convex to rectangular shape, and low degree of inheritance. Inter-column sediment consists of peloidal and intraclastic grainstone and layers of carbonate mudstone.

Hemispherical. These forms (Fig.10C, D) range from a few centimetres to several decimetres in diameter, and are commonly oblate to flat-topped. They are either laterally linked within biostromes or occur as isolated bioherms, and commonly alternate or intergrade with wavy forms. The stromatoids have a smooth to irregular shape and a moderate degree of inheritance.

Wavy. These forms have a wavy shape reminiscent of an egg-carton (Fig.10B, D). They are predominantly biostromal, and either cap columnar and hemispherical forms or alternate with undulose forms and beds of rippled grainstone. The stromatoids have a smooth sinusoidal shape, a wavelength of 5–10 cm and amplitude of 2–3 cm, and high degree of inheritance. A variant of this form has a linear ripple-like shape which mimics and perpetuates the shape of underlying current or wave ripples.

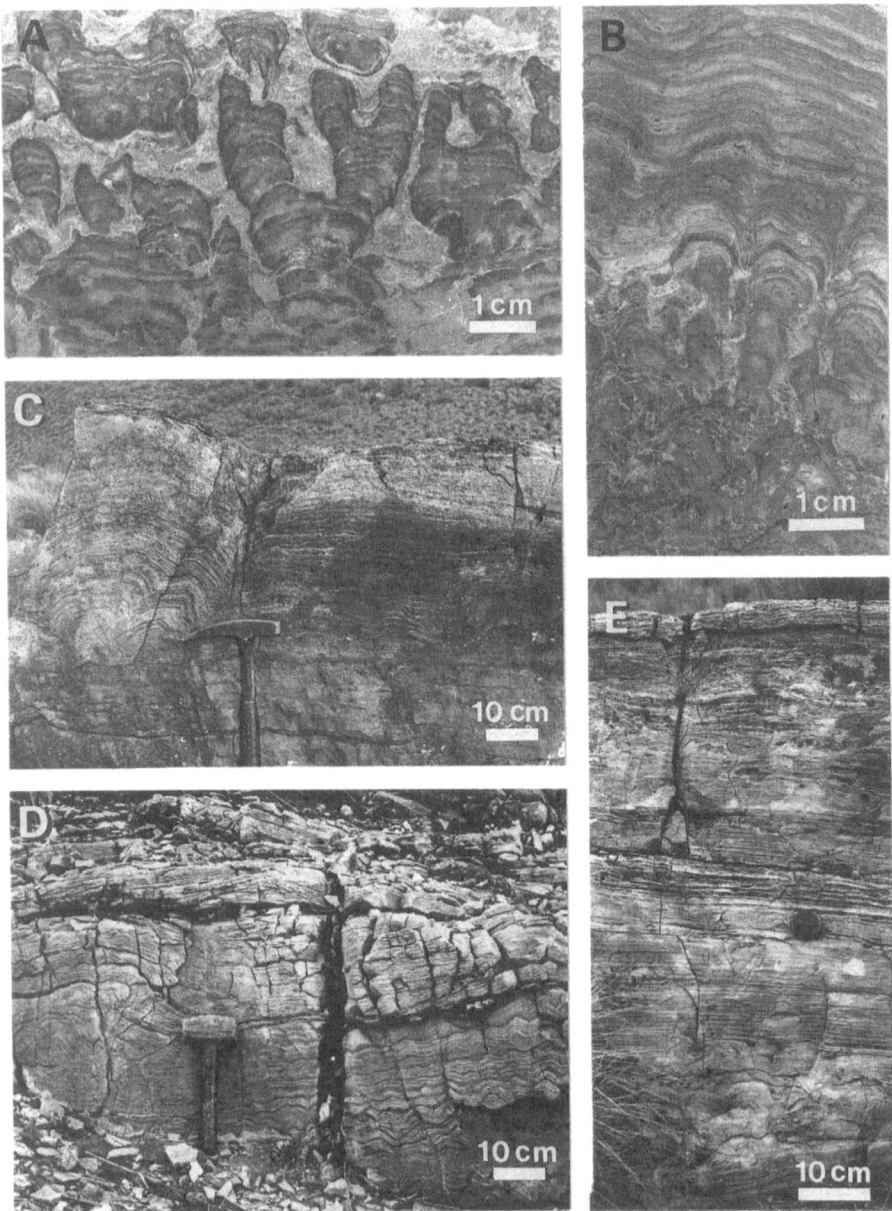

Figure 10 : Mesostructure of stromatolites. A) Columnar stromatolite (*Madiganites mawsoni*, Walter, 1972) with inter-column peloid-intraclast packstone. Polished slab. B) Columnar grading up to wavy stromatolite. Polished slab. C) Linked hemispherical stromatolite capped by undulose stromatolite. D) Wavy grading up to hemispherical (left of hammer) and undulose stromatolite. E) Alternating wavy and planar stromatolite capped by vuggy undulose stromatolite.

Planar. Planar forms comprise smooth stromatoids of minimal relief and considerable lateral continuity (Fig.10E). They are predominantly biostromal, and alternate with wavy and undulose forms.

Undulose. These forms comprise irregular undulose to pustular stromatoids of low relief (typically 1 cm or less), low degree of inheritance, and limited lateral continuity. They are exclusively biostromal and cap or alternate with wavy and planar forms (Fig.10E), or cap thrombolites. The laminae are commonly arched above fenestrae to form discrete blister-like features, and are disrupted by chert nodules, silicified evaporite pseudomorphs and desiccation cracks.

Microstructure

Three types of microstructures are recognised: 1) vermiform microstructure within columnar stromatolites, 2) banded microstructures (two varieties) within hemispherical, wavy and planar stromatolites, and 3) irregular streaky microstructure within undulose stromatolites.

Vermiform Microstructure. Vermiform microstructure (Walter, 1972; originally defined by Gürich (1906) as "vermicular" microstructure), consists of a network of fine sinuous tubules composed of clear microcrystalline carbonate, separated by darker coloured cryptocrystalline carbonate (Fig.11). The tubules branch and coalesce, are 20–60 μm wide, several hundred microns long (maximum 2 mm), and have a random reticulate to preferred prostrate orientation. Vermiform laminae are 300–3000 μm thick, and are either successively stacked and somewhat poorly differentiated, or episodically alternate with massive cryptocrystalline and microcrystalline laminae of comparable thickness (Fig.11B). Minor quartz silt commonly occurs within these alternating massive laminae. Vermiform and massive laminae are locally disrupted by metazoan burrows. With slight neomorphism, the vermiform microstructure intergrades with grumous microstructure (Walter, 1972) in which micro-clots of dark cryptocrystalline carbonate are separated by irregular patches of clear microcrystalline carbonate (Fig.11D).

The sinuous tubules are interpreted as the moulds of relatively large filaments between which micritic sediment was precipitated or trapped. Walter (1972, p. 87) suggested that upward gliding trichomes, a consequence of positive phototaxis, could generate a network of abandoned mucilaginous sheaths, the subsequent oxidation or calcification of which would result in a vermiform microstructure. In the samples examined in the Shannon Formation and elsewhere (Kennard, 1989), however, there is no evidence of a calcified wall around the sinuous tubules; the tubules are thus interpreted as moulds of oxidised sheaths and trichomes. Micrite between the tubules has most probably been precipitated in mucilages surrounding the filaments, and lacks definitive detritus. The alternating massive micritic laminae have a distinct appearance and density to the

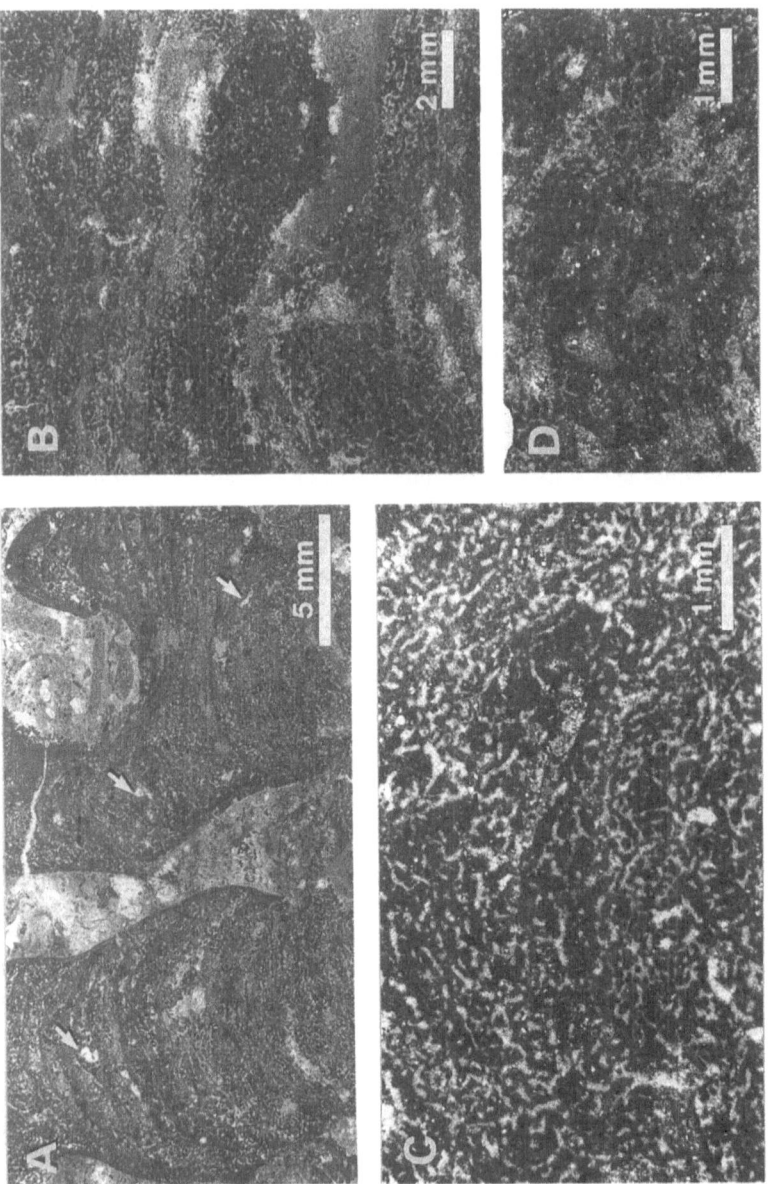

Figure 11 : Microstructure of columnar stromatolites. A) Columns composed of convex vermiform stromatoid layers and cryptocrystalline wall. Layers are disrupted by metazoan burrows (arrows). Inter-column peloid-intraclast grainstone. B) Alternating vermiform and massive cryptocrystalline stromatoid layers. C) Vermiform microstructure with reticulate network of sinuous tubules. D) Vermiform microstructure intergradational to grumous microstructure.

micrite between the tubules (Fig.12B), and are interpreted as layers of carbonate mud that periodically blanketed the filamentous mats. Similar filament moulds occur in modern stromatolites (Davies, 1970; Monty, 1976; Hardie and Ginsburg, 1977).

Banded Microstructure. Two varieties of banded microstructure (Walter, 1972) are recognised: a) composite repetitive, microcrystalline and cryptocrystalline laminae (Fig.12A), and b) simple repetitive microcrystalline laminae. The laminae are very distinct and generally continuous, and have a smooth to slightly crenulate shape. Microcrystalline laminae comprise equigranular, xenotopic calcite or dolomite (neomorphic microspar, generally 10–30 µm crystal size), are generally 300–2000 µm thick, and show marked crestal thickening within hemispherical and wavy forms. They locally pinch out laterally, and frequently contain minor silt-sized peloids and quartz silt. Cryptocrystalline laminae are generally uniformly thin (10–100 µm).
The precise origin of these banded microstructures is uncertain; microcrystalline laminae probably represent layers of trapped (partially neomorphosed) carbonate mud, peloids and quartz silt, and cryptocrystalline laminae may be calcified organic-rich layers. Although the composition of the formative microbial community cannot be determined, such uniformly laminated fabrics are generally constructed by filamentous rather than coccoid-dominated communities (Hofmann, 1973; Gebelein, 1974; Awramik, 1984). Modern banded microstructures occur in the "smooth mats" of Gladstone Embayment, Shark Bay (Davies, 1970), and Abu Dhabi, Persian Gulf (Kinsman and Park, 1976; Monty, 1976, fig. 11), the supratidal and intertidal mats of eastern Andros Island (Monty, 1967), and the flat-topped and conical stromatolite columns within hot springs in Yellowstone National Park, Wyoming (Walter et al., 1976). These modern banded microstructures are all constructed by filament-dominated microbial communities. Banding within the Andros Island mats variously results from the alternation of: 1) layers of erect and prostrate calcified filaments of the same species, 2) layers of calcified (erect or prostrate) filaments and non-calcified (prostrate or erect) filaments of a different species, and 3) layers of non-calcified filaments (organic-rich layers) and layers rich in grains trapped by these filaments (sediment-rich layers) (Monty, 1967, 1976). In the case of the Yellowstone National Park examples, Walter et al. (1976) demonstrated that the banding results from the alignment of filaments either parallel to laminae (conical forms) or alternately parallel and perpendicular to laminae (flat-topped forms), and that filament alignment is a direct result of their response to light   (phototaxis).   Banded   microstructures   within the hemispherical, wavy and planar stromatolites of the Shannon Formation were thus similarly probably constructed by filament-dominated microbial communities.

Irregular Streaky Microstructure. This form of streaky microstructure (Walter, 1972) consists of irregular, relatively poorly defined alternations of cryptocrystalline and inequigranular (10–100 µm crystal size) hypidiotopic carbonate (Fig.12B). The laminae

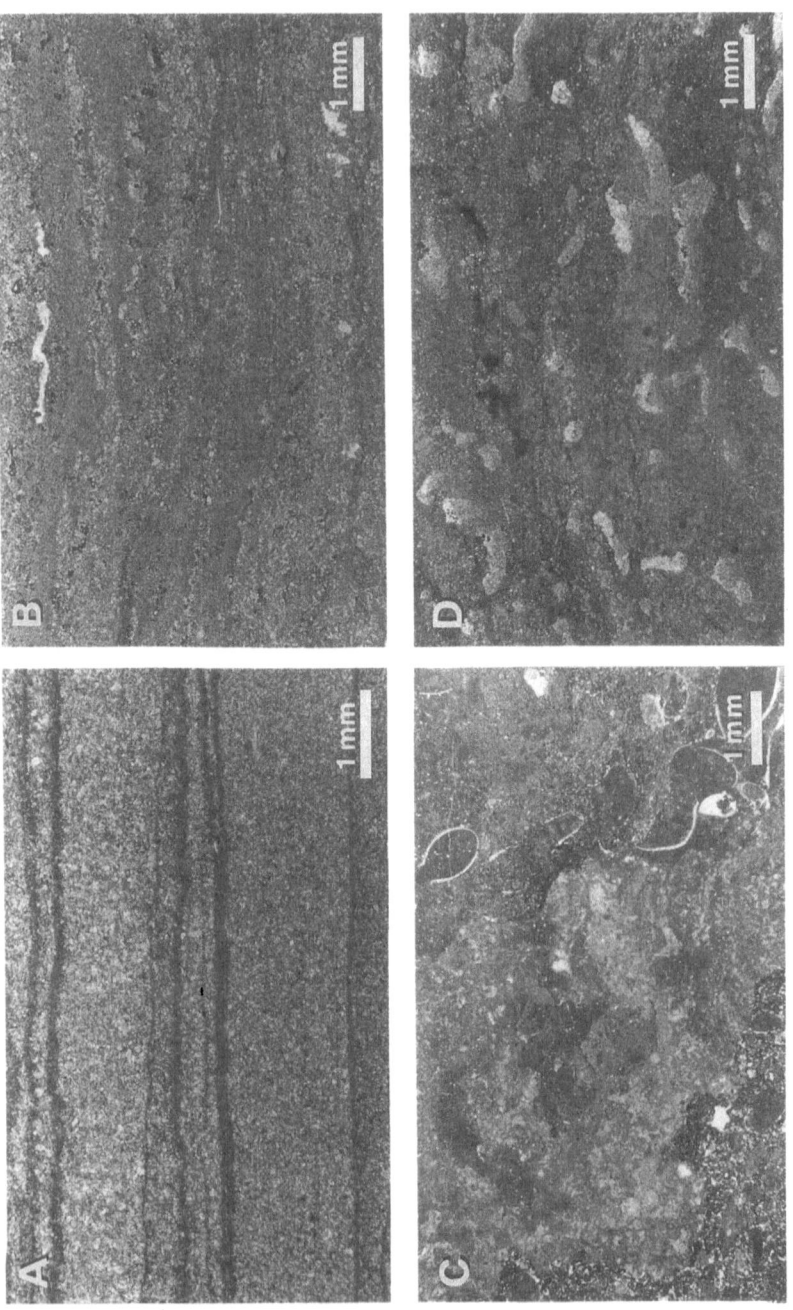

Figure 12 : A) Banded microstructure of planar stromatolite. B) Irregular streaky microstructure of undulose stromatolite. C) Cryptic microbialite, showing crudely laminated digitate column with mottled and diffuse vermiform grading to grumous microstructure. Note abundant mollusc fragments in inter-column sediments and embayed, probably bioeroded, margins of column. D) Extensively bioturbated cryptic microbialite.

have a diagnostic pustular to mamillate shape, are 200–3000 μm thick, and are commonly discontinuous. They are frequently disrupted by spar-filled fenestrae and silicified evaporite pseudomorphs (halite, gypsum or anhydrite).

Although definitive evidence is lacking, these irregular streaky microstructures were probably constructed by the *in situ* calcification of coccoid-dominated communities. This interpretation is consistent with: 1) the absence of trapped detrital grains, such as quartz silt and peloids, and 2) the observation that modern mats with similar shaped pustular and mamillate laminae are commonly dominated by coccoid rather than filamentous microbes (Hofmann, 1973; Gebelein, 1974; Golubic, 1983; Awramik, 1984; Pentecost and Riding, 1986).

OTHER MICROBIALITES

Some of the bioherms in the upper unit of the formation have irregular mottled, vaguely clotted or crudely laminated fabrics in which distinct mesoscopic constituents (thromboids, stromatoids and inter-framework sediments) cannot be differentiated. These bioherms are designated cryptomicrobial boundstones (Kennard and James, 1986b) or cryptic microbialites (Burne and Moore, 1987). They generally comprise irregular to digitate patches of mottled or crudely laminated microcrystalline and cryptocrystalline carbonate, locally with poorly preserved vermiform grading to grumous and lobate microstructures, and patches of silty peloid-bioclast grainstone, packstone and wackestone (Fig.12C). They are commonly extensively bioturbated (Fig.12D), and contain various metazoan debris (molluscs, pelmatozoans, brachiopods, trilobites). These microbialites were evidently constructed by an internally poorly differentiated, mixed filamentous and coccoid microbial community, the composition and structure of which has been largely obscured by extensive bioturbation, degradation and neomorphism.

INTERPRETATION

Each thrombolite and stromatolite type observed within the carbonate cycles is considered to have been constructed by a benthic microbial community of specific composition and or sediment-forming activity, restricted to a particular depositional environment. The sequence of stromatolites and thrombolites within each cycle is interpreted as an ecologic succession of benthic communities in response to sea-level fluctuations and shoaling sedimentation.

The observed relative frequency of transition between the various thrombolite and stromatolite types is shown in Figure 13. Within the lower unit of the formation, the most common sequence is one of decreasing synoptic relief from: 1) hemispherical, to 2) wavy, and 3) undulose stromatolites. The stromatolites were established on wave-rippled and lenticular bedded, peloid or ooid tidal sands and carbonate mud, and were commonly disrupted by evaporites in the upper part of the sequence. The banded

THROMBOLITES   STROMATOLITES

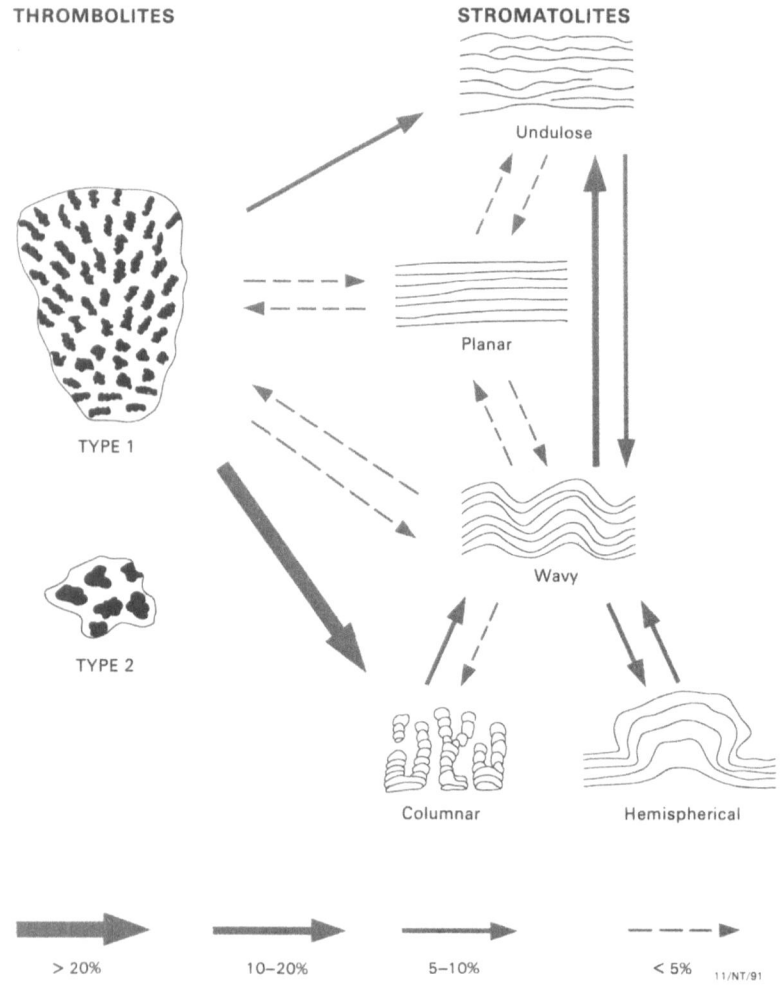

Figure 13. Relative frequency of sequence transitions between thrombolite and stromatolite types. 126 transitions recorded.

(?filamentous) wavy and planar forms are comparable to the smooth, filament-dominated (*Schizothrix*), mats observed in lower intertidal zones in Hamelin Pool, Shark Bay, Western Australia (Logan et al., 1974; Golubic, 1976; Hoffman, 1976); equivalent hemispherical forms, however, do not occur in Hamelin Pool. Undulose forms are interpreted to have been formed by ?coccoid mats in upper intertidal zones where they were disrupted by evaporites.

The most common sequence in the upper unit of the formation is also one of decreasing synoptic relief from: 1) Type 1 thrombolite (or undifferentiated cryptic microbialites), to 2) columnar stromatolite, and locally, 3) wavy stromatolite, and 4) planar stromatolite. The thrombolites were established on a foundation of peloid-ooid-intraclast subtidal sand, are flanked by poorly sorted peloid-intraclast sand, and pass laterally into ribbon and nodular bedded, peloidal and bioclastic subtidal carbonate mud. The capping columnar stromatolites are flanked by intraclast conglomerate, and are buried by wave-rippled, peloid-ooid intertidal sand and desiccated carbonate mud. The initial coccoid (thrombolite-forming) community grew in a moderately turbulent shallow subtidal environment, probably within a few to several metres of water, and was inhabited by grazing molluscs, scavenging trilobites and soft-bodied burrowing metazoans. Upward growth and progressive shoaling of the thrombolites culminated in the establishment of column-forming filamentous mats across their crests, probably within very shallow subtidal or lower intertidal environments. The only record of metazoans in this environment is minor bioturbation. These mats were in turn locally succeeded by banded (?filamentous) wavy and planar intertidal mats which show no evidence of metazoan activity.

This succession of microbial communities is broadly comparable to the sequence of subtidal and intertidal communities observed at Flagpole Landing, Carbla Point and Booldah Well in Hamelin Pool, Shark Bay, Western Australia (Playford and Cockbain, 1976; Burne and James, 1986); the thrombolites corresponding to subtidal club-shaped stromatolites built by colloform mats (a diverse community of coccoid and filamentous cyanobacteria, eucaryotic algae and many small metazoans), and the capping vermiform (filamentous) columnar forms and banded (?filamentous) wavy and planar forms corresponding to smooth filamentous (*Schizothrix*) mats in intertidal environments.

Each carbonate cycle corresponds to a period of reduced or nil siliciclastic influx, and is abruptly terminated either by the renewed influx and marine deposition of siliciclastic mud, or less commonly, subaerial exposure and karst erosion. Fourier analysis of cycle thicknesses (unpublished data) suggests that the cycles may have a common extrinsic cause, probably climatically controlled (Milankovitch band) sea-level oscillations.

DISCUSSION

Although it is generally agreed that stromatolites and thrombolites are organosedimentary structures formed by the interaction between benthic microbial communities and the environment, a great deal of uncertainty exists as to the relative importance of biological and environmental factors in controlling their morphology (Burne, 1987; Burne and Moore, 1987). The recognition of cyclic ecologic successions of biologically distinct microbial communities within the Shannon Formation, enables an evaluation of the relative contributions of biological and environmental factors in determining the microstructure, mesostructure and megastructure of these stromatolites and thrombolites.

Three major processes are involved in the formation of stromatolites and thrombolites (Burne and Moore, 1987): 1) trapping and binding of detrital particles, 2) biologically influenced calcification, and 3) inorganic calcification (cementation). Trapping and binding of sediment is governed by the ability of a microbial community to trap or bind detrital sediment, and the availability and rate of supply of suitable detrital sediment in a particular environment (Monty, 1967, 1972, 1976). Biologically influenced calcification is also controlled by the interaction of biological and environmental factors (Monty, 1972). Cyanobacteria commonly participate in processes of calcification, but such calcification is only partly within the influence of the organism; it also depends on suitable environmental conditions (Monty, 1967; Pentecost and Riding, 1986; Burne and Moore, 1987). Similarly bacteria, important constituents of modern and ancient microbial communities, also induce the precipitation of calcium carbonate (Lalou, 1957; Oppenheimer, 1961; McCallum and Guhathakurta, 1970; Monty, 1972; Deelman, 1975; Krumbein, 1974, 1979; Krumbein and Cohen, 1977; Krumbein et al., 1977; Chafetz and Folk, 1984; Chafetz, 1986), but again such calcification depends on suitable environmental, or micro-environmental, factors. Inorganic calcification, on the other hand, solely depends on environmental factors.

The microstructure of each stromatolite and thrombolite type provides a direct record of the sediment-forming processes of the microbial community, and commonly directly reflects the internal organisation and composition of that community. Thus the variegated saccate lobate, spherulitic lobate and massive microstructure of Types 1 and 2 thrombolites, directly reflects the morphology of the constituent coccoid colonies. Vermiform microstructure within columnar stromatolites mimics the architecture of successive layers of filamentous microbes, now represented as sinuous moulds. Finally, although the precise origins of banded and streaky microstructures are uncertain, their basic architecture is a direct consequence of the periodic growth, calcification and or sediment-trapping activities of smooth laterally continuous (?filament-dominated) and pustular laterally discontinuous (?coccoid-dominated) microbial mats, respectively. Furthermore, banded microstructure occurs within various stromatolite forms (hemispherical, wavy and planar) in inferred upper subtidal, lower intertidal and middle intertidal environments, yet corresponding microstructural variation were not observed within these forms. A significant non-biological contribution to banded and streaky microstructures, however, is diagenesis; stromatoid microstructure appears to have been obscured by neomorphism, stromatoids are locally disrupted by desiccation cracks or evaporites, and fenestrae are occluded by carbonate cement.

Thus the basic architecture of each type of microstructure recognised within these stromatolites and thrombolites is primarily controlled by the structure and composition of the formative microbial community; apart from desiccation and evaporitic features, there appears to be little direct environmental contribution to microstructure. Monty (1972, 1976) also showed this to be the case for modern stromatolites.

The mesostructure of the stromatolites and thrombolites appears to be controlled by a complex interaction of several biological and environmental factors: 1) shape of the microbial community or colonies, 2) coherence of the microbial community, 3) metazoan bioturbation, 4) physical turbulence and erosion, 5) production and accumulation of autochthonous sediments, and 6) influx and accumulation of allochthonous sediments. The basic shape and size of the thromboids and stromatoids largely reflects the growth pattern, coherence and lateral continuity of the formative microbial community (cf. Walter, 1972, p. 92). Thus prostrate, amoeboid and arborescent shaped thromboids within Type 1 thrombolites closely reflect the growth pattern of the constituent calcified coccoid colonies, which is in turn controlled by the relative rates of microbial growth, calcification and bacterial degradation. Their shape is also further modified by physical erosion, metazoan grazing and bioturbation. Columnar stromatoids result from the vertical stacking of discontinuous, button-like, filamentous mats. Mat discontinuity, column branching and column coalescence could be controlled by several factors (see discussion by Walter, 1972, p. 93–96): uneven growth of the microbial community, or localised physical erosion, bioturbation, and detrital sediment accumulation. Hemispherical and wavy stromatoids result from the successive stacking of laterally continuous mats which selectively thicken across the crests of underlying protuberances, such as ripples or eroded surfaces. Crestal thickening is probably a consequence of phototaxis (cf. Gebelein, 1969), or variations in sediment trapping ability due to depositional slope. These mats were sufficiently coherent to withstand mild currents that deposited interlayered rippled peloid sands. Planar and undulose stromatoids directly reflect the smooth and undulose-pustular shape, respectively, of laterally extensive mats. They were also sufficiently coherent to withstand mild currents.

Specific stromatolite and thrombolite types generally exhibit a range of megastructural forms both within a single horizon and from horizon to horizon. Conversely, microstructurally and mesostructurally distinct types sometimes exhibit similar megastructures. That is, megastructure is largely independent of microstructure and mesostructure, and is considered to be principally controlled by environmental rather than biological factors. The overall relief and size of a bioherm or biostrome is clearly limited by the depth of water in which it formed, and its shape and lateral continuity are probably principally controlled by the form of the substrate, the presence or absence of erosive currents, and the relative rates of microbial growth and influx, production and accumulation of detrital sediment (cf. Gebelein, 1969). Early (sea-floor) lithification is also obviously an important factor in determining megastructure. It is achieved by two processes, biologically influenced calcification and inorganic calcification, the relative biological and environmental contributions to which have been discussed above. Large club-shaped thrombolite bioherms are generally associated with coarse inter-biohermal grainstones and abundant reworked thrombolite fragments, their flanks are commonly eroded, and they are buried by intraclastic conglomerate and wave-rippled ooid-peloid grainstone. Early calcification and lithification of these forms enabled them to form relatively high relief bioherms in a relatively turbulent, wave and current-washed, subtidal environment. In contrast, small irregular thrombolite bioherms occur within ribbon and nodular bedded packstone and wackestone, their margins show no evidence of erosion, and they are extensively bioturbated; they formed low relief structures in a relatively tranquil subtidal environment. Both of these megastructural forms, however, were constructed by essentially identical, coccoid-dominated, microbial communities.

Similarly, different megastructural forms of columnar stromatolites (domed and discoidal bioherms, and domed and tabular biostromes) have identical microstructures and were thus probably constructed by a single, filament-dominated, microbial coenose. Early calcification enabled oxidised filaments to be preserved as sinuous cement-filled voids, and undoubtedly imparted a certain degree of rigidity and wave and current resistance to these structures. These stromatolites show no sign of abrasion or breakage, however, and formed moderate to low relief structures in mildly agitated shallow subtidal or lower intertidal environments.

Hemispherical, wavy and planar stromatolites also exhibit a variety of megastructural forms which have the same type of microstructure; that is, they were probably constructed by one microbial association.

The observed megastructural forms of these thrombolites and stromatolites are thus concluded to be primarily controlled by environmental factors, and biological contributions are relatively minor.

CONCLUSIONS

1. Several types of stromatolites and thrombolites are recognised within carbonate cycles in the Shannon Formation, each constructed by a benthic microbial community of specific composition and or sediment-forming activity.

2. Sequences of stromatolites and thrombolites within each carbonate cycle record an ecologic succession of these benthic microbial communities within laterally adjacent environments. Repeated sea-level rises and progradation generated the observed cyclic vertical sequences.

3. Cycles in the lower portion of the formation are relatively thin, exclusively dolostone, and are dominated by intertidal stratiform stromatolites of low synoptic relief. The most common microbial succession is from smoothly banded (?)filamentous communities which constructed hemispherical and wavy stromatolites in lower intertidal environments, to irregularly laminated pustular and mamillate (?)coccoid communities which constructed undulose stromatolites within middle and upper

intertidal environments. These latter mats were frequently disrupted by the precipitation of halite and anhydrite\gypsum crystals.

4. Cycles in the upper portion of the formation are thicker, predominantly limestone with a thin dolostone cap, and are dominated by subtidal thrombolite bioherms of relatively high synoptic relief. The thrombolites were constructed by variously degraded and calcified coccoid microbial colonies within a turbulent subtidal environment, and were inhabited by grazing, scavenging and burrowing metazoans. This community was typically succeeded by filamentous, column-building mats in shallower subtidal environments, and locally banded (?)filamentous mats which constructed wavy and planar stromatolites within intertidal environments.

5. The interpreted microbial successions show some similarities to those observed in Hamelin Pool, Shark Bay, Western Australia.

6. The microstructure of each stromatolite and thrombolite type provides a direct record of the sediment-forming processes of the microbial community, and commonly indicates if that community was dominated by filamentous or coccoid microbes. Apart from diagenesis, there appears to be little direct environmental contribution to the architecture of these microstructures.

7. The mesostructure of the stromatolites and thrombolites is controlled by a complex interaction of biological and environmental factors.

8. The megastructure of the stromatolites and thrombolites is primarily controlled by environmental factors.

ACKNOWLEDGMENTS

This study benefited substantially from discussions and constructive criticisms of earlier manuscripts by R.V. Burne, P.N. Southgate, M.R. Walter and J. Bertrand-Sarfati and C.L.V. Monty.

REFERENCES

Aitken, J.D. (1967) "Classification and environmental significance of cryptalgal limestones and dolostones, with illustrations from the Cambrian and Ordovician of southwestern Alberta", Journal of Sedimentary Petrology, 37, 1163-1178.

Awramik, S.M. (1984) "Ancient stromatolites and microbial mats", in Y. Cohen, R.W. Castenholz, and H.O. Halvorson (eds.), Microbial Mats, Alan R. Liss, New York, pp. 1-22.

Buchbinder, B. (1981) "Morphology, microfabric and origin of stromatolites of the Pleistocene precursor of the Dead Sea, Israel", in C. Monty (ed.), Phanerozoic Stromatolites, Springer, Berlin, pp. 181-196.

Burne, R.V. (1987) "Creation of an international collaboration programme: Stromatolites - IGCP Project 261", in C.L.V. Monty (ed.), Stomatolite Newsletter 13.

Burne, R.V., and Moore, L.S. (1987) "Microbialites: organosedimentary deposits of benthic microbial communities", Palaios, 2, 241-254.

Burne, R.V., and James, N.P. (1986) "Subtidal origin of club-shaped stromatolites, Shark Bay", 12th International Sedimentological Congress, Canberra, Australia, Abstracts, p. 49.

Chafetz, H.S. (1986) "Marine peloids: a product of bacterially induced precipitation of calcite", Journal of Sedimentary Petrology, 56, 812-817.

Chafetz, H.S., and Folk, R.L. (1984) "Travertines: depositional morphology and the bacterially constructed constituents", Journal of Sedimentary Petrology, 54, 289-316.

Davies, G.R. (1970) "Algal-laminated sediments, Gladstone Embayment, Shark Bay, Western Australia", American Assosiation of Petroleum Geologists Memoir 13, 169-205.

Deelman, J.C. (1975) "Two mechanisms of microbial carbonate precipitation", Naturwissenschaften, 62, 484-485.

Gebelein, C.D. (1969) "Distribution, morphology, and accretion rate of Recent subtidal algal stromatolites, Bermuda", Journal of Sedimentary Petrology, 39, 49-69.

Gebelein, C.D. (1974) "Biologic control of stromatolite microstructure: implications for Precambrian time stratigraphy", American Journal of Science, 274, 575-598.

Golubic, S. (1976) "Organisms that build stromatolites", in M.R. Walter (ed.), Stromatolites, Elsevier, Amsterdam, pp. 113-126.

Golubic, S. (1983) "Stromatolites, fossil and Recent", in P. Westbroek and E.W. de Jong (eds.), Biomineralization and Biological Metal Accumulation, D. Reidel, pp. 313-326.

Gürich, G (1906) "Les Spongiostromes du Viséen de la Province de Namur", Extraits des Mémoires du Musées D'Histoire Naturelle de Belgigue, T. II, 55p.

Hardie, L.A., and Ginsburg, R. (1977) "Layering: the origin and environmental significance of lamination and thin bedding", in L.A. Hardie (ed,), Sedimentation on the Modern Carbonate Tidal Flats of Northwest Andros Island, Bahamas, Johns Hopkins University Press, Baltimore, pp. 50-123.

Hoffman, P. (1976) "Stromatolite morphogenesis in Shark Bay", in M.R. Walter (ed.), Stromatolites, Elsevier, Amsterdam, pp. 261-271.

Hofmann, H.J. (1973) "Stromatolite characteristics and utility", Earth Science Reviews, 9, 339-373.

Horodyski, R.J., and Vonder Haar, S.P. (1975) "Recent calcareous stromatolites from Laguna Mormona (Baja California), Mexico", Journal of Sedimentary Petrology, 45, 894-906.

Kennard, J.M. (1989) "The structure and origin of Cambro-Ordovician thrombolites, western Newfoundland", PhD thesis, Memorial University of Newfoundland, St. John's, Newfoundland, Canada.

Kennard, J.M., and James, N.P. (1986a) "Early Palaeozoic thrombolites: an integrated approach to their structure, origin and palaeoecology", 12th International Sedimentological Congress, Canberra, Australia, Abstracts, p. 160.

Kennard, J.M., and James, N.P. (1986b) "Thrombolites and Stromatolites: two distinct types of microbial structures", Palaios, 1, 492-503.

Kennard, J.M., and Lindsay, J.F. (1991) "Sequence stratigraphy of the latest Proterozoic-Cambrian Pertaoorrta Group, northern Amadeus Basin, central Australia", in R.J. Korsch and J.M. Kennard (eds.), Geological and Geophysical Studies in the Amadeus Basin, Central Australia", Bureau of Mineral Resources, Geology and Geophysics, Australia, Bulletin 236, pp. 171-194.

Kennard, J.M., Nicoll, R.S., and Owen, M. (1986) "Late Proterozoic and Early Palaeozoic depositional facies of the northern Amadeus Basin, central Australia", 12th International Sedimentological Congress, Canberra, Australia, Field Excursion 25B, 125p.

Kinsman, D.J.J., and Park, R.K. (1976) "Algal belt and coastal sabkha evolution, Trucial coast, Persian Gulf", in M.R. Walter (ed.), Stromatolites, Elsevier, Amsterdam, pp. 421-433.

Korsch, R.J., and Kennard, J.M. (eds.) (1991) Geological and Geophysical Studies in the Amadeus Basin, Central Australia, Bureau of Mineral Resources, Geology and Geophysics, Australia, Bulletin 236, 594 p.

Korsch, R.J., and Lindsay, J.L. (1989) "Relationships between deformation and basin evolution in the intracratonic Amadeus Basin, central Australia", Tectonophysics, 158, 5-22.

Krumbein, W.E. (1974) "On the precipitation of aragonite on the surface of marine bacteria", Naturwissenschaften, 64, 167.

Krumbein, W.E. (1979) "Photolithotropic and chemoorganotrophic activity of bacteria and algae as related to beachrock formation and degradation (Gulf of Aqaba, Sinai)", Geomicrobiology Journal, 1, 139-203.

Krumbein, W.E., and Cohen, Y. (1977) "Primary production, mat formation and lithification: contribution of oxygenic and facultative anoxygenic cyanobacteria", in E. Flugel (ed.), Fossil Algae, Springer-Verlag, Berlin, pp. 37-56.

Krumbein, W.E., Cohen, Y., and Shilo, M. (1977) "Solar Lake (Sinai). 4. Stromatolitic cyanobacterial mats", Limnology and Oceanography, 22, 635-656.

Lalou, C. (1957) "Studies on bacterial precipitation of carbonate in seawater", Journal of Sedimentary Petrology, 27, 190-195.

Lindsay, J.F., and Korsch, R.J. (1989) "Interplay of tectonics and sea-level changes in basin evolution: an example from the intracratonic Amadeus Basin, central Australia", Basin Research, 2, 3-25.

Logan, B.W., Hoffman, P., and Gebelein, C.D. (1974) "Algal mats, cryptalgal fabrics and structures, Hamelin Pool, Western Australia", American Association of Petroleum Geology Memoir 22, 140-194.

McCallum, M.F., and Guhathakurta, K. (1970) "The precipitation of calcium carbonate from seawater by bacteria isolated from the Bahama Bank sediments", Journal of Applied Bacteriology, 33, 649-655.

Monty, C.L.V. (1967) "Distribution and structure of Recent stromatolitic algal mats, eastern Andros Island, Bahamas", Annuales Geologique de Belgique, 90, 55-100.

Monty, C.L.V. (1972) "Recent algal stromatolite deposits, Andros Island, Bahamas", Geologische Rundschau, 61, 742-783.

Monty, C.L.V. (1976) "The origin and development of cryptalgal fabrics", in M.R. Walter (ed.), Stromatolites, Elsevier, Amsterdam, pp. 193-249.

Oppenheimer, C.H. (1961) "Note on the formation of spherical aragonite bodies in the presence of bacteria from the Bahama Bank", Geochemica et Cosmochimica Acta, 23, 295-296.

Playford, P.E., and Cockbain, A.E. (1976) "Modern algal stromatolites at Hamelin Pool, a hypersaline barred basin in Shark Bay, Western Australia", in M.R. Walter (ed.), Stromatolites, Elsevier, Amsterdam, pp. 389-411.

Pentecost, A., and Riding, R. (1986) "Calcification in cyanobacteria", in R. Riding, and B.S.C. Leadbearer (eds.), Biomineralization in the Lower Plants and Animals, Systematics Association Special Volume, pp. 73-90.

Shergold, J.H. (1986) "Review of the Cambrian and Ordovician palaeontology of the Amadeus Basin, central Australia", Bureau of Mineral Resources, Geology and Geophysics, Australia, Report 276, 21 p.

Southgate, P.N. (1982) "Cambrian skeletal halite crystals and experimental analogues", Sedimentology, 29, 391-407.

Stolz, J.F. (1983) "Fine structure of the stratified community at Laguna Figueroa, Baja California, Mexico: 1 Methods of an in situ study of the laminated sediment", Precambrian Research, 20, 479-492.

Walter, M.R. (1972) "Stromatolites and the biostratigraphy of the Australian Precambrian and Cambrian", Palaeontological Association of London Special Papers in Palaeontology No. 11, 190 p.

Walter, M.R., Bauld, J., and Brock, T.D. (1976) "Microbiology and morphogenesis of columnar stromatolites (*Conophyton, Vacerrilla*) from hot springs in Yellowstone National Park", in M.R. Walter (ed.), Stromatolites, Elsevier, Amsterdam, pp. 273-310.

Wells, A.T., Forman, D.J., Ranford, L.C., and Cook, P.J. (1970) "Geology of the Amadeus Basin, central Australia", Bureau of Mineral Resources, Geology and Geophysics, Australia, Bulletin 100, 222 p.